Highly Selective Neurotoxins

Contemporary Neuroscience

Highly Selective Neurotoxins

Basic and Clinical Applications

Edited by

Richard M. Kostrzewa

East Tennessee State University, Johnson City, TN

Humana Press 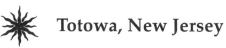 Totowa, New Jersey

© 1998 Humana Press Inc.
999 Riverview Drive, Suite 208
Totowa, New Jersey 07512

For additional copies, pricing for bulk purchases, and/or information about other Humana titles, contact Humana at the above address or at any of the following numbers: Tel.: 201-256-1699; Fax: 201-256-8341; E-mail: humana@mindspring.com or visit our Website: http://humanapress.com

This publication is printed on acid-free paper. ∞
ANSI Z39.48-1984 (American Standards Institute) Permanence of Paper for Printed Library Materials.

Cover illustration:

Cover design by Patricia F. Cleary.

Printed in the United States of America. 10 9 8 7 6 5 4 3 2 1

Library of Congress Cataloging-in-Publication Data

Preface

The noted physiologist Walter B. Cannon expounded a precept that the function of a nerve can often be deduced by observing the specific impairments produced subsequent to its destruction. Historically, denervation was ordinarily accomplished by either destroying the nucleus of origin or transecting the axonal trunk and waiting until anterograde degeneration effectively removed innervation to target tissues. Many principles of neurophysiology were uncovered in this way. Nevertheless, there are a number of inherent disadvantages to this approach, including the invasive surgical procedure with anesthetic effects and inadvertent transection of adjacent nerves containing other neurotransmitters. When perikarya are electrolytically destroyed or when their nerve trunks are surgically transected, particularly within the central nervous system, extensive secondary damage occurs to adjoining brain, spinal, or other structural regions. Also, when tissues receive innervation from more than one nerve trunk, surgical axotomy is incomplete. Although there has long existed a battery of chemicals able to destroy nerves, most substances nonselectively damage many kinds of nerves or tissues and produce a spectrum of adverse effects.

In the 1950s Dr. Rita Levi-Montalcini began a series of pioneering studies that established the importance of a protein nerve growth factor (NGF), as an essential neurotrophic substance for both survival and guided growth of the sympathetic neuropil. This Nobel Laureate's elegant studies with a specific antibody for NGF demonstrated that extensive destruction of sympathetic noradrenergic nerves could be produced in the absence of change in sensory nerve morphology, development, and growth. Anti-NGF represented the first highly selective neurotoxin, producing desired selective neuronal destruction in the virtual absence of damage to other kinds of nerves and with relatively few adverse effects.

Several years after these reports, during a search for electrophilic markers for monoaminergic nerves, Drs. Hans Thoenen and J. P. Tranzer made the serendipitous discovery that the norepinephrine isomer 6-hydroxydopamine (6-OHDA) produced overt and selective destruction of sympathetic noradrenergic nerves. Later, it was shown that dopaminergic or noradrenergic nerves in brain could be targeted for selective destruction by 6-OHDA if animals were pretreated with appropriate transport inhibitors. Anti-NGF and 6-OHDA have been used in more than 5000 published studies, attesting to their value as research tools. Over the past 25 years there have been a succession of discoveries of other neurotoxins that have high selectivity for specific neurotransmitter systems. These have been invaluable for uncovering facets of basic neuronal function, including processess related to ion channel opening, receptor dynamics, neuronal metabolism, axonal transport, axonal growth, and nerve development. Neurotoxins have proven useful for studying presynaptic receptor localization and postsynaptic receptor proliferation and sensitization.

Beacuse of their ability to destroy specific kinds of nerves, neurotoxins are used to produce animal models of many neurological and psychiatric disorders. This includes models of Parkinsonism, schizophrenia, tardive dyskinesia, mania and depression, attention deficit hyperactivity disorder, Tourette syndrome, Lesch-Nyhan syndrome, Alzheimer's disease, Huntington's chorea, dystonias, and other clinical entities. Some of the neurotoxins described are themselves drugs of abuse or have been consumed inadvertently by humans, with dire consequences. Neurotoxins have shed new light on biological mechanisms that underlie spontaneous degeneration of nerves during aging, as well as uncovering mechanisms associated with apoptosis (programmed cell death). Neurotoxins are essential for studies on nerve degeneration, regeneration, and sprouting, and in probing the neural adaptations that occur after injury. Through the use of neurotoxins, much knowledge on the neuropharmacological actions of drugs has been realized.

Highly Selective Neurotoxins: Basic and Clinical Applications is intended to be a compendium of the most common and most important highly selective neurotoxins. Although the term neurotoxin often includes those substances that temporarily impair neuronal function, the term neurotoxin is used in this book in the strictest sense, referring specifically to substances that produce overt destruction of neurons. Each chapter is intended to provide a description of the most important actions of a single neurotoxin or mechanistically similar group of neurotoxins. Biological actions, mechanisms, and experimental uses for each neurotoxin are summarized, along with projected newer uses, if applicable. All contributors have made significant discoveriers on one or more of the neurotoxins, thus helping to create new vistas in neuroscience research.

Richard M. Kostrzewa

Dedication
Richard E. Heikkila

I first met Dick Heikkila at a Catecholamine Symposium in 1975 and one week later joined him and his wife Dawn for dinner at the International Union of Pharmacology meeting. These were the first of many instances when Dick and I would get together at meetings—the last time being Amsterdam during another International Congress of Pharmacology. It was always a pleasure to meet with Dick, in whom ideas arose as fast as he spoke.

Having trained with Dr. David G. Cornwell in the Department of Physiological Chemistry at Ohio State University, Dick changed his research focus toward catecholamine metabolism, and rapidly rose to prominence during his postdoctoral training with Dr. Gerald Cohen at Mt. Sinai Medical School in New York. Gerald and Dick discovered the preeminent role for free radicals in mediating effects of neurotoxins. This revolutionary concept formed the foundation for research into mechanisms of action of neurotoxins and potentially toxic xenobiotics. In the span of five years they published 29 peer-reviewed papers in *Science, J. Biol. Chem., Experientia, J. Pharm. Exp. Ther., Eur. J. Pharm., Biochem. Pharm., Mol. Pharm., J. Neurochem., Brain Res.*, and other fine journals. They continued to publish together even after Dick formed his own lab in Dr. Roger Duvoisin's Department of Neurology at the University of Medicine and Dentistry of New Jersey, Rutgers Medical School. Here, Dick's work with Larry Manzino, Steve Youngster, Pat Sonsalla, and others gradually evolved toward determination of the mechanisms of action of 1-methyl-4-phenyl-1,2,3,6-tetrahydropyridine (MPTP) in destroying substantia nigra dopamine-containing nerves, an effect that is accompanied by development of an irreversible parkinsonian syndrome in primates.

Dick Heikkila's work stands at the forefront of research into mechanisms of action of neurotoxins and in advancing our understanding of basic cellular/neuronal oxygen metabolism. Only 16 years after receiving his PhD degree, Dick published his one hundredth paper; by this time he was recognized as an international authority. Dick was always cordial, friendly, and helpful to all, no matter their stage of development. He would instantly provide ideas and suggest experiments to anyone, including those who might be labeled as "competitors." His early death was a loss to all who knew him and to all in the neuroscience community that benefited from his findings. A busy schedule did not preclude Dick's involvement in church, community, and civic activities, as well as coaching year-round sports for youths.

His friendliness, openness, sincerity, and humble ways are models for all to follow. It is to his memory that this book is dedicated.

Contents

Contributors

C. LEROY BLANK, *Department of Chemistry and Biochemistry, University of Oklahoma, Norman, OK*

CHARLES R. BREESE, *Department of Psychiatry, University of Colorado Health Sciences Center, Denver, CO*

GEORGE R. BREESE, *Neurosciences Center, Departments of Psychiatry and Pharmacology, School of Medicine, University of North Carolina, Chapel Hill, NC*

NEAL CASTAGNOLI, JR., *Department of Chemistry, College of Arts and Sciences, Virginia Polytechnic Institute, Blacksburg, VA*

STANISLAW J. CZUCZWAR, *Department of Pharmacology and Toxicology, Lublin University Academy of Medicine; and Department of Clinical Toxicology, Institute of Agricultural Medicine, Lublin, Poland*

ANDRZEJ DEKUNDY, *Department of Clinical Toxicology, Institute of Agricultural Medicine, Lublin, Poland*

GLENN DRYHURST, *Department of Chemistry and Biochemistry, University of Oklahoma, Norman, OK*

GAYLORD ELLISON, *Department of Psychology, University of California at Los Angeles, CA*

GUILLERMO JAIM-ETCHEVERRY, *Departamento de Celular e Histologia, Facultad de Medicina, University of Buenos Aires, Argentina*

KRISTEN A. KEEFE, *Department of Pharmacology and Toxicology, University of Utah, Salt Lake City, UT*

ZDZISLAW KLEINROK, *Department of Pharmacology and Toxicology, Lublin University Academy of Medicine, Lublin, Poland*

RICHARD M. KOSTRZEWA, *Department of Pharmacology, Quillen College of Medicine, East Tennessee State University, Johnson City, TN*

J. WILLIAM LANGSTON, *The Parkinson's Institute, Sunnyvale, CA*

ROLAND E. LEHR, *Department of Chemistry and Biochemistry, University of Oklahoma, Norman, OK*

R. LEW, *Department of Pharmacological and Physiological Sciences, The University of Chicago, IL*

RUSSELL J. LEWIS, *Department of Chemistry and Biochemistry, University of Oklahoma, Norman, OK*

xi

J. E. MALBERG, *Department of Pharmacological and Physiological Sciences, The University of Chicago, IL*

SUSAN M. POND, *Department of Chemistry, College of Arts and Sciences, Virginia Polytechnic Institute, Blacksburg, VA*

PAMELA E. POTTER, *Department of Anesthesiology, Albert Einstein College of Medicine, Montefiore Medical Center, Bronx, NY*

MICHAEL R. PRANZATELLI, *Departments of Pharmacology, Neurology, and Pediatrics, George Washington University, Washington, DC. Current Address: The National Pediatric Myoclonus Center, Washington, DC*

SUNITA RAJDEV, *Department of Neurology, University of California at San Francisco and VA Medical Center, San Francisco, CA*

GEORGE A. RICUARTE, *Department of Neurology, Johns Hopkins Medical Institutions, Baltimore, MD*

JOYCE E. ROYLAND, *The Parkinson's Institute, Sunnyvale, CA*

LEWIS S. SEIDEN, *Department of Pharmacological and Physiological Sciences, The University of Chicago, IL*

FRANK R. SHARP, *Department of Neurology, University of California at San Francisco and VA Medical Center, San Francisco, CA*

ARPAD SZALLASI, *Department of Pharmacology, Karolinska Institute, Stockholm, Sweden. Current Address: Department of Anatomy and Neurobiology, Washington Unversity School of Medicine, St. Louis, MO*

TAHEREH TABATABAIE, *Free Radical Biology and Aging Research Program, Oklahoma Medical Research Foundation, Oklahoma City, OK*

WALDEMAR A. TURSKI, *Department of Pharmacology and Toxicology, Lublin University Academy of Medicine; and Department of Clinical Toxicology, Institute of Agricultural Medicine, Lublin, Poland*

EWA M. URBANSKA, *Department of Pharmacology and Toxicology, Lublin University Academy of Medicine; and Department of Clinical Toxicology, Institute of Agricultural Medicine, Lublin, Poland*

ETSUKO USUKI, *Department of Chemistry, College of Arts and Sciences, Virginia Polytechnic Institute, Blacksburg, VA*

CORNELIS J. VAN DER SCHYF, *Department of Chemistry, College of Arts and Sciences, Virginia Polytechnic Institute, Blacksburg, VA*

THOMAS J. WALSH, *Department of Psychology, Rutgers University, New Brunswick, NJ*

MICHAEL J. ZIGMOND, *Department of Neuroscience, University of Pittsburgh, PA*

6-Hydroxydopamine and Related Catecholaminergic Neurotoxins
Molecular Mechanisms

C. LeRoy Blank, Russell J. Lewis, and Roland E. Lehr

1. INTRODUCTION

6-Hydroxydopamine (6-HDA) (Fig. 1) first acquired neurochemical importance when it was reported as an in vivo metabolite of injected dopamine by Senoh et al. *(1)* Its peripheral noradrenergic depletion effects *(2,3)* were shown by Tranzer and Thoenen *(4,5)*, using electron microscopy, to correspond to neuronal degeneration, and it was soon demonstrated to elicit neurodegeneration in noradrenergic and dopaminergic projections in the central nervous system (CNS) as well *(6,7)*. Although immunosympathectomy *(8)* had preceded the appearance of 6-HDA by more than a decade as an alternative to simple surgical ablation in eliciting neurodegeneration, the broader applicability and considerable selectivity of 6-HDA quickly led to its extensive employment for such purposes. However, 6-HDA is not totally specific or complete in its degeneration of catecholaminergic neurons or associated projections even when employed under optimal conditions. The selectivity and potency of 6-HDA depend on the nature and age of the animal species, the environment of targeted neuronal entities, the dose, the manner of administration, and many other factors *(9)*. The history, biological effects, selectivity, potency, biological significance, and clinical relevance of 6-HDA and its congeners have been extensively covered elsewhere *(9–14)*, including Chapters 2–4. As such, we will only briefly mention such items in the current discussion.

A number of agents structurally similar to the original 6-HDA have been demonstrated, or are strongly suspected, to possess neurodegenerative properties. These include:

1. The 16 trisubstituted phenethylamines or α-methylphenethylamines in which the trisubstitution pattern on the ring is of the 2,3,5- or 2,4,5- variety with respect to the side chain, and involves either trihydroxy or aminodihydroxy functional group entities *(15–18)*;

From: Highly Selective Neurotoxins: Basic and Clinical Applications *Edited by: R. M. Kostrzewa.* Humana Press Inc., Totowa, NJ

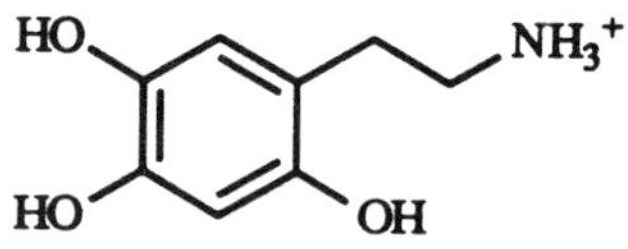

Fig. 1. 6-Hydroxydopamine (6-HDA).

2. 6-Hydroxydopa, which may be viewed as an amino acid precursor to 6-HDA (*9,10,* and Chapter 4);
3. Tetrahydroxyphenethylamines, which may be viewed as compounds from the first group containing an extra hydroxyl group on the ring *(18)*;
4. Compounds from the first group in which one of the hydroxy or amino ring substituents is replaced by a hydrogen, methyl, or methoxy group *(15,17,19,20)*;
5. Compounds from the first group in which alkyl groups other than methyl or cyclic alkyl groups are added to the side chain of the phenethylamine *(21)*;
6. The three ring methylated derivatives of 6-HDA, called 2-Me-, 5-Me-, and 2,5-diMe- 6-hydroxydopamines *(22)*; and
7. Compounds similar to those in the first group in which the side chain has been covalently attached to the benzene ring, such as 1,2-dimethyl-4,6,7-trihydroxy-1,2,3,4-tetrahydroisoquinoline *(23)*.

The 6-HDA-related agents outlined in the above paragraph, as well as similar species that lack neurodegenerative properties, have contributed to our knowledge concerning the mode of action of such agents. Investigations into the mode of action, which is commonly assumed to be at least similar for all such agents, have most frequently incorporated one or more of the subgroup of eight 2,4,5-trisubstituted compounds taken from the first group mentioned above. Members of this subgroup, which includes the parent 6-HDA, are shown collectively in Fig. 2, where R_α = H or Me and R_2, R_4, and R_5 constitute trihydroxy or amino and dihydroxy substituents. The compounds represented in Fig. 2 exhibit a dilemma which has arisen in the nomenclature of 6-HDA and related neurodegenerative agents. In fact, numbering of the ring substituents for 6-HDA and other compounds in this subgroup begins, as indicated, at the side chain and more properly proceeds clockwise around the benzene ring to provide the lowest numerical combination for ring substituents. Thus, 6-HDA is more properly identified as 2,4,5-trihydroxyphenethylamine. However, 6-HDA was originally viewed as a derivative of dopamine, or 3,4-dihydroxyphenethylamine, in which the corresponding ring substituents are properly numbered in a counterclockwise fashion to obtain the lowest numerical combination. Subsequently introduced 6-HDA analogs, with the exception of 6-aminodopamine (6-ADA), have most commonly been named using the more proper clockwise approach for substituents. The extensive prior use in the literature of the names 6-HDA and 6-ADA, instead of the more proper 2,4,5-trihydroxyphenethylamine and

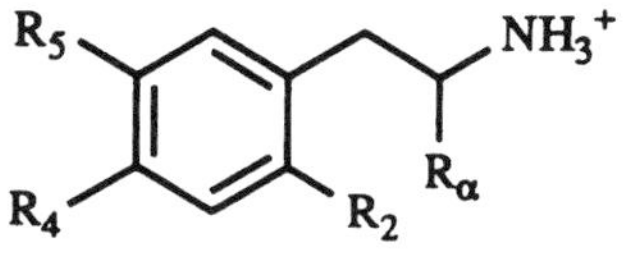

Fig. 2. 2,4,5-Trisubstituted phenethylamines or α-Me-phenethylamines.

2-amino-4,5-dihydroxyphenethylamine, respectively, for the two most commonly encountered compounds in this subgroup, however, precludes any correction in their nomenclature at this juncture. Thus, we will continue to employ this dichotomous nomenclature here, referring to 6-HDA, 6-ADA, and their α-methyl derivatives by their common names and the remainder of agents by their more correct names.

Considering the many structural analogs of 6-HDA that have been examined for neurodegenerative properties, it is somewhat remarkable that none has been found to be so innately superior as to displace the use of this original agent. Of course, "superior" requires clarification with regard to the animal species, targeted neurons or projections, potency, selectivity, and general lethality. For example, in regard to CNS noradrenergic effects, the seven other derivatives implicated in Fig. 2 have been shown equipotent or more potent than the standard 6-HDA while simultaneously being equally or more selective *(13,15–17,19,24,25)*. However, many of these agents are also more lethal than 6-HDA, which would obviously impede any claim of superiority. An exception among these congeners is afforded by 5-amino-2,4-dihydroxy-α-methylphenethylamine (α-5-ADA), which provides both slightly improved potency and selectivity in noradrenergic neurodegeneration with decreased lethal effects compared to 6-HDA *(16,26)*. The α-5-ADA is particularly more potent in its noradrenergic effects in the hypothalamus, and thus may be useful for studies concerned with this and other noradrenergic CNS regions having projections originating from the ventrotegmental nuclei. On the other hand, none of the analogs examined to date has provided substantially greater selectivity or potency with regards to dopaminergic systems than 6-HDA.

2. FUNDAMENTALS OF MODE OF ACTION

Chemical properties and biological events potentially related to the mode of action of 6-HDA and similar agents have been investigated by many different workers. However, no single such property or event has been unequivocally demonstrated to be **the** cause of degeneration. Previously proposed mechanistic possibilities remain viable today, and existing reviews on the subject are, thus, entirely pertinent *(9,10,12–14)*. The complete mode of action can be fairly cleanly separated into two distinct and sequential steps: (1) interaction with the uptake transport site on the cellular membrane leading to concentration of the agent intraneuronally and (2) subsequent intracellular events leading to degeneration. Uptake of the agent is well accepted to be a necessary, but not sufficient, requirement for neurodegeneration; thus, many exclude this component from their definition of the mode of action or mechanism of action. However, the pertinent intraneuronal degenerative properties or events associated with or constituting the mechanism of action are, by contrast, more speculative in nature. Experimental dissection and temporal isolation of the neurodegeneration process into individual events, with observation in vivo of the outcome of each, are not possible at the current time. Some potentially important properties and intraneuronal events, often interrelated, include:

1. Ease of oxidation of the agent;
2. Formation of cytotoxic oxygen species;
3. Reactions of oxidized forms of the agent with vital cell constituents;
4. Depletion of cellular oxygen;
5. Depletion of reducing components;
6. Disruption of oxidative phosphorylation or other mitochondrial functions;
7. Intraneuronal release of endogenous transmitters from vesicular stores;
8. Disruption of Ca^{2+} homeostasis;
9. Other electrolyte or pH disturbances, and
10. Triggering of apoptosis.

3. UPTAKE BY TRANSPORT PROTEINS

Transportation and intracellular concentration of the agent by the targeted neuronal projection are firmly established prerequisites to subsequent degeneration by 6-HDA-related species. Early investigations of 6-HDA demonstrated blockade of neurodegeneration by uptake inhibitors like desmethylimipramine *(27–30)*. From these observations, the selectivity of a particular agent has frequently been construed to be a combination of the preferential transport by the targeted neuronal uptake site and the relative lack of transport by nontargeted sites *(10,11)*. This concept was initially supported by results that yielded K_i values of 22 and 37 μM for 6-HDA blockade of ^{3}H-NE uptake into hypothalamic and striatal preparations, representing noradrenergic and dopaminergic transporter interactions, respectively *(70)*. However, more recent results *(16)*, obtained using ^{3}H-NE and ^{3}H-DA, respectively, yielded K_i values for 6-HDA of 51 and 12 μM. The latter results are clearly in opposition to the observed, substantially greater noradrenergic selectivity found following intraventricular 6-HDA administration. Also, even the earlier results do not support the degree of selectivity observed for 6-HDA. Thus, quantitative assessment of the affinity of an agent for various transmitter uptake sites does not, *a priori,* allow precise prediction of the relative selectivity of the agent. Other aspects, such as locally achieved extraneuronal concentrations, are obviously involved in the quantitative outcome from intraventricular treatments. However, it is also quite reasonable to conclude from these results that variations in the intraneuronal degenerative actions of individual agents may contribute significantly to their selectivity.

Selective interaction of the agents with the cellular uptake proteins notably leads to substantial intraneuronal concentrations. Indeed, Sachs and Jonsson *(14,31)* have estimated that 6-HDA intraneuronal concentrations of approx 30–100 mM must be obtained to produce subsequent degeneration.

Once inside the neuron, 6-HDA-related agents also may interact with the vesicular transporter, gaining access to the intravesicular region, promoting intraneuronal release of endogenous transmitters, and becoming available for release as false transmitters *(14)*. A recent report *(34)* indicates that the interaction of 6-HDA with the dopamine vesicular transporter, e.g., is quite potent, having a K_i value of 1.35 µM. However, although increased intraneuronal, extravesicular concentrations of endogenous transmitter may partially inhibit radical reactions *(vide infra)*, the impact of vesicular transport of an agent appears to be relatively minimal in the overall mode of action.

In contrast to the in vivo situation, transport of 6-HDA-related agents by the cell membrane-uptake system may not be essential to elicit neurodegeneration or toxicity in cell-culture-preparations, as shown in a bovine chromaffin cell culture *(32)*. However, the extraneuronal-environments in cell-culture situations are quite different from those found in vivo. Also, uptake inhibitors have produced at least some attenuation in 6-HDA toxicity using PC12 cell cultures *(33)*.

4. EASE OF OXIDATION

Investigations into the mode of action of 6-HDA and its analogs have repeatedly implicated the ease of oxidation of these compounds in the intraneuronal milieu *(9,14)*. Autoxidation of 6-HDA is summarized below in Fig. 3, along with associated redox interactions. The oxidation process and the associated molecular forms for 6-HDA under physiological conditions have a few notable characteristics that tend to be overlooked. First, the side chain amino group is assumed to have a pK_a of 9–10; thus, both the reduced and oxidized forms predominantly exist in the protonated form. Second, the oxidized form of 6-HDA possesses an enolic proton at position 4 on the ring, using the 2,4,5 nomenclature, which has a pK_a value of 4–5. Thus, the enolic—OH group will be predominantly unprotonated at pH 7.4, leading to an oxidized form that is a resonance structure somewhere between the *ortho*-quinone and the *para*-quinone. Further, the existence of the enolic proton on the oxidized form provides an oxidation pathway from the hydroquinone to the quinone form that involves one more proton than the number of electrons, i.e., 3 H$^+$ and 2 e$^-$.

Although Fig. 3 shows a reasonable overview of autoxidation for 6-HDA, the many mechanistic steps not shown substantially increase the complexity of the processes involved *(vide infra)*. Nonetheless, oxidation of 6-HDA and related agents by molecular oxygen generally leads to production of the superoxide radical, hydrogen peroxide, and the highly reactive hydroxyl radical. These three related oxygen species are considered, both individually and together, to be cytotoxic. Production of these oxygen-derived cytotoxins, particularly the

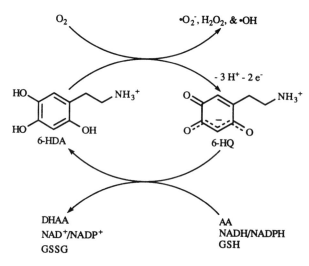

Fig. 3. Autoxidation of 6-HDA at physiological pH and associated recycling reactions.

hydroxyl radical, via interaction with 6-HDA-related agents, constitutes what most consider to be the primary neurodegenerative mechanism of action, a hypothesis initially put forward by Heikkila and Cohen *(35–39)*.

Thermodynamically, the ease of oxidation of 6-HDA analogs is derived from their low formal reduction potentials. Using a hanging mercury drop electrode, the $E^{\circ\prime}$ values at pH 7.40 for the eight 2,4,5-trisubstituted compounds represented in Fig. 2 were found to fall between –0.212 (6-HDA) and –0.123 V (α-Me-6-aminodopamine or α-Me-6-ADA) vs SCE *(16)*. These values correspond, respectively, to +0.029 and +0.118 V vs NHE. Combination of these $E^{\circ\prime}$ values with the value for molecular oxygen ($E^{\circ\prime}$ = 0.762 V vs NHE for O_2 + 4H+ + 4e– $\rightarrow$ 2H_2O at pH 7.40) shows that the ratio of the concentration of the oxidized quinoid form to that of the reduced hydroquinone or aminophenol form should be between 3×10^{23} (6-HDA) and 4×10^{20} (α-Me-6-ADA) when assuming P_{O2} = 0.21 atm. Employing P_{O2} values 100 times lower only decreases the calculated ratios by a factor of 10. Thus, for reasonable levels of O_2, all the 6-HDA analogs are enormously driven thermodynamically to their oxidized forms.

Oxygen and neurodegenerative agents, however, do not encounter one another in vivo in a redox vacuum. The oxidation/reduction condition of a particular locality within a tissue is normally controlled by redox components having substantial endogenous concentrations; additionally, the redox state is clearly dependent on the time of examination. Temporarily ignoring the locality and timing concerns, an alternative thermodynamic analysis to that given above might reasonably consider an "integrated" CNS redox state. It is easy to argue that such a redox state would fundamentally be controlled, or at least reflected, by the ascorbate half-reaction owing to the relatively high concentrations of its associated redox components. Using this approach, we employ the measured extraneuronal redox "buffer" value of approx -0.200 V vs SCE, as reported by McCreery et al. *(40)*. Substitution of this value into the appropriate Nernst Equation expressions for 6-HDA analogs yields predicted ratios of the oxidized to the reduced forms of the agents, which vary between 0.003 and 2. Thus, using this approach, one can at least say that the analogs would be sufficiently driven thermodynamically to provide a significant to substantial amount of their oxidized forms. For comparison, employing the same redox buffer value to obtain the ratio of oxidized to reduced forms for the endogenous catecholamines yields values between 2×10^{-10} and 4×10^{-12}, or the endogenous catecholamines are roughly 10 million to 1 trillion times less driven thermodynamically to their oxidized forms than are the neurodegenerative analogs.

5. PATHWAYS FOR AUTOXIDATION AND ASSOCIATED RATES

Although thermodynamic aspects of the oxidation of 6-HDA and its analogs are fairly straightforward to assess, many of the associated reactions do not proceed to the extent predicted within a reasonable period of time because of kinetic complications. As such, kinetics of the individual processes become considerably more relevant in the production of significant levels of potentially toxic or degenerative byproducts. Since degeneration is generally presumed not to be derived directly from the initial, reduced form of the agents, rates of formation of toxic or pretoxic products may well be the second most important determinant of selectivity, after transportation by the uptake site. Unfortunately, rates for many of the pertinent reactions involved in the "simple" in vitro autoxidation of even the best-known member of this group, i.e., 6-HDA, are largely unknown. Unraveling the impact of kinetics on the in vivo mode of action notably includes two linked items of importance: (1) identification of pertinent products and (2) determination of the rates of formation of these products. We believe that progress in the future pursuit of these linked kinetic aspects will be the most critical in obtaining complete understanding of the mode of action of such neurodegenerative compounds and, likewise, in the successful design of novel agents.

Rates of interaction of 6-HDA analogs with oxygen have been primarily examined in a qualitative or semiquantitative manner *(39,44–46)*. Such interactions generate both oxygen-derived cytotoxins and the oxidized (semiquinone, *p*-quinone, semiquinoneimine, and *p*-quinoneimine) forms of the analogs. Mechanistically, these processes may be represented as:

$$QH_2 + O_2 \rightarrow QH\bullet + \bullet O_2^- + H^+ \tag{1}$$

$$\text{or } QH_2 + \bullet O_2^- + H^+ \rightarrow QH\bullet + H_2O_2 \tag{2}$$

$$\text{and } QH\bullet + \bullet O_2^- \to Q + H_2O_2 \tag{3}$$

$$\text{or } QH\bullet + O_2 \to Q + \bullet O_2^- + H^+ \tag{4}$$

where at least some of these reactions are catalyzed by metal ions or complexed metal ions. In the case of 6-HDA *(37,41,43)*, a substantial initial lag period is observed in the presence of selected metal chelators and modest amounts of superoxide dismutase (SOD), which catalyzes disproportionation of $\bullet O_2^-$ according to:

$$2\ \bullet O_2^- + 2\ H^+ \to O_2 + H_2O_2 \tag{5}$$

Relatively large amounts of SOD alone also evoke an initial lag in the autoxidation of 6-HDA, although such large amounts of the SOD protein appear to have metal-chelating properties in addition to the normal catalytic properties, as shown by Davison et al. *(44–46)*. Combined, these data indicate that metal ion catalysis involving a ternary (QH_2–metal–O_2) complex probably represents the major initial oxidative pathway for 6-HDA under normal autoxidative circumstances at physiological pH *(46)*. However, the absence of metal ions, or effective complexation of metal ions with agents like desferrioxamine reveals an oxidation pathway that is $\bullet O_2^-$-dependent and, thus, readily disrupted by SOD. Also, the self-catalytic nature of the autoxidation of 6-HDA, in which the rate of oxidation accelerates after initiation, indicates subsequent involvement of more than the initial, metal ion-catalyzed pathway alone. Taken together, these results indicate that reaction (1) is quite sluggish, but reactions (2) and (4), by comparison, become relatively rapid, constituting effective free radical chain propagation steps *(39)*. Reactions (1) and (3) offer an alternative pathway to the fully oxidized Q species and H_2O_2, but Q formation thusly achieved incurs a radical chain termination. Formation of Q and H_2O_2 can similarly be accomplished through a combination of reactions (1) and (2), followed by disproportionation of the intermediate semiquinone:

$$2\ QH\bullet \rightleftharpoons QH_2 + Q \tag{6}$$

After the reaction progresses, producing a substantial concentration of Q, the reverse of reaction (6) may become important in the overall oxidation pathway by providing a pathway to the semiquinone independent of both reactions (1) and (2).

Regardless of the individual mechanistic steps taken among those outlined above, the overall oxidation of the agent to the quinone or quinoneimine form can be summarized as:

$$QH_2 + O_2 \to Q + H_2O_2 \tag{7}$$

Furthermore, inclusion of H_2O_2 into our consideration of the in vitro autoxidation engenders a considerable increase in the level of complexity for the process. For example, direct oxidation of 6-HDA by H_2O_2, although generally considered quite slow, is possible *(47)*, and this may partially explain inhibition of autoxidation on addition of catalase, which effects disproportionation of H_2O_2 as:

$$2\ H_2O_2 \rightarrow 2H_2O + O_2 \tag{8}$$

The role of H_2O_2 in such autoxidations also requires consideration of the interaction of this species with metal ions, such as Fe^{2+}:

$$Fe^{2+} + H_2O_2 \rightarrow Fe^{3+} + \bullet OH + OH^- \tag{9}$$

$$Fe^{3+} + \bullet O_2- \rightarrow Fe^{2+} + O_2 \tag{10}$$

where reaction (9) is the Fenton reaction and reactions (9) and (10) together constitute the iron-catalyzed Haber-Weiss reaction *(48)*. The potential importance of iron in the overall autoxidative process and the neurodegeneration observed in vivo is bolstered by the observation that 6-HDA induces release of this metal ion from ferritin *(49)*. Reduction of Fe^{3+}, as shown in reaction (10), could also be efficiently achieved by reducing agents, such as ascorbate. Reactions (9) and (10) represent the chemical pathway by which most believe the hydroxyl radical, the presumed primary cytotoxin associated with 6-HDA-related compounds, is generated. In addition to its cytotoxic or neurodegenerative actions, $\bullet OH$, a potent oxidizing agent, could simply accelerate autoxidation directly through oxidation of QH_2 or $QH\bullet$ or, indirectly, through interaction with QH_2 leading to tetra- or penta-hydroxy derivatives. Such perhydroxylated derivatives would be thermodynamically easier to oxidize than the original agent and would, perhaps, undergo the equivalent of reaction (1) more rapidly, enhancing production of the superoxide radical and, thus, the overall autoxidation process.

In short, metal ion catalysis and a superoxide-mediated radical chain reaction appear to play the predominant roles in the in vitro autoxidation of 6-HDA. Production of the presumed primary toxic or neurodegenerative species, the hydroxyl radical, is achieved via the Fenton, Haber-Weiss, or modified forms of these reactions. The above discussion, however, hopefully illustrates some of the multiple reactions and side reactions that must be quantitatively investigated to obtain a comprehensive picture of the relatively simple in vitro autoxidation of such an agent considering only the primary reactants and metal ions. Incorporation of additional reagents found in vivo, however, increases the complexity even further. For example, oxidative stress has been shown to increase production of nitric oxide, another physiologically relevant free radical *(50)*. Nitric oxide readily reacts with $\bullet O_2-$ as:

$$\bullet O_2- + \bullet NO \rightarrow ONOO- \tag{11}$$

where the peroxynitrite formed is relatively stable in its anionic form. However, ONOO⁻ is the conjugate base of the considerably less stable peroxynitrous acid, HONOO. This weak acid has a pK_a of 6.8, so that, under physiological conditions (pH 7.40), approx 20% of the conjugate acid/base pair would exist in the acid form. Also, HONOO is reported to have a half-life of <1 s under these conditions, decaying according to:

$$HONOO \rightarrow \bullet OH + \bullet NO_2 \tag{12}$$

Thus, reactions (11) and (12) provide another pathway to formation of the hydroxyl radical *(51)*, and represent some of the many alternative chemical and enzymatic interactions that must be incorporated into an understanding of the autoxidation of the neurodegenerative agents in vivo.

To complicate matters further, it is highly probable that each of the 6-HDA-related agents experiences a quite different set of predominant pathways for autoxidation both in vitro and in vivo from that discussed for 6-HDA. The seemingly quite similar 6-ADA, for example, exhibits a number of anomalies when compared to 6-HDA. The initial rate of autoxidation of 6-ADA is much faster than that of 6-HDA *(41,43)*, although this process is thermodynamically less favorable by a factor of approx 1000 *(vide supra)*. Autoxidation of 6-HDA is considerably more stimulated by the addition of $Fe(EDTA)^{2-}$ than is that of 6-ADA. Further, both the metal-catalyzed and innate autoxidations of 6-ADA are virtually unaffected by catalase or SOD, indicating the probable rapid nature of reaction (1) above for 6-ADA, as opposed to what is observed with 6-HDA. Unraveling the in vitro and in vivo autoxidation processes for the most important agent in this group, i.e., 6-HDA, may thus provide us with relevant information concerning its neurodegenerative mode of action. However, such information will not necessarily be directly applicable to understanding the neurodegenerative mode(s) of action of the structural analogs of 6-HDA.

6. REDOX CYCLING AND PROTECTION FROM AUTOXIDATION

Endogenous cellular components and enzymes afford some protection from the potentially destructive processes arising from autoxidation of 6-HDA and related agents. Autoxidation of 6-HDA in the presence of ascorbic acid leads to the relatively innocuous ascorbate free radical *(39)*; likewise, endogenous catecholamines like norepinephrine may effectively trap free radicals *(14)*. Thus, ascorbate, transmitters, and other endogenous species may afford protection through radical trapping mechanisms. The oxidized (Q) forms of 6-HDA analogs can also be reduced back to the original forms (QH_2) by ascorbic acid *(36)* and, perhaps, other reducing agents like NADPH, as depicted above in Fig. 3. In a similar fashion, the enzymes catalase, Cu/Zn SOD (cytosolic), Mn SOD (mitochondrial), GSH peroxidase, GSH *S*-transferase, and GSSG reductase are all capable of providing protection from cytotoxic byproducts of autoxidation *(39)*.

GSH peroxidase is probably the primary enzyme responsible for H_2O_2 removal intraneuronally in the CNS; this enzyme transforms GSH into GSSG and H_2O_2 into H_2O, thus consuming GSH in the process. The GSSG produced may be transformed back to GSH by GSSG reductase; however, NADPH is oxidized to NADP+ in the process. Thus, as summarized in Fig. 3, both chemical and enzymatic interactions can lead to reduction of the initially produced quinone or quinoneimine with concomitant loss of endogenous reducing capacity.

A consequence of multiple chemical and enzymatic reductions of the oxidized quinoid species is a substantial amplification in production of the cytotoxic oxygen byproducts. This process is called redox cycling and is presumed a significant factor in the mode of action of these agents. Indeed, accelerated oxygen consumption in the autoxidation of 6-HDA and 6-ADA has been demonstrated in vitro using ascorbic acid *(36,41,43,52)*. Given the modestly high intraneuronal concentration of this reducing agent, approx 2 mM, studies of the autoxidation of neurodegenerative agents in the presence of this and other intracellular redox active components may be more revealing than the relatively pure autoxidation studies outlined above.

Redox cycling could, thus, be viewed to be a primary factor in the mode of action of 6-HDA analogs through three separate means:

1. Hypoxia owing to accelerated O_2 depletion;
2. General loss of reductive capacity in the neuron through depletion of ascorbic acid, GSH, NADPH, and related reducing agents; and
3. Enhanced production of toxic components like •O_2^-, H_2O_2, and •OH.

7. NUCLEOPHILIC ATTACK ON THE QUINONE OR QUINONEIMINE

The quinones or quinoneimines produced in autoxidation of 6-HDA analogs are susceptible to attack by nucleophiles. Tranzer and Thoenen *(4,5)* first proposed that the primary mode of action of 6-HDA involved the ability of the oxidized form of this agent to undergo intramolecular interactions to form indoles and intermolecular interactions with other nucleophiles, e.g., the thiol groups of proteins, to cause a loss of membrane integrity. Early studies did, indeed, prove that 6-HDA and 6-ADA both intracyclized on oxidation to yield the same oxidized 5,6-dihydroxyindoline, which, on rearrangement, gave 5,6-dihydroxyindole, 5,6-DHI, a melanin precursor (Fig. 4) *(21,53)*.

These studies not only confirmed *(54)* formation of indoles from 6-HDA analogs under physiological conditions (pH 7.40, 37°C), but also provided an obvious potential link to the serotonergic dihydroxytryptamine neurotoxins *(12)*. Additionally, this work introduced the concept that a viable neurotoxin must have a balanced reactivity. The original form of the agent must be sufficiently stable to reach the site of action, while simultaneously it must be sufficiently unstable to effect degeneration on reaching the site of action. The kinetic half-lives for these intracyclization processes, 38 min for 6-HQ and 28 s for 6-AQI,

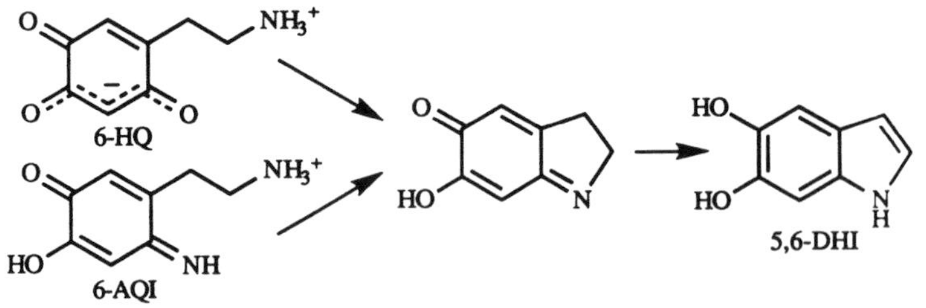

Fig. 4. Intracyclization of 6-HQ and 6-AQI leading to 5,6-DHI.

are relatively slow, leading to expected extraneuronal half-lives, assuming no alternative chemical reactions, of 1.9 and 0.3 h, respectively. This approach provides a surprisingly reasonable, although predictably optimistic, estimate of the measured extraneuronal half-life for 6-HDA in rat CNS of approx 30 min *(40)*.

Intraneuronally, however, nucleophilic attack of the quinone or quinoneimine species by thiol groups (GSH, cysteine, protein-SH) is anticipated to be much more rapid than intracyclization. Liang et al. *(55)*, for example, have shown the in vitro reaction of 6-HQ with GSH to have a half-life of approx 80 s in the presence of 2 mM GSH. Also, they isolated the 6-HDA/GSH adduct from the CNS of animals injected with 6-HDA, verifying its in vivo formation. Convincing evidence for the importance of such nucleophilic reactions, particularly attack by thiols, in the mechanism of action of such agents is provided by the observation that neurodegeneration by 6-HDA is always accompanied by a substantial amount of covalent binding to large mol-weight cellular components *(56)*. Thus, reaction of the oxidized form of an agent with nucleophiles is often viewed as equally, if not more, important than the production of cytotoxic oxygen-derived species in the mode of action of these agents.

An attempt to distinguish between the relative importance of the two primary accepted modes of action for 6-HDA analogs, i.e., cytotoxicity from oxygen byproducts and nucleophilic attack with associated loss of vital cellular function or membrane integrity, was undertaken by Borchardt et al. *(22)*, who synthesized and tested the two monomethyl- and the one dimethyl-ring-substituted derivatives of 6-HDA. Unfortunately, unanticipated complications in other aspects of these compounds, particularly transport by the cellular uptake site, precluded positive and singular identification of the relative importance of either of these two proposed mechanisms.

It should also be noted that attack of the quinone or quinoneimine by a thiol like GSH does not necessarily inactivate the agent with respect to further involvement in oxidative pathways. The initial attack, and even subsequent chemical reactions, should lead to derivatives that are as easy, or perhaps easier, to oxidize thermodynamically or kinetically than is the original agent. Thus, thiol adducts could also participate in redox cycling schemes similar to those shown in Fig. 3, independently leading to enhanced formation of reactive oxygen species, O_2 depletion, and depletion of reductive capacity. Indeed, such thiol adducts may be toxic in their own right *(57)*.

On the other hand, the covalent bonding of 6-HDA analogs to macromolecules and consumption of reducing agents via redox cycling intraneuronally may simply indicate a chemical/reactivity "barrier" that must be overcome before neurodegeneration may proceed. A crude estimation of the accessible intraneuronal concentration of nucleophiles and reducing agents might be on the order of 20–30 mM. Thus, the requirement for an agent to exceed an intraneuronal concentration of 30–100 mM to elicit neurodegeneration *(14,31)* could be interpreted to indicate that such nucleophiles and reducing agents must initially be eliminated through reactions with the agent before the excess "free" agent can accomplish its degenerative task.

8. MITOCHONDRIAL INTERACTIONS

The reduced agents or their oxidized forms could interact with mitochondria in a variety of ways leading to neurodegeneration. Wagner and Trendelenburg *(58)* were the first to suggest that 6-HDA produced neurodegeneration via uncoupling of oxidative phosphorylation. This hypothesis has been questioned by many, including Thakar and Hassan *(59)*, who showed no inhibition of oxidative phosphorylation at 6-HDA concentrations of 0.05 mM; however, mitochondria treated with 0.5 mM 6-HDA subsequently did exhibit some decreases in the respiratory control ratio and ATP synthesis. More recently, Glinka et al. *(60,61)* have shown inhibition of mitochondrial complexes I and IV by 6-HDA in the 10–50 μM range. Their results indicate that it is the reduced QH_2 form of the agent that is responsible for this inhibition, and thus bring into question the necessity for autoxidation of the agent in neurodegeneration.

9. LIPID PEROXIDATION AND LOSS
OF MEMBRANE INTEGRITY

Another possible contributor to the mode of action of 6-HDA-related species is disruption of the normal membrane character or function. Direct attack on nucleophilic components of vital membrane proteins *(62)* could quite easily be imagined to disrupt the normal function of such proteins. Interruption of oxidative phosphorylation, mentioned above, could lead to a depletion of ATP and, thus, contribute to neurodegeneration through repression of the Na+/K+-ATPase and associated loss of ability to maintain the cellular membrane potential. Similar effects on other proteins dependent on ATP could also be easily implicated in the neurodegenerative process as well. Direct lipid peroxidation, initiated by oxygen and semiquinone free radicals generated during autoxidation, is another viable pathway to degeneration *(43)*. The latter possibility is supported by the recent report of Kumar et al., who not only showed decreases in GSH, SOD activity, and glutathione peroxidase activity following 6-HDA treatment, but also showed a decreases in phospholipids and membrane fluidity *(63)*.

10. INDUCTION OF APOPTOSIS

6-HDA and other agents could effect degeneration by triggering a programmed cell death or apoptosis. Walkinshaw and Waters *(33)*, for example, have shown that cell death observed in PC12 cultured cells treated with 6-HDA was mediated by an induced apoptosis. Likewise, Marti et al. *(64)* have recently reported that 6-HDA treatment of neonates elicits apoptotic cell death in dopaminergic nigral cells, as visualized by a tyrosine hydroxylase immunostaining procedure combined with a Nissl counterstain to identify the characteristic intranuclear chromatin clumps.

11. DISRUPTIONS IN Ca^{2+} HOMEOSTASIS

Increases in the intraneuronal concentration of Ca^{2+} may result from dysfunctional membrane ion pumps, as discussed above. Likewise, mitochondrial release of Ca^{2+} is evoked by 6-HDA-related agents *(63,65,66)*. This increased intracellular, extramitochondrial Ca^{2+} could independently lead to neurodegeneration *(67,68)*, and such a mode of action has been strongly implicated in connection with many excitotoxic species *(69)*. However, although disruption in Ca^{2+} homeostasis may effect degeneration by itself or in concert with the other processes, this phenomena, like many previously discussed, may be no more than a result of the ongoing neurodegeneration.

12. DISRUPTIONS IN INTRANEURONAL pH

As seen above in Fig. 3, autoxidation of 6-HDA and related 2,4,5-trisubstituted agents in which an amine group occupies the 5-position on the ring involves the loss of three protons and two electrons. The third proton is lost at physiological pH because the pK_a of the enolic OH group is approx 4–5. If the 30–100 mM intraneuronal concentrations of such agents required to produce degeneration are substantially converted to their oxidized forms, the resulting 30–100 mM concentrations of excess H$^+$ incurred would certainly strain the buffering capacity of the intraneuronal system, which is estimated to have an effective concentration of approx 30–50 mM. Other intraneuronal processes triggered by autoxidative stress could also be envisioned to add to this acidification, eliciting an intraneuronal acidosis with obvious repercussions. Additionally, such a decrease in pH would further accelerate destructive reactions like (11) and (12) outlined above.

REFERENCES

1. Senoh, S., Witkop, B., Creveling, C. R., and Udenfriend, S. (1959) 2,4,5-Trihydroxyphenethylamine, a new metabolite of 3,4-dihydroxyphenethylamine. *J. Am. Chem. Soc.* **81,** 1768,1769.

2. Porter, C. C., Totaro, J. A., and Stone, C. A. (1963) Effect of 6-hydroxydopamine and some other compounds on the concentration of norepinephrine in the hearts of mice. *J. Pharmacol. Exp. Ther.* **140**, 308–316.

3. Laverty, R. and Taylor, K. M. (1970) Effects of intraventricular 2,4,5-trihydroxyphenylethylamine (6-hydroxydopamine) on rat behaviour and brain catecholamine metabolism. *Br. J. Pharmacol.* **40**, 836–846.

4. Tranzer, J. P. and Thoenen, H. (1967) Ultramorphologische veranderungen der sympatischen nervendigunden der katze nach vorbehandlung mit 5- and 6-hydroxydopamin. *Naunyn-Schmiedeberg's Arch. Pharmacol. Exp. Pathol.* **2257**, 343,344.

5. Tranzer, J. P. and Thoenen, H. (1968) An electron microscopic study of selective, acute degeneration of sympathetic nerve terminals after administration of 6-hydroxydopamine. *Experientia* **24**, 155,156.

6. Ungerstedt, U. (1968) 6-Hydroxydopamine induced degeneration of central monoamine neurons. *Eur. J. Pharmacol.* **5**, 107–110.

7. Ungerstedt, U. (1971) Histochemical studies on the effect of intracerebral and intraventricular injections of 6-hydroxydopamine on monoamine neurons in the rat brain, in *6-Hydroxydopamine and Catecholamine Neurons* (Malmfors, T. and Thoenen, H., eds.), North-Holland Elsevier, Amsterdam, pp. 101–128.

8. Levi-Montalcini, R. and Angeletti, P. U. (1966) Immunosympathectomy. *Pharmacol. Rev.* **18**, 619–628.

9. Kostrzewa, R. M. (1989) Neurotoxins that affect central and peripheral catecholamine neurons, in *Neuromethods*, vol. 12, *Drugs as Tools in Neurotransmitter Research* (Boulton, A. A., Baker, G. B., and Juorio, A. V., eds.), Humana, Clifton, NJ, pp. 1–48.

10. Kostrzewa, R. M. and Jacobowitz, D. M. (1974) Pharmacological actions of 6-hydroxydopamine. *Pharmacol. Rev.* **26**, 199–288.

11. Baumgarten, H. G. and Zimmerman, B. (1992) Neurotoxic phenylalkylamines and indolealkylamines. *Handbook Exp. Pharmacol.* **102**, 225–291.

12. Breese, G. R. (1975) Chemical and immunochemical lesions by specific neurotoxic substances and antisera, in *Handbook of Psychopharmacology,* vol. 1 (Iversen, L. L., Iversen, S. D., and Snyder, S. H., eds.), Plenum, New York, pp. 137–189.

13. Jonsson, G. (1983) Chemical sympathectomy agents, in *Handbook of Chemical Neuroanatomy*, vol. 1 (Bjorklund, A. and Hokfelt, T., eds.), Elsevier, New York, pp. 463–480.

14. Sachs, C. and Jonsson, G. (1975) Mechanisms of action of 6-hydroxydopamine. *Biochem. Pharmacol.* **24**, 1–8.

15. Tranzer, J. P. and Thoenen, H. (1973) Selective destruction of adrenergic nerve terminals by chemical analogues of 6-hydroxydopamine. *Experientia* **29**, 314,315.

16. Ma, S., Lin, L., Rhagavan, R., Cohenour, P., Lin, P. Y. T., Bennet, J., Lewis, R. J., Kostrzewa, R., Lehr, R. E., and Blank, C. L. (1995) In vivo and in vitro studies on the neurotoxic potential of 6-hydroxydopamine analogs. *J. Med. Chem.* **38**, 4087–4097.

17. Cheng, A. C. and Castagnoli, N., Jr. (1984) Synthesis and physicochemical and neurotoxicity studies of 1-(4-substituted-2,5-dihydroxyphenyl)-2-aminoethane analogues of 6-hydroxydopamine. *J. Med. Chem.* **27**, 513–520.

18. Lundstrom, J., Ong, H., Daly, J., and Creveling, C. R. (1973) Isomers of 2,4,5-trihydroxyphenethylamine (6-hydroxydopamine). Long-term effects of the accumulation of (3H)-norepinephrine in mouse heart in vivo. *Mol. Pharmacol.* **9**, 505–513.

19. Blank, C. L., Murrill, E., and Adams, R. N. (1972) Central nervous system effects of 6-aminodopamine and 6-hydroxydopamine. *Brain Res.* **45**, 635–637.

20. Ho, B. T., Meyer, A. L., and Taylor, D. (1973) Selective depletion of dopamine following *O*-methylation of 6-hydroxydopamine. *Res. Commun. Chem. Pathol. Pharmacol.* **6**, 47–56.

21. Blank, C. L., McCreery, R. L., Wightman, R. M., Chey, W., Adams, R. N., Reid, J. R., and Smissman, E. E. (1976) Intracyclization rates of 6-hydroxydopamine and 6-aminodopamine analogs under physiological conditions. *J. Med. Chem.* **19**, 178–180.

22. Borchardt, R. T., Burgess, S. K., Reid, J. R., Liang, Y. O., and Adams, R. N. (1977) Effects of 2- and/or 5-methylated analogues of 6-hydroxydopamine on norepinephrine- and dopamine-containing neurons. *Mol. Pharmacol.* **13**, 805–818.

23. Azevedo, I. and Osswald, W. (1977) Adrenergic nerve degeneration induced by condensation products of adrenaline and acetaldehyde. *Naunyn-Schmiedeberg's Arch. Pharmacol. Exp. Pathol.* **300**, 139–144.

24. Jacob, P., III, Kline, T., and Castagnoli, N., Jr. (1979) Chemical and biological studies of 1-(2,5-dihydroxy-4-methylphenyl)-2-aminopropane, an analogue of 6-hydroxydopamine. *J. Med. Chem.* **22**, 662–671.

25. Kostrzewa, R. M., Fukushima, H., Morrow, A., Cohenour, P., Hsi, T., Lehr, R. E., and Blank, C. L. (1980) α-Methyl-6-aminodopamine: depletion of catecholamines in mouse brain and peripheral tissues. *Life Sci.* **27**, 2245–2250.

26. Jarry, H., Lookingland, K. J., Palmer, J. R., and Moore, K. E. (1986) Neurochemical characterization of the actions of 5-amino-2,4-dihydroxy-α-methylphenethylamine (5-ADMP): a selective neurotoxin to central noradrenergic neurons. *J. Pharmacol. Exp. Ther.* **239**, 55–62.

27. Jonsson, G. and Sachs, C. (1971) Uptake and accumulation of ^{3}H-6-hydroxydopamine in adrenergic nerves. *Eur. J. Pharmacol.* **16**, 55–62.

28. Malmfors, T. and Sachs, Ch. (1968) Degeneration of adrenergic nerves produced by 6-hydroxydopamine. *Eur. J. Pharmacol.* **3**, 89–92.

29. Stone, C. A., Porter, C. C., Stavorski, J. M., Ludden, C. T., and Totaro, J. A. (1964) Antagonism of certain effects of catecholamine-depleting agents by antidepressant and related drugs. *J. Pharmacol. Exp. Ther.* **144**, 196–204.

30. Breese, G. R. and Traylor, T. D. (1971) Depletion of brain noradrenaline and dopamine by 6-hydroxydopamine. *Br. J. Pharmacol.* **42**, 88–99.

31. Jonsson, G. (1976) Studies on the mechanisms of 6-hydroxydopamine cytotoxicity. *Med. Biol.* **54**, 406–420.

32. Abad, F., Maroto, R., Lopez, M. G., Sanchez-Garcia, P., and Garcia, A. G. (1995) Pharmacological protection against the cytotoxicity induced by 6-hydroxydopamine and H_2O_2 in chromaffin cells. *Eur. J. Pharmacol.* **293**, 55–64.

33. Walkinshaw, G. and Waters, C. M. (1994) Neurotoxin-induced cell death in neuronal PC12 cells is mediated by induction of apoptosis. *Neuroscience* **63**, 975–987.

34. Vaccari, A. and Saba, P. (1995) The tyramine-labelled vesicular transporter for dopamine: a putative target of pesticides and neurotoxins. *Eur. J. Pharmacol.* **292**, 309–314.

35. Heikkila, R. and Cohen, G. (1971) A mechanism for toxic effects of 6-hydroxydopamine. *Science* **172**, 1257,1258.

36. Heikkila, R. and Cohen, G. (1972) Further studies on the generation of hydrogen peroxide by 6-hydroxydopamine: potentiation by ascorbic acid. *Mol. Pharmacol.* **8,** 241–248.

37. Heikkila, R. E. and Cohen, G. (1973) 6-Hydroxydopamine: evidence for superoxide radical as an oxidative intermediate. *Science* **181,** 456,457.

38. Cohen, G., Heikkila, R. E., and MacNamee, D. (1974) The generation of hydrogen peroxide, superoxide radical, and hydroxy radical by 6-hydroxydopamine, dialuric acid, and related cytotoxic agents. *J. Biol. Chem.* **249,** 2447–2452.

39. Cohen, G. and Werner, P. (1994) Free radicals, oxidative stress, and neurodegeneration, in *Neurodegenerative Diseases* (Calne, D. B., ed.), W. B. Saunders, Philadelphia, PA, pp. 139–161.

40. McCreery, R. L., Dreiling, R., and Adams, R. N. (1974) Voltammetry in brain tissue: the fate of injected 6-hydroxydopamine. *Brain Res.* **73,** 15–21.

41. Sullivan, S. G. and Stern, A. (1981) Effects of superoxide dismutase and catalase on catalysis of 6-hydroxydopamine and 6-aminodopamine autoxidation by iron and ascorbate. *Biochem. Pharmacol.* **30,** 2279–2285.

42. Heikkila, R. E. and Cohen, G. (1975) Cytotoxic aspects of the interaction of ascorbic acid with alloxan and 6-hydroxydopamine. *Ann. NY Acad. Sci.* **258,** 221–230.

43. Borg, D. C., Schaich, K. M., Elmore, J. J., Jr., and Bell, J. A. (1978) Cytotoxic reactions of free radical species of oxygen. *Photochem. Photobiol.* **28,** 887–907.

44. Gee, P. and Davison, A. J. (1984) 6-Hydroxydopamine does not reduce molecular oxygen directly, but requires a coreductant. *Arch. Biochem. Biophys.* **231,** 164–168.

45. Davison, A. J. and Gee, P. (1984) Redox state of cytochrome C in the presence of the 6-hydroxydopamine/oxygen couple: oscillations dependent on the presence of hydrogen peroxide or superoxide. *Arch. Biochem. Biophys.* **233,** 761–771.

46. Bandy, B. and Davison, A. J. (1987) Interactions between metals, ligands, and oxygen in the autoxidation of 6-hydroxydopamine: mechanisms by which metal chelation enhances inhibition by superoxide dismutase. *Arch. Biochem. Biophys.* **259,** 305–315.

47. Liang, Y. O., Wightman, R. M., and Adams, R. N. (1976) Competitive oxidation of 6-hydroxydopamine by oxygen and hydrogen peroxide. *Eur. J. Pharmacol.* **36,** 455–458.

48. Walling, C. (1975) Fenton's reagent revisited. *Accounts Chem. Res.* **8,** 125–131.

49. Monteiro, H. P. and Winterbourn, C. C. (1989) 6-Hydroxydopamine releases iron from ferritin and promotes ferritin-dependent lipid peroxidation. *Biochem. Pharmacol.* **38,** 4177–4182.

50. Malinski, T., Bailey, F., Zhang, Z. G., and Chopp, M. (1993) Nitric oxide measured by a porphyrinic microsensor in rat brain after transient middle cerebral artery occlusion. *J. Cereb. Blood Flow Metab.* **13,** 355–358.

51. Beckman, J. S. (1994) Peroxynitrite versus hydroxyl radical: the role of nitric oxide in superoxide-dependent cerebral injury. *Ann. NY Acad. Sci.* **738,** 69–75.

52. Nappi, A. J. and Vass, E. (1994) The effects of glutathione and ascorbic acid on the oxidations of 6-hydroxydopa and 6-hydroxydopamine. *Biochim. Biophys. Acta* **1201,** 498–504.

53. Blank, C. L., Kissinger, P. T., and Adams, R. N. (1972) 5,6-Dihydroxyindole formation from oxidized 6-hydroxydopamine. *Eur. J. Pharmacol.* **19,** 391–394.

54. Harley-Mason, J. (1953) Melanin and its precursors. Part VI. Further syntheses of 5:6-dihydroxyindole and its derivatives. *J. Chem. Soc.* **1953,** 200–203.

55. Liang, Y. O., Plotsky, P. M., and Adams, R. N. (1977) Isolation and identification of an in vivo reaction product of 6-hydroxydopamine. *J. Med. Chem.* **20,** 581–583.

56. Creveling, C. R., Rotman, A., and Daly, J. W. (1975) Interactions of 6-hydroxydopamine and related compounds with proteins; a model for the mechanism of cytotoxity, in *Chemical Tools in Catecholamine Research,* vol. I (Jonsson, G., Malmfors, T., and Sachs, C., eds.), North-Holland, Amsterdam, pp. 23–32.

57. Monks, T. J. and Lau, S. S. (1992) Toxicology of quinone thioethers. *Crit. Rev. Toxicol.* **22,** 243–270.

58. Wagner, K. and Trendelenburg, U. (1971) Effect of 6-hydroxydopamine on oxidative phosphorylation and on monoamine oxidase activity. *Naunyn-Schmiedeberg's Arch. Pharmacol. Exp. Pathol.* **269,** 110–116.

59. Thakar, J. H. and Hassan, M. N. (1987) Effects of 6-Hydroxydopamine on oxidative phosphorylation of mitochondria from rat striatum, cortex, and liver. *Can. J. Physiol. Pharmacol.* **66,** 376–379.

60. Glinka, Y. Y. and Youdim, M. B. H. (1995) Inhibition of mitochondrial complexes I and IV by 6-hydroxydopamine. *Eur. J. Pharmacol.* **292,** 329–332.

61. Glinka, Y., Tipton, K. F., and Youdim, M. B. H. (1996) Nature of inhibition of mitochondrial respiratory complex I by 6-hydroxydopamine. *J. Neurochem.* **66,** 2004–2010.

62. Saner, A. and Thoenen, H. (1971) Model experiments on the molecular mechanism of action of 6-hydroxydopamine. *Mol. Pharmacol.* **7,** 147–154.

63. Kumar, R., Agarwal, A. K., and Seth, P. K. (1995) Free radical-generated neurotoxicity of 6-hydroxydopamine. *J. Neurochem.* **64,** 1703–1707.

64. Marti, M. J., James, C. J., Oo, T. F., Kelly, W. J., and Burke, R. E. (1996) Striatal injection of 6-hydroxydopamine induces apoptotic cell death of nigral dopaminergic neurons in neonatal rats. *Soc. Neurosci. Abstracts* **22,** 41.

65. Frei, B. and Richter, C. (1986) *N*-Methyl-4-phenylpyridine (MMP+) together with 6-hydroxydopamine or dopamine stimulates Ca^{2+} release from mitochondria. *FEBS Lett* **198,** 99–102.

66. Reichman, N., Porteous, C. M., and Murphy, M. P. (1994) Cyclosporin A blocks 6-hydroxydopamine-induced efflux of Ca^{2+} from mitochondria without inactivating the mitochondrial inner-membrane pore. *Biochem. J.* **297,** 151–155.

67. Jewell, S. A., Bellomo, G., Thor, H., Orrenius, S., and Smith, M. T. (1982) Bleb formation in hepatocytes during drug metabolism is caused by disturbances in thiol and calcium homeostasis. *Science* **217,** 1257–1259.

68. Bellomo, G., Jewell, S. A., Thor, H., and Orrenius, S. (1982) Regulation of intracellular calcium compartmentation: studies with isolated hepatocytes and t-butyl hydroperoxide. *Proc. Natl. Acad. Sci. USA* **79,** 6842–6846.

69. Choi, D. W. (1988) Glutamate neurotoxicity and diseases of the nervous system. *Neuron* **1,** 623–634.

70. Iversen, L. L. (1970) Inhibition of catecholamine uptake by 6-hydroxydopamine in rat brain. *Eur. J. Pharmacol.* **10,** 408–410.

2

The Use of Neurotoxins to Lesion Catecholamine-Containing Neurons to Model Clinical Disorders

Approach for Defining Adaptive Neural Mechanisms and Role of Neurotrophic Factors in Brain

Charles R. Breese and George R. Breese

1. INTRODUCTION

The compound 6-hydroxydopamine (6-OHDA) was the first neurotoxin documented to produce long-term depletion and destruction of norepinephrine in peripheral sympathetic neurons without affecting central catecholamine content *(1–3)*. Subsequently, it was found that this neurotoxic compound would produce a selective destruction of brain catecholamine-containing neurons if administered directly into brain *(4–7)*. In addition to the use of 6-OHDA, the discovery and development of a second compound, 1-methyl-4-phenyl-1,2,3,6-tetrahydropyridine (MPTP), which selectively destroys central dopaminergic neurons, will be reviewed *(8,9)*. The various techniques utilized to provide effective lesions of catecholamine-containing neurons with these neurotoxins will be described *(see 10,11)*, as well as how the use of these neurotoxic compounds has made significant contributions to our knowledge of the structural organization, physiological role, and neural mechanisms of central catecholamine-containing systems. Recently initiated work to investigate the action and role of neurotrophic factors in relation to the growth, maintenance, and recovery of central catecholamine-containing neurons after neurotoxin treatment will be examined. Subsequently, we will review how these neurotoxin treatments are being used to model two clinical disorders with neuropathology to dopamine-containing systems in brain, parkinsonism and Lesch-Nyhan disease. Finally, we will describe the various efforts undertaken with the neurotoxin-treated animals to enhance dopaminergic function in vivo and the application of these results to the restoration of dopaminergic function in Parkinson's patients.

From: Highly Selective Neurotoxins: Basic and Clinical Applications *Edited by: R. M. Kostrzewa.* Humana Press Inc., Totowa, NJ

2. DESTRUCTION OF CENTRAL CATECHOLAMINE-CONTAINING NEURONS WITH 6-OHDA ADMINISTRATION INTO BRAIN

2.1. Administration of 6-OHDA into Brain of Adult Animals

Because 6-OHDA does not cross the blood–brain barrier *(1–3)*, methods were developed to administer 6-OHDA directly into the central nervous system (CNS) in order to attain destruction of central catecholamine-containing fibers. One approach was to administer 6-OHDA into the cerebroventricular system either by intracisternal or intraventricular (icv) injections *(5–7,11–14)*. Another approach was to administer 6-OHDA directly into the medial forebrain bundle or other regions with catecholamine-containing cell bodies *(4,15–17)*. Early studies were critical of microinjecting 6-OHDA directly into brain because of a proven degree of nonspecific damage *(18,19)*; however, this view is not universal, and microinjection of 6-OHDA into specific regions of brain continues to be widely used to produce site-specific lesions of catecholaminergic neurons *(10,17)*.

Following icv administration of 6-OHDA into brain, norepinephrine (NE) was gradually reduced over time. In contrast, the content of dopamine (DA) in the brain was initially shown to increase during the first 24 h after 6-OHDA treatment, followed by a reduction in DA content by 48 h *(6,20)*. Early biochemical studies found the degree of destruction by 6-OHDA was greater for NE than that for DA-containing neurons *(6,7)*. Subsequent evaluations of DA loss in relation to the reduction of tyrosine-hydroxylase levels in brain indicated that dopaminergic neurons were destroyed (*see 11,21*). Likewise, the loss of dopamine beta-hydroxylase (DBH) activity supported the view that 6-OHDA-induced reduction of NE was owing to a destruction of noradrenergic neurons *(22)*. However, after icv injection, there was some sparing of dopaminergic neurons in the pars compacta of the substantia nigra (SNR). This was likely the result of the inability of 6-OHDA to reach this area of brain following icv administration of 6-OHDA *(6,7)*. For this reason, an approach was sought to maximize the action of6-OHDA on DA-containing neurons when using this approach. Breese and Traylor *(6)* found that pretreatment with pargyline (50 mg/kg) and other monoamine oxidase (MAO) inhibitors enhanced the destructive action of6-OHDA on both noradrenergic and dopaminergic neurons *(6,11)*. The degreeof depletion produced by 6-OHDA administered icv to adult rats was further enhanced by a second dose of 6-OHDA with the MAO inhibitor *(6,11)*.

2.2. Administration of 6-OHDA to Neonates to Destroy Brain Catecholaminergic Neurons

In addition to the protocol to reduce brain catecholamine content in adult rats, a protocol was developed to destroy catecholamine-containing neurons in neonatal rats *(23–27)*. Treatment of neonates with 6-OHDA produced the greatest destruction of central catecholaminergic neurons of all approaches tested

(23). Neonatal treatment with 6-OHDA is accomplished by a single intracisternal injection of 6-OHDA at approx 3 d of age *(25,28,29)*. Unlike the adult rat, no pretreatment with pargyline or MAO inhibitors was required to produce near-complete destruction of catecholamine-containing neurons.

2.3. Procedures for Selective Destruction of Brain Noradrenergic Neurons

As implied above, early investigations of the effect of 6-OHDA on the content of individual catecholamines in brain noted that central 6-OHDA administration into the ventricular system produced a greater reduction of NE than DA *(6,7)*. In order to gain greater specificity with 6-OHDA against noradrenergic systems, three approaches were undertaken to destroy brain noradrenergic neurons without destruction of brain dopaminergic neurons. One approach was the icv administration of multiple small doses of 6-OHDA to adult and neonatal rats *(11,21,30,31)*. However, in spite of the specificity of this approach, not all noradrenergic terminals were destroyed *(11,21,31,32)*. Another approach to reducing NE-containing neurons was to lesion the locus cerulus (LC) by direct microinjection of 6-OHDA into this brain region *(33)*; however, this approach only reduced NE levels at brain sites innervated by the LC. Regardless of the disadvantages, both methods were sufficient to result in supersensitive noradrenergic receptors, resulting in accentuated responses to noradrenergic agonists *(34)*. Finally, there are limited reports that have used pretreatment with a drug acting on dopaminergic uptake sites prior to administration of a higher dose of 6-OHDA (i.e., higher than usually used to reduce only NE) to neonatal rats. This treatment protocol enhanced the overall destruction of noradrenergic neurons in these animals *(35,36)*. The only disadvantage was a moderate reduction of DA in cortical regions without an effect on DA occurring in other regions of brain *(35)*. In spite of this latter reported success using a DA uptake inhibitor to increase the selectivity of 6-OHDA on NE-containing neurons, this approach has received limited use. Although this strategy should be useful to induce selective destruction of noradrenergic neurons in adult animals, this technique appears not to have been used for this purpose.

2.4. Selective Destruction of Brain Dopaminergic Neurons with 6-OHDA or MPTP

2.4.1. Selective Reduction of Dopamine-Containing Neurons with 6-OHDA

Based on previous knowledge that some antidepressants block NE uptake *(1,37)*, desipramine and protriptyline were administered prior to 6-OHDA to adult rats and shown to attain selective destruction of dopaminergic neurons without destruction of NE-containing neurons after 6-OHDA administration *(21,38)*. The specific reduction of brain DA content by 6-OHDA, when administered with an uptake inhibitor of NE to adult rats, was accentuated by the addition of pargyline *(11,21)*. The same pretreatment with desipramine (10 mg/kg) to neonatal 3-d old rat pups prior to intracisternal administration of 100 μg of 6-OHDA also produced a major reduction of dopaminergic neurons without affecting the content of NE *(25,39)*. DA-containing neurons could also be selectively reduced in brain by microinjection of 6-OHDA bilaterally into the pars compacta, with or without desipramine pretreatment *(33,40)*.

2.4.2. 1-Methyl-4-Phenyl-1,2,3,6-Tetrahydropyridine

MPTP was discovered as a contaminant in synthesized meperidine, and was found to be responsible for Parkinson's-like symptoms of individuals who ingested the narcotic contaminated with MPTP *(8,9)*. Systemically administered MPTP to primates was demonstrated to cause severe motor abnormalities, producing the majority of cardinal motor symptoms associated with Parkinson's disease, with the exception of tremor *(41–44)*. These effects were attributed to a loss of dopaminergic neurons that follows MPTP treatment *(45,46)*. An advantage of this neurotoxin is that it can be given systemically to primates to destroy dopaminergic neurons. However, unlike the persistent and often worsening symptomatology in parkinsonism, the Parkinson syndrome produced by MPTP in primates recovered with time *(46–49)*. Subsequent to defining the effects of MPTP in primates, administration of MPTP to specific strains of mice (C57/BL6, for example) was shown to produce similar neuropathological changes in dopaminergic neurons like those observed in primates *(50–55)*; however, this destructive process has not been seen in all mouse strains or other species of rodents. 2-Methyl MPTP was shown to be more effective than MPTP in C57/BL6 mice *(56)*, but produced no behavioral deficiencies and less neurotoxicity when given to marmosets *(57)*.

2.5. Neurotoxic Mechanisms of 6-OHDA and MPTP on Catecholaminergic Neurons

Based on observations that drugs that blocked NE or DA transporters on catecholamine-containing neurons prevented neurotoxicity of 6-OHDA, it is apparent that the 6-OHDA must enter the neuron before neural toxicity can be attained *(6,35)*. In a series of manuscripts, Heikkila and Cohen *(58–60)* provided evidence that hydrogen peroxide and superoxide radical generation by 6-OHDA were responsible for the neurotoxicity once 6-OHDA entered the neuron, although the exact intracellular components affected by 6-OHDA to cause subsequent degeneration of catecholamine-containing neurons remain largely unknown. Given the availability of tools to investigate apoptosis vs necrosis, the mechanism by which 6-OHDA affects destruction of catecholamine-containing neurons may be a worthwhile exercise for additional assessment. In this regard, Jeon et al. *(61)* recently suggested that 6-OHDA does not act by an apoptotic mechanism.

The neurotoxic effect of MPTP is believed to be related to its metabolism by MAO-B in glial cells to produce MPDP+ and subsequently MPP+ *(62–64)*, which then accumulates within dopaminergic neurons *(65,66)*. Presumably, the specificity of MPTP for DA-containing neurons is related to its metabolites being taken up into dopaminergic neurons by way of the uptake transporter *(66)*. It is believed that the ultimate toxicity of MPP+ is owing to uptake in mitochondria *(67,68)* and subsequent inhibition of the complex of the respiratory chain *(69,70)*. It has been reported that transgenic mice that overexpress superoxide dismutase (SOD), which protect cells against superoxide radicals *(71)*, are resistant to MPTP *(72)*. This latter observation suggests a possible role of oxygen radicals in the action of MPTP. It has also been found that nitrous oxide (NO) synthetase inhibitors were effective antagonists of MPTP toxicity of dopaminergic neurons *(73)*. Likewise, mice lacking NO synthetase are resistant to MPTP neurotoxicity *(73)*. Therefore, although considerable progress has been made, the exact mechanism of the neurodegenerative action of MPTP has yet to be fully established.

3. FUNCTIONAL CONSEQUENCES OF SELECTIVE REDUCTION OF CATECHOLAMINE-CONTAINING NEURONS WITH NEUROTOXINS

3.1. Acute Physiological Consequences of Neurotoxin Administration

Immediately following and a few days after administration of 6-OHDA, the negative effect of reducing brain catecholamines on physiological functions becomes apparent. Rats appear sedated during the first few hours after treatment and resemble to some degree animals treated with reserpine *(6,11,21,31,74–76)*. This was also accompanied by a reduction in body temperature *(77–80)*. Following acute administration of 6-OHDA into the pars compacta or icv to adult rats, animals exhibit severe aphagia and adipsia for several days, which recovers with time *(6,23,31,81–84)*. A similar reaction to 6-OHDA appears to occur in monkeys *(85,86)* and cats *(87)*. The severity of this effect in rats and primates is such they often require intubation of nutrients for up to 2 wk postlesion to maintain life *(81,84,85,88)*. Ungerstedt *(81)* and others *(82–87)* likened these physiological changes to those seen in the lateral-hypothalamic syndrome following bilateral electrolytic lesions *(89–91)*. In spite of an appearance of recovery in 6-OHDA-lesioned rats, chronic deficits in ingestive behavior persisted when lesioned rats were challenged with various pharmacological agents that alter ingestive behaviors *(31,32,84)*. For example, "recovered" 6-OHDA-lesioned rats, like lateral-hypothalamic-lesioned rats, exhibited reduced feeding to hypoglycemia *(82,84)*, consumed less saline after DOCA treatment *(25,84)*, failed to increase water consumption to angiotensin *(92)*, and had a delayed onset of drinking to isoproterenol *(93)*. Additionally, 6-OHDA-treated rats failed to increase fluid intake when sucrose was added to the water *(11,31,84,94,95)*. Similar acute changes in function appear to accompany the administration of MPTP, but the effects of MPTP are not as severe as the changes observed after acute treatment with 6-OHDA.

Motor movements are impaired in animals lesioned with 6-OHDA, since persistent akinesia and rigidity are observed *(11,23,33,88)*. As will be discussed subsequently, these latter deficits have allowed rodents and primates treated with 6-OHDA to serve as an animal model of Parkinson's disease *(11,23,96,97)*. Nonetheless, the degree of recovery of motor function with time may be sufficient that the neurotoxin-lesioned animals appear to be near normal *(33,79,94,95,97,98)*. The mechanisms by which recovery occurs in 6-0HDA-lesioned rats and MPTP-treated animals appear to include a recovery of "injured" neurons *(11,48)*, an increase in the turnover of DA *(11,48,88,98)*, and supersensitivity of catecholamine receptors *(11,99–101)*. After recovery from MPTP is apparent, DA turnover has been shown to be elevated in the caudate of these animals, as reflected by an improved HVA-to-DA ratio *(48,102)*. Several months after MPTP administration, nucleus accumbens levels of DA were found to return nearly to normal *(48)*. Interestingly, the recovery of mesolimbic DA content appears to parallel the behavioral recovery, more so than the increased rate of DA turnover observed in the caudate-putamen *(48)*.

In addition to these acute physiological changes following neurotoxin treatment, 6-OHDA-treated animals exhibit a variety of deficits in operant behaviors, including impairment of conditioned avoidance responses, self-stimulation, and spatial learning tasks *(11,31,98,103–106)*. Generally, deficits in operant behaviors were most apparent when first exposed to the chamber following the lesion. Responding was shown to improve with successive exposures to the chamber until normal response rates often were observed *(98,103,104,106,107)*. Subsequently, administration of a relatively low dose of α-methyltyrosine, a tyrosine hydroxylase inhibitor, having virtually no effect in unlesioned rats, was shown to precipitate the operant deficits seen early in testing of the 6-OHDA-lesioned animals *(11,31,98)*, whereas inhibitors of DBH that blocked NE synthesis were without effect on these measures *(11,31)*. These latter observations provided convincing evidence that recovery of operant responding was related to a compensatory increase in the synthesis of DA. Of additional interest to adaptive changes reported after a 6-OHDA lesion, Stodgell et al. *(96)* showed that discrimination training could reverse the 6-OHDA-induced striatal DA deficiency in rats lesioned as neonates. Although such a phenomenon has not previously been encountered, this work could have incredible implications if reproduced.

3.2. Functional Consequences of Reducing Brain NE

Several reports based on investigations with 6-OHDA have suggested possible roles of the central noradrenergic systems in physiological function. As noted above, acute 6-OHDA administration produced hypothermia *(77–80,108)*, which was accentuated when rats were placed in the cold after injection *(79)*. Since this reduction in temperature was absent in animals previously treated with 6-OHDA to destroy catecholamine-containing neurons in brain, it was concluded that the hypothermia was the result of an acute release of catecholamines *(11)*. Subsequent efforts in rats with preferential reduction of NE by 6-OHDA revealed that the hypothermia produced by acute administration of 6-OHDA was dependent on release of endogenous NE *(79,80)*. For example, preferential reduction of NE, but not DA content, not only blocked the hypothermic response to acutely administered 6-OHDA, but also blocked the hypothermic response observed following acute administration of other phenylethylamines *(79,80)*. In contrast to the acute hypothermia produced by central administration of 6-OHDA, rats with chronic destruction of central catecholamine-containing neurons had no difficulty adapting to cold or heat *(77,80,93)*.

In addition to temperature control, a number of other physiological consequences of noradrenergic systems have been reported following 6-OHDA treatment. Of these, central noradrenergic control of cardiovascular and respiratory function was most prominent *(109–111)*. Particularly apparent were the accentuated physiological changes in blood pressure, heart rate, and respiratory rate after degeneration of noradrenergic pathways *(110)*. In addition, noradrenergic function has also been proposed to control sleep patterns *(112)*. The destruction of noradrenergic neurons has been implicated in aggressive behavior and irritability *(113–115)*, and learning impairments *(116)*. Investigators have implicated noradrenergic systems in obesity and overeating following lesioning of noradrenergic fibers *(117)*, and hypothalamic noradrenergic systems have been implicated in the modulation of intestinal motility *(118)*. Finally, the consequences of destroying noradrenergic function led to a theoretical position that altered noradrenergic function was involved in the core symptoms of schizophrenia *(see* Section 5.1.3.; *119–121)*. Although many key observations concerning physiological involvement of central noradrenergic neurons were made over two decades ago, these findings have since received little attention.

3.3. Functional Consequences of Reducing Brain DA in Adult 6-OHDA-Lesioned Rats

Studies in rats with selective lesions to either noradrenergic or dopaminergic neurons revealed that the acute and chronic deficits in ingestive behaviors were dependent on loss of dopaminergic neurons following 6-OHDA treatment *(11,25,81,84,85,95)*. Following acute administration of 6-OHDA that reduced only DA content, a major disruption of ingestive behavior was observed, just as when both catecholamine-containing systems were disrupted *(6,33,95)*. Presumably, this documentation indicated that the major physiological consequences of the lateral hypothalamic syndrome are related to the destruction of the dopaminergic innervation, which courses through this brain region in the medial forebrain bundle *(33)*.

The deficits in operant responding produced by 6-OHDA treatment were also observed after the preferential reduction of DA with 6-OHDA. Of the operant tasks, results obtained on self-stimulation reward following 6-OHDA treatments that preferentially reduced brain DA or NE content had the greatest theoretical impact. Even though self-stimulation and other operant responding had long been believed to be dependent on functioning noradrenergic neurons *(119,122)*, preferential reduction DA-containing neurons reduced this response, whereas the destruction of noradrenergic neurons was without effect *(31,32,105,123,124)*. The recovery in self-stimulation and operant responding observed with time in rats with a preferential reduction of DA could be disrupted by pharmacological treatments that reduced the synthesis of DA *(11,31,105,124)* or antagonists of D_2-DA receptors *(88,104,125,126)*. Furthermore, since DBH inhibitors, which block synthesis of NE, had no effect on these behaviors, a compensatory increase of dopaminergic synthesis was proposed to account for the time-dependent recovery in operant responding in lesioned animals *(31,98,103–105)*. Another contributing factor to recovery from the lesioning of dopaminergic neurons was the supersensitivity to DA agonists shown by6-OHDA-lesioned animals *(11,99–101)*. For example, L-DOPA or apomorphine administered to rats with bilateral lesions of dopaminergic neurons show a marked increase in locomotor responsiveness, accompanied by stereotypical behaviors, including intense sniffing, paw treading, and head bobbing, reflecting supersensitivity of DA receptors *(29,101)*. In contrast, unilaterally lesioned rats show a stereotypical increase in turning behavior when challenged with drugs acting on DA receptors *(99)*.

3.4. Responses in Neonatally 6-OHDA-Lesioned Rats: Comparison to Effects Seen in Adult 6-OHDA-Lesioned Rats

The initial motivation for reducing DA content during development was to determine if destruction of dopaminergic neurons early in development would illustrate distinctive biological consequences based on the age at which the lesioned occurred. As noted with other types of brain injury during development *(127)*, lesioning of dopaminergic neurons in young animals resulted in less impairment in ingestive behavior and motor function than in rats lesioned as adults, even though loss of DA content approached 99% *(23,125,128–131)*. Since neonates did not show the acute deficits observed when rats were lesioned as adults *(29,132)*, these data suggested that there was a compensatory mechanism in rats lesioned as neonates, which adjusts for the neonatal loss of dopaminergic neurons *(29,132)*.

The loss of body weight in neonatal 6-OHDA-lesioned rats has been attributed to the ingestive changes produced by 6-OHDA destruction of dopaminergic neurons *(23,25,81,84,131)*. In support of this latter view, neonatally lesioned rats with only DA reduced showed a persistent deficit in sucrose intake similar to that seen in lesioned adults *(25)*. Further, other selective challenges would reveal that motor function was not fully compensated in the neonatally lesioned rats *(126)*. Drug treatment revealed major differences in the responses to specific D_1-DA and D_2-DA agonists between neonate and adult lesioned rats *(28,29,132)*. The neonatally lesioned rats were more sensitive to D_1-DA agonists than D_2-DA agonists. Conversely, adult 6-OHDA-lesioned rats were more sensitive to D_2-DA than to D_1-DA agonists *(39,132)*.

Other differences were found to exist in behavioral responses to pharmacological treatments of adult rats lesioned as neonates compared to rats lesioned with 6-OHDA as adults. This was the appearance of self-injurious behavior (SIB) when L-DOPA was administered to rats lesioned as neonates. This behavior was not observed following L-DOPA treatment in the adult lesioned rats *(28,29,132)*. In addition, various behavioral measures were found to be different based on the age dependency of the lesion *(28,29,132,133)*. Head nodding and paw treading were prominent in the adult-lesioned rats, but were absent in the neonatally lesioned animals after L-DOPA administration *(29)*. Subsequently, SIB induced by L-DOPA *(28,29)* was found to be dependent on D_1-DA receptors, with SIB blocked by a D_1-DA receptor antagonists *(133)*. Another instance of a difference between rats lesioned as neonates and those lesioned as adults was observed when these animals were given DA antagonists *(125,126)*. This comparison revealed that rats lesioned as adults exhibited severe akinesia after D_2-DA antagonist, whereas the rats lesioned as neonates showed virtually no akinesia at doses producing virtual total immobility in nonlesioned control animals *(126)*. Finally, when administered a D_1-DA agonist, the pattern of 2-deoxyglucose accumulation in brain of the neonatally lesioned rats was found to be distinct from that in adult lesioned rats (unpublished data).

4. IDENTIFICATION OF SPECIFIC DOPAMINERGIC MECHANISMS WITH THE USE OF ANIMALS TREATED WITH NEUROTOXINS

4.1. Documentation of the Mechanism of Action of Centrally Acting Stimulants

The preferential destruction of dopaminergic or noradrenergic neurons with 6-OHDA led to our understanding of the mechanism by which *d*-amphetamine and similar phenylethylamines induced locomotion at moderate doses and stereotypes at higher doses. These studies demonstrated without ambiguity that the stimulant actions of these drugs were dependent on the release of DA in brain *(105,134–137)*. Ungerstedt *(137)* used turning behavior in rats with unilateral 6-OHDA lesions to reach the conclusion that intact dopaminergic neurons were essential for the actions of *d*-amphetamine. Since this early work, lesioning of dopaminergic neurons has been a popular method to investigate the actions of drugs thought to act by releasing central DA. This work has included derivatives of amphetamine, as well as evaluations of the action of cocaine *(138–143)*. In addition to defining the action of drugs that result in release of DA *(11)*, 6-OHDA-treated animals with specific lesions to dopaminergic neurons demonstrated that drugs could be identified that acted directly on dopaminergic receptors. In the latter case, the motor response of such direct-acting DA agonists was enhanced in the lesioned rats because of the presence of supersensitive DA receptors *(99,101)*. After it was discovered that a distinct D_1-DA receptor existed in brain *(144–146)*, the latter adaptive change in receptor sensitivity in the lesioned animals became particularly important for defining the role of the differing DA receptor subtypes in brain function.

4.2. Discovery of "Priming" D₁-DA Receptor Supersensitivity in Neonatally Lesioned Rats

When testing the action of L-DOPA to produce SIB in neonatally lesioned rats *(28,29)*, a second L-DOPA treatment was found to be more effective in producing SIB than the initial dose of L-DOPA *(29)*. Since it had been demonstrated that this and other behavior abnormalities produced by L-DOPA in the neonatally lesioned animals were related to D_1-DA receptor activation *(132,133)*, it was tested whether repeated administration of a D_1-DA agonist, SKF-38393, was required to reach maximal behavioral supersensitivity to this specific agonist *(147)*. When the D_1-DA agonist was administered at weekly intervals, a progressive increase in responsiveness was observed until a maximal behavioral response was present by the fourth dose of the D_1-DA agonist *(147)*. This sensitization phenomenon was referred to as "priming of D_1-DA receptor sensitivity." Such a sensitization process was not observed after repeated administration of D_2-DA agonist *(147)*. Further, a D_2-DA antagonist did not block the priming sensitization caused by a D_1-DA agonist *(147)*. Johnson et al. *(148)* demonstrated that A-68930, a full D_1-DA agonist, also induced priming in neonatally lesioned rats. A disadvantage of A-68930 was that at a large dose, D_1-DA receptor activation was desensitized although at reasonable doses, the full agonist was more potent than SKF-38393. Subsequent evaluation of priming of D_1-DA receptor supersensitivity demonstrated that this phenomenon was long-lasting and possibly permanent *(147)*. Another important finding was that sensitization of D_1-DA receptors was blocked if each dose of the D_1-DA agonist was preceded with the NMDA antagonist, MK-801 *(149)*.

The D_1-DA receptor priming in the neonatally lesioned rats and pharmacological treatments that blocked this sensitization inspired subsequent studies to define the basis of cocaine and *d*-amphetamine sensitization with repeated treatment *(150)*. In fact, discovery of this priming of D_1-DA receptor responses led to the discovery that the sensitization that accompanies repeated treatment with cocaine and amphetamines is based on "priming" of D_1-DA receptors *(151–153)*. Furthermore, the sensitization of the behavioral response to cocaine and amphetamine-related compounds *(154)* has been shown to be blocked by administration of a D_1-DA antagonist into the ventral tegmental area (VTA; *155*). The VTA seems particularly relevant to this phenomenon because microinjection of SKF-38393 into this brain region substituted for repeated injections of cocaine *(156)*. Likewise, it is known that systemic administration of NMDA antagonists, as well as their microinjection into the VTA, prevents the behavioral sensitization to psychostimulants *(157–159)*. The current work on psychostimulants suggests that SKF-38393 microinjected into the VTA of rats lesioned as neonates should facilitate priming of D_1-DA receptors, and administration of MK-801 into this site should block priming of the response to systemically administered D_1-DA agonists in neonatally lesioned rats. Given the similarities to cocaine and amphetamine sensitization, studying the sensitization phenomenon related to D_1-DA receptors in neonatally lesioned rats may be a means to understand the basis of psychostimulant abuse, without the complication of neurochemical changes related to the other pharmacological actions of this drug class.

4.3. Discovery of a D_1/D_2 Dopamine Receptor Interaction

In spite of the priming of D_1-DA receptor sensitivity in rats lesioned as neonates, administration of a D_1-DA agonist rarely produced SIB in most neonatally lesioned rats, even though D_1-DA antagonists blocked the self-injurious behavioral response to L-DOPA. This latter finding suggested the possibility that D_2-DA receptor activation, which would also occur with L-DOPA administration, was needed for maximal expression of the behavioral response, even though D_1-DA receptors were absolutely essential for SIB in the rats lesioned as neonates *(133)*. In support of this hypothesis, combining a D_1 with a D_2-DA agonist produced SIB in the neonatally lesioned rats exhibiting SIB to L-DOPA administration *(133)*, even though SKF-38393 alone did not produce this response. This phenomenon was interpreted as being a D_1/D_2-DA receptor "coupling," an interaction not previously recognized. It would seem reasonable that coupling of the D_1-DA receptor to the D_2-DA receptor accounts for a D_1-DA receptor antagonist blocking the action of D_2-DA agonist in normal unlesioned rats, while not having this action in lesioned rats *(132,160)*. Since the original descriptions of the D_1/D_2-DA receptor coupling *(132,160)* and the interaction between D_1/D_2-DA receptors *(133)*, several other investigators have subsequently encountered this coupling phenomenon associated with D_1-DA and D_2-DA receptors *(161–163)*.

4.4. Adaptive Changes in Brain Peptides After 6-OHDA Treatment

Enkephalin and tachykinin peptides have been shown to be present in high concentrations in the basal ganglia *(164)*. The striatopallidal pathway expresses enkephalin, and the striatonigral pathway expresses tachykinins *(164)*. Enkephalin-containing cell bodies have been identified in the SNR, which were not eliminated by 6-OHDA, indicating that this peptide was not in dopaminergic neurons *(165)*. Ultrastructural studies have shown that dopaminergic neurons directly synapse on enkephalin- and substance P-containing neurons in the rat *(166–168)*. Lesioning of dopaminergic neurons in the rat with 6-OHDA causes adaptive changes in these families of peptides in the striatum. After a unilateral 6-OHDA lesion in rats, met- and leu-enkephalin were increased *(169–174)*, and substance P was decreased in the striatum ipsilateral to the lesion *(169,170,175–177)*. When analyzed by electron microscopy following unilateral destruction of nigrostriatal DA neurons, the enkephalin-immunoreactive synaptic bouton profiles in the striatum were enhanced by the lesion, without affecting the boutons on postsynaptic targets *(178)*. In rats, lesioning of dopaminergic neurons was also accompanied by an increase in mRNA for enkephalin *(179–185)* and a decrease in preprotachykinin mRNA *(173,180,182,183,185)*. Since dopaminergic transplants reversed the increases in enkephalin levels in the striatum *(186,187)*, it was hypothesized that enkephalin levels in the striatum are controlled by dopaminergic input from the SNR, a conclusion consistent with the changes seen after lesioning dopaminergic neurons.

Although bilateral lesions in both neonate and adult animals consistently increase both protein and mRNA levels for enkephalin *(176,177,188–192)*, only the neonatal lesion consistently resulted in a decrease in tachykinin peptides in the striatum *(175–177,190)*. Snyder-Keller *(193)* found that neonatal administration of 6-OHDA did not affect the patchy distribution of substance P immunostaining nor did this treatment affect the distribution of enkephalin in the matrix of the striatum. In addition, neonatal destruction of dopaminergic neurons was found to enhance the release of acetylcholine, but had no effect on the action of substance P-induced acetylcholine release *(194)*.

Because of variable results in primates, the effects of a unilateral lesion with 6-OHDA on enkephalin and substance P levels were examined in the marmoset and found that enkephalin immunoreactivity was consistently increased, as seen in the rat, but variable results were obtained on the measure for substance P *(195)*. This study suggested that unlike the changes seen in rat, lesioning of dopaminergic neurons has little effect on substance P function in this primate. Nonetheless, in human Parkinson patients with greater than an 80% reduction in DA content, substance P was reduced in putamen, and met-enkephalin content was reduced in striatum *(196)*. The reduced enkephalin content in striatum of humans with Parkinson's disease contrasts with results obtained in lesioned rats and marmosets where enkephalin was increased when brain DA content was reduced. This suggests that destruction of enkephalinergic neurons may also be associated with the neurodegenerative progression of dopaminergic neuronal loss associated with parkinsonism.

In addition to enkephalin and tachykinin peptides, lesioning of the nigrostriatal pathway induces cholecystokinin and neurotensin mRNA in the rat striatum *(197,198)*. The cholecystokinin mRNA expression was restricted to the dorsolateral quadrant of the ipsilaterally denervated striatum *(197)*. The increase in cholecystokinin could be considered surprising because this peptide has been shown to colocalize with tyrosine hydroxylase in the pars compacta and VTA, although not all dopaminergic neurons contain this peptide *(199)*. In addition, dynorphin mRNA expression in patch neurons showed an average 75% reduction, whereas matrix neurons showed no change in this peptide after lesioning of dopaminergic neurons *(179)*. Finally, neonatal lesions had no effect on preprosomatostatin mRNA *(191)*, but was found to increase neurokinin B, a member of the tachykinin family, on the ipsilateral side of a 6-OHDA lesion *(200)*.

The effect of 6-OHDA lesion to dopaminergic neurons on the levels of enkephalin and substance P has also been examined in nucleus accumbens and SNR in rat. Enkephalin and its mRNA levels in the SNR, measured by immunohistochemistry and *in situ* hybridization, have been shown to be enhanced by 6-OHDA *(165,179)*. Following lesions of ascending dopaminergic neurons to the nucleus accumbens, contents of mRNA for enkephalin and peptides for leu- and met-enkephalin were increased *(171,184)*, and mRNA for dynorphin and substance P was decreased more in rostral parts than in caudal regions of the nucleus *(171,184)*. Destruction of dopaminergic neurons augmented behavioral response to μ-, but not δ-opioid receptor agonists in nucleus accumbens *(201)*, suggesting that the receptors for these agonists were not on presynaptic dopaminergic neurons. Neurotensin was increased and substance P was reduced in the SNR following 6-OHDA lesions *(171)*. Somatostatin mRNA was increased in the entopeduncular nucleus (internal pallidum) by a unilateral 6-OHDA lesion *(202)*. Thus, the changes in these peptide levels appear generally to reflect the peptide and mRNA changes seen in the striatum.

Kaakkola *(203)* found that circling behavior induced by an enkephalin analog microinjected into the SNR was antagonized by a 6-OHDA lesion, suggesting that the response was mediated by the dopaminergic nigrostriatal pathway. Subsequently, Stefano et al. *(204)* found a high-affinity opiate binding site on presynaptic dopaminergic neurons in the mollusc, a finding consistent with studies in rats *(203,205)*. In contrast, microinjection of *d*-ala-met-enkephalamide (DALA) into the nucleus accumbens was found to be unaffected by destruction of mesolimbic DA system with 6-OHDA, indicating that this action was independent of an intact dopaminergic system *(205)*. Administration of DA agonists has also implicated dopaminergic control of peptides localized to striatum. When unilateral 6-OHDA-lesioned rats were treated chronically with either a D_1-DA or a D_2-DA agonist, several peptides were affected, but in differing ways. Systemic administration of quinpirole, a D_2-DA agonist, blocked the elevation of striatal mRNA for enkephalin in 6-OHDA-lesioned rats *(182,198)*. Systemic administration and infusion of quinpirole for 7 d to lesioned rats elevated somatostatin and neuropeptide Y in the striatum, but reduced the lesion-induced increase in neurotensin and enkephalin *(182,198)*. After D_1-DA agonist treatment, enkephalin content was unaffected *(198)*; however, the D_1-DA agonist reduced somatostatin and neuropeptide Y, and increased neurotensin and dynorphin in the striatum of the lesioned animals *(198)*. These latter studies suggest a reciprocal relationship between the function of some of the peptides and the DA receptor subtypes.

Other work has indicated that cortical input and cholinergic function can affect content of the peptides in striatum after 6-OHDA lesion. For example, frontal cortical transections after 6-OHDA lesion reversed the lesion-induced increase in preproenkephalin mRNA in the medial striatum *(173)*, but had no effect on the upregulation in the lateral striatum. The downregulation of the preprotachykinin mRNA in the striatum was unaffected by the interruption of the corticostriatal pathway *(173)*. Systemic administration of scopolamine blocked the elevation of striatal mRNA for enkephalin in 6-OHDA-lesioned rats *(182)*. Likewise, very high doses of scopolamine prevented 6-OHDA lesion-induced decrease in striatal substance P mRNA, which may result from an action of scopolamine in other brain areas *(182)*. Zeng et al. *(185)* have found a modest increase in GAD mRNA in the denervated striatum. GAD mRNA was increased in the globus pallidus ipsilateral to the 6-OHDA lesion *(202)*. In addition to cortical influences, peptide changes induced by a unilateral 6-OHDA lesion have also been shown to be blocked by lesioning the subthalamic nucleus *(202)*, suggesting that the subthalamic nucleus can influence peptide function in the striatum through polysynaptic pathways. The basis of this conclusion is that the subthalamic nucleus lesion affected peptide changes induced by the 6-OHDA lesion in brain regions that do not receive a direct input from the subthalamus.

4.5. Activation of Immediate Early Genes by DA Agonists in 6-OHDA-Lesioned Rats

Early studies demonstrated that the cellular proto-oncogene c-Fos could be a metabolic marker of neuronal activity if activated by appropriate neurotransmitters *(206–209)*. An important discovery associated with lesioning of dopaminergic neurons was that the L-DOPA-induced rotation in rats with unilateral 6-OHDA lesions to the pars compacta was accompanied by a large increase in Fos-like immunoreactivity (Fos-LI) on the side of the lesion *(210–214)*; or in both striata when bilateral lesions were made *(148,215)*. The Fos-LI activated by L-DOPA was not affected by the administration of a D_2-DA antagonist, even though locomotor activation was induced by a D_2-DA agonist, eliminating the role of locomotion *per se* in the Fos response observed after L-DOPA administration *(148,211,216)*. Subsequent work demonstrated that indeed D_1-DA, but not D_2-DA agonists would activate Fos in striatum of 6-OHDA-lesioned rats *(148,214–217)*. The changes induced by a D_1-DA agonist were most prominent in the dorsal lateral region of the striatum compared to the ventral medial aspects of the striatum in rats with bilateral lesions of dopaminergic neurons *(216)*. In this latter investigation, electrophysiological responses to the D_1-DA agonist, SKF-38393, were enhanced in the lateral portion of the striatum, but not the basal portion or nucleus accumbens *(216)*.

Because of the relative absence of Fos-LI in unlesioned rats, measurement of the neurons expressing Fos-LI has proven to be a method by which to document D_1-DA receptor supersensitivity *(148,215)*. Although it is not clear what subsequent role the expression of Fos-LI has in relation to modifying dopaminergic or other neurotransmitter functions *(218)*, it certainly provides a valuable approach for measuring and mapping sites within specific regions of brain influenced directly or indirectly by supersensitive D_1-DA receptors. In spite of the supersensitive Fos response to SKF-38393, no change in D_1-activated adenylate cyclase activity or change in binding to D_1-DA receptors was observed *(216)*. Robertson et al. *(219)* documented that the neurons containing Fos were largely found in medium-size striatal cell bodies that innervated the substantia nigra reticulata. In contrast to the lack of effect of a D_2-DA agonist on Fos expression in the striatum, quinpirole caused an expression of Fos in the globus pallidus (GP) after a 6-OHDA lesion *(220)*. Even though a D_1-DA agonist accentuated this change in the GP by quinpirole, a D_1-DA antagonist had no effect on the Fos expression in the GP induced by the D_2-DA agonist *(220)*. The latter observation is another example of a D_1/D_2-DA receptor interaction (*see* Section 4.3.).

Recent investigations have focused on understanding the basis of the expression of Fos by D_1-DA agonists in lesioned rats. For example, it has been observed that blockade of NMDA receptors would reduce, but not block, the activation of c-Fos by L-DOPA in striatum *(221,222)*. In addition, cholinergic mechanisms were found to influence the expression of Fos in striatum. Administration of scopolamine was found to enhance D_1-DA agonist-induced Fos and the behavioral response in lesioned rats *(213)*. Interestingly, scopolamine did not affect the behavioral response to a D_2-DA agonist *(223)*. In another investigation, it was found that cholinergic interneurons in the striatum did not express Fos-LI after a D_1-DA agonist, even though this treatment has been shown to increase the release of acetylcholine *(224)*. However, it was noted in this latter investigation that the 6-OHDA lesion reduced the level of choline acetylase in neurons in the basal forebrain and lateral dorsal tegmental nucleus *(224)*. Since neurons in these brain regions exhibited Fos expression after a D_1-DA agonist, these cholinergic neurons presumably contain DA_1-DA receptors that are being activated directly by the D_1-DA agonist *(224)*.

Cenci and Björklund *(172)* demonstrated that transection of corticostriatal afferents reduced apomorphine-induced striatal Fos expression by 33–66%. In addition, it has been found that infusion of a glutamate antagonist into the SNR of 6-OHDA-lesioned rats significantly increased the expression of Fos induced in the striatum by SKF-38393 *(225)*. Such enhancement of this action of the D_1-DA agonist was also seen when muscimol was infused into the SNR *(225)*. Since these effects to enhance Fos in the striatum were accompanied by an increase in the behavioral response to the agonist, these data indicate that the activity of SNR efferent neurons is inhibitory on the action of a D_1-DA agonist in the striatum to increase Fos. Of additional interest is the observation that MK-801 can induce Fos in the cortex, but not in the striatum of adult rats lesioned as neonates *(221)*. This change in Fos expression by MK-801 was reduced, but not blocked, by a D_1-DA antagonist *(221)*. Collectively, these series of results emphasize the complex circuitry involved in the control of Fos in striatum.

4.6. Neural Hyperinnervation in the Striatum and Brainstem After Selective 6-OHDA Lesions

4.6.1. NE Hyperinnervation

Systemic administration of 6-OHDA induces an elevation of NE content in brainstem structures *(27,226–231)*. Subsequently, icv administration of a low dose of 6-OHDA into brain of neonates *(30)* resulted in an approximate twofold increase in the accumulation of NE in brainstem structures. This hyperinnervation appears to be limited to brain regions in the lower brainstem and cerebellum *(232)*, since this treatment selectively reduced telencephalic and upper brainstem noradrenergic content *(30)*. An immunohistochemical evaluation of the cerebellum in adult rats after intracisternal administration as 1-d-old neonates revealed that the increase in cerebellar NE accumulation was attributable to an invasion of noradrenergic processes into the cerebellum and brainstem *(30)*, which could be attributed to an accumulation of amine content in collateral processes not affected by 6-OHDA, or to secondary outgrowth of nerve terminals *(30,227,231,233,234)*.

4.6.2. Serotonergic Hyperinnervation

In addition to the noradrenergic hyperinnervation of brainstem structures following after a low dose of 6-OHDA given to developing rats, neonatal treatment with larger doses of 6-OHDA has been shown to elevate serotonin content markedly in the striatum of neonatally lesioned rats *(28,29)*. This change has not been reported in adult lesioned rats given 6-OHDA icv, but has been observed in adult rats that had dopaminergic neurons in the SNR lesioned by direct local infusion *(235)*. Requiring a near-maximal lesion of dopaminergic neurons *(235,236)*, the serotonergic hyperinnervation is presumed to be owing to an increase in collateral sprouting and a proliferation of serotonergic nerve terminals originating in the dorsal raphe nucleus in response to the loss of DA *(36,236,237–240)*. This increase is found throughout the striatum, but is most prominent in the rostral portions of the neostriatum *(241)*, and is accompanied by an increased capacity for neurotransmitter release *(238)*, as well as an increase in the 5HT-1B and 5HT-2A serotonin receptor subtypes *(242,243)*. Interestingly, treatment of neonatally 6-OHDA-lesioned animals with serotonergic antagonists to most of the serotonergic receptor subtypes failed to change stereotypical locomotion or SIB, indicating that the serotonergic hyperinnervation was not directly involved with these abnormalities following neonatal 6-OHDA lesions *(236,244)* although recent evidence suggests that the 5HT-1C receptor subtype may mediate some aspects of dopaminergic supersensitivity in the neonatally lesioned rat *(244)*.

4.6.3. Relationship of Innervation to Defining Growth Factors Relevant to Specific Neurotransmitter Systems

Although several trophic agents have been shown to promote survival and growth of dopaminergic neurons in culture, the endogenous neurotrophic factors that control the development, migration, and maturation of catecholaminergic and serotonergic neurons are still largely unknown *(see 245)*. The 6-OHDA-induced hyperinnervation of the cerebellum by noradrenergic neurons and the serotonergic hyperinnervation to the striatum produced by neonatal 6-OHDA treatments *(29)* would seem logical choices for experimental systems for identifying growth-promoting factors specific for initiating and maintaining the growth and integrity of the hyperinnervation of these specific neuronal systems. Based on the degree of hyperinnervation of the noradrenergic and serotonergic fibers in these models, it could be hypothesized that there are increases in yet to be defined specific endogenous neurotrophic factors, which may be directly associated with the adaptive hyperinnervation of these neurotransmitter systems after the distinct neonatal 6-OHDA lesion. In addition, it is interesting to note that in a study using neonatally lesioned animals, lesion of the noradrenergic neurons produced an abnormal development of cortical neurons innervated by the LC *(246)*, providing evidence that catecholamine-containing neurons may also provide trophic support for target neurons during development.

4.7. Neurotrophic Factors and Dopaminergic Neuronal Survival After Neurotoxin Treatment

In addition to the growth-promoting properties of neurotrophic agents, recent evidence suggests that several of the peptide neurotrophic factors may act as neuroprotective agents to inhibit neuronal degeneration and cell death of catecholamine-containing neurons *(247)*, as well as having actions on glia and other neurotransmitter systems in the CNS *(248)*. For example, a number of trophic agents have been shown to promote the survival and growth of dopaminergic neurons in culture *(245,249–254)*, and produce an improvement in functional and behavioral measures when infused into the brain of animals with lesioned dopaminergic neurons *(251,255–258)*. It has also been shown that extracts from lesioned striatum demonstrate trophic actions on dopaminergic neurons in culture *(259)*. With both 6-OHDA and MPTP available to model the dopaminergic neuronal loss in Parkinson's disease, research has recently focused on the potential role of trophic agents in either the degenerative neuropathology (*see* Section 5.1.1.) or in their ability to reduce or even reverse the neuronal loss caused by this neurodegenerative disorder (*see* Section 5.2.2.). However, owing to the rapidly expanding numbers of trophic factors, considerable effort will be required to understand fully the role of these agents in normal catecholaminergic function, and in the CNS neuropathology of dopaminergic neurons and other neurotransmitter systems. Reversal of the neurotoxic action of 6-OHDA and MPTP by the neurotrophin factors is expected to contribute to this understanding.

The neurotrophins (nerve growth factor [NGF], NT-3, NT4/5, and brain-derived neurotrophic factor [BDNF]) have been one family of neurotrophic factors extensively studied as possible agents to reverse neurodegenerative cell loss. Perhaps the most promising of the peptides within this family for reversing deficits in DA function following lesion has been BDNF *(260)*. BDNF has been shown to be expressed in dopaminergic neurons *(261)*, as well as found to protect dopaminergic neurons against MPTP and 6-OHDA toxicity in vitro *(262,263)*, and against 6-OHDA and MPP+ lesions in vivo *(258,264–269)*. The role for BDNF as a potential dopaminergic trophic factor is further supported by the finding that the trkB receptor for BDNF is found within the SNR *(270)*. Although NT-3 failed to provided protection to dopaminergic neurons following 6-OHDA lesions *(265)*, NT-3 prevented the loss of noradrenergic neurons following 6-OHDA lesion *(270,271)*. It has also been shown that NT-4/5 may provide protection for dopaminergic neurons following 6-OHDA treatment *(270,272)*. Interestingly, NGF does not appear to have a substantial impact on central noradrenergicand dopaminergic neurons following either 6-OHDA or MPTP lesions *(30,270,273,274)*; however, NGF has been shown to sustain the viability of adrenal catecholamine-containing cells *(259,275–278)*, which have been tested as a potential source of DA when transplanted into brain of Parkinson's patients *(247)*.

The fibroblast growth factors (bFGF and aFGF) have also been shown to promote survival of dopaminergic neurons in culture *(279–281)*, as well as following MPTP lesions *(282)*. The expression of bFGF has been shown to increase within the lesioned striatum following MPTP lesions in black C/57 mice *(283,284)*. It has been shown that infused bFGF can protect dopaminergic neurons in vivo from the effects of MPTP lesions *(282,285)*, and improve behavioral measures and tyrosine hydroxylase staining levels in 6-OHDA-lesioned animals *(286)*. aFGF has also been shown to induce the recovery of brain DA levels in young, but not older mice treated with MPTP *(287)*. bFGF and epidermal growth factor (EGF) have been shown to protect dopaminergic neurons produced by glutamate toxicity and axotomy *(252)*. Other factors that have been shown either to stimulate or promote survival of catecholaminergic neurons in culture systems include the insulin-like growth factors (IGF-1 and IGF-2), EGF, and NT-3 *(279,281,288,289)*. Ciliary neurotrophic factor (CNTF) and transforming growth factor (TGF) may also prevent axotomy *(290)* or MPTP- *(291)* induced degeneration, respectively, of adult SNR pars compacta neurons. Substance P has also been shown to have a protective action on noradrenergic neurons and to counteract the cerebellar hyperinnervation following neonatal 6-OHDA treatment *(292)*. One factor found to influence the survival of dopaminergic neurons after neurotoxin treatment or other insults to the dopaminergic system is the ganglioside GM-1 *(293,294)*. Perhaps more importantly, cotreatment of GM-1 with a variety of trophic agents appears to increase the degree of observed improvement compared to neurotrophic treatment alone *(295–297)*.

The most promising of the trophic agents to be investigated recently with respect to survival of dopaminergic neurons, both in culture and in vivo, is glial cell-line-derived neurotrophic factor (GDNF). GDNF has been shown to promote survival of embryonic dopaminergic neurons in culture *(254,298)* and stimulate dopamine production *(299)*. GDNF has also been shown to protect dopaminergic neurons from 6-OHDA lesion *(256,300)*, MPTP lesion *(301)*, and axotomy of dopaminergic neurons *(302)*. GDNF treatment can also reduce the delayed degeneration caused by 6-OHDA treatment *(303)*. Studies have clearly demonstrated that treatment of lesioned animals with GDNF, even weeks following the lesion, can induce an improvement in behaviors and function of the dopaminergic system *(264,304)*. The expressional pattern of GDNF suggested that this agent might play a role in the development of dopaminergic neurons in the pars compacta of the SNR *(305,306)*. However, in contrast to this latter view, recent data in the GDNF-deficient mouse found that GDNF was not the critical neurotrophic factor for dopaminergic or noradrenergic neurons during embryogenesis *(307)*. Rather, GDNF was essential for the development of the enteric nervous system, kidney, and subpopulations of sensory and sympathetic neurons *(307)*. Further, these investigators noted the similarity in phenotype between the cRet-deficient and GDNF-deficient mice, suggesting a close relationship of these two proteins *(307,308)*. Nonetheless, these data do not negate the findings that GDNF can clearly facilitate the survival of dopaminergic neurons in vitro and in vivo, and antagonize the effects of lesioning DA neurons.

5. MODELING OF SPECIFIC DISEASE STATES WITH NEUROTOXINS: BASIC APPROACHES TO BETTER TREATMENT STRATEGIES

5.1. Clinical Disorders with Brain DA Reduced

5.1.1. Parkinson's Disease

The first of the clinical neurodegenerative disorders found to have reduced DA and its metabolite, homovanillic acid, in brain and CSF was Parkinson's disease *(309)*. Later, dopaminergic neurons were shown to be reduced using PET analysis with labeled agents that either bind to DA uptake sites or enter dopaminergic neurons through the uptake sites, such as fluoro-DOPA *(310–312)*. Therefore, the availability of 6-OHDA, and later MPTP, allowed specific lesions of dopaminergic neurons in animals to model the DA deficiency in Parkinson's disease *(313)*. As discussed earlier, modeling of Parkinson's disease in animals has allowed evaluation of adaptive changes in various neurotransmitter systems following lesioning of dopaminergic neurons. These models have also provided a means to screen and define drugs that could be potentially useful in treating this disease. Furthermore, in primates, MPTP induces the formation of Lewy bodies, a neuropathological finding in parkinsonism *(314)*. Thus, there are important similarities between the changes induced by MPTP and 6-OHDA, and those seen in Parkinson's patients. This latter observation has prompted investigators to continue their search for the destructive mechanism by which dopaminergic neurons are destroyed by MPTP, as well as by 6-OHDA, as a source of understanding of the pathogenesis of dopaminergic neurons in Parkinsonism.

A number of the peptide neurotrophic factors discussed earlier have been shown to be expressed in the basal ganglia and undergo changes in response to 6-OHDA and MPTP-induced lesions to dopaminergic neurons (*see* Sections 4.4. and 4.7.). aFGF has been reported to be reduced in 6-OHDA-lesioned animals *(315)*, but increased, along with bFGF, in the MPTP-lesioned mouse *(283)*. NGF has been reported to increase in the brain of 6-OHDA-lesioned animals *(316)*, with BDNF shown to be reduced following 6-OHDA lesions *(317)*. In Parkinson's disease, there have been reports of reduced serum NGF *(318)* and brain mRNA levels of bFGF *(319,320)*, with increases seen in CSF levels of TGFβ1 and in EGF, TGFα, and interleukin 1 (IL-1) *(321,322)*. There are also reports of an abnormal production of interluekin 2 (IL-2) in Parkinson's patients *(323)*, which might suggest a possible immunological abnormality in this disease. Although it has been suggested that Parkinson's disease may result from an abnormality in growth factor response with aging, additional research will be necessary to examine possible changes in neurotrophic factor expression during the progression of Parkinson's disease *(324)*.

Although it is unknown if a loss of trophic support underlies the etiology or contributes to the neuropathology observed in Parkinson's disease *(248,325)*, the usefulness of the neurotoxins to lesion dopaminergic neurons in adult animals as a model of Parkinson's disease has allowed the investigation of strategies using trophic factors and transplantation to determine whether these agents may facilitate recovery and provide protection against the loss of dopaminergic neurons by neurotoxins *(326)*. Although trophic factor treatment has often improved behavioral measures following the DA lesion in animal models of parkinsonism, it has not always been clear by what mechanism these improvements were accomplished. In several studies, although no major improvement was observed in the overall numbers of dopaminergic neurons or fibers, significant improvements were often seen in behavioral measures *(see 247)*. This latter observation might suggest that the primary involvement of these trophic agents is to modulate the function of surviving dopaminergic neurons *(255)*. Therefore, these findings should increase our knowledge of the modulatory role of the various neurotrophic factors that affect dopaminergic neurons, and may provide an additional means of manipulating dopaminergic neurotransmitter levels and function in parkinsonism. This strategy may be particularly important for Parkinson's patients, since much of the functional dopaminergic system may have been lost by the time these patients present with symptoms characteristic of this disease. In addition, as will be discussed, the use of various transplantation and gene therapy technologies holds further promise and has been investigated in patients with Parkinson's disease *(248,327–332)*.

5.1.2. Lesch-Nyhan Disease

In addition to parkinsonism, the developmental disorder referred to as Lesch-Nyhan disease, which has a genetic deficiency of the enzyme hypoxanthine phospho-ribothransferase (HPRT) *(333,334)*, is a second clinical disorder in humans with a documented reduction in central DA. Initially, the DA deficiency was found at autopsy of patients with Lesch-Nyhan disease. *(335)*. Because the motor symptoms were distinct from those in parkinsonism, there was a degree of skepticism about the DA deficiency being critical to the symptoms of this developmental disorder. Recently, the uncertainty about the DA deficiency in this disorder was resolved with two clinical reports using PET technology to evaluate the integrity of dopaminergic neurons in Lesch-Nyhan disease. Wong et al. *(336)* found that a labeled marker for the DA transporter was drastically reduced in brain of patients with Lesch-Nyhan disease compared to either normal controls and patients with Rett's syndrome. Ernst et al. *(337)* found that labeled fluoro-DOPA accumulation was reduced in brain of patients with Lesch-Nyhan disease.

A model of the enzyme deficiency in Lesch-Nyhan syndrome is the HPRT-deficient mouse *(338,339)*. This mouse model has a reduction of DA of approx 50% *(340,341)*, providing evidence that the developmental deficiency of HPRT results in a disruption of dopaminergic function, even though the DA reduction in Lesch-Nyhan patients is far greater than in the HPRT-deficient mouse strains. The major drawback for using HPRT-deficient mice to model directly the neurological deficiencies seen in Lesch-Nyhan syndrome, is the lack of an adequate loss of DA content to allow the behavioral manifestations and central adaptive mechanisms of the clinical disorder to be studied. Until the depletion of DA can be further reduced in HPRT-deficient mice to the levels of deficiency seen in Lesch-Nyhan disease, the most widely accepted model to emulate the neonatal destruction of DA-containing neurons in Lesch-Nyhan disease is the neonatal 6-OHDA lesion model using desipramine pretreatment to reduced brain DA selectively *(28,29,39)*.

Rats treated at this early age with 6-OHDA have biological responses to D_1-DA receptor agonist treatment, which is in many ways distinct from that seen in rats lesioned as adults (*see* Sections 3.3. and 3.4.). In particular, neonatally lesioned rats given 100 mg/kg of ₁-DOPA with a decarboxylase inhibitor show distinctive stereotypic responses, followed by self-biting behaviors *(28,29,39)*. Since SIB is a cardinal symptom of Lesch-Nyhan disease *(334)*, neonatal lesioning of dopaminergic neurons with 6-OHDA has provided an approach to delineate neural systems that contribute to the abnormal behavioral symptoms observed in this disorder. This use of the neonate 6-OHDA-lesioned rat model may result in better treatment regimens for the behavioral symptoms of patients with the Lesch-Nyhan syndrome. However, since HPRT deficiency reduces adenosine levels, it should be noted that that adenosine A2-receptors regulate gene expression of many of the previously described peptides in the striatopallidal and striatonigral pathways *(342–344)*. Consequently, there are likely to be important interactions between the reduction of DA and the altered purine mechanisms in Lesch-Nyhan disease.

5.1.3. Use of Neonatal 6-OHDA Lesions to Model Specific Deficiencies Found in Schizophrenia

Although dopaminergic systems have been scrutinized in an attempt to demonstrate neuropathological changes that account for the psychopharmacological role of dopaminergic antagonist treatment in schizophrenia, brain regions associated with dopaminergic neurotransmission have often been found to be relatively normal in schizophrenic brains *(345,346)*. Nevertheless, it has been suggested that specific symptoms of schizophrenia, particularly those related to the dysfunction of DA innervation to the prefrontal cortex, could be modeled by 6-OHDA lesioning of DA input into this region *(347)*. This view has received further support from the observation that rats lesioned neonatally with 6-OHDA have alterations in sensory gating that are comparable to those seen in schizophrenics *(348,349)*. It should be noted that neonatal 6-OHDA-lesioned rats are resistant to the effects of classical antipsychotics *(126)*, which could be relevant to modeling schizophrenic patients who show extreme resistance to the beneficial effects of classical antipsychotic agents. Likewise, it should be noted that the response to MK-801 in neonatally lesioned rats is less dependent on DA release than in control or adult-lesioned rats *(149)*. Thus, blockade of the behavioral and functional responses to MK-801 in the neonatally lesioned rat may provide a means to define new treatment approaches that would alleviate the symptoms of schizophrenic patients that are resistant to the available antipsychotics.

Coyle and Johnston *(350)* suggested that some of the symptoms of schizophrenics may be caused by a hyperinnervation of NE, which can be duplicated following low-dose 6-OHDA treatment of neonates. In this regard, the consequences of destroying noradrenergic function led to a theoretical position that altered noradrenergic function was involved in the core symptoms of schizophrenia *(120,121)*. Furthermore, based on early theoretical positions of self-stimulation reward to the deficiencies in schizophrenia *(119)*, it would seem reasonable to reassess the importance of self-stimulation responding to the symptoms in schizophrenia, which are treated with dopaminergic antagonists. In this latter case, the major response patterns of self-stimulation were shown to be related to intact dopaminergic function (*see* Section 3.2.).

5.2. Use of Neurotoxin Models to Define Therapeutic Strategies for Clinical Diseases with Dopamine Reduced

5.2.1. Screening for New Drug Treatments

The improved motor function following administration of D_2-DA agonists to 6-OHDA-lesioned rats provided a means to screen drugs for possible therapeutic effects in parkinsonism *(33,99,160,137)*. To screen for drugs that may be useful in parkinsonism, the treatment of rats unilaterally treated with 6-OHDA into the pars compacta of the SNR *(33)* has proven to be the most widely used, although bilateral lesions can be made by microinjecting 6-OHDA into pars compacta on both sides or by administering the 6-OHDA intracisternally after administration of pargyline as noted earlier *(11)*. The major advantage of the unilateral lesioning protocol is that the rats do not show an extended period of aphagia and adipsia like that seen when dopaminergic neurons are destroyed on both sides of brain *(31,81,84)*. In the case of animals with bilateral lesions of dopaminergic neurons, locomotor activity or various behaviors are measured to assess the supersensitivity produced by administering DA agonists following nearly complete loss of dopaminergic function *(101,132,133,160)*. When drugs are given that activate dopaminergic receptors in unilaterally lesioned rats, the contralateral turning in response to direct-acting DA agonists can be easily measured and used to quantitate the efficacy of drug treatments *(137,351)*. However, it should be mentioned that a recent study found that accurate quantification of drug effects on rotational behaviors following unilateral 6-OHDA lesion could only be attained if a maximal reduction of DA was attained in the lesioned pars compacta of the SNR *(351)*.

5.2.2. Transplantation of Dopaminergic Neurons to Neurotoxin-Treated Rats: Possible Complementary Use of Gene Therapy

One important use of animals that have received 6-OHDA or MPTP lesions as a model of Parkinson's disease has been the investigation of various aspects of functional restoration of dopaminergic neurotransmission through the use of transplantation and trophic factor treatments *(248,327–332,352)*. To accomplish this, adrenal medullary cells *(275,353–361)*, or dissociated fetal dopaminergic neurons *(362–370)*, have been transplanted or grafted into basal ganglia in order to replace lost dopaminergic function. Initial studies were performed in rodent and primate animal models of Parkinson's disease using 6-OHDA and MPTP lesions, with some degree of success *(371–374)*. In many of these studies, a significant degree of functional improvement was observed in rotational behaviors, Fos expression, and tyrosine hydroxylase levels in the lesioned animals *(375–386)*. One recent report examined the possibility of functional repopulation of the SNR with fetal DA neurons following 6-OHDA lesion. Animals received transplants of fetal dopaminergic cells directly into the lesioned SNR. These animals demonstrated improved behavioral measures compared to striatally implanted animals and, most importantly, demonstrated axonal ingrowth into the striatum and a much improved degree of reinnervation of striatal targets than animals receiving striatal implants *(387,388)*.

Based on the success of preclinical studies in animals *(389)*, transplantation of adrenal medullary cells and dissociated fetal dopaminergic neurons were performed in humans with either Parkinson's disease *(327–332,348,352)* or in subjects who had ingested MPTP *(363)*. The early trials primarily used adrenal medullary tissue placed into the striatum or putamen of Parkinson's patients *(275,353–361)*. The success of these initial efforts was followed by another technique to treat parkinsonism, transplantation of fetal dopaminergic neurons into the basal ganglia *(362–370)*. However, with both types of transplants, although some patients in these trials did show modest improvements during the early period following transplant *(365,370,390–392)*, in many patients, improvements were often short-lived and failed to show marked improvement at later time-points *(247,332,355,357,393–397)*.

Although transplantation has shown great potential as a treatment for the restoration of dopaminergic function in patients suffering from Parkinson's disease, there are still a number of technical and ethical problems related to transplantation and grafting, which have yet to be fully resolved. These issues include transplant rejection by the host, the low degree of survival of neurons in transplanted material, and ethical issues surrounding human fetal tissue and the use of xenografts for transplantation *(398–400)*. The ethical issues will not be discussed, but have been previously reviewed *(401–405)*. To date, only a limited number of studies have examined the role of immunological rejection in relation to either grafting or cross-species xenografting *(406,407)*; however, Galpern et al. *(406)* reported transplantation of porcine fetal ventral mesencephalic neurons into a rats lesioned with 6-OHDA and found functional recovery in the rats treated with cyclosporin to prevent rejection, suggesting that xenografts may be an alternative to human fetal tissue in neurotransplantation in parkinsonism *(406–410)*. However, the most persistent finding following transplantation in both animal models and patients who have been examined postmortem has been the low degree of survival of the transplanted material *(379,411,412)*.

In order to improve the health and increase the survival rate of cells within the transplants, two similar approaches have been undertaken related to neurotrophic factor cotreatment with the transplant. In one case, neurotrophic factors have been directly infused into brain, or placed into sponges and cotransplanted with the cells, in order to promote survival of the grafted or transplanted cells *(248,325,326,413)*. The other relates to the use of astrocytes, fibroblasts, and other cell types *(277,414–417)* that have been genetically engineered to express various trophic agents for adrenal cells or dopaminergic neurons. These cells are then cografted with the transplants to promote neuronal survival *(418)*. NGF has been extensively investigated as a potential agent to provide trophic support for adrenal chromaffin cells *(275–278,414,419)*. Another approach to provide NGF to adrenal chromaffin cells has been the use of peripheral nerve cografted with the adrenal chromaffin cells *(420,421)*, which resulted in an increased the survival of the graft. In addition to NGF, it has been noted that the peripheral nerves also contain a number of other growth factors, including IGF-1 *(419,422)* and CNTF *(423)*, suggesting that these factors may have contributed to the increased viability of the grafted medullary cells.

Trophic agents that have been shown to provide support for dopaminergic neuronal survival in culture have been investigated as possible cotreatment regimens for fetal dopaminergic transplants, including bFGF *(424)*, NT4/5 *(425,426)*, and BDNF *(255,427)*. In addition, cell lines genetically altered to overexpress these trophic agents have also been developed, and shown to produce marked improvements in dopaminergic function or behaviors when implanted into lesioned animals *(267–269)*. For example, BDNF was shown not only to improve the survival of dopaminergic cells in the grafted material, but also to enhance dopaminergic reinnervation of the striatum, as shown by an increase in tyrosine hydroxylase staining in the striatum *(427)*. It has also been shown in animal models that the regions surrounding the grafts often contain a large degree of gliosis *(379)*. Although the effect of the gliosis that accompanies grafting is currently unknown, it has been suggested that the glia may provide a source of endogenous trophic support for transplants, as well as surviving neurons *(428)*.

Whereas growth factor therapy for neurodegenerative disorders may be an attainable goal within the foreseeable future, the primary difficulty of attaining entrance into the human brain must first be resolved. One approach to supply peptide neurotrophic factors to the CNS may include various forms of gene therapy that have proven to be successful in animal models, as described above. Another use of gene therapy has been the use of encapsulated cells to deliver dopamine directly to target areas *(429–434)*. Collectively, in spite of the short-comings related to approaches to supply trophic factors alone and to transplanted tissue in brain, clinical studies aimed at restoring dopaminergic function for patients suffering from Parkinson's disease offer hope for the future that such technology will be available to ameliorate the symptoms of these diseases.

An additional treatment approach being considered for parkinsonism is gene delivery of tyrosine hydroxylase or aromatic decarboxylase into brain with gene therapy. The tyrosine hydroxylase gene would be expected to increase L-DOPA and rely on the endogenous decarboxylation of the L-DOPA for increasing the production of DA. The introduction of the aromatic decarboxylase gene, which decarboxylates L-DOPA, would take into consideration that dopaminergic and serotoninergic neurons are both reduced in the striatum of Parkinson's patients *(309)*. The simultaneous reduction of both serotonin- and DA-containing neurons has been demonstrated in animals to result in a dramatic reduction in the decarboxylation of L-DOPA *(101)*. In parkinsonism, reduced 5-HIAA in cerebrospinal fluid (CSF) has been associated with a loss of effectiveness of L-DOPA to treat symptoms, presumably resulting from a reduced capacity to decarboxylate the L-DOPA *(435,436)*. With the addition of the decarboxylase enzyme with gene therapy, the treatment dose of L-DOPA could be titrated to each patient for any variability in the amount of the decarboxylase enzyme expression following delivery of the gene to brain. Of these approaches, replacement of the decarboxylase in Parkinson's patients would appear to have the greatest potential.

6. CONCLUSION

The overview of neurotoxins affecting dopaminergic and noradrenergic neurons clearly documents the usefulness of these agents. The use of these agents has allowed numerous discoveries regarding how these neurotransmitter systems control specific physiological functions and specific behavioral responses. Further, the adaptive changes resulting from the loss of dopaminergic neurons revealed receptor supersensitivity, D_1/D_2-DA receptor interaction, and D_1-DA priming of supersensitivity, neural mechanisms that appear to be relevant to the clinical disorders modeled by the neurotoxins. The loss of dopaminergic neurons following neurotoxin treatment in adult animals is widely accepted as a model of the DA reduction in Parkinson's disease. Likewise, the neonatal 6-OHDA treatment has been accepted as a model of the DA reduction in Lesch-Nyhan disease. More recently, the reduction of DA following neurotoxin treatment has proven useful in testing whether growth factors would minimize the functional deficits resulting from neurotoxin lesions. The property of specific types of lesions in neonates with 6-OHDA, resulting in regionally specific hyperinnervation, would appear to be another method of relating neurotrophic factor systems elaborated in brain to specific neurotransmitter systems. The knowledge gained from transplanting tissue containing dopaminergic neurons to reverse deficits in lesioned animals has been directly applied to Parkinson's disease, may provide alternative treatments for the motor and behavioral abnormalities seen in Lesch-Nyhan disease, and may serve as a model system for the treatment of other neurodegenerative diseases in the future *(437,438)*. Likewise, information concerning the importance of neurotrophic factors in neural survival is being applied to clinical treatment of parkinsonism. Consequently, it would appear that this class of neurotoxins will continue to play an important role in developing and testing new therapies for clinical disorders with brain DA reduced.

REFERENCES

1. Porter, C., Totara, J., and Stone, C. (1963) Effect of 6-hydroxydopamine and some other compounds on the concentration of norepinephrine in the heart of mice. *J. Pharmacol. Exp. Ther.* **140,** 308–316.
2. Laverty, R., Sharman, D., and Vogt, M. (1965) Action of 2,4,5-tri-hydroxyphenylethylamine on the storage and release of noradrenaline. *Br. J. Pharmacol.* **24,** 549–560.
3. Thoenen, H. and Tranzer, J. P. (1968) Chemical sympathectomy by selective destruction of adrenergic nerve endings with 6-hydroxydopamine. *Naunyn-Schmiedeberg's Arch. Pharmacol.* **261,** 271–288.
4. Ungerstedt, U. (1968) 6-Hydroxydopamine induced degeneration of central monoamine neurons. *Eur. J. Pharmacol.* **5,** 107–110.
5. Bloom, F. E., Algeri, S., Croppetti, A., Revuelta, A., and Costa, E. (1969) Lesions of Central norepinephrine terminals with 6-OH-dopamine: Biochemistry and fine structure. *Science* **166,** 1284–1286.
6. Breese, G. R. and Traylor, T. D. (1970) Effects of 6-hydroxydopamine on brain norepinephrine and dopamine: Evidence for selective degeneration of catecholamine neurons. *J. Pharmacol. Exp. Ther.* **174,** 413–420.
7. Uretsky, N. J. and Iversen, L. L. (1970) Effects of 6-hydroxydopamine on catecholamine neurones in the rat brain. *J. Neurochem.* **17,** 269–278.
8. Davis, G. C., Williams, A. C., Markey, S. P., Ebert, M. H., Caine, E. D., Reichert, C. M., and Kopin, I. J. (1979) Chronic parkinsonism secondary to intravenous injection of meperidine analogues. *Psychiatr. Res.* **1,** 249–254.
9. Langston, J. W., Ballard, P., Tetrud, J. W., and Irwin, I. (1983) Chronic parkinsonism in humans due to a product of meperidine-analog synthesis. *Science* **219,** 979,980.
10. Kostrzewa, R. and Jacobowitz, D. (1974) Pharmacological actions of 6-hydroxydopamine. *Pharmacol. Rev.* **26,** 199–288.

11. Breese, G. R. (1975) Chemical and immunochemical lesions by specific neurotoxic substances and antisera, in *Handbook of Psychopharmacology*, vol. 1 (Iversen, L. L., Iversen, S. D., and Snyder, S. H., eds.), Plenum, New York, pp. 137–189.

12. Iversen, L. L. and Uretsky, N. J. (1970) Regional effects of 6-hydroxydopamine on catecholamine containing neurons in rat brain and spinal cord. *Brain. Res.* **24,** 364–367.

13. Jacks, B. R., DeChamplain, J., and Cordeau, J. P. (1972) Effects of 6-hydroxydopamine on putative transmitter substances in the central nervous system. *Euro. J. Pharmacol.* **18,** 353–360.

14. Jalfre, M. and Haefely, W. (1971) Effects of some centrally acting agents in rats after intraventricular injections of 6-hydroxydopamine, in *6-Hydroxydopamine and Catecholamine Neurons* (Malmfors, T. and Thoene, H., eds.), North Holland Publishing, Amsterdam, The Netherlands, pp. 333–346.

15. Ungerstedt, U. (1971) Stereotaxic mapping of th monoamine pathways in the rat brain. *Acta. Physiol. Scand.* **367,** 1–48.

16. Frigyesi, T. L., Ige, A., Iulo, A., and Schwartz, R. (1971) Denigration and sensorimotor disability induced by ventral tegmental injection of 6-hydroxydopamine in the cat. *Exp. Neurol.* **33,** 78–87.

17. Hökfelt, T. and Ungerstedt, U. (1973) Specificity of 6-hydroxydopamine induced degeneration of central monoamine neurons: An electron and fluorescence microscopy study with special reference to intracerebral injection on the nigrostriatal dopamine system. *Brain Res.* **60,** 269–298.

18. Butcher, L. L., Eastgate, S. M., and Hodge, G. K. (1974) Evidence that punctuate intracerebral administration of 6-hydroxydopamine fails to produce selective neuronal degeneration. Comparison with copper sulfate and factors governing the deportment of fluids injected into brain. *Naunyn-Schmeideberg's Arch. Pharmacol.* **235,** 31–70.

19. Poirier, L. J., Langlier, P., Roberge, A., Boucher, R., and Kitsikis, A. (1972) Nonspecific histopathological changes induced by the intracerebral injection of 6-hydroxy-dopamine (6-OH-DA). *J. Neurol. Sci.* **16,** 401–416.

20. Bell, L. J., Iversen, L. L., and Uretsky, N. J. (1970) Time course of the effects of 6-hydroxydopamine on catecholamine-containing neurones in rat hypothalamus and striatum. *Br. J. Pharmacol.* **40,** 790–799.

21. Breese, G. R. and Traylor, T. D. (1971) Depletion of brain noradrenaline and dopamine by 6-hydroxydopamine. *Br. J. Pharmacol.* **42,** 88–99.

22. McGeer, E. G., Fibiger, H. C., McGeer, P. L., and Brooke, S. (1973) Temporal changes in amine synthesizing enzymes of rat extrapyramidal structures after hemitransections or 6-hydroxydopamine administration. *Brain Res.* **52,** 289–300.

23. Breese, G. R. and Traylor, T. D. (1972) Developmental characteristics of brain catecholamines and tyrosine hydroxylase in the rat: Effects of 6-hydroxydopamine. *Br. J. Pharmacol.* **44,** 210–222.

24. Lytle, L., Shoemaker, W., Cottman, K., and Wurtman, R. (1972) Long-term effects of postnatal 6-hydroxydopamine treatment on tissue catecholamine levels. *J. Pharmacol. Exp. Ther.* **183,** 56–64.

25. Smith, R. D., Cooper, B. R., and Breese, G. R. (1973) Growth and behavioral changes in developing rats treated intracisternally with 6-hydroxydopamine: Evidence for involvement of brain dopamine. *J. Pharmacol. Exp. Ther.* **185,** 609–619.

26. Sachs, C. and Jonsson, G. (1975) Effects of 6-hydroxydopamine on central noradrenaline neurons during ontogeny. *Brain Res.* **99,** 277–291.

27. Sachs, C., Pycok, P., and Jonsson, G. (1974) Altered development of central noradrenaline neurons during ontogeny by 6-hydroxydopamine. *Med. Biol.* **52,** 55–65.

28. Breese, G. R., Baumeister, A. A., McCown, T. J., Emerick, S. G., Frye, G. D., and Mueller, R. A. (1984) Neonatal-6-hydroxydopamine: Model of susceptibility for self-mutilation in the Lesch-Nyhan Syndrome. *Pharmacol. Biochem. Behav.* **21,** 459–461.

29. Breese, G. R., Baumeister, A. A., McCown, T. J., Emerick, S. G., Frye, G. D., and Mueller, R. A. (1984) Behavioral differences between neonatal and adult 6-hydroxy-dopamine-treated rats: Relevance to neurological symptoms in clinical syndromes with reduced dopamine. *J. Pharmacol. Exp. Ther.* **231,** 343–354.

30. Konkol, R. J., Benedeich, E. G., and Breese, G. R. (1978) A biochemical and morphological study of the altered growth pattern of central catecholamine neurons following 6-hydroxydopamine. *Brain Res.* **140,** 125–135.

31. Breese, G. R., Cooper, B. R., and Smith, R. D. (1973) Biochemical and behavioral alterations following 6-hydroxydopamine administration into brain, in *Frontiers in Catecholamine Research* (Usdin, E. and Snyder, S., eds.), Pergamon, UK, pp. 701–706 (Reprint: *Biochem. Pharmacol.* **23 Suppl. Part 2,** 574–579.)

32. Breese, G. R., Cooper, B. R., and Hollister, A. S. (1974) Relationship of biogenic amines to behavior, in *Catecholamines and Their Enzymes in the Neuropharmacology of Schizophrenia* (Kety, S. S. and Mattysee, S., eds.), North Holland Publishing, Amsterdam, The Netherlands, Suppl. to *J. Psychiat. Res.* **11,** 125–134.

33. Ungerstedt, U. (1971) Use of intracerebral injections of 6-hydroxdopamine as a tool for morphological and functional studies on central catecholamine neurons, in *6-Hydroxydopamine and Catecholamine Neurons* (Malmfors, T. and Thoenen, H., eds.), North-Holland, Amsterdam, pp. 315–332.

34. Palmer, G. C. (1972) Increased cyclic AMP response to norepinephrine in rat brain following 6-hydroxydopamine. *Neuropharmacology* **11,** 145–149.

35. Teicher, M. H., Barber, N. I., Reichheld, J. H., Baldess, R. J., and Finklestein, S. P. (1986) Selective depletion of cerebral norepinephrine with 6-hydroxydopamine and GBR-12909 in neonatal rat. *Dev. Brain Res.* **30,** 124–128.

36. Molina-Holgado, E., Dewar, K. M., Descarries, L., and Reader, T. A. (1994) Altered dopamine and serotonin metabolism in the dopamine-denervated and serotonin-hyper-innervated neostriatum of adult rat after neonatal 6-hydroxydopamine. *J. Pharmacol. Exp. Ther.* **270,** 713–721.

37. Stone, C., Porter, C., Stavorski, J., Ludden, C., and Totaro, J. (1964) Antagonism of certain effects of catecholamine-depleting agents by antidepressant and related drugs. *J. Pharmacol. Exp. Ther.* **144,** 196–204.

38. Evetts, K. D. and Iversen, L. L. (1970) Effects of protriptyline on the depletion of catecholamines induced by 6-hydroxydopamine in the brain of the rat. *J. Pharm. Pharmacol.* **22,** 540–542.

39. Breese, G. R., Criswell, H. E., Johnson, K. B., O'Callaghn, J. P., Duncan, G. E., Jensen, K. F., Simson, P. E., and Mueller, R. A. (1994) Neonatal destruction of dopaminergic neurons. *Neurotoxicology* **15,** 149–160.

40. Agid, Y., Javoy, F., Glowinski, J., Bouvet, D., and Sotelo, C. (1973) Injection of 6-hydroxydopamine into the substantia nigra of the rat. II. Diffusion and specificity. *Brain Res.* **58,** 291–301.
41. Burns, R. S., Chiueh, C. C., Markey, S. P., Ebert, M. H., Jacobowitz, D. M., and Kopin, I. J. (1983) A primate model of Parkinsonism: Selective destruction dopaminergic neurons in the pars compacta of the substantia nigra by *N*-methyl-4-phenyl-1,2,3,6-tetrahydropyridine. *Proc. Natl. Acad. Sci. USA* **80,** 4546–4550.
42. Langston, J. W., Forno, L. S., Rebert, C. S., and Irwin, I. (1984) Selective nigral toxicity after systemic administration of 1-methyl-4-1,2,5,6-tetrahydropyridine (MPTP) in the squirrel monkey. *Brain Res.* **292,** 390–394.
43. Jenner, P., Rupniak, N. M. Y., Rose, S., Kelly, E., Kilpatrick, G., Lees, A., and Marsden, C. D. (1984) 1-Methyl-4-phenyl-1,2,3,6-tetrahydropyridine-induced parkinsonism in the common marmoset. *Neurosci. Lett.* **50,** 85–90.
44. Kitt, C. A., Cork, L. C., Eidelburg, F., Joh, T. H., and Price, D. L. (1986) Injury of nigral neurons exposed to 1-methyl-4-1,2,3,6-tetrahydropyridine: A tyrosine hydroxylase immunocytochemical study in monkey. *Neuroscience* **17,** 1089–1103.
45. Weihmuller, F. B., Hadjiconstantinou, M., and Bruno, J. P. (1989) Dissociation between biochemical and behavioral recovery in MPTP-treated mice. *Pharmacol. Biochem. Behav.* **34,** 113–117.
46. Jenner, P. G., Marsden, C. D., Costall, B., and Marsden, C. D. (1986) MPTP and MPP+ induced toxicity in rodents and the common marmoset as experimental models for investigating Parkinson's disease, in: *MPTP: A Neurotoxin Producing a Parkinsonian Syndrome* (Markey, S. P., Castagnoli, N., Trevor, A. J., and Kopin, I. J., eds.), Academic, New York, pp. 45–68.
47. Eidelberg, E., Brooks, B. A., Morgan, W. W., Walden, J. G., and Kokemoor, R. H. (1986) Variability and functional recovery in the *N*-methyl-4-phenyl-1,2,3,6-tetrahydropyridine model of parkinsonism in monkeys. *Neuroscience* **18,** 817–822.
48. Rose, S., Nomoto, M., Kelly, E., Kilpatrick, G., Jenner, P., and Marsden, C. D. (1989) Increased caudate dopamine turnover may contribute tot he recovery of motor function in marmosets treated with the dopaminergic neurotoxin MPTP. *Neurosci. Lett.* **101,** 305–310.
49. Ueki, A., Chong, P. N., Albanese, A., Rose, S., Chivers, J. K., Jenner, P., and Marsden, C. D. (1989) Further treatment with MPTP does not produce parkinsonism in marmosets showing behavioral recovery from motor deficits induced by an earlier exposure to the toxin. *Neuropharmacology* **28,** 1089–1097.
50. Heikkila, R. E., Cabbat, F. S., Manzino, L., and Duvoisin, R. C. (1984) Effects of 1-methyl-4-phenyl 1,2,5,6-tetrahydropyridine on neostriatal dopamine in mice. *Neuropharmacology* **23,** 711–713.
51. Heikkila, R. E., Hess, A., and Duvoisn, R. C. (1984) Dopaminergic neurotoxicity of 1-methyl-4-phenyl 1,2,3,6-tetrahydropyridine (MPTP) in mice. *Science* **224,** 1451–1453.
52. Heikkila, R. E. (1985) Differential neurotoxicity of 1-methyl-4-phenyl 1,2,3,6-tetrahydropyridine (MPTP) in Swiss-Webster mice from different sources. *Eur. J. Pharmacol.* **117,** 131–133.

53. Heikkila, R. E., Sieber, B-E., Manzino, L., and Sonsalla, P. K. (1989) Some features of the nigrostriatal dopaminergic neurotoxin 1-methyl-4-phenyl 1,2,3,6-tetrahydropyrdine (MPTP) in the mouse. *Mol. Chem. Neuropathol.* **10,** 171–183.
54. Sonsalla, P. K. and Heikkila, R. E. (1986) The influence of dose and dosing internal on MPTP-induced dopaminergic neurotoxicity in mice. *Eur. J. Pharmacol.* **129,** 339–345.
55. Sonsalla, P. K., Youngster, S. K., Kindt, M. V., and Heikkila, R. E. (1987) Characteristics of 1-methyl-4-(2'methylphenyl)-1,2,3,6-tetrahydropyridine-induced neurotoxicity in the mouse. *J. Pharmacol. Exp. Ther.* **242,** 850–857.
56. Youngster, S. K., Duvoisin, R. C., Hess, A., Sonsalla, P. K., Kindt, M. V., and Heikkila, R. E. (1986) 1-Methyl-4-(2'-methylphenyl)- 1,2,3,6-tetrahydropyridine (2'-CH3-MPTP) is a more potent dopaminergic neurotoxin than MPTP in mice. *Eur. J. Pharmacol.* **122,** 283–287.
57. Rose, S., Nomoto, M., Jackson, E. A., Gibb, W. R. G., Jenner, P., and Marsden, C. D. (1990) 1-Methyl-4-phenyl 1,2,3,6-tetrahydropyridine (2'-methyl-MPTP) is less neurotoxic than MPTP in the common marmoset. *Eur. J. Pharmacol.* **181,** 97–103.
58. Heikkila, R. and Cohen, G. (1971) Inhibition of biogenic amine uptake by hydrogen peroxide: A mechanism for toxic effects of 6-hydroxydopamine. *Science* **172,** 1257–1258.
59. Heikkila, R. and Cohen, G. (1972) Further studies on the generation of hydrogen peroxide by 6-hydroxydopamine:Potentiation by ascorbic acid. *Mol. Pharmacol.* **8,** 241–248.
60. Heikkila, R. E. and Cohen, G. (1973) 6-Hydroxydopamine: Evidence for superoxide radical as an oxidative intermediate. *Science* **181,** 456–457.
61. Jeon, B. S., Jackson-Lewis, V., and Burke, R. E. (1995) 6-Hydroxydopamine lesion of the rat substantia nigra: time course and morphology of cell death. *Neurodegeneration.* **4,** 131–137.
62. Chiba, K., Trevor, A., and Castagonoli, N., Jr. (1984). Metabolism of the neurotoxic tertiary amine, MPTP, by brain monoamine oxidase. *Biochem. Biophys. Res. Commun.* **120,** 574–578.
63. Salach, J. I., Singer, T. P., Castagnoli, N., and Trevor, A., Jr. (1984) Oxidation of the neurotoxic amine 1-methyl-4-phenyl-1,2,3,6-tetrahydropyridine (MPTP) by monoamine oxidases A and B and suicide inactivation of the enzymes by MPTP. *Biochem. Biophys. Res. Commun.* **125,** 831–835.
64. Heikkila, R. E., Manzino, L., Cabbat, F. S., and Duvoisin, R. C. (1985) Studies on the oxidation of the dopaminergic neurotoxin 1-methyl-4-phenyl 1,2,3,6-tetrahydropyridine by monoamine oxidase B. *J. Neurochem.* **45,** 1049–1054.
65. Javitch, J. A., D'Amato, R. J., Strittmatter, S. M., and Snyder, S. H. (1985) Parkinsonism-inducing neurotoxin, N-methyl-4-phenylpyridine by dopamine: Uptake of the metabolite N-methyl-4-phenylpyridine dopamine neurons explains selective toxicity. *Proc. Natl. Acad. Sci. USA* **82,** 2173–2177.
66. Mayer, R. A., Kindt, M. V., and Heikkila, R. E. (1986) Prevention of the nigrostriatal toxicity of 1-methyl-4-phenyl-1,2,3,6-tetrahydropyridine by inhibitors of 3,4-dihhydroxphenylethylamine transport. *J. Neurochem.* **47,** 1073–1079.
67. Ramsay, R. R. and Singer, T. P. (1986) Energy-dependent uptake of N-methyl-phenylpyridinium, the neurotoxic metabolite of 1-mehtyl-4-phenyl-1,2,3,6-tetrahydroppyridine, by mitochondria. *J. Biol. Chem.* **261,** 7585–7587.

68. Ramsay, R. R., Salach, J. I., and Singer, T. P. (1986) Uptake of the neurotoxin 1-methyl-4-phenylpyridine (MPP+) by mitochondria and its relation to the inhibition of the mitochondrial oxidation of NAD+-linked substrates by MPP+. *Biochem Biophys. Res. Commun.* **134,** 743–748.

69. Nicklas, W. J., Vyas, I., and Heikkila, R. E. (1985) Inhibition of NADH-linked oxidation brain mitochondria by 1-methyl-4-phenyl 1,2,3,6-tetrahydropyridine. *Life Sci.* **36,** 2503–2508.

70. Vyas, I., Heikkila, R. E., and Nicklas, W. J. (1986) Studies on the neurotoxicity of 1-methyl-4-phenyl 1,2,3,6-tetrahydropyridine: Inhibition of NAD-linked substrate oxidation by its metabolite, 1-methyl-4-phenylpyridinium. *J. Neurochem.* **46,** 1501–1507.

71. Halliwell, B. and Gutteridge, J. M. (1991) *Free Radicals in Biology and Medicine.* Clarendon, Oxford.

72. Przedborski, S., Kostic, V., Jackson-Lewis, V., Naini, A. B., Simonetti, S., Fahn, S., Carlson, E., Epstein, C. J., and Cadet, J. L. (1992) Transgenic mice with increased Cu/Zn-superoxide dismutase activity are resistant to *N*-methyl-4-phenyl-1,2,3,6-tetrahydropyridine-induced neurotoxicity. *J. Neurosci.* **12,** 1658–1667.

73. Przedborski, S., Jackson-Lewis, V., Yokoyama, R., Shibata, T., Dawson, V. L., and Dawson, T. M. (1996) Role of neuronal nitric oxide in 1-methyl-4-phenyl-1,2,3,6-tetrahydropyridine (MPTP)-induced dopaminergic neurotoxicity. *Proc. Natl. Acad. Sci. USA* **93,** 4565–4571.

74. Laverty, R. and Taylor, K. (1970) Effects of intraventricular 2,4,5-trihydrophenylethylamine on the storage and release of noradrenaline. *Br. J. Pharmacol.* **40,** 836–846.

75. Herman, Z. S., Kmieciak-Kolada, K., and Brus, R. (1972) Behavior of rats and biogenic amine level in brain after 6-hydroxydopamine. *Psychopharmacologia* **24,** 407–416.

76. Fibiger, H. C., Lonsbury, B., Cooper, H. P., and Lytle, L. D. (1972) Early behavioral effects of intraventricular administration of 6-hydroxydopamine in rat. *Nature New Biol.* **236,** 209–211.

77. Simmonds, M. A. and Uretsky, N. J. (1970) Central effects of 6-hydroxydopamine and 1-5-hydroxytryptophan on body temperature in the rat. *Br. J. Pharmacol.* **40,** 630–638.

78. Nakamura, K. and Thoenen, H. (1971) Hypothermia induced by intraventricular administration of 6-hydroxydopamine in rats. *Eur. J. Pharmacol.* **16,** 46–54.

79. Breese, G. R. and Howard, J. L. (1971) Effect of central catecholamine alterations on the hypothermic response to 6-hydroxydopamine in desipramine-treated rats. *Br. J. Pharmacol.* **43,** 671–674.

80. Breese, G. R., Moore, R. A., and Howard, J. L. (1972) Central actions of 6-hydroxydopamine and other phenylethylamine derivatives on body temperature in the rat. *J. Pharmacol. Exp. Ther.* **180,** 591–602.

81. Ungerstedt, U. (1971) Adipsia and aphagia after 6-hydroxydopamine induced degeneration of the nigro-striatal dopamine system in the rat brain. *Acta Physiol. Scand.* **367,** 95–122.

82. Zigmond, M. J. and Stricker, E. M. (1972) Deficits in feeding behavior after intraventricular injection of 6-hydroxydopamine in rats. *Science* **177,** 1211–1214.

83. Fibiger, H. C., Zis, A. P., and McGeer, E. G. (1973) Feeding and drinking deficits after 6-hydroxdopamine administration in the rat: Similarities to the lateral hypothalamic syndrome. *Brain Res.* **53,** 135–148.

84. Breese, G. R., Smith, R. D., Cooper, B. R., and Grant, L. D. (1973) Alterations in consummatory behavior following intracisternal injection of 6-hydroxydopamine. *Pharmacol. Biochem. Behav.* **1,** 319–328.

85. Breese, G. R., Prange, A. J., Jr., Howard, J. L., Lipton, M. A., McKinney, W. T., Bowman, R. E., and Bushnell, P. (1972) 3-Methoxy-4-hydroxyphenylglycol excretion and behavioral changes in rat and monkey after central sympathectomy with 6-hydroxydopamine. *Nature New Biol.* **240,** 286–287.

86. Redmond, D. E., Jr., Hinrichs, R. L., Maas, J. W., and Kling, A. (1973) Behavior of free-ranging macaques after intraventricular 6-hydroxydopamine. *Science* **181,** 1256–1258.

87. Howard, J. L. and Breese, G. R. (1974) Physiological and behavioral effects of centrally-administered 6-hydroxydopamine in cats. *Pharmacol. Biochem. Behav.* **2,** 651–661.

88. Cooper, B. R., Grant, L. D., and Breese, G. R. (1973) Comparison of the behavioral depressant effects of biogenic amine depleting and neuroleptic agents following various 6-hydroxydopamine treatments. *Psychopharmacologia* **31,** 95–109.

89. Teitelbaum, P., Cheng, M. F., and Rozin, P. (1969) Stages of recovery and development of lateral hypothalamic control of food and water intake. *Ann. NY Acad. Sci.* **157,** 849–860.

90. Marshall, J. F. and Teitelbaum, P. (1973) A comparison of the eating in response to hypothermic and glucoprivic challenges after nigral 6-hydroxydopamine and lateral hypothalamic electrolytic lesions in rats. *Brain Res.* **55,** 229–233.

91. Marshall, J. F., Richardson, J. S., and Teitelbaum, P. (1974) Nigrostriatal bundle damage and the lateral hypothalamic syndrome. *J. Comp. Physiol. Psychol.* **87,** 808–830.

92. Fitzsimons, J. T. (1972) Thirst. *Physiol. Rev.* **52,** 468–561.

93. Stricker, E. M. and Zigmond, M. J. (1974) Effects of homeostasis of intraventricular injection of 6-hydroxydopamine in rats. *J. Comp. Pyschol.* **86,** 973–994.

94. Cooper, B. R. and Breese, G. R. (1974) Relationship of dopamine neural systems to the behavioral alterations produced by 6-hydroxydopamine administration into brain, in *Neuropsychopharmacology of Monoamines and Their Regulatory Enzymes* (Usdin, E., ed.), Raven, New York, pp. 353–368.

95. Cooper, B. R., Howard, J. L., Grant, L. D., Smith, R. D., and Breese, G. R. (1974) Alteration of avoidance and ingestive behavior after destruction of central catecholamine pathways with 6-hydroxydopamine. *Pharmacol. Biochem. Behav.* **2,** 639–649.

96. Stodgell, C. J., Schroeder, S. R., and Tessel, R. E. (1996) FR discrimination training reverses 6-hydroxydopamine-induced striatal dopamine depletion in a rat model of Lesch-Nyhan syndrome. *Brain Res.* **713,** 246–252.

97. Zigmond, M. J. and Stricker, E. M. (1984) Parkinson's disease: Studies with an animal model. *Life Sci.* **35,** 5–18.

98. Cooper, B. R., Breese, G. R., Howard, J. L., and Grant, L. D. (1972) Enhanced behavioral depressant effects of reserpine and alpha-methyltyrosine after 6-hydroxydopamine treatment. *Psychopharmacologia* **27,** 99–110.

99. Ungerstedt, U. (1971) Post synaptic supersensitivity after 6-hydroxydopamine induced degeneration of the nigrostriatal dopamine system. *Acta Physiol. Scand.* **367,** 69–93.

100. Uretsky, N. J. and Schoenfeld, R. I. (1971) Effect of L-DOPA on the locomotor activity of rats pretreated with 6-hydroxydopamine. *Nature New Biol.* **234,** 157–159.

101. Hollister, A. S., Breese, G. R., and Mueller, R. A. (1979) Role of monoamine neural systems in L- dihydroxyphenylalanine stimulated activity. *J. Pharmacol. Exp. Ther.* **208,** 37–43.

102. Jenner, P. (1992) MPTP-induced parkinsonism: Recovery, species differences, and relevance to Parkinson's Disease, in *Progress in Parkinson's Disease Research* (Hefti, F. and Weiner, W. J., eds.), Futura, Mt. Kisco, NY, pp. 15–38.

103. Cooper, B. R., Breese, G. R., Howard, J. L., and Grant, L. D. (1972) Effect of central catecholamine alterations by 6-hydroxydopamine on shuttle box avoidance acquisition. *Physiol. Behav.* **9,** 727–731.

104. Cooper, B. R., Breese, G. R., Grant, L. D., and Howard, J. L. (1973) Effects of 6-hydroxydopamine treatments on active avoidance responding: Evidence for involvement of brain dopamine. *J. Pharmacol. Exp. Ther.* **185,** 358–370.

105. Cooper, B. R., Cott, J. M., and Breese, G. R. (1974) Effects of catecholamine-depleting drugs and amphetamine on self-stimulation of brain following various 6-hydroxydopamine treatments. *Psychopharmacologia* **37,** 235–248.

106. Howard, J. L., Grant, L. D., and Breese, G. R. (1974) Effects of intracisternal 6-hydroxydopamine treatment acquisition and performance of rats in a double-T-Maze. *J. Comp. Physiol. Psychol.* **86,** 995–1007.

107. Schoenfeld, R. I. and Uretsky, N. J. (1972) Operant behavior and catecholamine-containing neurons: Prolonged increase in lever-pressing after 6-hydroxydopamine. *Eur. J. Pharmacol.* **20,** 357–362.

108. Howard, J. L., Leahy, J. P., and Breese, G. R. (1971) Some physiological and behavioral consequences of acute and chronic injection of 6-hydroxydopamine (6-OHDA). *Fed. Proc.* **30,** 541.

109. Gupta, P. P., Drimai, R. C., and Dhawan, B. N. (1972) Central cardiovascular effects of 6-hydroxydopamine. *Eur. J. Pharmacol.* **20,** 215–223.

110. Bolme, P., Fuxe, K., Nygren, L. G., Olson, L., and Sachs, C. (1974) Enhanced sensitivity to the noradrenaline receptor stimulating agent, clonidine, following degeneration of noradrenaline pathways: Studies on arterial pressure, heart rate and respiration, in *Dynamics of Degeneration and Growth in Neurons* (Fuxe, K., Olson, L., and Zotterman, Y., eds.), Pergamon, New York, pp. 597–602.

111. Howard, J. L., Smith, R. D., Mueller, R. A., and Breese, G. R. (1974) Cardiovascular changes following DOPA/NaCl or conditioning in 6-hydroxydopamine-treated rats. *Pharmacol. Biochem. Behav.* **2,** 537–543.

112. Hartman, N., Chung, R., Draskoczy, P. R., and Schildkraut, J. J. (1971) Effects of 6-hydroxydopamine on sleep in the rat. *Nature* **233,** 425–427.

113. Eichelman, B. S., Jr., Thoa, N. B., and Ng, K. Y. (1972) Facilitated aggression in the rat following 6-hydroxydopamine administration. *Physiol. Behav.* **8,** 1–3.

114. Nakamura, K. and Thoenen, H. (1972) Increased irritability: A permanent behavior change induced in rat by intraventricular administration of 6-hydroxydopamine. *Psychopharmacolgia* **24,** 359–372.

115. Jimerson, D. and Reis, D. J. (1973) Effects of intrahypothalamic injection of 6-hydroxydopamine on predatory aggression in rat. *Brain Res.* **61,** 141–152.

116. Anlezark, G. M., Crow, T. J., and Greenway, A. P. (1973) Impaired learning and decreased cortical norepinephrine after bilateral locus ceruleus lesions. *Science* **181,** 682–684.

117. Ahlskog, J. E. and Hoebel, G. G. (1973) Overeating and obesity from damage to a noradrenergic system in brain. *Science* **182,** 166–169.

118. Bonaz, B., Martin, L., Beurriand, E., Hostein, J., and Feuerstein, C. (1995) Brain noradrenergic systems modulate the ceco-colonic myoelectric activity in rats. *Neurogastroenterol. Motil.* **7,** 101–110.

119. Stein, L. and Wise, C. (1971) Possible etiology of schizophrenia: Progressive damage to the noradrenergic reward system of 6-hydroxydopamine. *Science* **171,** 1032–1037.

120. Antelman, S. M., Lippa, A. S., and Fisher, A. E. (1972) 6-hydroxydopamine, noradrenergic reward and schizophrenia. *Science* **175,** 919–920.

121. Bowers, M. G., Jr. and VanWoert, M. H. (1972) 6-Hydroxydopamine, noradrenergic reward and schizophrenia. *Science* **175,** 920–921.

122. Stein, L. (1964) Self-stimulation of the brain and the central stimulant action of amphetamine. *Fed. Proc.* **23,** 836–850.

123. Breese, G. R., Howard, J. L., and Leahy, J. P. (1971) Effects of 6-hydroxydopamine on electrical self-stimulation of the brain. *Br. J. Pharmacol.* **43,** 255–257.

124. Breese, G. R. and Cooper, B. R. (1975) Relationship of dopamine neural systems to the maintenance of self-stimulation, in *Neurotransmitter Balances Regulating Behavior.* (Domino, E. F. and Davis, J. M., eds.), Ann Arbor, MI, pp. 37–56.

125. Bruno, J. P., Stricker, E. M., and Zigmond, M. J. (1985) Rats given dopamine-depleting brain lesions as neonates are subsensitive to dopaminergic antagonists as adults. *Behav. Neurosci.* **99,** 771–775.

126. Duncan, G., Criswell, H., McCown, T. J., Paul, I., Mueller, R. A., and Breese, G. R. (1987) Behavioral and neurochemical response to haloperidol and SCH-23390 in rats treated neonatally or as adults with 6-hydroxydopamine. *J. Pharmacol. Exp. Ther.* **243,** 1027–1034.

127. de Brabander, J. M., de Bruin, M. P. C., and van Eden, C. G. (1991) Comparison of the effects of neonatal and adult medial prefrontal cortex lesions on food hoarding and spatial delayed alternation. *Behav. Brain Res.* **42,** 67–75.

128. Bruno, J. P., Snyder, A. M., and Stricker, E. M. (1984) Effect of dopamine-depleting brain lesions on suckling and weaning in rats. *Behav. Neurosci.* **98,** 156–161

129. Bruno, J. P., Zigmond, M. J., and Stricker, E. M. (1986) Rats given dopamine-depleting brain lesions as neonates do not respond to acute homeostatic imbalances as adults. *Behav. Neurosci.* **100,** 125–128.

130. Weihmuller, F. B. and Bruno, J. P. (1989) Age-dependent plasticity in the dopaminergic control of sensorimotor development. *Behav. Brain Res.* **35,** 95–109.

131. Weihmuller, F. B. and Bruno, J. P. (1989) Drinking behavior and motor function in rat pups depleted of brain dopamine during development. *Dev. Psychobiol.* **22,** 101–113.

132. Breese, G. R., Napier, T. C., and Mueller, R. A. (1985) Dopamine agonist-induced locomotor activity in rats treated with 6-hydroxydopamine at differing ages: Functional supersensitivity of D-1 dopamine receptors in neonatally lesioned rats. *J. Pharmacol. Exp. Ther.* **234,** 447–455.

133. Breese, G. R., Baumeister, A., Napier, T. C., Frye, G. D., and Mueller, R. A. (1985) Evidence that D-1 dopamine receptors contribute to supersensitive behavioral

responses induced by L-dihydroxyphenylalanine in rats treated neonatally with 6-hydroxydopamine. *J. Pharmacol. Exp. Ther.* **235,** 287–295.

134. Creese, I. and Iversen, S. D. (1972) Amphetamine response in rat after dopamine neurone destruction. *Nature New Biol.* **238,** 247–248.

135. Creese, I. and Iversen, S. D. (1973) Blockade of amphetamine induced motor stimulation and stereotype in the adult rat following neonatal treatment with 6-hydroxdopamine. *Brain Res.* **55,** 369–382.

136. Hollister, A. S., Breese, G. R., and Cooper, B. R. (1974) Comparison of tyrosine hydroxylase and dopamine-beta-hydroxylase inhibition with the effects of various 6-hydroxydopamine treatments on *d*-amphetamine induced motor activity. *Psychopharmacologia* **36,** 1–16.

137. Ungerstedt, U. (1971) Striatal dopamine release after amphetamine or nerve degeneration revealed by rotational behavior. *Acta Physiol Scand.* **367,** 49–68.

138. Evetts, K. D., Uretsky, N. J., Iversen, L. T., and Iversen, S. D. (1970) Effects of 6-hydroxydopamine on CNS catecholamines, spontaneous motor activity and amphetamine induced hyperactivity in rats. *Nature* **225,** 961–962.

139. Von Voigtlander, P. and Moore, K. (1973) Turning behavior of mice with unilateral 6-hydroxydopamine lesions in the striatum: Effects of apomorphine, L-DOPA, amantadine, amphetamine and other psychomotor stimulants. *Neuropharmacology* **12,** 451–462.

140. Christie, J. E. and Crow, T. J. (1971) Turning behaviour as an index of the action of amphetamines and ephedrines on central dopamine-containing neurones. *Br. J. Pharmacol.* **53,** 658–667.

141. Fibiger, H. C., Fibiger, H. P., and Zis, A. P. (1973) Attenuation of amphetamine-induced motor stimulation and stereotypy by 6-hydroxydopamine in the rat. *Br. J. Pharmacol.* **47,** 683–692.

142. Marsden, C. A. and Guldberg, H. D. (1973) The role of monoamines in rotation induced or potentiated by amphetamine after nigral, raphe and mesencephalic reticular lesions in the rat brain. *Neuropharmacology* **12,** 195–211.

143. Peterson, D. W. and Sparber, S. B. (1974) Increased fixed ratio performance and differential *d*- and (–)-amphetamine action following norepinephrine depletion by intraventricular 6-hydroxydopamine. *J. Pharmacol. Exp. Ther.* **191,** 349–357.

144. Dearry, A., Gingrich, J. A., Falardeau, P., Fremeau, R. T., Jr., Bates, M. D., and Caron, M. G. (1990) Molecular cloning and expression of the gene for a human D1 dopamine receptor. *Nature* **347,** 72–76.

145. Gingrich, J. A., Dearry, A., Falardeau, P., Fremeau, R. T., Jr., Bates, M. D., and Caron, M. G. (1991) Molecular characterization of G-protein coupled receptors: isolation and cloning of a D1 dopamine receptor. *J. Receptor Res.* **11,** 521–534.

146. Fremeau, R. T., Jr., Duncan, G. E., Fornaretto, M. G., Dearry, A., Gingrich, J. A., Breese, G. R., and Caron, M. G. (1991) Localization of D1 dopamine receptor mRNA in brain supports a role in cognitive, affective, and neuroendocrine aspects of dopaminergic neurotransmission. *Proc. Natl. Acad. Sci. USA* **88,** 3772–3776.

147. Criswell, H. E., Breese, G. R., and Mueller, R. A. (1989) Priming of D1-dopamine receptor responses: Long lasting supersensitivity to D1-dopamine agonist following repeated administration to neonatal 6-hydroxydopamine lesioned rats. *J. Neurosci.* **9,** 125–133.

148. Johnson, K. B., Criswell, H. E., Jensen, K. F., Simson, P. E., Mueller, R. A., and Breese, G. R. (1992) Comparison of the D_1-dopamine agonists SKF-38393 and A-68930 in neonatal 6-OHDA-lesioned rats: Behavioral effects and induction of c-Fos immunoreactivity. *J. Pharmacol. Exp. Ther.* **262,** 855–865.

149. Criswell, H. E., Mueller, R. A., and Breese, G. R. (1990) Long-term D_1-dopamine receptor sensitization in neonatal-6-OHDA-lesioned rats is blocked by an NMDA antagonist. *Brain Res.* **512,** 284–290.

150. Paulson, P. E., Camp, D. M., and Robinson, T. E. (1991) Time course of transient behavioral depression and persistent behavioral sensitization in relation to regional brain monoamine concentrations during amphetamine withdrawal in rats. *Psychopharmacology* **103,** 480–492.

151. Ujike, H., Onoue, R., Akiyama, K., Hamamura, T., and Otsuki, S. (1989) Effects of selective D-1 and D-2 dopamine antagonists on development of methamphetamine-induced behavioral sensitization. *Psychopharmacology* **98,** 89–92.

152. Drew, K. L. and Glick, S. D. (1990) Role of D-1 and D-2 receptor stimulation in sensitization to amphetamine-induced circling behavior and in expression and extinction of the Pavlovian conditioned response. *Psychopharmacol. (Berl.)* **101,** 465–471.

153. Vezina, P. (1996) D_1 Dopamine receptor activation is necessary for the induction of sensitization by amphetamine in the ventral tegmental area. *J. Neurosci.* **16,** 2411–2420.

154. Robinson, T. E., Jurson, P. A., Bennett, J. A., and Bentgen, K. M. (1988) Persistent sensitization of dopamine neurotransmission in ventral striatum (nucleus accumbens) produced by prior experience with (+)-amphetamine: a microdialysis study in freely moving rats. *Brain Res.* **462,** 211–222.

155. Stewart, J. and Vezina, P. (1989) Microinjections of SCH-23390 into the ventral tegmental area and substantia nigra pars reticulata attenuate the development of sensitization to the locomotor activating effects of systemic amphetamine. *Brain Res.* **495,** 401–406.

156. Pierce, R. C., Born, B., Adams, M., and Kalivas, P. W. (1996) Repeated intra-ventral Tegmental area administration of SKF-38393 induces behavioral and neurochemical sensitization to a subsequent cocaine challenge. *J. Pharmacol. Exp. Ther.* **278,** 384–392.

157. Karler, R., Calder, L. D., Chaudhry, I. A., and Turkanis, S. A. (1989) Blockade of reverse tolerance to cocaine and amphetamine by MK-801. *Life Sci.* **45,** 599–606.

158. Kalivas, P. W. and Alesdatter, J. E. (1993) Involvement of NMDA receptor stimulation in the VTA and amygdala in behavioral sensitization to cocaine. *J. Pharmacol. Exp. Ther.* **267,** 486–495.

159. Wolf, M. E., White, F. J., and Hu, X-T. (1994) MK-801 prevents alterations in the mesoaccumbens dopamine system associated with behavioral sensitization to amphetamine. *J. Neurosci.* **14,** 1735–1745.

160. Breese, G. R. and Mueller, R. A. (1985) SCH-23390 antagonism of a D-2 dopamine agonist depends upon catecholaminergic neurons. *Eur. J. Pharmacol.* **113,** 109–114.

161. Walters, J. R., Bergstrom, D. A., Carlson, J. H., Chase, T. N., and Braun, A. R. (1987) D-1 dopamine receptor activation required for postsynaptic expression of D-2 agonist effects. *Science* **236,** 719–722.

162. Jackson, D. M., Ross, S. B., and Hashizume, M. (1987) Dopamine-mediated behaviors produced in naive mice by bromocriptine plus SKF-38393. *J. Pharm. Pharmacol.* **40,** 221–222.

163. Longoni, R., Spina, L., and Di Chiara, G. (1987) Permissive role of D-1 receptor stimulation for the expression of D-2 mediated behavioral responses: A quantitative phenomenological study in rats. *Life Sci.* **41,** 2135–2145.

164. Graybiel, A. M. and Ragsdale, C. W., Jr. (1983) Biochemical anatomy of the striatum, in *Chemical Neuroanatomy* (Emson, P. C., ed.), Raven, New York, pp. 427–504.

165. Vankova, M., Arluison, M., Boyer, P. A., Bourgoin, S., and Quignon, M. (1991) Enkephalin-containing nerve cell bodies in the substantia nigra of the rat: demonstration by immunohistochemistry and *in situ* hybridization. *Brain Res. Bull.* **27,** 19–27.

166. Kubota, Y., Inagaki, S., and Kito, S. (1986) Innervation of substance P neurons by catecholaminergic terminals in the neostriatum. *Brain Res.* **375,** 163–167.

167. Kubota, Y., Inagaki, S., Kito, S., Takagi, H., and Smith, A. D. (1986) Ultrastructural evidence of dopaminergic input to enkephalinergic neurons in rat neostriatum. *Brain Res.* **367,** 374–378.

168. Kawai, Y., Takagi, H., Kumoi, Y., Shiosaka, S., and Tohyama, M. (1987) Nigrostriatal dopamine neurons receive substance P-ergic inputs in the substantia nigra: application of the immunoelectron microscopic mirror technique to fluorescent double-staining for transmitter-specific projections. *Brain Res.* **401,** 371–376.

169. Voorn, P., Roest, G., and Groenewegen, H. J. (1987) Increase of enkephalin an decrease of substance P immunoreactivity in the dorsal and ventral striatum of the rat after midbrain the 6-hydroxydopamine-lesions. *Brain Res.* **412,** 391–396.

170. Engber, T. M., Susel, Z., Kuo, S., Gerfen, C. R., and Chase, T. N. (1991) Levodopa replacement therapy alters enzyme activities in striatum and neuropeptide content in striatal output regions of the 6-hydroxydopamine-lesioned rats. *Brain Res.* **552,** 113–118.

171. Taylor, M. D., De Ceballos, M. L., Rose, S., Jenner, P., and Marsden, C. D. (1992) Effects of a unilateral the 6-hydroxydopamine-lesion and prolonged L-3, 4-dihydroxyphenylalanine treatment on peptidergic system in rat basal ganglia. *Eur. J. Pharmacol.* **219,** 183–192.

172. Cenci, M. A. and Björklund, A. (1993) Transection of corticostriatal afferents reduces amphetamine- and apomorphine-induced striatal Fos expression and turning behaviour in unilaterally 6-hydroxydopamine-lesioned rats. *Eur. J. Neurosci.* **5,** 1062–1070.

173. Campbell, K. and Björklund, A. (1994) Prefrontal corticostriatal afferents maintain increased enkephalin gene expression in the dopamine-denervated rat striatum. *Eur. J. Neurosci.* **6,** 1371–1383.

174. Thal, L. J., Sharpless, N. S., Hirschhorn, I. D., Horowitz, S. G., and Makman, M. H. (1983) Striatal met-enkephalin concentration increases following nigrostriatal denervation. *Biochem. Pharmacol.* **32,** 3297–3301.

175. Sivam, S. P. (1989) D1 dopamine receptor mediated substance P depletion in the striatonigral neurons of rats subjected to neonatal dopaminergic denervation: implications for self-injurious behavior. *Brain Res.* **500,** 119–130.

176. Sivam, S. P., Breese, G. R., Krause, J. E., Napier, T. C., Mueller, R. A., and Hong, J. S. (1987) Neonatal and adult-6-hydroxydopamine-induced lesions differentially alter tachykinin and enkephalin gene expression. *J. Neurochem.* **49,** 1623–1633.

177. Sivam, S. P., Krause, J. E., Breese, G. R., and Hong, J. S. (1991) Dopamine-dependent postnatal development of enkephalin and tachykinin neurons of rat basal ganglia. *J. Neurochem.* **56,** 1499–1508.
178. Ingham, C. A., Hood, S. H., and Arbuthnott, G. W. (1991) A light and electron microscopical study of enkephalin-immunoreactive structures in the rat neostriatum after removal of the nigrostriatal dopaminergic pathway. *Neuroscience* **42,** 715–730.
179. Gerfen, C. R., McGinty, J. F., and Young, W. S. (1991) Dopamine differentially regulates dynorphin, substance P, and enkephalin expression in striatal neurons: *in situ* hybridization histochemical analysis. *J. Neurosci.* **11,** 1016–1031.
180. Campbell, K., Wictorin, K., and Björklund, A. (1992) Differential regulation of neuropeptide mRNA expression in intrastriatal striatal transplants by host dopaminergic afferents. *Proc. Natl. Acad. Sci. USA* **89,** 10,489–10,493.
181. Cenci, M. A., Campbell, K., and Björklund, A. (1993) Neuropeptide messenger RNA expression in the 6-hydroxydopamine-lesioned rat striatum reinnervated by fetal dopaminergic transplants: differential effects of the grafts on preproenkephalin, preprotachykinin and prodynorphin messenger RNA levels. *Neuroscience* **57,** 275–296.
182. Nisenbaum, L. K., Kitai, S. T., and Gerfen, C. R. (1994) Dopaminergic and muscarinic regulation of striatal enkephalin and substance P messenger RNAs following striatal dopamine denervation: effects of systemic and central administration of quinpirole and scopolamine. *Neuroscience* **63,** 435–439.
183. Nisenbaum, L. K., Kitai, S. T., Crowley, W. R., and Gerfen, C. R. (1994) Temporal dissociation between changes in striatal enkephalin and substance P messenger RNAs following striatal dopamine depletion. *Neuroscience* **60,** 927–937.
184. Voorn, P., Docter, G. J., Jongen-Relo, A. L., and Jonker, A. J. (1994) Rostrocaudal subregional differences in the response of enkephalin, dynorphin and substance P synthesis in rat nucleus accumbens to dopamine depletion. *Eur. J. Neurosci.* **6,** 486–496.
185. Zeng, B. Y., Jolkkonen, J., Jenner, P., and Marsden, C. D. (1995) Chronic L-DOPA treatment differentially regulates gene expression of glutamate decarboxylase, preproenkephalin and preprotachykinin in the striatum of 6-hydroxdopamine-lesioned rat. *Neuroscience* **66,** 19–28.
186. Abrous, D. N., Manier, M., Mennicken, F., Feuerstein, C., Le Moal, M., and Herman, J. P. (1993) Intrastriatal transplants of embryonic dopaminergic neurons counteract the increase of striatal enkephalin immunostaining but not serotoninergic sprouting elicited by a neonatal lesion of the nigrostriatal dopaminergic pathway. *Eur. J. Neurosci.* **5,** 128–136.
187. Bal, A., Savasta, M., Chritin, M., Mennicken, F., Abrous, D. N., Le Moal, M., Feuerstein, C., and Herman, J. P. (1993) Transplantation of fetal nigral cells reverses the increase of preproenkephalin mRNA levels in the rat striatum caused by 6-OHDA lesion of the dopaminergic nigrostriatal pathway: a quantitative *in situ* hybridization study. *Brain Res. Mol.* **18,** 221–227.
188. Sivam, S. P., Breese, G. R., Napier, T. C., Mueller, R. A., and Hong, J. S. (1986) Dopaminergic regulation of prenkephalin-A gene expression in the basal ganglia. *NIDA Res. Monogr.* **75,** 389–392.
189. Kurumaji, A., Takashima, M., Watanabe, S., and Takahashi, K. (1988) An increase in striatal Met-enkephalin-like immunoreactivity in neonatally dopamine-depleted rats. *Neurosci. Lett.* **87,** 109–113.

190. Sivam, S. P. and Krause J. E. (1990) The adaptation of enkephalin, tachykinin and monoamine neurons of the basal ganglia following neonatal dopaminergic denervation is dependent on the extent of dopamine depletion. *Brain Res.* **536,** 169–175.

191. Cimino, M., Zoli, M., and Weiss, B. (1991) Differential ontogenetic expression and regulation of proenkephalin and preprosomatostatin mRNAs in rat caudate-putamen as studied by *in situ* hybridization histochemistry. *Brain Res. Dev. Brain Res.* **60,** 115–122.

192. Soghomonian, J. J. (1994) Differential regulation of glutamate decarboxylase and preproenkephalin mRNA levels in the rat striatum. *Brain Res.* **640,** 146–154.

193. Synder-Keller, A. M. (1991) Developmental striatal compartmentalization following pre- or postnatal dopamine depletion. *J. Neurosci.* **11,** 810–821.

194. Perez-Navarro, E., Alberch, J., and Marsal, J. (1993) Postnatal development of functional dopamine, opioid and tachykinin receptors that regulate acetylcholine release from rat neostriatal slices. Effect of 6-hydroxydopamine lesion. *Int. J. Dev. Neurosci.* **11,** 701–708.

195. Roeling, T. A., Docter, G. J., Voorn, P., Melchers, B. P., Wolters, E. C., and Groenewegen, H. J. (1995) Effects of unilateral 6-OHDA lesions on neuropeptide immunoreactivity in the basal ganglia of the common marmoset, Callithrix Jacchus, a quantitative immunohistochemical analysis. *J. Chem. Neuroanat.* **9,** 155–164.

196. Sivam, S. P. (1991) Dopamine dependent decrease in enkephalin and substance P levels in basal ganglia regions of postmortem parkinsonian brains. *Neuropeptides* **18,** 201–207.

197. Schiffmann, S. N. and Vanderhaeghen, J. J. (1992) Lesion of the nigrostriatal pathway induces cholecystokinin messenger RNA expression in the rat striatum. *Neuroscience* **50,** 551–557.

198. Engber, T. M., Boldry, R. C., Kuo, S., and Chase, T. N. (1992) Dopaminergic modulation of striatal neuropeptides: differential effects of D1 and D2 receptor stimulation on somatostatin, neuropeptide Y, neurotensin, dynorphin and enkephalin. *Brain Res.* **581,** 261–268.

199. Savasta, M., Ruberte, E., Palacious, J. M., and Mengod, G. (1989) The co-localization of cholecystokinin and tyrosine hydroxylase mRNAs in mesencephalic dopaminergic neurons in the rat brain examined by *in situ* hybridization. *Neuroscience* **29,** 363–369.

200. Burgunder, J. M. and Young, W. S. (1989) Distribution, projection and dopaminergic regulation of the neurokinin B mRNA-containing neurons of the rat caudate-putamen. *Neuroscience* **32,** 323–335.

201. Churchill, L. and Kalivas, P. W. (1992) Dopamine depletion produces augmented behavioral responses to a mu-, but not a delta-opioid receptor agonist in the nucleus accumbens: lack of a role for receptor upregulation. *Synapse* **11,** 47–57.

202. Delfs, J. M., Ciaramitaro, V. M., Parry, T. J., and Chesselet, M. F. (1995) Subthalamic nucleus lesions: widespread effects on changes in gene expression induced by nigrostriatal dopamine depletion in rats. *J. Neurosci.* **15,** 6562–6575.

203. Kaakkola, S. (1980) Contralateral circling behaviour induced by intranigral injection of morphine and enkephalin analogue FK 33-824 in rats. *Acta Pharmacol. Toxicol.* **47,** 385–393.

204. Stefano, G. B., Surkin, R. S., and Kream, R. M. (1982) Evidence for the presynaptic localization of a high affinity opiate binding site on dopamine neurons in the pedal ganglia of Mytilus edulis (Bivalvia). *J. Pharmacol. Exp. Ther.* **222,** 759–764.

205. Kalivas, P. W., Widerlove, E., Stanley, D., Breese, G., and Prange, A. J., Jr. (1983) Enkephalin action on the mesolimbic system: a dopamine-dependent and a dopamine-independent increase in locomotor activity. *J. Pharmacol. Exp. Ther.* **227,** 229–237.

206. Arenander, A. T., de Vellis, J., and Herschman, H. R. (1989) Induction of c-Fos and TIS genes in cultured rat astrocytes by neurotransmitters. *J. Neurosci. Res.* **24,** 107–114.

207. Chang, S. L., Squinto, S. P., Harlan, R. E. (1988) Morphine activation of c-Fos expression in rat brain. *Biochem. Biophys. Res. Commun.* **157,** 698–704.

208. Sonnenberg, J. L., Mitchelmore, C., Macgregor-Leon, P. F., Hempstead, J., Morgan, J. I., and Curran, T. (1989) Glutamate receptor agonists increase the expression of Fos, Fra, and AP-1 DNA binding activity in the mammalian brain. *J. Neurosci. Res.* **24,** 72–80.

209. Gubits, R. M., Smith, T. M., Fairhurst, J. L., and Yu, H. (1994) Adrenergic receptors mediate changes in c-Fos mRNA levels in brain. *Brain Res. Mol. Brain Res.* **18,** 39–45.

210. Robertson, G. S., Herrera, D. G., Dragunow, M., and Robertson, H. A. (1989) L-DOPA activates c-Fos in the striatum ipsilateral to a 6-hydroxydopamine lesion of the substantia nigra. *Eur. J. Pharmacol.* **159,** 99–100.

211. Robertson, H. A., Peterson, M. R., Murphy, K., and Robertson, G. S. (1989) D1-Dopamine receptor agonists selectively activate striatal c-Fos independent of rotational behaviour. *Brain Res.* **503,** 346–349.

212. Dragunow, M., Leah, J. D., and Faull, R. L. (1991) Prolonged and selective induction of Fos-related antigen(s) in striatal neurons after 6-hydroxydopamine lesions of the rat substantia nigra pars compacta. *Brain Res. Mol. Brain Res.* **10,** 355–358.

213. Morelli, M., Cozzolino, A., Pinna, A., Fenu, S., Carta, A., and Di Chiara, G. (1993) L-DOPA stimulates c-Fos expression in dopamine denervated striatum by combined activation of D1-and D2-receptors. *Brain Res.* **623,** 334–336.

214. Cole, D. G., Growdon, J. H., and Difiglia, M. (1993) Levodopa induction of Fos immunoreactivity in rat brain following partial and complete lesions of the substantia nigra. *Exp. Neurol.* **120,** 223–232.

215. Mueller, R. A., Grimes, L. M., Criswell, H. E., Carter, L. S., McGimsey, W. C., Stumpf, W., and Breese, G. R. (1989) D1 Dopamine agonist induces c-Fos protein in the striatum of 6-OHDA-lesioned rats. *Soc. Neurosci. Abstracts* **15,** 430.

216. Simson, P. E., Johnson, K. B., Jurevics, H. A., Criswell, H. E., Napier, T. C., Duncan, G. E., Mueller, R. A., and Breese, G. R. (1992) Augmented sensitivity of D1-dopamine receptors in lateral but not medial striatum after 6-hydroxydopamine-induced lesions in the neonatal rat. *J. Pharmacol. Exp. Ther.* **263,** 1454–1463.

217. Paul, M. L., Graybiel, A. M., David, J. C., and Robertson, H. A. (1992) D1-like and D2-like dopamine receptors synergistically activate rotation and c-Fos expression in the dopamine-depleted striatum in a rat model of Parkinson's disease. *J. Neurosci.* **12,** 3729–3742.

218. Pennypacker, K. R., Hong, J. S., and McMillian, M. K. (1995) Implications of prolonged expression of Fos-related antigens. *Trends Pharmacol. Sci.* **16,** 317–321.

219. Robertson, G. S., Vincent, S. R., and Fibiger, H. C. (1990) Striatonigral projection neurons contain D1 dopamine receptor-activated c-Fos. *Brain Res.* **523,** 288–290.

220. Marshall, J. F., Cole, B. N., and LaHoste, G. J. (1993) Dopamine D2 receptor control of pallidal Fos expression: comparisons between intact and 6-hydroxydopamine-treated hemispheres. *Brain Res.* **632,** 308–313.

221. Criswell, H. E., Johnson, K. B., Mueller, R. A., and Breese, G. R. (1993) Evidence for involvement of brain dopamine and other mechanisms in the behavioral action of the *N*-methyl-D-aspartic acid antagonist MK-801 in control and 6-hydroxydopamine-lesioned rats. *J. Pharmacol. Exp. Ther.* **265,** 1001–1010.

222. Morelli, M., Pinna, A., Fenu, S., Carta, A., Cozzolino, A., and DiChiara, G. (1994) Differential effect of MK-801 and scopolamine on c-Fos expression induced by L-DOPA in the striatum of 6-hydroxydopamine lesioned rats. *Synapse* **18,** 288–293.

223. Morelli, M., Fenu, S., Cozzolino, A., Pinna, A., Carta, A., and Di Chiara, G. (1993) Blockade of muscarinic receptors potentiates D1 dependent turning behavior and c-Fos expression in 6-hydroxydopamine-lesioned rats but does not influence D2 mediated responses. *Neuroscience* **53,** 673–678.

224. Robertson, G. S. and Staines, W. A. (1994) D1 dopamine receptor agonist-induced Fos-like immunoreactivity occurs in basal forebrain and mesopontine tegmentum cholinergic neurons and striatal neurons immunoreactive for neuropeptide Y. *Neuroscience* **59,** 375–387.

225. Fenu, S., Carta, A., and Morelli, M. (1995) Modulation of dopamine D1-mediated turning behavior and striatal c-Fos expression by the substantia nigra. *Synapse* **19,** 233–240.

226. Singh, B. and de Champlan, J. (1972) Altered ontogenesis of central noradrenergic neurons following neonatal treatment of 6-hydroxydopamine. *Brain Res.* **48,** 432–437.

227. Jonsson, G., Pycock, C., Fuxe, K., and Sachs, C. H. (1974) Changes in the development of central noradrenaline neurons following neonatal administration of 6-hydroxydopamine. *J. Neurochem.* **22,** 419–426.

228. Loizou, L. A. (1972) The postnatal ontogeny of monoamine-containing neurons in the central nervous system of the albino rat. *Brain Res.* **40,** 395–418.

229. Pappas, B. A. and Sobrian, S. K. (1972) Neonatal sympathectomy by 6-hydroxydopamine in the rat: no effects on behavior but changes in endogenous brain norepinephrine. *Life Sci.* **11,** 653–659.

230. Pappas, B. A., Peters, D. A. V., Sobrian, S. K., Blouin, A., and Drew, B. (1975) Early behavioral and catecholaminergic effects of 6-hydroxydopamine and quanethidine in the neonatal rat. *Pharmacol. Biochem. Behav.* **3,** 681–685.

231. Jonsson, G., Wiesel, F. A., and Hallman, H. (1979) Developmental plasticity of central noradrenaline neurons after neonatal damage-changes in transmitter functions. *J. Neurobiol.* **10,** 337–353.

232. Gustafson, E. L. and Moore, R. Y. (1987) Noradrenaline neuron plasticity in developing rat brain: effects of neonatal 6-hydroxydopamine demonstrated by dopamine-beta-hydroxylase immunocytochemistry. *Brain Res.* **465,** 143–155.

233. Bendeich, E., Konkol, R. J., Krigman, M. R., and Breese, G. R. (1978) Morphological evidence for 6-hydroxydopamine-induced sprouting or noradrenergic neurons in the cerebellum. *J. Neurol. Sci.* **38,** 47–57.

234. Hemmendinger, L. M. and Moore, R. Y. (1986) Synaptic reorganization in the motor trigeminal nucleus of the rat following neonatal 6-hydroxydopamine treatment. *J. Comp. Neurol.* **250,** 462–468.

235. Zhou, F. C., Bledsoe, S., and Murphy, J. (1991) Serotonergic sprouting is induced by dopamine-lesion in substantia nigra of adult rat brain. *Brain Res.* **556**, 108–116.

236. Towle, A. C., Maynard, E. H., Criswell, H. E., Lauder, J. M., Joh, T. H., Mueller, R. A., and Breese, G. R. (1989) Serotonergic innervation of the rat caudate following a neonatal-6-hydroxydopamine lesion: An anatomical, biochemical and pharmacological study. *Pharmacol. Biochem. Behav.* **34**, 367–374.

237. Luthman, J., Bolioli, B., Tsutsumi, T., Verhofstad, A., and Jonsson, G. (1987) Sprouting of striatal serotonin nerve terminals following selective lesions of nigrostriatal dopamine neurons in neonatal rat. *Brain Res. Bull.* **19**, 269–274.

238. Jackson, D. and Abercrombie, E. D. (1992) *In vivo* neurochemical evaluation of striatal serotonergic hyperinnervation in rats depleted of dopamine at infancy. *J. Neurochem.* **58**, 890–897.

239. Berger, T. W., Kaul, S., Stricker, E. M., and Zigmond, M. J. (1985) Hyperinnervation of the striatum by dorsal raphe afferents after dopamine-depleting brain lesions in neonatal rats. *Brain Res.* **336**, 354–358.

240. Snyder, A. M., Zigmond, M. J., and Lund, R. D. (1986) Sprouting of serotoninergic afferents into striatum after dopamine-depleting lesions in infant rats: a retrograde transport and immunocytochemical study. *J. Comp. Neurol.* **245**, 274–281.

241. Mrini, A., Soucy, J. P., Lafaille, F., Lemoine, P., and Descarries, L. (1995) Quantification of the serotonin hyperinnervation in adult rat neostriatum after neonatal 6-hydroxydopamine lesion of nigral dopamine neurons. *Brain Res.* **669**, 303–308.

242. Numan, S., Lundgren, K. H., Wright, D. E., Herman, J. P., and Seroogy, K. B. (1995) Increased expression of 5HT2 receptor mRNA in rat striatum following 6-OHDA lesions of the adult nigrostriatal pathway. *Brain Res. Mol.* **29**, 391–396.

243. Radja, F., Descarries, L., Dewar, K. M., and Reader, T. A. (1993) Serotonin 5-HT1 and 5-HT2 receptors in adult rat brain after neonatal destruction of nigrostriatal dopamine neurons: a quantitative autoradiographic study. *Brain Res.* **606**, 273–285.

244. Kostrzewa, R. M., Gong, L., and Brus, R. (1993) Serotonin (5-HT) systems mediate dopamine (DA) receptor supersensitivity. *Acta Neurobiol. Exp.* **53**, 31–41.

245. Beck, K. D. (1994) Functions of brain-derived neurotrophic factor, insulin-like growth factor-I and basic fibroblast growth factor in the development and maintenance of dopaminergic neurons. *Prog. Neurobiol.* **44**, 497–516.

246. Felten, D. L., Hallman, H., and Jonsson, G. (1982) Evidence for a neurotropic role of noradrenaline neurons in the postnatal development of rat cerebral cortex. *J. Neurocytol.* **11**, 119–135.

247. Date, I. (1996) Parkinson's disease, trophic factors, and adrenal medullary chromaffin cell grafting: Basic and clinical studies. *Brain Res. Bull.* **40**, 1–19.

248. Lewin, G. R. and Barde, Y.-A. (1996) Physiology of the neurotrophins. *Annu. Rev. Neurosci.* **19**, 289–317.

249. Kupsch, A. and Oertel, W. H. (1994) Neural-transplantation, trophic factors and Parkinson's disease. *Life Sci.* **55**, 2083–2095.

250. Akaneya, Y., Takahashi, M., and Hatanaka, H. (1995) Selective acid vulnerability of dopaminergic neurons and its recovery by brain-derived neurotrophic factor. *Brain Res.* **704**, 175–183.

251. Hou, J. G., Lin, L. F. H., and Mytilineou, C. (1996) Glial Cell Line-derived neurotrophic factor exerts neurotrophic effects on dopaminergic neurons *in vitro* and

promotes their survival and regrowth after damage by 1-methyl-4-phenylpyridinium. *J. Neurochem.* **66,** 74–82.

252. Casper, D. and Blum, M. (1995) Epidermal growth factor and basic fibroblast growth factor protect dopaminergic neurons from glutamate toxicity in culture. *J. Neurochem.* **65,** 1016–1026.

253. Engele, J., Rieck, H., Choi-Lundberg, D., and Bohn, M. C. (1996) Evidence for a novel neurotrophic factor for dopaminergic neurons secreted from mesencephalic glial cell lines. *J. Neurosci. Res.* **43,** 576–586.

254. Clarkson, E. D., Zawada, W. M., and Freed, C. R. (1996) GDNF reduces apoptosis in dopaminergic neurons *in vitro. Neuroreport* **7,** 145–149.

255. Sauer, H., Fischer, W., Nikkhah, G., Wiegand, S. J., Brundin, P., Lindsay, R. M., and Björklund, A. (1993) Brain-derived neurotrophic factor enhances function rather than survival of intrastriatal dopamine cell-rich grafts. *Brain Res.* **626,** 37–44.

256. Shults, C. W., Kimber, T., and Martin, D. (1996) Intrastriatal injection of GDNF attenuates the effects of 6-hydroxydopamine. *Neurorep.* **7,** 627–631.

257. Shults, C. W., O'Connor, D. T., Baird, A., Hill, R., Goetz, C. G., Watts, R. L., Klawans, H. L., Carvey, P. M., Bakay, R. A., Gage, F. H., et al. (1991) Clinical improvement in parkinsonian patients undergoing adrenal to caudate transplantation is not reflected by chromogranin A or basic fibroblast growth factor in ventricular fluid. *Exp. Neurol.* **111,** 276–281.

258. Kirschner, P. B., Jenkins, B. F., Schulz, J. B., Finkelstein, S. P., Matthews, R. T., Rosen, B. R., and Beal, M. F. (1996) NGF, BDNF and NT-5, but not NT-3 potent against MPP+ toxicity and oxidative stress in neonatal animals. *Brain. Res.* **713,** 178–185.

259. Niijima, K., Araki, M., Ogawa, M., Nagatsu, I., Sato, F., Kimura, H., and Yoshida, M. (1990) Enhanced survival of cultured dopamine neurons by treatment with soluble extracts from chemically deafferentiated striatum of adult rat brain. *Brain Res.* **528,** 151–154.

260. Hyman, D., Hofer, M., Barde, Y. A., Juhasz, M., Yancopoulous, G. D., Squinto, S. P., and Lindsay, R. M. (1991) BDNF in a neurotrophic factor for dopaminergic neurons of the substantia nigra. *Nature* **350,** 230–232.

261. Seroogy, K. B., Lundgren, K. H., Tran, T. M., Guthrie, K. M., Isackson, P. J., and Gall, C. M. (1994) Dopaminergic neurons in rat ventral midbrain express brain-derived neurotrophic factor and neurotrophin-3 mRNAs. *J. Comp. Neurol.* **342,** 321–334.

262. Spina, M. B., Hyman, C., Squinto, S., and Lindsay, R. M. (1992) Brain-derived neurotrophic factor protects dopaminergic cells from 6-hydroxydopamine toxicity. *Ann. NY Acad. Sci.* **648,** 348–350.

263. Spina, M. B., Squinto, S. P., Miller, J., Lindsay, R. M., and Hyman, C. (1992) Brain-derived neurotrophic factor protects dopamine neurons against 6-hydroxydopamine and *N*-methyl-4-phenylpyridinium ion toxicity: involvement of the glutathione system. *J. Neurochem.* **59,** 99–106.

264. Hoffer, B. J., Hoffman, A., Bowenkamp, K., Huettl, P., Hudson, J., Martin, D., Lin, L. F. H., and Gerhardt, G. A. (1994) Glial cell line-derived neurotrophic factor reverses toxin-induced injury to midbrain dopaminergic neurons *in vivo. Neurosci. Lett.* **182,** 107–111.

265. Altar, C. A., Boylan, C. B., Fritsche, M., Jones, B. E., Jackson, C., Wiegand, S. J., Lindsay, R. M., and Hyman, C. (1994) Efficacy of brain-derived neurotrophic factor

and neurotrophin-3 on neurochemical and behavioral deficits associated with partial nigrostriatal dopamine lesions. *J. Neurochem.* **63**, 1021–1032.

266. Shults, C. W., Kimber, T., and Altar, C. A. (1995) BDNF attenuates the effects of intrastriatal injection of 6-hydroxydopamine. *Neuroreport* **6**, 1109–1112.

267. Levivier, M., Przedborski, S., Bencsics, C., and Kang, U. J. (1995) Intrastriatal implantation of fibroblasts genetically engineered to produce brain-derived neurotrophic factor prevents degeneration of dopaminergic neurons in a rat model of Parkinson's disease. *J. Neurosci.* **15**, 7810–7820.

268. Lucidi-Phillipi, C. A., Gage, F. H., Shults, C. W., Jones, K. R., Reichardt, L. F., and Kang, U. J. (1995) Brain-derived neurotrophic factor-transduced fibroblasts: production of BDNF and effects of grafting to the adult rat brain. *J. Comp. Neurol.* **354**, 361–376.

269. Yoshimoto, Y., Lin, Q., Collier, T. J., Frim, D. M., Breakefield, X. O., and Bohn, M. C. (1995) Astrocytes retrovirally transduced with BDNF elicit behavioral improvement in a rat model of Parkinson's disease. *Brain Res.* **691**, 25–36.

270. Sauer, H., Wong, V., and Björklund, A. (1995) Brain-derived neurotrophic factor and neurotrophin-4/5 modify neurotransmitter-related gene expression in the 6-hydroxy-dopamine-lesioned rat striatum. *Neuroscience* **65**, 927–933.

271. Arenas, E. and Persson, H. (1994) Neurotrophin-3 prevents the death of adult central noradrenergic neurons *in vivo. Nature* **367**, 368–371.

272. Hynes, M. A., Poulsen, K., Armanini, M., Berkemeier, L., Phillips, H., and Rosenthal, A. (1994) Neurotrophin-4/5 is a survival factor for embryonic midbrain dopaminergic neurons in enriched cultures. *J. Neurosci. Res.* **37**, 144–154.

273. Lewis, M. E., Brown, R. M., Brownstein, M. J., Hart, T., and Stein, D. G. (1979) Nerve growth factor: effects on D-amphetamine-induced activity and brain monoamines. *Brain Res.* **176**, 297–310.

274. Bergdall, V. K. and Becker, J. B. (1994) Effects of nerve growth factor infusion on behavioral recovery and graft survival following intraventricular adrenal medulla grafts in the unilateral 6-hydroxydopamine lesioned rat. *J. Neural Trans. Plast.* **5**, 163–167.

275. Stromberg, I., Herrera-Marschitz, M., Ungerstedt, U., Ebendal, T., and Olson, L. (1985) Chronic implants of chromaffin tissue into the dopamine-denervated striatum. Effects of NGF on graft survival, fiber growth and rotational behavior. *Exp. Brain. Res.* **60**, 335–349.

276. Pezzoli, G., Fahn, S., Dwork, A., Truong, D. D., De Yebenes, J. G., Jackson-Lewis, V., Herbert, J., and Cadet, J. L. (1988) Non-chromaffin tissue plus nerve growth factor reduces experimental parkinsonism in aged rats. *Brain Res.* **459**, 398–403.

277. Cunningham, L. A., Short, M. P., Breakefield, X. O., and Bohn, M. C. (1994) Nerve growth factor released by transgenic astrocytes enhances the function of adrenal chromaffin cell grafts in a rat model of Parkinson's disease. *Brain Res.* **658**, 219–231.

278. Olson, L., Backlund, E. O., Ebendal, T., Freedman, R., Hamberger, B., Hansson, P., Hoffer, B., Lindblom, U., Meyerson, B., Stromberg, I., et al. (1991) Intraputaminal infusion of nerve growth factor to support adrenal medullary autografts in Parkinson's disease. One-year follow-up of first clinical trial. *Arch. Neurol.* **48**, 373–381.

279. Knusel, B., Michel, P. P., Schwaber, J. S., and Hefti, F. (1990) Selective and nonselective stimulation of central cholinergic and dopaminergic development *in vitro* by nerve growth factor, basic fibroblast growth factor, epidermal growth factor, insulin and the insulin-like growth factors I and II. *J. Neurosci.* **10**, 558–570.

280. Mena, M. A., Casarejos, M. J., Gimenez-Gallego, G., and Garcia de Yebenes, J. (1995) Fibroblast growth factors: structure-activity on dopamine neurons *in vitro. J. Neural Trans. Parkinson's Disease Dementia* **Section 9,** 1–14.

281. Nakao, N., Odin, P., Lindvall, O., and Brundin, P. (1996) Differential trophic effects of basic fibroblast growth factor, insulin-like growth factor-1, and neurotrophin-3 on striatal neurons in culture. *Exp. Neurol.* **138,** 144–157.

282. Otto, D. and Unsicker, K. (1990) Basic FGF reverses chemical and morphological deficits in the nigrostriatal system of MPTP-treated mice. *J. Neurosci.* **10,** 1912–1921.

283. Leonard, S., Luthman, D., Logel, J., Luthman, J., Antle, C., Freedman, R., and Hoffer, B. (1993) Acidic and basic fibroblast growth factor mRNAs are increased in striatum following MPTP-induced dopamine neurofiber lesion: assay by quantitative PCR. *Brain Res. Mol. Brain Res.* **18,** 275–284.

284. Chadi, G., Moller, A., Rosen, L., Janson, A. M., Agnati, L. A., Goldstein, M., Ogren, S. O., Pettersson, R. F., and Fuxe, K. (1993) Protective actions of human recombinant basic fibroblast growth factor on MPTP-lesioned nigrostriatal dopamine neurons after intraventricular infusion. *Exp. Brain Res.* **97,** 145–158.

285. Chadi, G., Cao, Y., Pettersson, R. F., and Fuxe, K. (1994) Temporal and spatial increase of astroglial basic fibroblast growth factor synthesis after 6-hydroxydopamine-induced degeneration of the nigrostriatal dopamine neurons. *Neuroscience* **61,** 891–910.

286. Matsuda, S., Saito, H., and Nishiyama, N. (1992) Basic fibroblast growth factor ameliorates rotational behavior of substantia nigral-transplanted rats with lesions of the dopaminergic nigrostriatal neurons. *Jpn. J. Pharmacol.* **59,** 365–370.

287. Date, I., Notter, M. F., Felten, S. Y., and Felten, D. L. (1990) MPTP-treated young mice but not aging mice show partial recovery of the nigrostriatal dopaminergic system by stereotaxic injection of acidic fibroblast growth factor (aFGF). *Brain Res.* **526,** 156–160.

288. Beck, K. D., Knusel, B., and Hefti, F. (1993) The nature of the trophic action of brain-derived neurotrophic factor, des(1-3)-insulin-like growth factor-1, and basic fibroblast growth factor on mesencephalic dopaminergic neurons developing in culture. *Neuroscience* **52,** 855–866.

289. Hwang, O. and Choi, H. J. (1995) Induction of gene expression of the catecholamine-synthesizing enzymes by insulin-like growth factor-I. *J. Neurochem.* **65,** 1988–1996.

290. Hagg, T. and Varon, S. (1992) Ciliary neurotrophic factor (CNTF) prevents axotomy-induced degeneration of adult rat substantia nigra dopaminergic neurons. *Soc. Neurosci. Abstracts* **18,** 390.

291. Krieglstein, K. and Unsicker, K. (1994) Transforming growth factor-β promotes survival of midbrain dopaminergic neurons and protects them against *N*-methyl-4-pheynypyridinimum ion toxicity. *Neuroscience* **63,** 1189–1196.

292. Jonsson, G. and Hallman, H. (1982) Substance P modifies the 6-hydroxydopamine induced alteration of postnatal development of central noradrenaline neurons. *Neuroscience* **7,** 2909–2918.

293. Hadjiconstantinou, M., Rosette, Z. L., Paxton, R. C., and Neff, N. H. (1986) Administration of GM1 ganglioside restores the dopamine content in striatum after chronic treatment with MPTP. *Neuropharmacology* **25,** 479–482.

294. Schneider, J. S. and Distefano, L. (1995) Response of the damaged dopamine system to GM1 and semisynthetic gangliosides: effects of dose and extent of lesion. *Neuropharmacology* **34,** 489–493.

295. Fadda, E., Negro, A., Facci, L., and Skaper, S. D. (1993) Ganglioside GM1 cooperates with brain-derived neurotrophic factor to protect dopaminergic neurons from 6-hydroxydopamine-induced degeneration. *Neurosci. Lett.* **159,** 147–150.

296. Fusco, M., Vantini, G., Schiavo, N., Zanotti, A., Zanoni, R., Facci, L., and Skaper, S. D. (1993) Gangliosides and neurotrophic factors in neurodegenerative diseases: from experimental findings to clinical perspectives. *Ann. NY Acad. Sci.* **695,** 314–317.

297. Iwashita, A., Hisajima, H., Notsu, Y., and Okuhara, M. (1996) Effects of fibroblast growth factor and ganglioside GM1 on neuronal survival in primary cultures and on eight-arm radial maze task in adult rats following partial fimbria transections. *Naunyn-Schmiedeberg's Arch. Pharmacol.* **353,** 342–348.

298. Lin, L. F., Doherty, D. H., Lile, J. D., Bektesh, S., and Collins, F. (1993) GDNF:a glial cell line-derived neurotrophic factor for midbrain dopaminergic neurons. *Science* **260,** 1130–1132.

299. Beck, K. D., Irwin, I., Valverde, J., Brennan, T. J., Langston, J. W., amd Hefti, F. (1996) GDNF induces a dystonia-like state in neonatal rats and stimulates dopamine and serotonin synthesis. *Neuron* **16,** 665–673.

300. Kearns, C. M. and Gash, D. M. (1995) GDNF protects nigral dopamine neurons against 6-hydroxydopamine *in vivo. Brain Res.* **672,** 104–111.

301. Tomac, A., Lindqvist, E., Lin, L-F. H., Ogren, S. O., Young, D., and Hoffer, B. J. (1995) Protection and repair of the nigrostriatal dopaminergic system by GDNF *in vivo. Nature* **373,** 335–339.

302. Beck, K. D., Valverde, J., Alexi, T., Poulsen, K., Moffat, B., Vandlen, R. A., Rosenthal, A., and Hefti, F. (1995) Mesencephalic dopaminergic neurons protected by GDNF from axotomy-induced degeneration in the adult brain. *Nature* **373,** 339–341.

303. Sauer, H., Rosenblad, C., and Björklund, A. (1995) Glial cell line-derived neurotrophic factor but not transforming growth factor beta 3 prevents delayed degeneration of nigral dopaminergic neurons following striatal 6-hydroxydopamine lesion. *Proc. Natl. Acad. Sci. USA* **92,** 8935–8939.

304. Bowenkamp, K. E., Hoffman, A. F., Gerhardt, G. A., Henry, M. A., Biddle, P. T., Hoffer, B. J., and Granholm, A. C. (1995) Glial cell line-derived neurotrophic factor supports survival of injured midbrain dopaminergic neurons. *J. Comp. Neurol.* **355,** 479–489.

305. Stromberg, I., Björklund, L., Johansson, M., Tomac, A., Collins, F., Olson, L., Hoffer, B., and Humpel, C. (1993) Glial cell line-derived neurotrophic factor is expressed in the developing but not adult striatum and stimulates developing dopamine neurons *in vivo. Exp. Neurol.* **124,** 401–412.

306. Schaar, D. G., Sieber, B. A., Dreyfus, C, F., and Black, I. B. (1993) Regional and cell-specific expression of GDNF in rat brain. *Exp. Neurol.* **124,** 368–371.

307. Moore, M. W., Klein, R. D., Farinas, I., Sauer, H., Armanin, M., Phillips, H., Reichardt, L. F., Ryan, A. M., Carver-Moore, K., and Rosenthal, A. (1996) Renal and neuronal abnormalities in mice lacking GDNF. *Nature* **382,** 76–79.

308. Durbec, P., Marcos-Gutierrez, C. V., Kilkenny, C., Grigoriou, M., Wartiowaara, K., Suvanto, P., Smith, D., Ponder, B., Costantini, F., Saarma, M., Sariola, H., and Pachnis, V. (1996) GDNF signaling through the Ret receptor tyrosine hydroxylase. *Nature* **381,** 789–793.

309. Hornykiewicz, O. (1975) Brain monoamines and parkinsonism. *NIDA: Res. Mon. Ser.* **3**, 13–21.

310. Eidelberg, D. (1992) Positron emission tomography studies in parkinsonism. *Neurol. Clin.* **10**, 421–433.

311. Eidelberg, D., Moeller, J. R., Ishikawa, T., Dhawan, V., Spetsieris, P., Chaly, T., Robeson, W., Dahl, J. R., and Margouleff, D. (1995) Assessment of disease severity in parkinsonism with flourine-18-fluorodeoxyglucose and PET. *J. Nuclear Med.* **36**, 378–383.

312. Shinotoh, H. and Calne, D. B. (1995) The use of PET in Parkinson's disease. *Brain Cogn.* **28**, 297–310.

313. Albin, R. L., Young, A. B., and Penny, J. B. (1989) The functional neuroanatomy of basal ganglia disorders. *TINS* **12**, 366–375.

314. Jenner, P. (1990) Parkinson's disease: Clues to the cause of cell death in substantia nigra. *Semin. Neurosci.* **2**, 117–126.

315. Bean, A. J., Elde, R., Cao, Y. H., Oellig, C., Tamminga, C., Goldstein, M., Pettersson, R. F., and Hökfelt, T. (1991) Expression of acidic and basic fibroblast growth factors in the substantia nigra of rat, monkey, and human. *Proc. Natl. Acad. Sci. USA* **88**, 10,237–10,241.

316. Nitta, A., Furukawa, Y., Hayashi, K., Hiramatsu, M., Kameyama, T., Hasegawa, T., and Nabeshima, T. (1992) Denervation of dopaminergic neurons with 6-hydroxy-dopamine increases nerve growth factor content in rat brain. *Neurosci. Lett.* **144**, 152–156.

317. Venero, J. L., Beck, K. D., and Hefti, F. (1994) 6-Hydroxydopamine lesions reduce BDNF mRNA levels in adult rat brain substantia nigra. *Neuroreport* **5**, 429–432.

318. Lorigados, L., Soderstrom, S., and Ebendal, T. (1992) Two-site enzyme immunoassay for beta NGF applied to human patient sera. *J. Neurosci. Res.* **32**, 329–339.

319. Tooyama, I., Kawamata, T., Walker, D., Yamada, T., Hanai, K., Kimura, H., Iwane, M., Igarashi, K., McGeer, E. G., and McGeer, P. L. (1993) Loss of basic fibroblast growth factor in substantia nigra neurons in Parkinson's disease. *Neurology* **43**, 372–376.

320. Tooyama, I., McGeer, E. G., Kawamata, T., Kimura, H., and McGeer, P. L. (1994) Retention of basic fibroblast growth factor immunoreactivity in dopaminergic neurons of the substantia nigra during normal aging in humans contrasts with loss in Parkinson's disease. *Brain Res.* **656**, 165–168.

321. Mogi, M., Harada, M., Kondo, T., Riederer, P., Inagaki, H., Minami, M., and Nagatsu, T. (1994) Interleukin-1 beta, interleukin-6, epidermal growth factor and transforming growth factor-alpha are elevated in the brain from parkinsonian patients. *Neurosci. Lett.* **180**, 147–150.

322. Mogi, M., Harada, M., Kondo, T., Narabayashi, H., Riederer, P., and Nagatsu, T. (1995) Transforming growth factor-beta 1 levels are elevated in the striatum and in ventricular cerebrospinal fluid in Parkinson's disease. *Neurosci. Lett.* **193**, 129–132.

323. Kluter, H., Vieregge, P., Stolze, H., and Kirchner, H. (1995) Defective production of interleukin-2 in patients with idiopathic Parkinson's disease. *J. Neurol. Sci.* **133**, 134–139.

324. Unsicker, K. (1994) Growth factors in Parkinson's disease. *Prog. Growth Factor Res.* **5**, 73–87.

325. Hefti, F., Michel, P. P., and Knusel, B. (1990) Neurotrophic factors and Parkinson's disease. *Adv. Neurol.* **53,** 123–127.

326. Lindsay, R. M., Altar, C. A., Cedarbaum, J. M., Hyman, C., and Wiegand, S. J. (1993) The therapeutic potential of neurotrophic factors in the treatment of Parkinson's disease. *Exp. Neurol.* **124,** 103–118.

327. Olson, L. (1990) Grafts and growth factors in CNS. Basic science with clinical promise. *Stereotactic Func. Neurosurg.* **54,55,** 250–267.

328. Freed, W. J., Poltorak, M., Takashima, H., LaMarca, M. E., and Ginns, E. I. (1991) Brain grafts and Parkinson's disease. *J. Cell. Biochem.* **45,** 261–267.

329. Lindvall, O. (1991) Prospects of transplantation in human neurodegenerative diseases. *TINS* **14,** 376–384.

330. Ahlskog, J. E. (1993) Cerebral transplantation for Parkinson's disease: current progress and future prospects. *Mayo Clin. Proc.* **68,** 578–591.

331. Lopez-Lozano, J. J. and Brera, B. (1993) Neural transplants in Parkinson's Disease. *Transplant. Proc.* **25,** 1005–1011.

332. Widner, H. and Rehncrona, S. (1993) Transplantation and surgical treatment of Parkinsonian syndromes. *Curr. Opinion Neurol. Neurosurg.* **6,** 344–349.

333. Sweetman, L. and Nyhan, W. L. (1967) Excretion of hypoxanthine and xanthine in a genetic disease of purine metabolism. *Nature* **215,** 859–860.

334. Nyhan, W. L. (1968) Clinical features of the Lesch-Nyhan syndrome: Introduction—clinical and genetic features. *Fed. Proc.* **27,** 1027–1033.

335. Lloyd, K. G., Hornykiewicz, O., Davidson, L., Shannak, K., Farley, I., Goldstein, M., Shibuya, M., Kelly, W. N., and Fox, I. H. (1981) Biochemical evidence of dysfunction of brain neurotransmitters in the Lesch-Nyhan syndrome. *N. Engl. J. Med.* **305,** 1106–1111.

336. Wong, D. F., Harris, J. C., Naidu, S., Yokoi, F., Marenco, S., Dannals, R. F., Ravert, H. T., Yaster, M., Evans, A., Rousset, O., Bryan, R. N., Gjedde, A., Kuhar, M. J., and Breese, G. R. (1996) Dopamine transporters are markedly reduced in Lesch-Nyhan disease *in vivo. Proc. Natl. Acad. Sci. USA* **93,** 5539–5543.

337. Ernst, M., Zametkin, A. J., Matochik, J. A., Pascualvaca, D., Jons, P. H., Hardy, K., Hankerson, J. G., Doudet, D. J., and Cohen, R. M. (1996) Presynaptic dopaminergic deficits in Lesch-Nyhan disease. *N. Engl. J. Med.* **334,** 1568–1572.

338. Fingers, S., Heavens, R. P., Sirinathsinghji, D. J., and Kuehn, M. R. (1988) Behavioral and neurochemical evaluation of a transgenic mouse model of Lesch-Nyhan syndrome. *J. Neurol. Sci.* **86,** 203–213.

339. Dunnett, S. B., Sirinathsinghji, D. J., Heavens, R., Rogers, D. C., and Kuehn, M. R. (1989) Monoamine deficiency in a transgenic (HPRT-) mouse model of Lesch-Nyhan syndrome. *Brain Res.* **501,** 401–406.

340. Jinnah, H. A., Langlais, P. J., and Friedmann, T. (1992) Functional analysis of brain dopamine systems in a genetic mouse model of Lesch-Nyhan syndrome. *J. Pharmacol. Exp. Ther.* **263,** 596–607.

341. Jinnah, H. A., Wojcik, B. E., Hunt, M., Narang, N., Lee, K. Y., Goldstein, M., Wamsley, J. K., Langlais, P. J., and Friedmann, T. (1994) Dopamine deficiency in a genetic mouse model of Lesch-Nyhan disease. *J. Neurosci.* **14,** 1164–1175.

342. Schiffmann, S. N. and Vanderhaeghen, J. J. (1993) Adenosine A2 receptors regulate the gene expression of striatopallidal and striatonigral neurons. *J. Neurosci.* **13,** 1080–1087.

343. Morelli, M., Pinna, A., Wardas, J., and DiChiara, G. (1995) Adenosine A2 receptors stimulate c-Fos expression in striatal neurons of 6-hydroxydopamine-lesioned rats. *Neuroscience* **67,** 49–55.

344. Pollack, A. E. and Fink, J. S. (1995) Adenosine antagonists potentiate D2 dopamine-dependent activation of Fos in the striatopallidal pathway. *Neuroscience* **68,** 721–728.

345. Joyce, J. N. (1993) The dopamine hypothesis of schizophrenia: limbic interactions with serotonin and norepinephrine. *Psychopharmacology* **112,** S16–34.

346. Halberstadt, A. L. (1995) The phencyclidine-glutamate model of schizophrenia. *Clin. Neuropharmacol.* **18,** 237–249.

347. Deutch, A. Y. (1994) The regulation of subcortical dopamine systems by the prefrontal cortex: interactions of central dopamine systems and the pathogenesis of schizophrenia. *J. Neural Trans.* **36,** 61–89.

348. Schwarzkopf, S. B., Mitra, T., and Bruno, J. P. (1992) Sensory gating in rats depleted of dopamine as neonates: potential relevance to findings in schizophrenic patients. *Biol. Psychiatr.* **31,** 759–773.

349. Bubser, M. and Koch, M. (1994) Prepulse inhibition of the acoustic startle response of rats is reduced by 6-hydroxydopamine lesions of the medial prefrontal cortex. *Psychopharmacol.* **113,** 487–492.

350. Coyle, J. T. and Johnston, M. V. (1980) Functional hyperinnervation of cerebral cortex by noradrenergic neurons results from fetal lesions: parallels with schizophrenia. *Psychopharmacol. Bull.* **16,** 27–29.

351. Hudson, J. L., van Horne, C. G., Stromberg, I., Brock, S., Clayton, J., Masserano, J., Hoffer, B. J., and Gerhardt, G. A. (1993) Correlation of apomorphine- and amphetamine-induced turning with nigrostriatal dopamine content in unilateral 6-hydroxydopamine lesioned rats. *Brain Res.* **626,** 167–174.

352. Lindvall, O. (1994) Clinical application of neuronal grafts in Parkinson's disease. *J. Neurol.* **242,** S54–56.

353. Backlund, E. O., Granberg, P. O., Hamberger, B., Knutsson, E., Martensson, A., Sevall, G., Seiger, A., and Olson, L. (1985) Transplantation of adrenal medullary tissue to striatum in parkinsonism: First clinical trials. *J. Neurosurg.* **62,** 169–173.

354. Freed, W. J., Morishisa, J. M., Spoor, H. E., Hoffer, B. J., Olson, L., Seiger, A., and Wyatt, R. J. (1981) Transplanted adrenal chromaffin cells in rat brain reduce lesion-induced rotational behavior. *Nature* **292,** 351,352.

355. Freed, W. J., Poltorak, M., and Becker, J. B. (1990) Intracerebral adrenal medulla grafts. A review. *Exp. Neurol.* **110,** 139–166.

356. Madrazo, I., Drucker-Colin, R., Diaz, V., Martinez-Mata, J., Torres, C., and Becerri, J. J. (1987) Open microsurgical autograft of adrenal medulla to the right caudate nucleus in two patients with intractable Parkinson's disease. *N. Engl. J. Med.* **316,** 831–834.

357. Allen, G. S., Burns, R. S., Tulipan, N. B., and Parker, R. A. (1989) Adrenal medullary transplantation to the caudate nucleus in Parkinson's disease. *Arch. Neurol.* **46,** 487–491.

358. Lindvall, O., Backlund, E. O., Farde, L., Sedvall, G., Freedman, R., and Hoffer, B. (1987) Transplantation in Parkinson's disease: Two cases of adrenal medullary grafts to the putamen. *Ann. Neurol.* **22,** 457–568.

359. Broggi, G., Pluchino, F., Gennari, L., Geminian, S., Tamma, F., and Caraceni, T. (1991) Adrenal medulla autograft in caudate nucleus as treatment for Parkinson's disease. *Acta. Neurochir. Suppl.* (Wien) **52,** 45–47.

360. Ben, R., Ji-Chang, F., Yao-Dong, B., Yie-Jian, L., and Yi-Fang, Z. (1991) Transplantation of fetal adrenal medullary tissue into the brain of Parkinsonian. *Acta. Neurochir. Suppl.* (Wien) **52,** 42–44.
361. Goetz, C. G., Stebbins, G. T., Klawans, H. L., Koller, W. C., Grossman, R. G., Bakay, R. A. E., and Penn, R. D. (1991) United Parkinson foundation neurotransplantation registry on adrenal medullary transplants: Presurgical, and 1-year and 2-year follow-up. *Neurology* **41,** 1719–1722.
362. Widner, H., Brundin, P., Rehncrona, S., Gustavii, B., Frackowiak, R., Leenders, K. L., Sawle, G., Rothwell, J. C., Marsden, C. D., Björklund, A., et al. (1991) Transplanted allogeneic fetal dopamine neurons survive and improve motor function in idiopathic Parkinson's disease. *Transplant. Proc.* **23,** 793–795.
363. Widner, H., Tetrud, J., Rehncrona, S., Snow, B., Grundin, P., Gustativii, B., Björklund, A., Lindvall, O., and Langston, J. W. (1992) Bilateral fetal mesencephalic granting in two patients with Parkinsonism induced by 1-methyl-4-phenyl-1,2,3,6-tetrahydropyridine (MPTP). *N. Engl. J. Med.* **327,** 1556–1563.
364. Lindvall, O., Brundin, P., Widner, H., Rehncrona, S., Gustavii, B., Frackowiak, R., Lenders, K. L., Sawle, G., Rothwell, J. C., Marsden, C. D., and Björklund, A. (1990) Grafts of fetal dopamine neurons survive and improve motor function in Parkinson's disease. *Science* **247,** 574–577.
365. Lindvall, O., Widner, H., Rehncrona, S., Brundin, P., Odin, P., Gustavii, B., Franckowiak, R., Leenders, K. L., Sawle, G., Rothwell, J. C., Björklund, A., and Marsden, C. D. (1992) Transplantation of fetal dopamine neurons in Parkinson's Disease: One year Clinical and Neurophysiological Observations in two patients with putaminal implants. *Ann. Neurol.* **31,** 155–165.
366. Sawle, G., Bloomfield, P. M., Björklund, A., Brooks, D. J., Brundin, P., Leenders, K. L., Lindvall, O., Marsden, C. D., Rehcrona, S., Widner, H., and Frackowiak, R. S. J. (1992) Transplantation of fetal dopamine neurons in Parkinson's disease: PET [18F]-6-L-Fluorodopa studies in two patients with putaminal implants. *Ann. Neurol.* **31,** 166–173.
367. Freed, C. R., Breeze, R. E., Rosenberg, N. L., and Schneck, S. A. (1993) Embryonic dopamine cell implants as a treatment for the second phase of Parkinson's disease. Replacing failed nerve terminals. *Adv. Neurol.* **60,** 721–728.
368. Redmond, D. E., Jr., Robbins, R. J., Naftolin, F., Marek, K. L., Vollmer, T. L., Leranth, C., Roth, R. H., Price, L. H., Gjedde, A., Bunney, B. S., et al. (1993) Cellular replacement of dopamine deficit in Parkinson's disease using human fetal mesencephalic tissue: preliminary results in four patients. *Res. Publications Assoc. Res. Nerv. Mental Dis.* **71,** 325–359.
369. Björklund, A. and Stenevi, U. (1979) Reconstruction of the nigrostriatal dopamine pathway by intracerebral nigral transplants. *Brain Res.* **177,** 555–560.
370. Freeman, T. B., Olanow, C. W., Hauser, R. A., Nauert, G. M., Smith, D. A., Borlongan, C. V., Sanberg, P. R., Holt, D. A., Kordower, J. H., Vingerhoets, F. J., et al. (1995) Bilateral fetal nigral transplantation into the postcommissural putamen in Parkinson's disease. *Ann. Neurol.* **38,** 379–388.
371. Freed, W. J., Perlow, M. J., Karoum, F., Seiger, A., Olson, L., Hoffer, B. J., and Wyatt, R. J. (1980) Restoration of dopaminergic function by grafting of fetal rat substantia

nigra to the caudate nucleus: long-term behavioral, biochemical, and histochemical studies. *Ann. Neurol.* **8,** 510–519.

372. Zhu, S. M., Kujirai, K., Dollison, A., Angulo, J., Fahn, S., and Cadet, J. L. (1992) Implantation of genetically modified mesencephalic fetal cells into the rat striatum. *Brain Res. Bull.* **29,** 81–93.

373. Annett, L. E., Martel, F. L., Rogers, D. C., Ridley, R. M., Baker, H. F., and Dunnett, S. B. (1994) Behavioral assessment of the effects of embryonic nigral grafts in marmosets with unilateral 6-OHDA lesions of the nigrostriatal pathway. *Exp. Neurol.* **125,** 228–246.

374. Annett, L. E., Torres, E. M., Ridley, R. M., Baker, H. F., and Dunnett, S. B. (1995) A comparison of the behavioural effects of embryonic nigral grafts in the caudate nucleus and in the putamen of marmosets with unilateral 6-OHDA lesions. *Exp. Brain Res.* **103,** 355–371.

375. Björklund, A., Dunnett, S. B., Stenevi, U., Lewish, M. E., and Iversen, S. D. (1980) Reinnervation of the denervated striatum by substantia nigra transplants: Functional consequences as revealed by pharmacological and sensorimotor testing. *Brain Res.* **199,** 307–333.

376. Taylor, J. R., Elsworth, J. D., Roth, R. H., Collier, T. J., Sladek, J. R., Jr., and Redmond, D. E., Jr. (1990) Improvements in MPTP-induced object retrieval deficits and behavioral deficits after fetal nigral grafting in monkeys. *Prog. Brain Res.* **82,** 543–559.

377. Taylor, J. R., Elsworth, J. D., Roth, R. H., Sladek, J. R., Jr., Collier, T. J., and Redmond, D. E., Jr. (1991) Grafting of fetal substantia nigra to striatum reverses behavioral deficits induced by MPTP in primates: a comparison with other types of grafts as controls. *Exp. Brain Res.* **85,** 335–348.

378. Collier, T. J. and Springer, J. E. (1991) Co-grafts of embryonic dopamine neurons and adult sciatic nerve into the denervated striatum enhance behavioral and morphological recovery in rats. *Exp. Neurol.* **114,** 343–350.

379. Blunt, S. B., Jenner, P., and Marsden, C. D. (1992) Motor function, graft survival and gliosis in rats with 6-OHDA lesions and foetal ventral mesencephalic grafts chronically treated with L-DOPA and carbidopa. *Exp. Brain Res.* **88,** 326–340.

380. Cenci, M. A., Kalen, P., Mandel, R. J., Wictorin, K., and Björklund, A. (1992) Dopaminergic transplants normalize amphetamine- and apomorphine-induced Fos expression in the 6-hydroxydopamine-lesioned striatum. *Neuroscience* **46,** 943–957.

381. Savasta, M., Mennicken, F., Chritin, M., Arbous, D. N., Feuerstein, C., LeMoal, M., and Herman, J. P. (1992) Intrastriatal dopamine-rich implants reverse in the changes in dopamine D2 receptor densities caused by 6-hydroxydopamine lesion of the nigrostriatal pathway in rats: An autoradiographic study. *Neuroscience* **46,** 729–738.

382. Rioux, L., Gagnon, C., Gaudin, D. P., Di Paolo, T., and Bedard, P. J. (1993) A fetal nigral graft prevents behavioral supersensitivity associated with repeated injections of L-DOPA in 6-OHDA rats. Correlation with D1 and D2 receptors. *Neuroscience* **56,** 45–51.

383. Zhu, Y. S., Jones, S. B., Burke, R. E., Franklin, S. O., and Inturrisi, C. E. (1993) Quantitation of the levels of tyrosine hydroxylase and preproenkephalin mRNAs in nigrostriatal sites after 6-hydroxydopamine lesions. *Life Sci.* **52,** 1577–1584.

384. Cenci, M. A. and Björklund, A. (1994) Transection of corticostriatal afferents abolishes the hyperexpression of Fos and counteracts the development of rotational overcompensation induced by intrastriatal dopamine-rich grafts when challenged with amphetamine. *Brain Res.* **665,** 167–174.

385. Wang, Y., Lin, J. C., Chiou, A. L., Liu, J. Y., Liu, J. C., and Zhou, F. C. (1995) Human ventromesencephalic grafts restore dopamine release and clearance in hemiparkinsonian rats. *Exp. Neurol.* **136,** 98–106.

386. Abrous, D. N., Desjardins, S., Sorin, B., Hancock, S. D., Le Moal, M., and Herman, J-P. (1996) Changes in striatal immediate early gene expression following neonatal dopaminergic lesion and effects of intrastriatal dopaminergic transplants. *Neuroscience* **73,** 145–159.

387. Nikkah, G., Cunningham, M. G., Cenci, M. A., McKay, R. D., and Björklund, A. (1995) Dopaminergic microtransplants into the substantia nigra of neonatal rats with bilateral 6-OHDA lesions. I. Evidence for anatomical reconstruction of the nigrostriatal pathway. *J. Neurosci.* **15,** 3548–3561.

388. Nikkhah, G., Cunningham, M. G., McKay, R., and Björklund, A. (1995) Dopaminergic microtransplants into the substantia nigra of neonatal rats with bilateral 6-OHDA lesions. II. Transplant-induced behavioral recovery. *J. Neurosci.* **15,** 3562–3570.

389. Dunnett, S. B. (1991) Towards a neural transplantation therapy for Parkinson's disease: experimental principles from animal studies. *Acta Neurochir.* **Suppl. 52,** 35–38.

390. Henderson, B. T., Kenny, B. G., Hitchcock, E. R., Hughes, R. C., and Clough, C. G. (1991) A comparative evaluation of clinical rating scales and quantitative measurements in assessment pre and post striatal implantation of human foetal mesencephalon in Parkinson's disease. *Acta Neurochirur.* **Suppl. 52,** 48–50.

391. Kordower, J. H., Freeman, T. B., Snow, B. J., Vingerhoets, F. J., Mufson, E. J., Sanberg, P. R., Hauser, R. A., Smith, D. A., Nauert, G. M., Perl, D. P., et al. (1995) Neuropathological evidence of graft survival and striatal reinnervation after the transplantation of fetal mesencephalic tissue in a patient with Parkinson's disease. *N. Engl. J. Med.* **332,** 1118–1124.

392. Lindvall, O., Sawle, G., Widner, H., Rothwell, J. C., Björklund, A., Brooks, D., Brundin, P., Frackowiak, R., Marsden, C. D., Odin, P., et al. (1994) Evidence for long-term survival and function of dopaminergic grafts in progressive Parkinson's disease. *Ann. Neurol.* **35,** 172–180.

393. Ahlskog, J. E., Kelly, P. J., van Heerden, J. A., Stoddard, S. L., Tyce, G. M., Windebank, A. J., Bailey, P. A., Bell, G. N., Blexrud, M. D., and Carmichael, S. W. (1990) Adrenal medullary transplantation into the brain for treatment of Parkinson's disease: clinical outcome and neurochemical studies. *Mayo Clin. Proc.* **65,** 305–328.

394. Hitchcock, E. R., Kenny, B. G., Henderson, B. T., Clough, C. G., Hughes, R. C., and Detta, A. (1991) A series of experimental surgery for advanced Parkinson's disease by foetal mesencephalic transplantation. *Acta Neurochir.* **Suppl. 52,** 54–57.

395. Hoffer, B. J., Leenders, K. L., Young, D., Gerhardt, G., Zerbe, G. O., Bygdeman, M., Seiger, A., Olson, L., Stromberg, I., and Freedman, R. (1992) Eighteen-month course of two patients with grafts of fetal dopamine neurons for severe Parkinson's disease. *Exp. Neurol.* **118,** 243–252.

396. Defer, G. L., Geny, C., Ricolfi, F., Fenelon, G., Monfort, J. C., Remy, P., Villafane, G., Jeny, R., Samson, Y., Keravel, Y., Gaston, A., Degos, J. D., Peschanski, M., Cesaro, P.,

and Nguyen, J. P. (1996) Long-term outcome of unilaterally transplanted parkinsonian patients. I. Clinical approach. *Brain* **119,** 41–50.

397. Diamond, S. G., Markham, C. H., Rand, R. W., Becker, D. P., and Treciokas, J. (1994) Four-year follow-up of adrenal-to-brain transplants in Parkinson's disease. *Arch. Neurol.* **51,** 559–563.

398. Folkerth, R. D. and Durso, R. (1996) Survival and proliferation of nonneural tissues, with obstruction of cerebral ventricles, in a parkinsonian patient treated with fetal allografts. *Neurology* **46,** 1219–1225.

399. Rosenstein, J. M. (1995) Why do neural transplants survive? An examination of some metabolic and pathophysiological considerations in neural transplantation. *Exp. Neurol.* **133,** 1–6.

400. Shinoda, M., Hudson, J. L., Stromberg, I., Hoffer, B. J., Moorhead, J. W., and Olson, L. (1995) Allogeneic grafts of fetal dopamine neurons: immunological reactions following active and adoptive immunizations. *Brain Res.* **680,** 180–195.

401. Fletcher, S. (1992) Innovative treatment or ethical headache? Fetal transplantation in Parkinson's Disease. *Prof. Nurse* **7,** 592–595.

402. Bain, L. (1993) Fetal tissue transplantation in Parkinson's disease. *Br. J. Nurs.* **2,** 1012–1016.

403. Boer, G. J. (1994) Ethical guidelines for the use of human embryonic or fetal tissue for experimental and clinical neurotransplantation and research. Network of European CNS Transplantation and Restoration (NECTAR). *J. Neurol.* **242,** 1–13.

404. Markowitz, M. S. (1993) Human fetal tissue: ethical implications for use in research and treatment. *AWHONNS Clin. Issues Perinat. Womens Health Nurs.* **4,** 578–588.

405. Ritschl, D. (1995) Use of embryonal CNS tissue in Parkinson disease from the medical ethics viewpoint. *Zentralbl. Neurochir.* **56,** 206–209.

406. Galpern, W. R., Burns, L. H., Deacon, T. W., Dinsmore, J., and Isacson, O. (1996) Xenotransplantation of porcine fetal ventral mesencephalon in a rat model of Parkinson's disease: Functional recovery and graft morphology. *Exp. Neurol.* **140,** 1–13.

407. Stromberg, I., Adams, C., Bygdeman, M., Hoffer, B., Boyson, S., and Humpel, C. (1995) Long-term effects of human-to-rat mesencephalic xenografts on rotational behavior, striatal dopamine receptor binding, and mRNA levels. *Brain Res. Bull.* **38,** 221–233.

408. Brudin, P., Milsson, O. G., Gage, F. H., and Björklund, A. (1985) Cyclosporin A increases survival of cross-species intrastriatal grafts of embryonic dopamine-containing neurons. *Exp. Brain Res.* **60,** 204–208.

409. Zhou, J., Date, I., Sakai, K., Yoshimoto, Y., Furuta, T., Asari, S., and Ohmoto, T. (1993) Xenogeneic dopaminergic grafts reverse behavioral deficits induced by 6-OHDA in rodents: effect of 15-deoxyspergualin treatment. *Neurosci. Lett.* **163,** 81–84.

410. Kondoh, T., Pundt, L. L., and Low, W. C. (1995) Development of human fetal ventral mesencephalic grafts in rats with 6-OHDA lesions of the nigrostriatal pathway. *Neurosci. Res.* **21,** 223–233.

411. Rosenfeld, J. V., Kilpatrick, T. J., and Bartlett, P. F. (1991) Neural transplantation for Parkinson's disease: a critical appraisal. *Aust. N. Zeal. J. Med.* **21,** 477–484.

412. Hoffer, B. J. and van Horne, C. (1995) Survival of dopaminergic neurons in fetal-tissue grafts. *N. Engl. J. Med.* **332,** 1163–1164.

413. Date, I. and Ohmoto, T. (1996) Neural transplantation and trophic factors in Parkinson's disease: Special reference to chromaffin cell grafting, NGF support from pre-

transeheral nerve, and encapsulated dopamine-secreting cell grafting. *Exp. Neurol.* **137,** 333–344.

414. Niijima, K., Chalmers, G. R., Peterson, D. A., Fisher, L. J., Patterson, P. H., and Gage, F. H. (1995) Enhanced survival and neuronal differentiation of adrenal chromaffin cells cografted into the striatum. *J. Neurosci.* **15,** 1180–1194.

415. Westermann, R., Hardung, M., Meyer, D. K., Ekhrhard, P., Otten, U., and Unsicker, K. (1988) Neurotrophic factors released by C6 glioma cells. *J. Neurochem.* **50,** 1747–1758.

416. Westlund, K. N., Lu, Y., Kadekaro, M., Harmann, P., Terrell, M. L., Pizzo, D. P., Hulse-bosch, C. E., Eisenberg, H. M., and Perez-Polo, J. R. (1995) NGF-producing trans-fected 3T3 cells: behavioral and histological assessment of transplants in nigral lesioned rats. *J. Neurosci. Res.* **41,** 367–373.

417. Kordower, J. H., Chen, E. Y., Mufson, E. J., Winn, S. R., and Emerich, D. F. (1996) Intrastriatal implants of polymer encapsulated cells genetically modified to secrete human nerve growth factor: Trophic effects upon cholinergic and noncholinergic stri-atal neurons. *Neuroscience* **72,** 63–77.

418. Collier, T. J. and Springer, J. E. (1994) Neural graft augmentation through co-grafting: implantation of cells as sources of survival and growth factors. *Prog. Neurobiol.* **44,** 309–311.

419. Frodin, M. and Gammeltoft, S. (1994) Insulin-like growth factors act synergistically with basic fibroblast growth factor and nerve growth factor to promote chromaffin cell proliferation. *Proc. Natl. Acad. Sci. USA* **91,** 1771–1775.

420. Kordower, J. H., Fiandaca, M. S., Notter, M. F. D., Hansen, J. T., and Gash, D. M. (1990) NGF-like trophic support from peripheral nerve for grafted rhesus adrenal chro-maffin cells. *J. Neurosurg.* **73,** 418–428.

421. Doering, L. C. (1992) Peripheral nerve segments promote consistent long terms sur-vival of adrenal medulla transplants in the brain. *Exp. Neurol.* **118,** 253–260.

422. Hansson, H. A., Dahlin, L. B., Danielsen, N., Fryklund, L., Nachemson, K., Polleryd, P., Rozell, B., Skottner, A., Stemme, S., and Lundborg, G. (1986) Evidence indicating trophic important of IGF-1 in regenerating peripheral nerves. *Acta. Physiol. Scand.* **126,** 609–614.

423. Manthorpe, M., Skaper, S. D., Williams, L. R., and Varon, L. (1986) Purification of adult rat sciatic nerve ciliary neurotrophic factor. *Brain. Res.* **367,** 282–286.

424. Takayama, H., Ray, J., Raymon, H. K., Baird, A., Hogg, J., Fisher, L. J., and Gage, F. H. (1995) Basic fibroblast growth factor increases dopaminergic graft survival and function in a rat model of Parkinson's disease. *Nature Med.* **1,** 53–58.

425. Haque, N. S., Hlavin, M. L., Fawcett, J., and Dunnett, S. B. (1994) The effect of neu-rotrophin-5 on the growth and survival of nigral grafts in a rat model of Parkinson's disease. *Gene Ther.* **1,** S60.

426. Haque, N. S., Hlavin, M. L., Fawcett, J. W., and Dunnett, S. B. (1996) The neu-rotrophin NT4/5, but not NT3, enhances the efficacy of nigral grafts in a rat model of Parkinson's disease. *Brain Res.* **712,** 45–52.

427. Yurek, D. M., Lu, W., Hipkens, S., and Wiegand, S. J. (1996) BDNF enhances the functional reinnervation of the striatum by grafted fetal dopamine neurons. *Exp. Neurol.* **137,** 105–118.

428. Sheng, J. G., Shirabe, S., Nishiyama, N., and Schwartz, J. P. (1993) Alterations in striatal glial fibrillary acidic protein expression in response to 6-hydroxydopamine-induced denervation. *Exp. Brain Res.* **95**, 450–456.

429. Freed, W. J., Geller, H. M., Poltorak, M., Cannon-Spoor, H. E., Cottingham, S. L., LaMarca, M. E., Schultzberg, M., Rehavi, M., Paul, S., and Ginns, E. I. (1990) Genetically altered and defined cell lines for transplantation in animal models of Parkinson's disease. *Prog. Brain Res.* **82**, 11–21.

430. Aebischer, P., Winn, S. R., Tresco, P. A., Jaeger, C. B., and Greene, L. A. (1991) Transplantation of polymer encapsulated neurotransmitter secreting cells: effect of the encapsulation technique. *J. Biomech. Eng.* **113**, 178–183.

431. Chen, L. S., Ray, J., Fisher, L. J., Kawaja, M. D., Schinstine, M., Kang, U. J., and Gage, F. H. (1991) Cellular replacement therapy for neurologic disorders: potential of genetically engineered cells. *J. Cell Biochem.* **45**, 252–257.

432. Gage, F. H., Fisher, L. J., Jinnah, H. A., Rosenberg, M. B., Tuszynski, M. H., and Friedmann, T. (1990) Grafting genetically modified cells to the brain: conceptual and technical issues. *Prog. Brain Res.* **82**, 1–10.

433. Winn, S. R., Tresco, P. A., Zielinski, B., Greene, L. A., Jaeger, C. B., and Aebischer, P. (1991) Behavioral recovery following intrastriatal implantation of microencapsulated PC12 cells. *Exp. Neurol.* **113**, 322–329.

434. Emerich, D. F., Winn, S. R., Christenson, L., Palmatier, M. A., Gentile, F. T., and Sanberg, P. R. (1992) A novel approach to neural transplantation in Parkinson's disease: use of polymer-encapsulated cell therapy. *Neurosci. Biobehav. Rev.* **16**, 437–447.

435. Gumpert, J., Sharpe, D., and Curzon, G. (1973) Amine metabolites in the cerebrospinal fluid in Parkinson's disease and the response to levodopa. *J. Neurol. Sci.* **19**, 1–12.

436. Korf, J., van Praag, H. M., Schut, D., Nienhuis, R. J., and Lakke, J. P. (1974) Parkinson's disease and amine metabolites in cerebrospinal fluid: implications for L-DOPA therapy. *Eur. J. Neurol.* **12**, 340–350.

437. Olson, L. (1993) Reparative strategies in the brain: treatment strategies based on trophic factors and cell transfer techniques. *Acta Neurochir. Suppl.* **58**, 3–7.

438. Redmond, D. E., Jr., Roth, R. H., Spencer, D. D., Naftolin, F., Leranth, C., Robbins, R. J., Marek, K. L., Elsworth, J. D., Taylor, J. R., Sass, K. J., et al. (1993) Neural transplantation for neurodegenerative diseases: past, present, and future. *Ann. NY Acad. Sci.* **695**, 258–266.

3

6-Hydroxydopamine as a Tool for Studying Catecholamines in Adult Animals

Lessons from the Neostriatum

Michael J. Zigmond and Kristen A. Keefe

1. INTRODUCTION

One of the oldest and most enduring approaches for exploring the functional significance of a specific portion of the nervous system has been that of ablation: removing the tissue in question and then comparing the capacities of the resulting animal to one that is intact. There are problems with this paradigm, of course. Lesions are never entirely specific. Moreover, after a lesion, the remaining tissue may be altered. Nonetheless, trying to unravel the functional organization of the nervous system by a process of subtraction remains a broadly applied and often useful approach. Our current understanding of the role of the sympathetic nervous system in homeostasis is a case in point, since it derives in large part from the pioneering studies of Walter B. Cannon, who examined the physiological impact of removing the sympathetic ganglionic chain of adult cats. However, removing the sympathetic chain is a relatively straightforward surgical process; the application of the ablation paradigm to the central nervous system (CNS) is more problematic. This is owing in large part to the tremendous heterogeneity of the CNS, almost every region of which contains multiple cell types as well as axons of passage. Thus, a great advance was made with the discovery of a tool—6-hydroxydopamine (6-OHDA)—that allowed one to destroy catecholamine-(CA) containing neurons selectively within the brain.

6-OHDA was first described in studies of the autonomic nervous system in which the drug produced a long-lasting depletion of norepinephrine (NE) and was accompanied by the selective degeneration of noradrenergic terminals *(1,2)*. Although 6-OHDA fails to cross the blood–brain barrier, this difficulty was soon overcome by administering the toxin directly into the brain *(3)*. In this way, 6-OHDA can be used to destroy catecholaminergic neurons with little or no apparent damage to other types of cells (Fig. 1).

From: Highly Selective Neurotoxins: Basic and Clinical Applications *Edited by: R. M. Kostrzewa.*
Humana Press Inc., Totowa, NJ

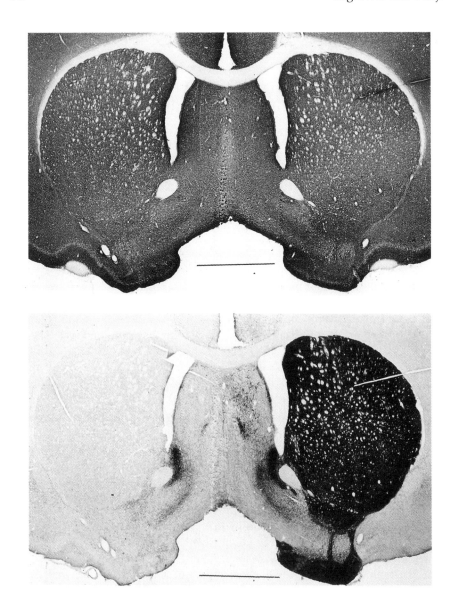

Fig. 1. Selective destruction of tyrosine hydroxylase immunoreactive nerve terminals with 6-OHDA. The animal received desipramine followed 30 min later by 6-OHDA (4 µg in 2 µL over 5 min) injected unilaterally into the left medial forebrain bundle at the level of the lateral hypothalamus (AP, −4.0 mm; ML, −1.8 mm; DV from surface of the skull, −8 mm). Shown is synaptophysin immunoreactivity (top) and tyrosine hydroxylase immunoreactivity (bottom) as seen several weeks postoperatively. Bar = 2 mm. (Courtesy of S. Castro, M. G. Scaffidi, M. Brady, and D. A. Lewis.)

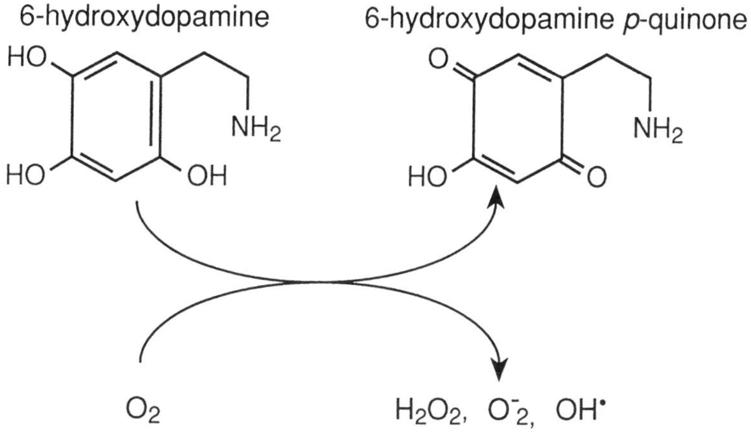

Fig. 2. Oxidation of 6-OHDA to form the reactive oxygen species 6-hydroxy-dopamine quinone, hydrogen peroxide (H_2O_2), superoxide radical (O_s^-), and hydroxyl radical (OH•).

The mechanism by which 6-OHDA causes the selective loss of CAs is discussed in some detail elsewhere in this volume (*see* Chapter 1). Briefly, 6-OHDA derives its value as a tool in the study of CAs by virtue of two features. First, 6-OHDA is highly electroactive and rapidly oxidizes at physiological pH to form 6-OHDA-quinone and hydrogen peroxide, the latter oxidizing still further to superoxide and hydroxyl radicals (Fig. 2). These compounds are highly reactive and can damage proteins, lipids, and DNA. Thus, not surprisingly, 6-OHDA is a potent cytotoxin and can probably destroy any cell type when the compound is available in sufficiently high concentration. However, 6-OHDA also is a structural analog of the CAs and, thus, a substrate for the high-affinity uptake systems present in the plasma membrane of catecholaminergic neurons. 6-OHDA can, therefore, be concentrated into CA-containing neurons, form a sufficient quantity of reactive oxygen species to overwhelm the antioxidant buffering capacity of these cells, and produce selective degeneration.

Studies of the impact of 6-OHDA-induced loss of CAs from brain have taught us a great deal about the role of CAs in brain functions; they also have illustrated the remarkable ability of the brain to compensate for damage, and this, in turn, has greatly improved our understanding about neuroplasticity within the affected systems. Finally, given the relationship between the loss of dopamine (DA) and parkinsonism, studies with 6-OHDA have had considerable clinical significance. In this review we will begin with the practical issue of how 6-OHDA is best used. Next, we will discuss its impact on the nervous system, first on CA neurons, and then on other types of neurons. Finally, we will discuss some of the behavioral effects of 6-OHDA treatment in rats. Because space does not permit a comprehensive review, we often will focus on the topic with which we have the most experience: the consequences of 6-OHDA-induced lesions of the dopaminergic projection of nigrostriatal bundle. (For additional reviews on the use of 6-OHDA in the CNS, *see 4–12*).

2. CRITICAL VARIABLES IN THE USE OF 6-OHDA

2.1. A General Protocol for Administering 6-OHDA

The exact protocol for administering 6-OHDA will depend on the effect one wishes to achieve. 6-OHDA usually is administered into the CNS via a narrow cannula (approx 30 gage) stereotaxically placed in the brain. (In the early days of 6-OHDA use, free-hand injections of 6-OHDA often were utilized; however, this is no longer common and should be avoided.) We recommend that no more than 4 μg of 6-OHDA be used when the toxin is injected directly into brain tissue, and no more than 250 μg when the injection is into the ventricular space. A standard vehicle for these injections is 0.9% NaCl and 0.1% ascorbic acid. We suggest that 6-OHDA be administered at a rate not >1–2 μL/min into the ventricles and not >0.1 μL/min into tissue. A Hamilton syringe connected to an infusion pump permits regulated and reproducible delivery. Once the injection is complete, the cannula should be left in place for several minutes to minimize the unwanted spread of the solution. In some cases, two or more injections can be used over several days to produce larger lesions without any apparent increase in nonspecific damage.

In the sections below, we will discuss in some detail these and related variables in the use of 6-OHDA.

2.2. Placement

6-OHDA is typically administered either into brain parenchyma *(3)* or into the ventricular spaces, usually the lateral ventricle *(13)*, but occasionally the third ventricle via the cisterna magna *(14)* (Figs. 3 and 4). By injecting 6-OHDA into parenchyma, it is possible to obtain a considerable degree of anatomical specificity. For example, one can damage a particular dopaminergic projection while leaving other pathways relatively intact. On the other hand, when injecting 6-OHDA directly into tissue, there is a danger of achieving high local concentrations of 6-OHDA and thus some nonspecific damage. Intracerebroventricular injection can be an effective way to destroy catecholaminergic neurons throughout the brain. This approach probably minimizes nonspecific damage owing to dilution of the toxin by cerebrospinal fluid (CSF) before it comes in contact with tissue, but is not, of course, an appropriate procedure for selectively damaging a particular aminergic pathway.

When 6-OHDA is placed unilaterally into brain tissue on one side, it sometimes is assumed that the contralateral side is unaffected by the toxin. Often this may be the case (e.g., *15*); however, contralateral changes have been reported (e.g., *16*), and thus, care must be taken to establish true baseline values

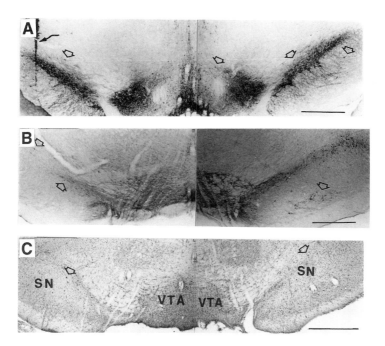

Fig. 3. Effect of 6-OHDA on the distribution of putative DA cells that were stained for TH immunoreactivity. **(A)** Normal distribution of numerous TH-immunoreactive cells (open arrows) in the ventral tegmental area (VTA) and the substantia nigra (SN). **(B)** The pattern of TH-immunoreactive cells in a rat with 6-OHDA (8 μg in 4 μL) injected unilaterally into the medial forebrain bundle. Few TH-positive cells were left in the lesion side, in contrast to numerous TH-positive cells stained in the nonlesion side. **(C)** A rat that had received pargyline followed by intraventricular 6-0HDA (250 μg). In this case, the loss of TH-positive cells in the midbrain was bilateral. Open arrows indicate TH-positive cells in the VTA and the SN. The curved arrow in A indicates the electrode track marked during extracellular recordings. Bars = 0.50 mm. (Courtesy of S.-P. Onn.)

against which the impact of 6-OHDA can be determined. When 6-OHDA is given intraventricularly, there is considerable distribution of the toxin. Nonetheless, the loss of CA may not be entirely symmetrical and certainly is not uniform throughout the CNS. To ensure bilateral symmetry with injections into the lateral ventricle, it is advisable to administer half of the toxin into each ventricle. Even then, however, the lesion will be more extensive in regions nearest the sites of injection (e.g., periventricular striatum vs far lateral striatum; Fig. 5) and particularly low in distant sites (e.g., frontal cortex).

Not all of the heterogeneity regarding 6-OHDA-induced lesions can be explained in terms of the distribution of the toxin following administration. Thus, for example, injections of 6-OHDA into the mesencephalon have a greater

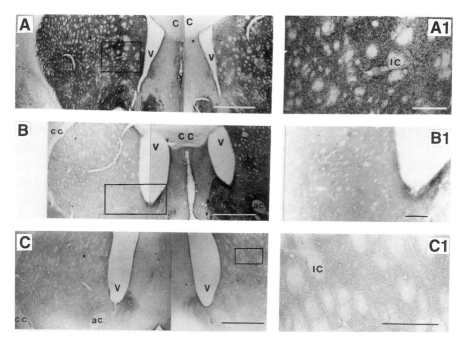

Fig. 4. Effect of 6-OHDA on the distribution of TH-immunoreactive terminals. (**A**) Low-power view showing the patchy appearance of TH-positive terminals characteristic of the intact striatum. (**A1**) An enlargement of the box in Panel A to illustrate better individual terminals stained positively for TH immunoreactivity. (**B**) Striata of a rat that had received a unilateral 6-OHDA injection (8 μg in 4 μL) into the left medial forebrain bundle. A unilateral loss of TH-positive terminals was identified in the lesion side, as opposed to relatively normal-appearing stains in the non-lesion striatum (right). (**B1**) An enlargement of the box in Panel B to show a nearly complete loss of such terminals with some remaining in the ventral striatum. (**C**) Striata of a rat that had been subjected to pargyline plus 250 μg 6-OHDA injected in the lateral ventricles (V). In this case, a bilateral loss of TH-containing terminals was observed with a somewhat more severe loss in the medial (**C1**) as compared to the lateral striatum (*see* Fig. 5). Bars = 1.0 mm (A, B, and C), 125 μm (A1, B1, and C1). (Courtesy of S.-P. Onn.)

effect on the cells of the substantia nigra than those of the ventral tegmental area, and thus cause a larger depletion of DA in the striatum than the nucleus accumbens *(17,18)*. In addition, injections of 6-OHDA into the nucleus accumbens have a greater impact on the core region of that structure than on the shell region *(19)*. Additional observations of regional differences in the efficacy of 6-OHDA have been reported *(13,19–22)*.

Fig. 5. Effect of 6-OHDA on the distribution of TH-immunoreactive terminals. 6-OHDA (250 µg) was delivered to the lateral ventricle with no pretreatments, thereby producing a somewhat smaller lesion than that observed in Fig. 4C. One month postoperative, few TH-immunoreactive terminals remained in the medial striatum, with an increasing density of residual terminals toward the lateral striatum. The lateral ventricle (V) is shown on the left, and the corpus callosum (CC) on the right. Bar = 125 µm. (Courtesy of S.-P. Onn.) (Reprinted from histology presented in ref. *79* with permission of *Brain Res.* and the authors.)

A variable that undoubtedly plays an important role in the regional distribution of the effects of 6-OHDA is the density of high-affinity uptake sites, since it is the uptake of 6-OHDA that determines the extent to which the toxin is accumulated intraneuronally. The density of uptake sites varies with the location along a given catecholaminergic neuron *(23)* and with the particular group of neurons *(24)*. This presumably explains the general observation that nerve terminals are more susceptible to the neurotoxic actions of 6-OHDA than are axons of passage, and that cell bodies are less susceptible still. Most if not all of the regional differences in the actions of 6-OHDA may be explained in a similar way. Another variable, possibly related to that of uptake, is that of catecholaminergic phenotype. Thus, NE neurons are more affected by 6-OHDA than are DA neurons *(13,14,20)*, and adrenergic neurons are the least sensitive *(25)*.

2.3. Dose, Volume, and Rate

There have been a number of studies of the effects of 6-OHDA as a function of the dose, volume, and/or rate of administration (e.g., *26–28*). The dose of 6-OHDA administered must be low enough to be rapidly and selectively accumulated by the high-affinity transporters present in DA and NE neurons, but high enough to exceed the buffering capacity of the catecholaminergic cell and cause degeneration once accumulated. Thus, not surprisingly, there were early claims that 6-OHDA had little or no selectivity *(29,30)*, and other reports that it produced no structural changes at all *(31)*. Within the proper range, however, a dose–response relation exists between the amount of 6-OHDA administered and the loss of CAs, with relatively little apparent damage to other types of neurons *(26,27,32–37)*. It is important to note, however, that the hypothesis of specificity is one that can only be proven in the negative, i.e., by showing that a particular measure is not affected by 6-OHDA. However, only a limited number of measures can be made in any given experiment. Moreover, certain variables have never been examined in 6-OHDA-treated animals. Finally, it should be kept in mind that the plasticity of the CNS can result in some types of damage being repaired or at least obscured by the time the tissue is examined (e.g., *38*). In short, the assumption that 6-OHDA has only affected catecholaminergic neurons should be treated with skepticism.

Coupled to the variable of dose are those of volume and rate of injection *(26,39)*. These three parameters must be adjusted in order to achieve the proper extracellular concentration and distribution of 6-OHDA. In addition, one must be careful to avoid mechanical damage as well as the possibility of fluid traveling back up the needle track. The rates of infusion that we have recommended (Section 2.1.) are considerably slower than those we and most other investigators have employed in the past. However, we have recently observed that a more rapid rate of injection will cause fluid to spread well beyond the region of interest. Thus, reports in which large amounts of 6-OHDA, large volumes, or rapid injection rates are employed must be considered with caution until the necessary examination for nonspecific effects has been made.

Virtually all studies with 6-OHDA involve either one or two injections of the toxin, each one being given over a few minutes. This approach is convenient and, when care is taken, can produce a relatively specific lesion. However, such injections fail to mimic the gradual neurodegenerative process presumed to underlie Parkinson's disease. Recently, osmotic minipumps have been employed to deliver 0.05–0.5 µg/h of 6-OHDA to the striatum over a period of up to 11 d *(39a)*. (Although 6-OHDA is highly reactive, the authors report that the compound remained relatively stable within the minipump for at least 8 d.) Interestingly, when 0.2 µg/h was delivered for 7 d, producing a 50% loss of striatal DA, no loss of TH-immunoreactive cells in the pars compacta was observed when this was examianed 5 wk postoperatively.

2.4. Vehicle

6-OHDA is sufficiently electroactive that it will begin to oxidize as soon as it is placed in solution. This must be minimized, since the selectivity of the toxin depends on it being taken up into CA neurons prior to oxidation. To reduce this problem, most investigators use a vehicle that contains a low concentration of ascorbic acid, keeping the solution on ice until use. In addition, the solution often is shielded from light. We know of no systematic studies of the impact of these conditions. However, care must be taken with the use of ascorbic acid, since this antioxidant also can act as a pro-oxidant *(40–42)* and can have neurotoxic effects of its own (e.g., *43*), some of which might even be confused with those of 6-OHDA (e.g., *44*). Moreover, acting as a pro-oxidant, ascorbic acid can increase the rate of free radical production by 6-OHDA *(45)* and also can block CA transport *(46)*. Both of these effects would reduce the value of 6-OHDA as a tool for the selective destruction of CA neurons, and may explain reports that ascorbic acid can increase nonspecific damage *(47,48)*.

2.5. Pretreatment

The DA and NE transporters are distinct proteins, and this fact can be exploited to produce still more selective lesions. For example, NE terminals are more sensitive than are other CA terminals to the neurotoxic effects of 6-OHDA (*see* Section 2.2.), and this may be because the NE transporter has a greater affinity for the neurotoxin. On the other hand, it often is possible to produce a selective destruction of DA neurons by pretreating animals with desipramine, a drug that inhibits the NE transporter at concentrations that have little effect on the DA transporter. In some applications, the desipramine is quite effective in producing a selective loss of DA *(49)*; under other conditions, however, it fails to protect *(52,53)*. Typically 25–30 mg/kg, ip, of desipramine are given 30–60 min prior to 6-OHDA *(49)*. However, investigators using desipramine should note that the drug can potentiate the gastrointestinal tract distension and accompanying morbidity that often occurs during chloral hydrate anesthesia. The effects of the anesthetic have been attributed to adynamic ileus, and it seems likely that desipramine exacerbates this condition by increasing sympathetic inhibition of gastrointestinal motility. The problem can be reduced by handling animals for several days prior to lesioning *(50)*, administering desipramine via intracerebral injection *(51)*, or using a different anesthetic.

Another common pretreatment is a monoamine oxidase inhibitor. For example, administration of 25–50 mg/kg, ip, of pargyline at least 30 min prior to the intracerebral injection of 6-OHDA increases the loss of DA *(14)*. This effect presumably occurs because the rate of metabolism of 6-OHDA is reduced. In addition, the inhibition of monoamine oxidase would change the nature of the reactive compounds formed from 6-OHDA, since oxidation via monoamine oxidase would lead to the formation of an aldehyde rather than a quinone.

2.6. Age, Gender, Species, and Strain

All of the comments made in this chapter refer to adult, male rats, generally of the Sprague-Dawley strain. The age of the animals can be critical. Thus, for example, as discussed elsewhere in this volume, both the behavioral and the neurobiological consequence of injections of 6-OHDA into very young rats can be quite different than those of injections into adults (*see* Chapter 2). The behavioral effects of 6-OHDA administered to aged rats also may be different from those seen in young adults; however, this does not appear to be owing to an altered neurodegenerative response of the animals to the toxin *(54)*, but rather to other age-related factors, such as a decreased capacity for compensation (*see* 55 for review). We are unaware of any systematic examination of gender, although 6-OHDA has frequently been used in female rats as well as males. In addition, 6-OHDA has frequently been used in a variety of species including, mouse *(56)*, cat *(57–59)*, dog *(60)*, and monkey *(61,62)*. Rat strain does not seem to have been explored with respect to 6-OHDA, although in the mouse the strain can have a profound impact on the effects of another selective neuro-toxin, 1-methyl-4-phenyl-1,2,3,6-tetrahydropyridine (MPTP).

3. EFFECTS OF 6-OHDA ON CA NEURONS IN BRAIN

3.1. Evidence of Degeneration of Catecholaminergic Neurons

The initial evidence that 6-OHDA caused the degeneration of CA-containing neurons came from the peripheral nervous system, where 6-OHDA was shown to cause a long-lasting depletion of NE, as well as the loss of other biochemical markers of NE neurons, including tyrosine hydroxylase (TH) and high-affinity NE uptake. Moreover, light and electron microscopic analyses strongly suggested the loss of peripheral noradrenergic terminals without the concomitant loss of cholinergic terminals. Subsequently, analogous biochemical evidence was provided for the effects of 6-OHDA on catecholaminergic neurons in the CNS *(3,13,14,20,63)*.

Often 6-OHDA appears to cause the degeneration of CA terminals without an equivalent loss of CA cells. For example, in the noradrenergic locus ceruleus system, one can destroy up to 80% of the NE terminals in hippocampus with an intraventricular injection of 6-OHDA without any apparent loss of cell bodies *(64)*. A discrepancy also exists in the nigrostriatal DA system, though it is less marked: An injection of 6-OHDA that destroys 80% of the DA terminals in striatum is accompanied by the loss of 40% of the DA cells of the substantia nigra *(65)*. One interpretation of such data is that 6-OHDA acts primarily on CA terminals, the site of the highest density of CA trans-porters, and the loss of cells depends on the extent to which the terminal axons of a given neuron are destroyed, thereby cutting the soma off from its supply of trophic factors. Thus, in the highly branched and diffuse locus ceruleus system, most cells continue to have a significant number of axons despite extensive lesions; this may be less true for the nigrostriatal DA system, which has less branching.

The exact time-course of the degenerative events depends on the route of administration and the region being examined. For example, after the intra-ventricular injection of 6-OHDA, there is a loss of transmitter content, high-affinity uptake, and TH activity. However, the time-course of these events depends on the brain region under study. For example, in the hypothalamus, these changes are complete within 2 h, whereas in the striatum and substantia nigra, they are not complete for several days and, indeed, are preceded by an initial elevation of striatal DA content *(66,67)*. Even slower time-courses would be expected when lesions are made by injecting 6-OHDA along an axonal trajectory, since the loss of catecholaminergic markers would occur almost exclusively by orthograde and retrograde degeneration, and in some distant brain regions, this could take many days *(see 68)*.

3.2. Damage to DA Neurons Not Associated with Degeneration

In low doses, 6-OHDA can cause transient depletions of CAs that are not accompanied by any apparent damage to the neuron itself; in such cases, 6-OHDA may simply be acting as a false transmitter *(5)*. However, some of the earliest studies of intracerebral 6-OHDA also reported long-lasting losses of CAs without attendant ultrastructural damage *(31)*. Moreover, there has been a recent report of long-lasting loss of DA-specific markers in brain without any apparent degeneration of the neurons themselves *(69)*. Together, these findings raise the possibilitythat under certain conditions, 6-OHDA can interfere with the expressin ofCA-specific genes. 6-OHDA also has been reported to cause morphological changes in CA neurons that do not degenerate. After administration of the neurotoxin, the surviving TH-positive substantia nigra cells appear smaller and more rounded *(70–73)*.

3.3. Changes in Residual DA Neurons

In most studies with 6-OHDA, some number of CA neurons remain intact, with the number of residual neurons depending on the specific protocol used to administer the toxin. These spared neurons may exhibit biochemical or morphological changes.

3.3.1. Changes in TH mRNA, Protein, and Enzyme Activity.

Although 6-OHDA causes a loss of each of the specific markers of noradrenergic markers in terminal fields, such as the hippocampus, the loss of TH activity often is significantly less than the loss of other markers. As a result, there can be an increase in the ratio of TH activity to NE. This appears to be a result of two events: an initial activation of existing TH molecules and the synthesis and delivery of new TH protein *(64,74,75)*.

A similar phenomenon may exist for the dopaminergic projections of the nigrostriatal bundle. Thus, we have frequently observed that large 6-OHDA-induced lesions are accompanied by significant increases in the ratios of TH activity and TH protein to DA in the striatum *(76–78)*. However, some other groups have failed to find these increased TH/DA ratios *(49,73,79)*. The basis for these discrepancies is not clear, although it may relate to the method of lesioning, the extent of the depletion, the assay for TH, or the postoperative time at which analyses are performed.

If the availability of TH is increased in remaining terminals, this does not appear to be owing to an increase in the synthesis of the protein, at least in the case of the nigrostriatal DA system. In this case, the accumulation of TH mRNA in surviving CA cells is actually reduced after 6-OHDA administration *(71–73)*, and the transcription of the mRNA is unchanged *(73)*. Thus, it seems likely that under certain conditions, the lesion-induced reduction in the number of terminals per residual cell results in a redirection of TH protein into those terminals that remain.

3.3.2. Increase in CA Turnover

The relatively few DA neurons that are spared by large doses of 6-OHDA exhibit an increase in DA turnover as indicated by an elevated ratio of DA metabolite to the parent compound or an increase in the rate of disappearance of radiolabeled DA after prelabeling *(20,73,77,80–84)*. This increase in DA turnover appears to develop gradually over several weeks *(78)*; Jackson and Zigmond, unpublished data). Moreover, it is accompanied by an increase in the delivery of DA from residual DA neurons as measured both in vitro *(85,86)* and in vivo *(87–89)*. The increased availability of DA appears to result from a passive loss of high-affinity DA uptake sites that accompanies the degeneration of DA terminals and an increase in DA release *(85,86,90)*.

3.4. Sprouting and Regeneration

Although regrowth of axons after damage is the rule rather than the exception in the peripheral nervous system, it generally is not a very significant factor in central studies of 6-OHDA. Some examples of the restoration of CA levels have been reported in the adult CNS following 6-OHDA-induced lesions (e.g., *71*), and this may represent sprouting following initial neurodegeneration. On the other hand, most lesion paradigms—including the widely used destruction of nigrostriatal DA neurons via intraventricular 6-OHDA—seem to produce little if any regeneration or sprouting of axon collaterals *(77)*.

4. ADDITIONAL CONSEQUENCES OF 6-OHDA-INDUCED LESIONS OF NIGROSTRIATAL DA NEURONS

4.1. Changes in DA Receptors

When the input to a target is removed, the responsiveness of that cell to the missing transmitter often is increased. Such "supersensitivity" occurs in striatal neurons that normally receive a DA input. This has been shown by looking at the effects of DA agonists on behavior *(91,92)*, and on striatal electrophysiology *(93)* and neurochemistry *(94–97)*. The increased responsiveness to D2 receptor agonists may be secondary to an increase in D2 receptors in the DA-depleted striatum, as numerous studies have revealed increased D2 receptor binding *(98,99)* and D2 receptor mRNA *(100,101)*. In contrast, there have been reports of increases *(102–105)*, decreases *(16,100,106–109)*, or no change *(110–113)* in D1 receptor binding and mRNA levels after injection of 6-OHDA. In addition, it has been argued that functional supersensitivity after treatment with 6-OHDA may not always be related to increased receptor number, because behavioral supersensitivity is seen in the absence of detectable differences in receptor binding *(114,115)*. Thus, mechanisms in addition to changes in receptor number, such as 6-OHDA-induced increases in the stimulatory G-proteins G_s and G_{olf}, may sometimes be responsible for functional supersensitivity after injections of 6-OHDA *(116,117)*.

4.2. Functional Alterations in Striatal Interneurons

The gross behavioral abnormalities produced by very large losses of striatal DA are accompanied by evidence of dysfunction within the striatum. It seems likely that many of the behavioral effects of 6-OHDA are secondary to these changes in the function of striatal neurons.

4.2.1. Somatostatin–Neuropeptide Y (NPY) Interneurons

One class of striatal interneuron coexpresses the neuropeptides somatostatin and NPY along with nitric oxide synthase, NADPH diaphorase, and possibly GABA. Treatment of rats with 6-OHDA results in a decrease in the number of striatal neurons expressing preprosomatostatin mRNA and a decrease in the intensity of the labeling for that message per cell *(118)*. However, no effect of 6-OHDA has been observed on the number of somatostatin immunoreactive neurons or in the amount of somatostatin in striatal tissue *(119,120)*. Interestingly, although NPY is expressed by the same population of neurons as is somatostatin, lesions induced by 6-OHDA are reported to increase the number of NPY-immunoreactive neurons in striatum *(121)*. This effect is mimicked by the D1 receptor antagonist SCH 23390, whereas the D2 receptor antagonist sulpiride tends to decrease the number of NPY-immunoreactive neurons *(122)*. These data suggest that the function of this population of striatal interneurons is regulated by DA via both the D1 and D2 families of DA receptors. Furthermore, DA seems to regulate the expression of somatostatin and NPY differentially in these neurons.

4.2.2. Cholinergic Interneurons

Interneurons containing acetylcholine (ACh) comprise about 1–2% of the neurons in striatum, and their activity also is regulated by DA via D2 and D1 receptors *(123–127)*. After large lesions of DA projections to striatum, there is an increase in ACh turnover and a fall in ACh content. Provided that at least 10% of the DA terminals remain, these changes are only present for a few days. However, even in such animals, acute administration of D2 antagonists still increases striatal ACh release and reduces ACh levels *(128,129)*, suggesting that a small number of DA terminals continue to maintain an inhibitory influence on ACh-containing neurons via D2 receptors. This hypothesis also is supported by in vitro studies of striatal slices. Three days after extensive destruction of striatal DA neurons with 6-OHDA, the effects of nomifensine and sulpiride on ACh release are reduced, although dopaminergic control over ACh release persists when lesions are somewhat smaller. Two months postsurgery, however, the effects of both drugs are restored unless striatal DA depletion exceeds 90% *(130)*. A comparable phenomenon has been observed for the inhibitory influence of DA on GABA efflux, which also is mediated via a D2 receptor *(131)*. It is not clear, however, whether this latter effect represents the action of DA on GABAergic interneurons or medium spiny efferents.

DA also exerts an excitatory influence on cholinergic function, which is mediated via a D1 receptor *(124,125)*. However, the effects of 6-OHDA on this interaction are less clear. The neurotoxin has been reported both to decrease *(97)* and to increase *(132)* the basal efflux of ACh. Other indices of cholinergic function also are affected by treatment of animals with 6-OHDA. Thus, there is a 20% increase in high-affinity choline uptake sites labeled with ^{3}H-hemicholinium in the lateral striatum *(107)*, and a decrease in striatal M1 and M2 muscarinic receptor binding *(107,133)*.

4.3. Functional Alterations in Striatal Efferent Neurons Produced by 6-OHDA

The majority of neurons in striatum are the medium spiny neurons, which propagate information from striatum to basal ganglia output nuclei. The behavioral impairments induced by 6-OHDA-induced depletions of striatal DA therefore are likely to arise as a consequence of alterations in the function of these cells. A wide variety of lesion-induced changes have been observed in these neurons. These include neurochemical and electrophysiological effects (reviewed in Sections 4.3.1. and 4.3.2.), as well as morphological changes *(134,135)*. All of the medium spiny neurons appear to use GABA as a primary neurotransmitter, but are roughly equally divided into two populations, the striatopallidal (indirect) and striatonigral (direct), on the basis of their primary axonal projections and the neuropeptides and DA receptors they express. Striatopallidal neurons project primarily to the external segment of the globus pallidus (globus pallidus in rodents), coexpress the neuropeptide enkephalin, and express predominately the D2 DA receptor *(136–139)*. Striatonigral neurons send their primary axonal projection to the internal segment of the globus pallidus (entopeduncular nucleus in rodents) and the substantia nigra pars reticulata, coexpress the neuropeptides substance P and dynorphin, and express predominately the D1 receptor *(137–141)*.

4.3.1. Differential Neurochemical Effects of 6-OHDA

The use of 6-OHDA has provided considerable insight into the regulation of striatal efferent neurons by DA. First, it appears that DA inhibits the activity of the striatopallidal pathway via a D2 receptor and stimulates the activity of the striatonigral pathway via a D1 receptor. This conclusion derives in part from two sets of observations. First, 6-OHDA-induced lesions increase local cerebral glucose utilization *(142–144)* and decrease the activity of neurons in globus pallidus *(145)*. Second, the lesions also increase the expression of immediate early genes in striatopallidal neurons and decrease expression in striatonigral neurons *(146,147)*.

The data also suggest that the sensitivity of these two pathways to DA loss is different. The loss of DA associated with injection of 6-OHDA into either striatum or substantia nigra leads to an increase in the expression of pre-proenkephalin mRNA and enkephalin immunoreactivity in the lesioned striatum

(148–150). In addition, the content of enkephalin in the striatum is elevated *(151)*. This increase requires at least a 90% depletion of striatal DA *(152)*, and is reversed by continuous administration of a D2 receptor agonist or systemic administration of the muscarinic antagonist scopolamine *(100,153,154)*. In contrast, the expression of preprotachykinin (substance P) mRNA in the striatum *(148,155)* and the levels of substance P in the striatum and substantia nigra *(150,156,157)* are reduced by the same treatment. These changes in substance P mRNA occur with DA depletions of 50–100% *(152)*, and are reversed by D1 receptor agonists.

Changes in neuropeptide levels in the striatum of 6-OHDA-lesioned animals are typically thought to reflect consequences of the loss of striatal DA. However, recent evidence suggests that 6-OHDA-induced loss of DA in other basal ganglia nuclei, as well as other system-level changes, may contribute to the observed alterations in striatal peptide expression. For example, the increases in enkephalin mRNA observed in the striatum after unilateral, 6-OHDA-induced DA depletions are not seen after bilateral injections of 6-OHDA *(157a)* and also are blocked by excitotoxic lesions of the subthalamic nucleus *(157b)*.

4.3.2. Electrophysiological Effects of 6-OHDA

In addition to being defined on the basis of their projections and neuropeptide expression, striatal efferent neurons have been divided into two different classes, Type I and Type II neurons, on the basis of their electrophysiological response to paired impulse stimulation of cortical afferents *(158,159)*. DA depletions induced by 6-OHDA increase the firing rate of both types of cells. However, the extent of DA depletion required to elicit these changes varies. The loss of 50% or more of the DA input to striatum is associated with an increase in the firing rate of Type I cells *(160)*, whereas the loss of at least 90% of this input is required in order to produce increases in the firing of Type II cells *(161)*. The effects of DA on Type I cells appear to be mediated by the D1 receptor subtype, whereas the effects of DA on Type II cells are mediated by the D2 receptor subtype *(12,161)*.

4.4. Behavioral Effects of 6-OHDA-Induced Loss of Striatal DA

Profound behavioral effects can be caused by 6-OHDA as long as very large depletions of striatal DA are produced. The behavioral effects of large depletions of striatal DA include akinesia, sensory neglect, and attendant aphagia and adipsia *(3,91,162–164)*; (for review, *see* 165). Although, there is little evidence for the regeneration of DA neurons after 6-OHDA-induced lesions of the nigrostriatal bundle *(77)*, these behavioral deficits usually are not permanent. As long as animals are kept alive by intragastric intubation during the initial period of aphagia and adipsia, and some minimum number of dopaminergic neurons remain, there is a gradual recovery of function, and after a few days or weeks, such animals often appear quite normal under basal laboratory conditions *(163,166)*.

Even moderate lesions produce behavioral deficits, and these deficits may not recover. One of the earliest such observations was an increase in the irritability of rats *(167)*. Subsequently, it was shown that 6-OHDA-lesioned animals failed to respond appropriately to a variety of regulatory challenges. For example, the rats did not eat in response to the administration of insulin or drink in response to an injection of hypertonic saline *(164,168–173)*. These deficits were accompanied by a generalized impairment of sensory and motor function, which also can be precipitated by a wide variety of other stressors, including cold and foot shock *(174)*. Although some stress-induced increase in DA release can still be elicited even after relatively large lesions *(175,176; see also 177)*, such deficits may represent the inability of a compensated dopaminergic system to respond adequately to an acute stress. Support for this hypothesis derives in part from the observation that many of the deficits seen when 6-OHDA-lesioned rats are exposed to stress can be reversed by DA agonists *(178,179)*. These stress-induced deficits must be contrasted with stress-induced improvement of function in animals with gross behavioral impairments *(180; see also 181)*, sometimes termed "paradoxical kinesia." Although the neurobiological basis for this phenomenon is unclear, it does not seem to result from an increase in the availability of DA *(182)*.

More subtle behavioral deficits after partial 6-OHDA-induced lesions include altered responding on an operant task *(92,183,184)*, impaired learning *(95)*, delayed reaction time *(185–187)*, and a reduction in skilled motor function *(188–192)*.

5. THE 6-OHDA-LESIONED RATS AS A MODEL FOR PARKINSONISM

5.1. Parkinson's Disease

In addition to offering a model with which to study some general principles of adaptation *(87–89)*, does 6-OHDA provide any insights into clinical states? If so, the most obvious candidate would be Parkinson's disease, a progressive neurodegenerative disorder of the basal ganglia, particularly the substantia nigra. Parkinson's disease is observed in approx 100 individuals/100,000, usually manifesting itself in the fifth or sixth decade of life. Commonly, the disease is detected as a mild resting tremor of one or more limbs, muscular rigidity, postural abnormalities, and bradykinesia. These abnormalities tend to be accompanied by other symptoms, including decreased food intake and concomitant weight loss, characteristic autonomic dysfunctions, depression, and a general slowing of intellectual processes. The severity of the symptoms is exacerbated by physical or emotional trauma, although there also are numerous reports of a temporary stress-induced improvement in the behavior of otherwise akinetic patients. Both neurological and psychiatric symptoms usually worsen slowly, but inexorably and lead to an end state of complete akinesia, often accompanied by dementia, within 10–15 yr of the initial diagnosis.

Postmortem examination of parkinsonian brains reveals a number of neuro-chemical and histological abnormalities (for reviews, *see 76,193–195*). Most striking is the degenerative loss of the DA neurons of the nigrostriatal projection. This is manifest as a loss of pigmented cells in the substantia nigra, and of DA in the caudate and putamen of the dorsal corpus striatum. However, the relationship between the loss of DA and neurological deficits is not linear. Indeed, the loss of as much as 80% of DA in caudate nucleus tissue and 90% of DA in putamen is accompanied by only mild neurological impairments, whereas marked clinical deficits require much larger DA-depleting brain lesions. In first observing that frank neurological deficits only occurred when the loss of DA was extensive, Hornykiewicz and his colleagues also noted that the concentrations of DA metabolites were less affected by the disease than were DA levels themselves, and hypothesized that a compensatory increase in DA release occurred from the residual DA neurons *(193,196–198)*. This hypothesis is similar to that suggested by our own group *(12,199; see 200)*.

5.2. Validity of the 6-OHDA Model

There are several obvious parallels between the 6-OHDA-treated animal model and the human parkinsonian state. First, rats with large depletions of striatal DA show hypokinesia, often manifesting as akinesia in animals with bilateral lesions *(201)*. This hypokinetic syndrome is associated with increased rigidity and tremor, as demonstrated by increased electromyographic activity *(102,202)*. In addition, as is the case in patients with parkinsonism, rats treated with 6-OHDA show impairments in food intake and weight loss. Thus, those hallmark features of Parkinson's disease that are amenable to study in rats are replicated in such animals when treated with 6-OHDA. Second, the neuropathology of parkinsonism also is well mimicked by the use of 6-OHDA. There clearly is a loss of the DA neurons innervating the striatum from the substantia nigra. As is the case in the human condition, this loss must reach at least 80–90% before marked symptomatology is observed *(163,203)*. In addition, the ratio of DA metabolites to DA, as discussed in Section 3.3., is increased, indicating an increase in DA turnover. Finally, some of the changes in neuropeptide and DA receptor gene expression seen in animals treated with 6-OHDA, such as decreased substance P levels and increased D2 receptor binding, are observed in human patients as well *(204–207)*.

5.3. What Have We Learned from 6-OHDA About Parkinsonism?

Given the validity of the 6-OHDA-treated animal as a model of Parkinson's disease, what has it taught us about the disorder? Although Hornykiewicz speculated that there was an increase in the activity of residual DA neurons in Parkinson's disease (*196,198*), it is the work completed in the 6-OHDA-treated rodent that has provided the actual evidence that such changes do indeed occur. Additionally, this work has provided considerable insight into the mechanisms by which this increase in DA turnover is realized and the implications of these adaptations for the emergence of parkinsonian symptomatology. Thus, the 6-OHDA model has served as a vehicle for confirming speculations based on postmortem studies in humans and for conducting mechanistic studies to understand better the neuropathology of Parkinson's disease.

The insights gained from studies on 6-OHDA-treated animals also have contributed to improved understanding of functional basal ganglia anatomy. This improved understanding, in turn, has led to new models of basal ganglia function in Parkinson's disease and treatment options. For example, 6-OHDA-induced decreases in GABA receptor binding in the globus pallidus and increases in the entopeduncular nucleus and substantia nigra *(208)* led to the novel hypothesis that DA inhibited striatopallidal neurons and excited striatonigral neurons. This hypothesis was further supported by the observations that 6-OHDA-induced depletions of striatal DA had opposite effects on the function of the striatopallidal and striatonigral pathways, as discussed above. In light of these findings and those obtained from MPTP-treated primates, a new model of basal ganglia function in Parkinson's disease was proposed *(209)*, which now serves as the foundation on which our understanding of and therapeutic approaches to the treatment of Parkinson's disease and other diseases of basal ganglia origin are based.

Finally, like DA, 6-OHDA is an electroactive molecule capable of oxidizing to form reactive oxygen species. Moreover, when injected into the striatum, DA can produce a specific degeneration of DA neurons *(210,211)*. For example, immunohistochemical studies indicate that the administration of DA into the striatum causes a loss of tyrosine hydroxylase and the DA transporter without any decrease in serotonin or synaptophysin. The apparent loss of DA terminals caused by DA can be reduced by the simultaneous injection of an equimolar amount of ascorbic acid or glutathione (Fig. 6) *(211)*. Thus, it is possible that studies with 6-OHDA have provided not only a model for parkinsonism, but also insight into the biological basis of the disease itself.

ACKNOWLEDGMENTS

The ideas presented in this chapter derive in large part from conversations with our many collaborators on studies of 6-OHDA over the years. Our special appreciation goes to Shao-pii Onn for the preparation of Figs. 3–5. Support has been provided in part by USPHS grants NS19608, MH00343, MH00058, MH18273, MH42217, MH30915, the American Parkinson's Disease Association, and the National Parkinson Foundation.

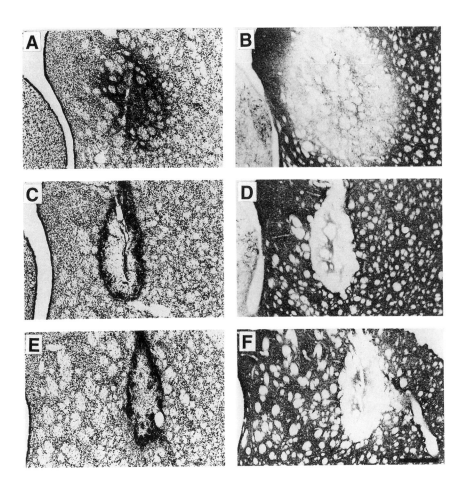

Fig. 6. Histological analysis of the effects of intrastriatal DA and the impact of antioxidants. Animals received intrastriatal injections of DA (**A** and **B**), DA and glutathione (GSH; **C** and **D**), or DA and ascorbic acid (**E** and **F**). In each case, 0.4 μmol of each substance was used. Adjacent sections through the caudate putamen were labeled for Nissl substance (A, C, and E) or TH immunoreactvity (B, D, F). Note the zone of decreased immunoreactivity in the animal injected with DA only (B) is much larger than in the animals that also received an antioxidant (D and F), whereas the impact of DA on Nissl substance was comparable in all three cases (A, C, and E). Bar = 500 μm. (From ref. *211*, reprinted with permission of the *Natl. Acad Sci. USA* and the authors.)

REFERENCES

1. Porter, C. C., Totaro, J. A., and Stone, C. A. (1963) Effect of 6-hydroxydopmine and some other compounds on the concentration of norepinephrine in the hearts of mice. *Journal of Pharmaco Exp. Thera.* **140**, 308.

2. Thoenen, H. and Tranzer, J. P. (1968) Chemical sympathectomy by selective destruction of adrenergic nerve endings with 6-Hydroxydopamine. *Naunyn-Schmiedebergs Arch. Pharmacol.* **261**, 271–288.

3. Ungerstedt, U. (1968) 6-Hydroxydopamine-induced degeneration of central monoamine neurons. *Eur. J. Pharmacol.* **5**, 107–110.

4. Malmfors, T. and Thoenen, H. (1971) *6-Hydroxydopamine and Catecholamine Neurons.* North-Holland, Amsterdam.

5. Thoenen, H. and Tranzer, J. P. (1973) The pharmacology of 6-hydroxydopamine. *Annu. Rev. Pharmacol.* **13**, 169–180.

6. Kostrzewa, R. M. and Jacobowitz, D. M. (1974) Pharmacological actions of 6-Hydroxydopamine. *Pharmacol. Rev.* **26**, 199–288.

7. Breese, G. R. (1975) Biochemical pharmacology of brain amine mechanisms. *Psychopharmacol. Bull.* **11**, 40–41.

8. Jonsson, G., Einarsson, P., Fuxe, K., and Hallman, H. (1975) Microspectrofluorimetric analysis of the formaldehyde induced fluorescence in midbrain raphe neurons. *Med. Biol.* **53**, 25–39.

9. Jonsson, G. (1980) Chemical neurotoxins as denervation tools in neurobiology. *Annu. Rev. Neurosci.* **3**, 169–187.

10. Schallert, T. and Wilcox, R. E. (1985) Neurotransmitter selective brain lesions, in *Neuromethods: Neurochemical Techniques,* vol. 1 (Boulton, A. A., and Baker, G. B., eds.), Humana, Clifton, NJ, pp. 343–387.

11. Zigmond, M. J. and Stricker, E. M. (1989) Animal models of parkinsonism using selective neurotoxins: clinical and basic implications. *Int. Rev. Neurobiol.* **31**, 1–79.

12. Zigmond, M., Abercrombie, E., Berger, T., Grace, A., and Stricker, E. (1993) Compensatory responses to partial loss of dopaminergic neurons: studies with 6-hydroxydopamine, in *Current Concepts in Parkinson's Disease Research,* (Schneider, J. and Gupta, M. eds.), Hogrefe & Huber Publishers, Toronto, pp. 99–140.

13. Uretsky, N. J. and Iversen, L. L. (1970) Effects of 6-hydroxydopamine on catecholamine containing neurones in the rat brain. *J. Neurochem.* **17**, 269–278.

14. Breese, G. R., and Traylor, T. D. (1970) Effect of 6-hydroxydopamine on brain norepinephrine and dopamine evidence for selective degeneration of catecholamine neurons. *J. Pharmacol. Exp. Ther.* **174**, 413–420.

15. Saavedra, J. M., Setler, P. E., and Kebabian, J. W. (1978) Biochemical changes accompanying unilateral 6-hydroxy-dopamine lesions in the rat substantia nigra. *Brain Res.* **151**, 339–352.

16. Berger, K., Prsedborski, S., and Cadet, J. L. (1991) Retrograde degeneration of nigrostriatal neurons induced by intrastriatal 6-hydroxydopamine injection in rats. *Brain Res. Bull.* **26**, 301–307.

17. Perese, D. A., Ulman, J., Viola, J., Ewing, S. E., and Bankiewicz, K. S. (1989) A 6-hydroxydopamine-induced selective parkinsonian rat model. *Brain Res.* **494**, 285–293.

18. Zahm, D. S. and Johnson, S. N. (1989) Asymmetrical distribution of neurotensin immunoreactivity following unilateral injection of 6-hydroxydopamine in rat ventral tegmental area (VTA). *Brain Res.* **483,** 301–311.

19. Zahm, D. S. (1991) Compartments in rat dorsal and ventral striatum revealed following injection of 6-hydroxydopamine into the ventral mesencephalon. *Brain Res.* **552,** 164–169.

20. Bloom, F. E., Algeri, S., Groppetti, A., Revuelta, A., and Costa, E. (1969) Lesions of central norepinephrine terminals with 6-OH-dopamine: biochemistry and fine structure. *Science* **166,** 1284–1286.

21. Iversen, L. L. and Uretsky, N. J. (1970) Regional effects of 6-hydroxydopamine on catecholamine containing neurones in rat brain and spinal cord. *Brain Res.* **24,** 364–367.

22. Pehek, E. A., Crock, R., and Yamamoto, B. K. (1992) Selective subregional dopamine depletions in the rat caudate-putamen following nigrostriatal lesions. *Synapse* **10,** 317–325.

23. Nirenberg, M. J., Vaughan, R. A., Uhl, G. A., Kuhar, M. J., and Pickel, V. M. (1996) The dopamine transporter is localized to dendritic and axonal plasma membranes of nigrostriatal dopamine neurons. *J. Neurosci.* **16,** 436–447.

24. Jones, S. R., Garris, P. A., and Wightman, R. M. (1995) Comparison of dopamine uptake in the basolateral amygdaloid nucleus, caudate putamen, and nucleus accumbens of the rat. *J. Neurochem.* **64,** 2581–2589.

25. Jonsson, G., Fuxe, K., Hokfelt, T., and Goldstein, M. (1976) Resistance of central phenylethanolamine-N-methyl transferase containing neurons to 6-hydroxydopamine. *Med. Biol.* **54,** 421–426.

26. Agid, Y., Javoy, F., Glowinski, J., Bouvet, D., and Sotelo, C. (1973) Injection of 6-hydroxydopamine into the substantia nigra of the rat. II. Diffusion and specificity. *Brain Res.* **58,** 291–301.

27. Sotelo, C., Javoy, F., Agid, Y., and Glowinski, J. (1973) Injection of 6-hydroxydopamine in the substantia nigra of the rat. I. Morphological study. *Brain Res.* **58,** 269–290.

28. Willis, G. L., Singer, G., and Evans, B. K. (1976) Intracranial injections of 6-OHDA. Comparison of catecholamine depleting effects of different volumes and concentrations. *Pharmacol. Biochem. Behav.* **5,** 207–213.

29. Butcher, L. L. (1975) Degenerative processes after punctate intracerebral administration of 6-hydroxydopamine. *J. Neural Transm.* **37,** 189–208.

30. Poirier, L. J. (1975) Histopathological changes associated with the intracerebral injection of 6-hydroxydopamine (6-OHDA) and peroxide (H2O2) in the cat and the rat. *J. Neural Transm.* **37,** 209–218.

31. Bartholini, G., Richards, J. G., and Pletscher, A. (1970) Dissociation between biochemical and ultrastructural effects of 6-hydroxydopamine in rat brain. *Experientia* **26,** 143.

32. Hedreen, J. C. and Chalmers, J. P. (1972) Neuronal degeneration in rat brain induced by 6 hydroxydopamine: a histological and biochemical study. *Brain Res.* **47,** 1–36.

33. Jacks, B. R., De Champlain, J., and Cordeau, J. P. (1972) Effects of 6-hydroxydopamine on putative transmitter substances in the central nervous system. *Eur. J. Pharmacol.* **18,** 353–360.

34. Hokfelt, T. and Ungerstedt, U. (1973) Specificity of 6-hydroxydopamine induced degeneration of central monoamine neurones: an electron and fluorescence microscopic study

with special reference to intracerebral injection on the nigro-striatal dopamine system. *Brain Res.* **60,** 269–297.

35. Javoy, F., Agid, Y., and Glowinski, J. (1975) Oxotremorine- and atropine-induced changes of dopamine metabolism in the rat striatum. *J. Pharm. Pharmacol.* **27,** 677–681.

36. Javoy, F. (1975) Partial selectivity of 6-OHDA induced neuronal degeneration after intraissular injection in the brain with special reference to the nigro-striatal dopamine system. *J. Neural Transm.* **37,** 219–227.

37. Harper, J. A., Labuszewski, T., and Lidsky, T. I. (1979) Substantia nigra unit responses to trigeminal sensory stimulation. *Exp. Neurol.* **65,** 462–470.

38. Wolf, G., Stricker, E. M., and Zigmond, M. J. (1978) Brain lesions: induction, analysis, and the problem of recovery of function, in *Recovery of Function from Brain Damage: Research and Theory* (Finger, S., ed.), Plenum, New York, pp. 91–112.

39. Meyers, R. D. (1966) Injection of solutions into cerbral tissues: relation between volume and diffusion. *Physiol. Behav.* **1,** 171–174.

39a. Jones, B. E., Boyland, C. B., Fritsche, M., Juhasz, M., Jackson, C., Wiegland, S. J., Hyman, C., Lindsay, R. M., and Altar, C. A. (1996) A continuous striatal infusion of 6-hydroxydopamine produces a terminal axotomy and delayed behavioral effects. *Brain Res.* **709,** 275–284.

40. Stadtman, E. R. (1991 Dec) Ascorbic acid and oxidative inactivation of proteins. *Am. J. Clin. Nutr.* **54,** 1125S–1128S.

41. Halliwell, B. (1992) Reactive oxygen species and the central nervous system. *J. Neurochem.* **59,** 1609–1623.

42. Halliwell, B. (1992) Oxygen radicals as key mediators in neurological disease: fact or fiction? *Ann. Neurol.* **32,** S10–S15.

43. Davison, A. J., Legault, A. N., and Donald, W. S. (1986) Effect of 6-hydroxydopamine on polymerization of tubulin- protection by superoxide dismutase, catalase, or anaerobic conditions. *Biochem. Pharmacol.* **35,** 1411–1417.

44. Waddington, J. L. and Crow, T. J. (1979) Drug-induced rotational behavior following unilateral intracerebral injection of saline-ascorbate solution: neurotoxicity of ascorbic acid and monoamine-independent circling. *Brain Res.* **161,** 371–376.

45. Heikkila, R. E. and Cohen, G. (1972) Further studies on the generation of hydrogen peroxide by 6-hydroxydopamine: potentiation by ascorbic acid. *Mol. Pharmacol.* **8,** 241–248.

46. Berman, S. B., Zigmond, M. J., and Hastings, T. G. (1996) Modification of dopamine transporter function: effect of reactive oxygen species and dopamine. *J. Neurochem.* **67,** 593–600.

47. Jonsson, G., Lohmander, S., and Sachs, C. (1974) 6-Hydroxydopamine induced degeneration of noradrenaline neurons in the scorbutic guinea-pig. *Biochem. Pharmacol.* **23,** 2585–2593.

48. Jonsson, G., Pycock, C., Fuxe, K., and Sachs, C. (1974) Changes in the development of central noradrenaline neurones following neonatal administration of 6-hydroxydopamine. *J. Neurochem.* **22,** 419–426.

49. Breese, G. R. and Traylor, T. D. (1971) Depletion of brain noradrenaline and dopamine by 6-hydroxydopamine. *Br. J. Pharmacol.* **42,** 88–99.

50. Saller, C. F. and Stricker, E. M. (1978) Gastrointestinal motility and body weight gain in rats after brain serotonin depletion by 5,7-hydroxytryptamine. *Neuropharmacology* **17,** 499–506.

51. Hernandez, L., and Hoebel, B. G. (1982) Overeating after midbrain 6-hydroxydopamine: prevention by central injection of selective catecholamine reuptake blockers. *Brain Res.* **245,** 333–343.

52. Commins, D. L., Shaughnessy, R. A., Axt, K. J., Vosmer, G., and Seiden, L. S. (1989) Variability among brain regions in the specificity of 6-hydroxydopamine (6-OHDA)-induced lesions. *J. Neural Transm.* **77,** 197–210.

53. King, D. and Finlay, J. M. (1997) Loss of dopamine terminals in the medial prefrontal cortex selectively increased the ratio of DOPAC to DA in tissue of the nucleus accumbens shell: role of stress. *Brain Res.* in press.

54. Ricaurte, G. A., DeLanney, L. E., Finnegan, K. T., Irwin, I., and Langston, J. W. (1988) The dopamine-depleting effect of 6-hydroxydopamine does not increase with aging. *Brain Res.* **438,** 395–398.

55. Zigmond, M. J., Berger, T. W., Grace, A. A., and Stricker, E. M. (1989) Compensatory responses to nigrostriatal bundle injury. Studies with 6-hydroxydopamine in an animal model of parkinsonism. *Mol. Chem. Neuropathol.* **10,** 185–200.

56. Mandel, R. J. and Randall, P. K. (1985) Quantification of lesion-induced dopaminergic supersensitivity using the rotational model in the mouse. *Brain Res.* **330,** 358–363.

57. Frigyesi, T. L., Ige, A., Iulo, A., and Schwartz, R. (1971) Denigration and sensorimotor disability induced by ventral tegmental injectin of 6-hydroxydopamine in the cat. *Exp. Neurol.* **33,** 78–87.

58. Beleslin, D. B., Samardzic, R., Krstic, S. K., Micic, D., and Terzic, B. (1981) Comparison of behavioral changes in cats treated with intracerebroventricular 6-hydroxydopamine and reserpine. *Brain Res. Bull.* **6,** 285–287.

59. Beleslin, D. B., Samardzic, R., and Stefanovic-Denic, K. (1981) 6-hydroxydopamine and aggression in cats. *Pharmacol. Biochem. Behav.* **14,** 29–32.

60. Van Woert, M. H., Ambani, L., and Bowers, M. B., Jr. (1972) Levodopa and cholinergic hypersensitivity in Parkinson's disease. *Neurology* **22 (Suppl.),** 86–93.

61. Maas, J. W., DeKirmenjian, H., Garner, D., Redmond, D. E. J., and Landis, D. H. (1972) Catecholamine metabolite excretion following intraventricular injection 6-OH-dopamine. *Brain Res.* **41,** 507–511.

62. Redmond, D. E. J., Hinrichs, R. L., Maas, J. W., and Kling, A. (1973) Behavior of free-ranging macaques after intraventricular 6-hydroxydopamine. *Science* **181,** 1256–1258.

63. Uretsky, N. J. and Iversen, L. L. (1969) Effects of 6-hydroxydopamine on noradrenaline-containing neurones in the rat brain. *Nature* **221,** 557–559.

64. Acheson, A. L. and Zigmond, M. J. (1981) Short and long term changes in tyrosine hydroxylase activity in rat brain after subtotal destruction of central noradrenergic neurons. *J. Neurosci.* **1,** 493–504.

65. Hollerman, J. R. and Grace, A. A. (1990) The effects of dopamine-depleting brain lesions on the electrophysiological activity of rat substantia nigra dopamine neurons. *Brain Res.* **533,** 203–212.

66. Bell, L. J., Iversen, L. L., and Uretsky, N. J. (1970) Time course of the effects of 6-hydroxydopamine on catecholamine-containing neurones in rat hypothalamus and striatum. *Br. J. Pharmacol.* **40,** 790–799.

67. McGeer, E. G., Fibiger, H. C., McGeer, P. L., and Brooke, S. (1973) Temporal changes in amine synthesizing enzymes of rat extrapyramidal structures after hemitransections or 6-hydroxydopamine administration. *Brain Res.* **52,** 289–300.

68. Zigmond, M. J., Chalmers, J. P., Simpson, J. R., and Wurtman, R. J. (1971) Effect of lateral hypothalamic lesions on uptake of norepinephrine by brain homogenates. *J. Pharmacol. Exp. Ther.* **179,** 20–28.

69. Bowenkamp, K. E., David, D., Lapchak, P. L., Henry, M. A., Granholm, A. C., Hoffer, B. J., and Mahalik, T. J. (1996) 6-Hydroxydopamine induced the loss of the dopaminergic phenotype in substantia nigra neurons of the rat—A possible mechanism for restoration of the nigrostriatal circuit mediated by glial cell line-derived neurotrophic factor. *Exp. Brain Res.* **111,** 1–7.

70. Pasinetti, G. M., Morgan, D. G., Johnson, S. A., Lerner, S. P., Myers, M. A., Poirier, J., and Finch, C. E. (1989) Combined in situ hybridization and immunocytochemistry in the assay of pharmacological effects on tyrosine hydroxylase mRNA concentration. *Pharmacological Research* **21,** 299–311.

71. Pasinetti, G. M., Lerner, S. P., Johnson, S. A., Morgan, D. G., Telford, N. A., and Finch, C. E. (1989) Chronic lesions differentially decrease tyrosine hydroxylase messenger RNA in dopaminergic neurons of the substantia nigra. *Brain Res.* **5,** 203–209.

72. Pasinetti, G. M., Osterburg, H. H., Kelly, A. B., Kohama, S., Morgan, D. G., Reinhard, J. F. J., Stellwagen, R. H., and Finch, C. E. (1992) Slow changes of tryosine hydroxylase gene expression in dopaminergic brain neurons after neurotoxin lesioning: a model for neuron aging. *Brain Res.* **13,** 63–73.

73. Sherman, T. G. and Moody, C. A. (1995) Alterations in tyrosine hydroxylase expression following partial lesions of the nigrostriatal bundle. *Mol. Brain Res.* **29,** 285–296.

74. Acheson, A. L., Zigmond, M. J., and Stricker, E. M. (1980) Compensatory increase in tyrosine hydroxylase activity in rat brain after intraventricular injections of 6-hydroxydopamine. *Science* **207,** 537–540.

75. Unnerstall, J. R. and Ladner, A. (1994) Deficits in the activation and phosphorylation of hippocampal tyrosine hydroxylase in the aged Fischer 344 rat following intraventricular administration of 6-hydroxydopamine. *J. Neurochem.* **63,** 280–290.

76. Zigmond, M. J., Acehson, A. L., Stachowiak, M. K., and Stricker, E. M. (1984) Neurochemical compensation after nigrostriatal bundle injury in an animal model of preclinical Parkinsonism. *Arch. Neurol.* **41,** 856–861.

77. Onn, S.-P., Berger, T. W., Stricker, E. M., and Zigmond, M. J. (1986) Effects of intraventricular 6-hydroxydopamine on the dopaminergic innervation of striatum: histochemical and neurochemical analysis. *Brain Res.* **376,** 8–19.

78. Liang, L. P., and Zigmond, M. J. (1993) Dopamine synthesis in neostriatal slices after intraventricular 6-hydroxydopamine. *Soc. Neurosci. Abstracts* **19,** 401.

79. Hefti, F., Dravid, A., and Hartikka, J. (1984) Chronic intraventricular injections of nerve growth factor elevate hippocampal choline acetyltransferase activity in adult rats with partial septo-hippocampal lesions. *Brain Res.* **293,** 305–311.

80. Uretsky, N. J., Simmonds, M. A., and Iversen, L. L. (1971) Changes in the retention and metabolism of 3 H-l-norepinephrine in rat brain in vivo after 6-hydroxydopamine pretreatment. *J. Pharmacol. Exp. Ther.* **176**, 489–496.

81. Agid, Y., Javoy, F., and Glowinski, J. (1973) Hyperactivity of remaining dopaminergic neurones after partial destruction of the nigro-striatal dopaminergic system in the rat. *Nature: New Biol.* **245**, 150–151.

82. Zigmond, M. J. and Stricker, E. M. (1977) Behavioral and neurochemical effects of central catecholamine depletion: a possible model for "subclinical" brain damage, in *Animal Models in Psychiatry and Neurology*, vol. 100. (Hanin, I. and Usdin, E., eds.), Pergamon, Oxford, pp. 415–429.

83. Hefti, F., Melamed, E., and Wurtman, R. J. (1980) Partial lesions of the dopaminergic nigrostriatal system in rat brain: biochemical characterization. *Brain Res.* **195**, 123–137.

84. Altar, C. A., Marien, M. R., and Marshall, J. F. (1987) Time course of adaptations in dopamine biosynthesis, metabolism, and release following nigrostriatal lesions: Implications for behavioral recovery from brain damage. *J. Neurochem.* **48**, 390–399.

85. Snyder, G. L., Keller, R. W. J., and Zigmond, M. J. (1990) Dopamine efflux from striatal slices after intracerebral 6 hydroxydopamine: Evidence for compensatory hyperactivity of residual terminals. *J. Pharmacol. Exp. Ther.* **253**, 867–876.

86. Stachowiak, M. K., Keller, R. W. J., Stricker, E. M., and Zigmond, M. J. (1987) Increased dopamine efflux from striatal slices during development and after nigrostriatal bundle damage. *J. Neurosci.* **7**, 1648–1654.

87. Abercrombie, E. D., Bonatz, A. E., and Zigmond, M. J. (1990) Effects of L-DOPA on extracellular dopamine in striatum of normal and 6-hydroxydopamine-treated rats. *Brain Res.* **525**, 36–44.

88. Robinson, T. E. and Whishaw, I. Q. (1988) Normalization of extracellular dopamine in striatum following recovery from a partial unilateral 6-OHDA lesion of the substantia nigra: a microdialysis study in freely moving rats. *Brain Res.* **450**, 209–224.

89. Zhang, W. Q., Tilson, H. A., Nanry, K. P., Hudson, P. M., Hong, J. S., and Stachowiak, M. K. (1988) Increased dopamine release from striata of rats after unilateral nigrostriatal bundle damage. *Brain Res.* **461**, 335–342.

90. Garris, P. A., Walker, Q. D., and Wightman, R. M. (1997) Dopamine release and uptake rates both decrease in the partially denervated striatum in proportion to the loss of dopamine terminals. *Brain Res.,* in press.

91. Ungerstedt, U. (1971) Postsynaptic supersensitivity after 6-hydroxy-dopamine induced degeneration of the nigro-striatal dopamine system. *Acta. Physiol. Scand. Suppl.* **367**, 69–93.

92. Schoenfeld, R. and Uretsky, N. (1972) Altered response to apomorphine in 6-hydroxydopamine-treated rats. *Eur. J. Pharmacol.* **19**, 115–118.

93. Feltz, P. and De Champlain, J. (1972) Enhanced sensitivity of caudate neurones to microiontophoretic injections of dopamine in 6-hydroxydopamine treated cats. *Brain Res.* **43**, 601–605.

94. Mishra, R. K., Gardner, E. L., Katzman, R., and Makman, M. H. (1974) *Proc. Natl. Acad. Sci. USA* **71**, 3883–3887.

95. Fibiger, H. C. and Grewaal, D. S. (1974) Neurochemical evidence for denervation supersensitivity: the effect of unilateral substantia nigra lesions on apomorphine-induced increases in neostriatal acetylcholine levels. *Life Sci.* **15,** 57–63.

96. Zigmond, M. J. and Stricker, E. M. (1980) Supersensitivity after intraventricular 6-hydroxydopamine: relation to dopamine depletion. *Experientia* **36,** 436–438.

97. Robertson, G. S., Hubert, G. W., Tham, C. S., and Fibiger, H. S. (1992) Lesions of the mesotelencephalic dopamine system enhance the effects of selective dopamine D1 and D2 receptor agonists on striatal acetylcholine release. *Eur. J. Pharmacol.* **219,** 323–325.

98. Creese, I., Burt, D. R., and Snyder, S. H. (1977) Dopamine receptor binding enhancement accompanies lesion-induced behavioral supersensitivity. *Science* **197,** 596–598.

99. Staunton, D. A., Wolfe, B. B., Groves, P. M., and Molinoff, P. B. (1981) Dopamine receptor changes following destruction of the nigrostriatal pathway: lack of a relationship to rotational behavior. *Brain Res.* **211,** 315–327.

100. Gerfen, C. R., Engber, T. M., Mahan, L. C., Susel, Z., Chase, T. N., Monsma, F. J., Jr., and Sibley, D. R. (1990) D1 and D2 dopamine receptor-regulated gene expression of striatonigral and striatopallidal neurons. *Science* **250,** 1429–1432.

101. Angulo, J. A., Coirini, H., Ledoux, M., and Schumacher, M. (1991) Regulation by dopaminergic neurotransmission of dopamine D2 mRNA and receptor levels in the striatum and nucleus accumbens of the rat. *Mol. Brain Res.* **11,** 161–166.

102. Bounamici, M., Maj, R., Pagani, F., Rossi, A. C., and Khazan, N. (1986) Tremor at rest episodes in unilaterally 6-OHDA-induced substantia nigra lesioned rats: EEG-EMG and behavior. *Neuropharmacology* **25,** 323–325.

103. Narang, N. and Wamsley, J. K. (1995) Time dependent changes in DA uptake sites, D1 and D2 receptor binding and mRNA after 6-OHDA lesions of the medial forebrain bundle in the rat brain. *J. Chem. Neuroanat.* **9,** 41–53.

104. Rioux, L., Gagnon, C., Gaudin, D. P., Di Paolo, T., and Bedard, P. J. (1993) A fetal nigral graft prevents behavioral supersensitivity associated with repeated injections of L-dopa in 6-OHDA-rats. Correlation with D1 and D2 receptors. *Neuroscience* **56,** 45–51.

105. Fornaretto, M. G., Caccia, C., Caron, M. G., and Fariello, R. G. (1993) Dopamine receptor status after unilateral nigral 6-OHDA lesion. Autoradiographic and in situ hybridization study in the rat brain. *Mol. Chem. Neuropathol.* **19,** 147–162.

106. Marshall, J. F., Navarrete, R., and Joyce, J. N. (1989) Decreased striatal D1 binding density following mesotelencephalic 6-hydroxydopamine injections: An autoradiographic analysis. *Brain Res.* **493,** 247–257.

107. Joyce, J. N. (1991) Differential response of striatal dopamine and muscarinic cholinergic receptor subtypes to the loss of dopamine. I. Effects of intranigral or intracerebroventricular 6-hydroxydopamine lesions of the mesostriatal dopamine system. *Exp. Neurol.* **113,** 261–276.

108. Blunt, S. B., Jenner, P., and Marsden, C. D. (1992) Autoradiographic study of striatal D1 and D2 dopamine receptors in 6-OHDA-lesioned rats receiving foetal ventral mesencephalic grafts and chronic treatment with L-Dopa and carbidopa. *Brain Res.* **582,** 299–311.

109. Stromberg, I., Adams, C., Bygdeman, M., Hoffer, B., Boyson, S., and Humpl, C. (1995) Long term effects of human-to-rat mesencephalic xenografts on rotational behavior, striatal dopamine receptor binding, and mRNA levels. *Brain Res. Bull.* **38,** 221–233.

110. Graham, W. C., Crossman, A. R., and Woodruff, G. N. (1990) Autoradiographic studies in animal models of hemi-Parkinsonism reveal dopamine D2 but not D1 receptor supersensitivity. I. 6-OHDA lesions of ascending mesencephalic dopaminergic pathways in the rat. *Brain Res.* **514**, 93–102.

111. Cadet, J. L., Last, R., Kostic, V., Przedborski, S., and Jackson-Lewis, V. (1991) Long-term behavioral and biochemical effects of 6-hydroxydopamine injections in rat caudate-putamen. *Brain Res. Bull.* **26**, 707–713.

112. Leslie, C. A. and Bennett, J. P. J. (1987) Striatal D1- and D2-dopamine receptor sites are separately detectable in vivo. *Brain Res.* **415**, 90–97.

113. Gnanalingham, K. K. and Robertson, R. G. (1994) The effects of chronic continuous versus intermittent levodopa treatments on striatal and extrastriatal D1 and D2 dopamine receptors and dopamine uptake sites in teh 6-hydroxydopamine lesioned rat—an autoradiographic study. *Brain Res.* **640**, 185–194.

114. LaHoste, G. J. and Marshall, J. F. (1991) Chronic eticlopride and dopamine denervation induce equal nonadditive increases in striatal D2 receptor density: Autoradiographic evidence against the dual mechanism hypothesis. *Neuroscience* **41**, 473–481.

115. Mileson, B. E., lewis, M. H., and Mailman, R. B. (1991) Dopamine receptor supersensitivity occurring without receptor up-regulation. *Brain Res.* **561**, 1–10.

116. Marcotte, E. R., Sullivan, R. M., and Mishra, R. K. (1994) Striatal G-proteins: effects of 6-hydroxydopamine lesions. *Neurosci. Lett.* **169**, 195–198.

117. Herve, D., Levi-Strauss, M., Marey-Semper, I., Verney, C., Tassin, J. P., Glowinski, J., and Girault, J. A. (1993) G(olf) and Gs in rat basal ganglia: possible involvement of G(olf) in the coupling of dopamine D1 receptor with adenylyl cyclase. *J. Neurosci.* **13**, 2237–2248.

118. Soghomonian, J.-J. and Chesselet, M.-F. (1991) Lesions of the dopaminergic nigrostriatal pathway alter preprosomatostatin messenger RNA in the striatum, the entopeduncular nucleus and the lateral hypothalamus of the rat. *Neuroscience* **42**, 49–59.

119. Salin, P., Kerkerian-Le Goff, L., Heidet, V., Epelbaum, J., and Nieoullon, A. (1990) Somatostatin-immunoreactive neurons in the rat striatum: effects of corticostriatal and nigrostriatal dopaminergic lesions. *Brain Res.* **521**, 23–32.

120. Beal, M. F. and Martin, J. B. (1983) Effects of lesions on somatostatin-like immunoreactivity in the rat striatum. *Brain Res.* **266**, 67–73.

121. Kerkerian, L., Bosler, O., Pelletier, G., and Nieoullon, A. (1986) Striatal neuropeptide Y neurons are under the influence of the nigrostriatal dopaminergic pathway: immunohistochemical evidence. *Neurosci. Lett.* **66**, 106–112.

122. Kerkerian, L., Salin, P., and Nieoullon, A. (1988) Pharmacological characterization of dopaminergic influence on expression of neuropeptide Y immunoreactivity by rat striatal neurons. *Neuroscience* **26**, 809–817.

123. Cubeddu, L. X. and Hoffmann, I. S. (1983) Frequency-dependent release of acetylcholine and dopamine from rabbit striatum: its modulation by dopaminergic receptors. *J. Neurochem.* **41**, 94–101.

124. DeBoer, P. and Abercrombie, E. D. (1996) Physiological release of striatal acetylcholine in vivo: Modulation by D1 and D2 dopamine receptor subtypes. *J. Pharmacol. Exp. Ther.* **277**, 775–783.

125. Damsma, G., Tham, C.-S., Robertson, G. S., and Fibiger, H. C. (1990) Dopamine D1 receptor stimulation increases striatal acetylcholine release in the rat. *Eur. J. Pharmacol.* **186**, 335–338.

126. Ajima, A., Yamaguchi, T., and Kato, T. (1990) Modulation of acetylcholine release by D1, D2 dopamine receptors in rat striatum under freely moving conditions. *Brain Res.* **518,** 193–198.

127. Bertorelli, R. and Consolo, S. (1990) D1 and D2 dopaminergic regulation of acetylcholine release from striata of freely moving rats. *J. Neurochem.* **54,** 2145–2148.

128. Grewaal, D. S., Fibiger, H. C., and McGeer, E. G. (1974) 6-Hydroxydopamine and striatal acetylcholine levels. *Brain Res.* **73,** 372–375.

129. Guyenet, P. G., Javoy, F., Agid, Y., Beaujouan, J. C., and Glowinski, J. (1975) Dopamine receptors and cholinergic neurons in the rat neostriatum. *Adv. Neurol.* **9,** 43–51.

130. MacKenzie, R. G., Stachowiak, M. K., and Zigmond, M. J. (1989) Dopaminergic inhibition of striatal acetylcholine release after 6-hydroxydopamine. *Eur. J. Pharmacol.* **168,** 43–52.

131. Harsing, L. G., Jr. and Zigmond, M. J. (1996) Dopaminergic inhibition of striatal GABA release after 6-hydroxydopamine. *Brain Res.* **738,** 142–145.

132. DeBoer, P., Abercrombie, E. D., Heeringa, M., and Westerink, B. H. (1993) Differential effect of systemic administration of bromocriptine and L-dopa on the release of acetylcholine from striatum of intact and 6-OHDA-treated rats. *Brain Res.* **608,** 198–203.

133. Dawson, T. M., Dawson, V. L., Gage, F. H., Fisher, L. J., Hunt, M. A., and Wamsley, J. K. (1991) Downregulation of muscarinic receptors in the rat caudate-putamen after lesioning of the ipsilateral nigrostriatal dopamine pathway with 6-hydroxydopamine (6-OHDA): Normalization by fetal mesencephalic transplants. *Brain Res.* **540,** 145–152.

134. Ingham, C. A., Hood, S. H., and Arbuthnott, G. W. (1989) Spine density of neostriatal neurones changes with 6-hydroxydopamine lesions and with age. *Brain Res.* **503,** 334–338.

135. Ingham, C. A., Hood, S. H., van Maldegem, B., Weenink, A., and Arbuthnott, G. W. (1993) Morphological changes in the rat neostriatum after unilateral 6-hydroxydopamine injections into the nigrostriatal pathway. *Exp. Brain Res.* **93,** 17–27.

136. Del Fiacco, M., Paxinos, G., and Cuello, A. C. (1982) Neostriatal enkephalinimmunoreactive neurones project to the globus pallidus. *Brain Res.* **231,** 1–17.

137. Kawaguchi, Y., Wilson, C. J., and Emson, P. C. (1990) Projection subtypes of rat neostriatal matrix cells revealed by intracellular injection of biocytin. *J. Neurosci.* **10,** 3421–3438.

138. Gerfen, C. R. and Young, W. S. (1988) Distribution of striatonigral and striatopallidal peptidergic neurons in both patch and matrix compartments: an in situ hybridization histochemistry and fluorescent retrograde tracing study. *Brain Res.* **460,** 161–167.

139. Le Moine, C., Normand, E., Guitteny, A. F., Fouque, B., Teoule, R., and Bloch, B. (1990) Dopamine receptor gene expression by enkephalin neurons in rat forebrain. *Proc. Natl. Acad. Sci. USA* **87,** 230–234.

140. Hong, J. S., Yang, H.-Y., Racagni, G., and Costa, E. (1977) Projections of substance P containing neurons from neostriatum to substantia nigra. *Brain Res.* **122,** 541–544.

141. Mroz, E. A., Brownstein, M. J., and Leeman, S. E. (1977) Evidence for substance P in the striato-nigral tract. *Brain Res.* **125,** 305–311.

142. Trugman, J. M. and Wooten, G. F. (1987) Selective D_1 and D_2 dopamine agonists differentially alter basal ganglia glucose utilization in rats with unilateral 6-hydroxydopamine substantia nigra lesions. *J. Neurosci.* **7,** 2927–2935.

143. Morelli, M., Pontieri, F. E., Lifante, I., Orzi, F., and DiChiara, G. (1993) Local cerebral glucose utilization after D1 receptor stimulation in 6-OHDA-lesioned rats: Effect of sensitization (priming) with a dopaminergic agonist. *Synapse* **13**, 264–269.

144. Kozlowski, M. R. and Marshall, J. F. (1980) Plasticity of [^{14}C]2-deoxy-d-glucose incorporation into neostriatum and related structures in response to dopamine neuron damage and apomorphine replacement. *Brain Res.* **197**, 167–183.

145. Pan, H. S. and Walters, J. R. (1988) Unilateral lesion of the nigrostriatal pathway decreases the firing rate and alters the firing pattern of globus pallidus neurons in the rat. *Synapse* **2**, 650–656.

146. Gerfen, C. R., Keefe, K. A., and Gauda, E. B. (1995) D1 and D2 dopamine receptor function in the striatum: Co-activation of D1- and D2-dopamine receptors on separate populations of neurons results in potentiated immediate early gene response in D1-containing neurons. *J. Neurosci.* **15**, 8167–8176.

147. Jian, M., Staines, W. A., Iadarola, M. J., and Robertson, G. S. (1993) Destruction of the nigrostriatal pathway increases Fos-like immunoreactivity predominately in striatopallidal neurons. *Mol. Brain Res.* **19**, 156–160.

148. Young, W. S., III, Bonner, T. I., and Brann, M. R. (1986) Mesencephalic dopamine neurons regulate the expression of neuropeptide mRNAs in the rat forebrain. *Proc. Natl. Acad. Sci. USA* **83**, 9827–9831.

149. Sivam, S. P., Breese, G. R., Napier, T. C., Mueller, R. A., and Hong, J. S. (1986) Dopaminergic regulation of proenkephalin-A gene expression in the basal ganglia. *NIDA Res. Monogr.* **75**, 389–392.

150. Voorn, P., Roest, G., and Groenewegen, H. J. (1987) Increase of enkephalin and decrease of substance P immunoreactivity in the dorsal and ventral striatum of the rat after midbrain 6-hydroxydopamine lesions. *Brain Res.* **412**, 391–396.

151. Thal, L. J., Sharpless, N. S., Hirshorn, I. D., Horowitz, S. G., and Makman, M. H. (1983) Striatal met-enkephalin concentration increases following nigrostriatal denervation. *Biochem. Pharmacol.* **44**, 3297–3301.

152. Nisenbaum, L. K., Kitai, S. T., Crowley, W. R., and Gerfen, C. R. (1994) Temporal dissociation between changes in striatal enkephalin and substance P messenger RNAs following striatal dopamine depletion. *Neuroscience* **60**, 927–937.

153. Nisenbaum, L. K., Kitai, S. T., and Gerfen, C. R. (1994) Dopaminergic and muscarinic regulation of striatal enkephalin and substance P messenger RNAs following striatal dopamine denervation: Effects of systemic and central administration of quinpirole and scopolamine. *Neuroscience* **63**, 435–449.

154. Pollack, A. E. and Wooten, G. F. (1992) D_2 dopaminergic regulation of striatal preproenkephaline mRNA levels is mediated at least in part through cholinergic interneurons. *Mol. Brain Res.* **13**, 35–41.

155. Gerfen, C. R., McGinty, J. F., and Young, W. S., III. (1991) Dopamine differentially regulates dynorphin, substance P, and enkephalin expression in striatal neurons: *In situ* hybridization histochemical analysis. *J. Neurosci.* **11**, 1016–1031.

156. Cruz, C. J. and Beckstead, R. M. (1989) Nigrostriatal dopamine neurons are required to maintain basal levels of substance P in the rat substantia nigra. *Neuroscience* **30**, 331–338.

157. Hanson, G. R., Alphs, L., Wolf, W., Levine, R., and Lovenberg, W. (1981) Haloperidol-induced reduction of nigral substance P-like immunoreactivity: A probe for the inter-

actions between dopamine and substance P neuronal systems. *J. Pharmacol. Exp. Ther.* **218**, 568–574.

157a. Salin, P., Hajji, M. D., and Kerkerian-Le Goff, L. (1996) Bilateral 6-hydroxydopamine-induced lesions of the nigrostriatal dopamine pathway reproduces the effects of unilateral lesion on substance P but not on enkephalin expression in rat basal ganglia. *Eur. J. Neurosci.* **8**, 1746–1757.

157b. Delfs, J. M., Ciaramitaro, V. M., Parry, T. J., and Chesselet, M.-F. (1995) Subthalamic nucleus lesions: widespread effects on changes in gene expression induced by nigro-striatal dopamine depletion in rats. *J. Neurosci.* **15(10)**, 6562–6575.

158. Berger, T. W., Nisenbaum, E. S., Stricker, E. M., and Zigmond, M. J. (1987) Evidence of two functionally distinct subpopulations of neurons within striatum and their differential sensitivity to dopamine, in *Neurophysiology of Dopaminergic Systems: Current Status and Clinical Perspectives.* (Chiodo, L. A. and Freeman, A. S., eds.), Lakeshore Publishing, Detroit, pp. 253–284.

159. Nisenbaum, E. S., Orr, W. B., and Berger, T. W. (1988) Evidence for two functionally distinct subpopulations of neurons within the rat striatum. *J. Neurosci.* **8**, 4138–4150.

160. Orr, W. B., Stricker, E. M., Zigmond, M. J., and Berger, T. W. (1987) Effects of dopamine depletion on the spontaneous activity of type 1 striatal neurons: Relation to local dopamine concentration and motor behavior. *Synapse* **1**, 461–469.

161. Orr, W. B., Gardiner, T. W., Stricker, E. M., Zigmond, M. J., and Berger, T. W. (1986) Short-term effects of dopamine-depleting brain lesions on spontaneous activity of striatal neurons: relation to local dopamine concentration and behavior. *Brain Res.* **376**, 20–28.

162. Fibiger, H. C., Phillips, A. G., and Zis, A. P. (1974) Deficits in instrumental responding after 6-hydroxydopamine lesions on nigro-neostriatal dopaminergic projection. *Pharmacol. Biochem. Behav.* **2**, 87–96.

163. Zigmond, M. J. and Stricker, E. M. (1973) Recovery of feeding and drinking by rats after intraventricular 6-hydroxydopamine or lateral hypothalamic lesions. *Science* **182**, 717–720.

164. Marshall, J. F. and Teitelbaum, P. (1973) A comparison of the eating in response to hypothermic and glucoprivic challenges after nigral 6-hydroxydopamine and lateral hypothalamic electrolytic lesions in rats. *Brain Res.* **55**, 229–233.

165. Stricker, E. M. and Zigmond, M. J. (1976) Recovery of function after damage to central catecholamine containing neurons: A neurochemical model for the lateral hypothalamic syndrome, in *Progress in Psychobiology and Physiological Psychology*, vol. 6 (Sprague, J. and Epstein, A. N., eds.), Academic, New York, pp. 121–188.

166. Ungerstedt, U. (1971) Adipsia and aphagia after 6-hydroxydopamine-induced degeneration of the nigrostriatal dopamine system. *Acta Physiol. Scand.* **Suppl. 367**, 95–122.

167. Evetts, K. D., Uretsky, N. J., Iversen, L. L., and Iversen, S. D. (1970) Effects of 6-hydroxydopamine on CNS catecholamines, spontaneous motor activity and amphetamine induced hyperactivity in rats. *Nature* **225**, 961–962.

168. Zigmond, M. J. and Stricker, E. M. (1972) Deficits in feeding behavior after intraventricular injection of 6-hydroxydopamine in rats. *Science* **177**, 1211–1214.

169. Breese, G. R., Smith, R. D., Cooper, B. R., and Grant, L. D. (1973) Alterations in consummatory behavior following intracisternal injection of 6-hydroxydopamine. *Pharmacol. Biochem. Behav.* **1,** 319–328.

170. Fibiger, H. C., Zis, A. P., and McGeer, E. G. (1973) Feeding and drinking deficits after 6-hydroxydopamine administration in the rat: similarities to the lateral hypothalamic syndrome. *Brain Res.* **55,** 135–148.

171. Fibiger, H. C., Phillips, A. G., and Zis, A. P. (1973) Deficits in instrumental responding after 6-hydroxydopamine lesions of the nigro-neostriatal dopaminergic projection. *Pharmacol. Biochem. Behav.* **2,** 87–96.

172. Marshall, J. F., Richardson, J. S., and Teitelbaum, P. (1974) Nigrostriatal bundle damage and the lateral hypothalamic syndrome. *J. Comp. Physiol. Psychol.* **87,** 808–830.

173. Stricker, E. M. and Zigmond, M. J. (1974) Effects on homeostasis of intraventricular injections of 6-hydroxydopamine in rats. *J. Comp. Physiol. Psychol.* **86,** 973–994.

174. Snyder, A. M., Stricker, E. M., and Zigmond, M. J. (1985) Stress-induced neurological impairments in an animal model of Parkinsonism. *Ann. Neurol.* **18,** 544–551.

175. Keefe, K. A., Stricker, E. M., Zigmond, M. J., and Abercrombie, E. D. (1990) Environmental stress increases extracellular dopamine in striatum of 6-hydroxydopamine-treated rats: in vivo microdialysis studies. *Brain Res.* **527,** 350–353.

176. Keefe, K. A., Zigmond, M. J., and Abercrombie, E. D. (1990) Excitatory amino acid receptor involvement in the regulation of striatal extracellular dopamine, in *Presynaptic Receptors and the Question of Autoregulation of Neurotransmitter Release,* vol. 604 (Kalsner, S. and Westfall, T. C., eds.), The New York Academy of Sciences, New York, pp. 614–616.

177. Hefti, F., Enz, A., and Melamed, E. (1985) Partial lesions of the nigrostriatal pathway in the rat. Acceleration of transmitter synthesis and release of surviving dopaminergic neurones by drugs. *Neuropharmacology* **24,** 19–23.

178. Marshall, J. F. and Ungerstedt, U. (1976) Apomorphine-induced restoration of drinking to thirst challenges in 6-hydroxydopamine-treated rats. *Physiol. Behav.* **17,** 817–822.

179. Marshall, J. F. and Gotthelf, T. (1979) Sensory inattention in rats with 6-hydroxydopamine-induced degeneration of ascending dopaminergic neurons: apomorphine-induced reversal of deficits. *Exp. Neurol.* **65,** 398–411.

180. Marshall, J. F., Levitan, D., and Stricker, E. M. (1976) Activation-induced restoration of sensorimotor functions in rats with dopamine-depleting brain lesions. *J. Comp. Physiol. Psychol.* **90,** 536–546.

181. Levitt, D. and Teitelbaum, P. (1975) Somnolence, akinesia, and sensory activation of motivated behavior in the lateral hypothalamic syndrome. *Proc. Natl. Acad. Sci. USA* **72,** 2819–2823.

182. Keefe, K. A., Salamone, J. D., Zigmond, M. J., and Stricker, E. M. (1989) Paradoxical kinesia in parkinsonism is not caused by dopamine release. Studies in an animal model. *Arch. Neurol.* **46,** 1070–1075.

183. Schoenfeld, R. I. and Uretsky, N. J. (1972) Operant behavior and catecholamine-containing neurons: prolonged increase in lever-pressing after 6-hydroxydopamine. *Eur. J. Pharmacol.* **20,** 357–362.

184. Schoenfeld, R. I. and Zigmond, M. J. (1973) Behavioral pharmacology of 6-hydroxydopamine, in *Frontiers in Catecholamine Research* (Usdin, E. and Snyder, S. H., eds.), Pergamon, New York. pp. 695–700.

185. Spirduso, W. W., Gilliam, P. E., Schallert, T., Upchurch, M., Vaughn, D. M., and Wilcox, R. E. (1985) Reactive capacity: a sensitive behavioral marker of movement initiation and nigrostriatal dopamine function. *Brain Res.* **335,** 45–54.

186. Amalric, M. and Koob, G. F. (1987) Depletion of dopamine in the caudate nucleus but not in nucleus accumbens impairs reaction-time performance in rats. *J. Neurosci.* **7,** 2129–2134.

187. Brown, V. J. and Robbins, T. W. (1991) Simple and choice reaction time performance following unilateral striatal dopamine depletion in the rat. *Brain* **114,** 513–525.

188. Whishaw, I. Q., O'Connor, W. T., and Dunnett, S. B. (1986) The contributions of motor cortex, nigrostriatal dopamine and caudate-putamen to skilled forelimb use in the rat. *Brain* **199,** 805–843.

189. Dunnett, S. B., Isacson, O., Sirinathsinghji, J. S., Clarke, D. J., and Björklund, A. (1988) Striatal grafts in rats with unilateral neostriatal lesions III. Recovery from dopamine-dependent motor asymmetry and deficits in skilled paw reaching. *Neuroscience* **20,** 813–820.

190. Mandel, R. J., Brundin, P., and Björklund, A. (1990) The importance of graft place-ment and task complexity for transplant induced recovery of simple and complex sen-sorimotor deficits in dopamine denervated rats. *Eur. J. Neurosci.* **2,** 888–894.

191. Olsson, M., Nikkhah, G., Bentlage, C., and Björklund, A. (1995) Forelimb akinesia in the rat parkinson model: Differential effects of dopamine agonists and nigral transplants as assessed by new stepping test. *J. Neurosci.* **15,** 3863–3875.

192. Lindner, M. D., Winn, S. R., Baetge, E. E., Hammang, J. P., Gentile, F. T., Doherty, E. D., McDermott, P. E., Frydel, B., Ullman, M. D., Schallert, T., and Emerich, D. F. (1995) Implantation of encapsulated catecholamine and DNF-producing cells in rats with unilateral dopamine depletions and parkinsonian symptoms. *Exp. Neurol.* **132,** 62–76.

193. Bernheimer, H., Birkmayer, W., Hornykiewicz, O., Jellinger, K., and Seitelberger, F. (1973) Brain dopamine and the syndromes of Parkinson and Huntington: Clinical, mor-phological, and neurochemical correlations. *J. Neurol. Sci.* **20,** 415–455.

194. Hornykiewicz, O. and Kish, S. J. (1987) Biochemical pathophysiology of Parkinson's disease. *Adv. Neurol.* **45,** 19–34.

195. Javoy-Agid, F., Ruberg, M., Hirsch, E., Cash, R., Raisman, R., Taquet, H., Epelbaum, J., Scatton, B., Duyckaerts, D., and Agid, Y. (1986) in *Recent Developments in Parkin-son's Disease* (Fahn, S., Marsden, C. D., Jenner, P., and Teychenne, P., eds.), Raven, New York. pp. 67–83.

196. Bernheimer, H. and Hornykiewicz, O. (1965) Decreased homovanillic acid concentra-tion in the brain in parkinsonian subjects as an expression of a disorder of central dopamine metabolism. *Klinische Wochenschrift* **43,** 711–715.

197. Lloyd, K. G., Davidson, L., and Hornykiewicz, O. (1975) The neurochemistry of Parkinson's disease: effect of L-dopa therapy. *J. Pharmacol. Exp. Ther.* **195,** 453–464.

198. Hornykeiwicz, O. (1966) Dopamine (3-hydroxytyramine) and brain function. *Pharma-col. Rev.* **18,** 925–964.

199. Zigmond, M. J. and Stricker, E. M. (1974) Ingestive behavior following damage to central dopamine neurons: implications for homeostasis and recovery of function, in *Neuropsychopharmacology of Monoamines and Their Regulatory Enzymes* (Usdin, E., ed.), Raven, New York. pp. 385–402.

200. Sharman, D. F., Poirier, L. J., Murphy, G. F., and Sourkes, T. L. (1967) Homovanillic acid and dihydroxyphenylacetic acid in the striatum of monkeys with brain lesions. *Can. J. Physiol. Pharmacol.* **45,** 57–62.

201. Apicella, P., Trouche, E., Nieoullon, A., Legallet, E., and Dusticier, N. (1990) Motor impairments and neurochemcial changes after unilateral 6-hydroxydopamine lesion of the nigrostriatal dopaminergic system in monkeys. *Neuroscience* **38,** 655–666.

202. Klockgether, T. and Turski, L. (1990) NMDA antagonists potentiate antiparkinsonian actions of L-Dopa in monoamine-depleted rats. *Ann. Neurol.* **28,** 539–546.

203. Zigmond, M. J. and Stricker, E. M. (1984) Parkinson's disease: Studies with an animal model. *Life Sci.* **35,** 5–18.

204. Antonini, A., Schwarz, J., Oertel, W. H., Beer, H. F., Madeja, U. D., and Leenders, K. L. (1994) [11C]raclopride and positron emission tomography in previously untreated patients with Parkinson's disease: Influence of L-dopa and lisuride therapy on striatal dopamine D2-receptors. *Neurology* **44,** 1325–1329.

205. Brooks, D. J., Ibanez, V., Sawle, G. V., Playford, E. D., Quinn, N., Mathias, C. J., Lees, A. J., Marsden, C. D., Bannister, R., and Frackowiak, R. S. (1992) Striatal D2 receptor status in patients with Parkinson's disease, striatonigral degeneration, and progressive supranuclear palsy, measured with 11C-raclopride and positron emission tomography. *Ann. Neurol.* **31,** 184–192.

206. Fernandez, A., de Ceballos, M. L., Jenner, P., and Marsden, C. D. (1992) Striatal neuropeptide levels in Parkinson's disease patients. *Neurosci. Lett.* **145,** 171–174.

207. Rinne, U. K., Rinne, J. O., Rinne, J. K., and Laakso, K. (1987) Chemical neurotransmission in the parkinsonian brain. *Med. Biol.* **65,** 75–81.

208. Pan, H. S., Penney, J. B., and Young, A. B. (1985) Gamma-aminobutyric acid and benzodiazepine receptor changes induced by unilateral 6-hydroxydopamine lesions of the medial forebrain bundle. *J. Neurochem.* **45,** 1396–1404.

209. Albin, R. L., Young, A. B., and Penney, J. B. (1989) The functional anatomy of basal ganglia disorders. *Trends Neurosci.* **12,** 366–375.

210. Filloux, F. and Townsend, J. J. (1993) Pre- and postsynaptic neurotoxic effects of dopamine demonstrated by intrastriatal injection. *Exp. Neurol.* **119,** 79–88.

211. Hastings, T. G., Lewis, D. A., and Zigmond, M. J. (1996) Role of oxidation in the neurotoxic effects of intrastriatal dopamine injections. *Proc. Natl. Acad. Sci. USA* **93,** 1956–1961.

4

6-Hydroxydopa, a Catecholamine Neurotoxin and Endogenous Excitotoxin at Non-NMDA Receptors

Richard M. Kostrzewa

1. HISTORICAL PERSPECTIVE

By the late 1960s, 6-hydroxydopamine (6-OHDA) was recognized as a substance that would produce long-lasting depletion of norepinephrine (NE) from the heart and other organs *(1–3)*, by virtue of overt destruction of sympathetic innervation to these tissues *(4,5)*. However, systemically administered 6-OHDA was restricted to noncentral compartments by the blood–brain barrier. 6-Hydroxydopa (6-OHDOPA) was synthesized specifically to overcome this limitation. Being an amino acid, 6-OHDOPA would be transported by facilitated diffusion into the brain, where abundant decarboxylase enzymes could convert this metabolic precursor to the active neurotoxic species, 6-OHDA, which then could produce damage to catecholamine- (CA) containing neurons in brain (Fig. 1).

Ong et al. *(6)* and Berkowitz and coworkers *(7)* validated this prediction by (1) showing that systemically administered 6-OHDOPA partly depleted NE from brain and (2) demonstrating that this effect was enhanced by coordinate administration of a peripherally restricted decarboxylase inhibitor. With these criteria established, 6-OHDOPA was ready for testing as a candidate neurotoxin. For the next 20 years, 6-OHDOPA was used in this way to study processes associated with damage to CA neurons and/or subsequent nerve regeneration and sprouting.

In 1990, 6-OHDOPA was discovered to be an excitotoxin via an action at non-*N*-methyl-D-aspartate (non-NMDA) receptors *(8)*. In that year, 6-OHDOPA was also found to be an endogenous cofactor for amine oxidases *(9)*, able to liberate reactive oxygen intra- and extraneuronally. 6-OHDOPA is the only endogenous neurotoxin acting by two disparate, but relatively selective mechanisms of action. The effect of 6-OHDOPA on CA-containing neurons is highlighted in this chapter. Its role as an excitotoxin is discussed more extensively in Chapter

From: Highly Selective Neurotoxins: Basic and Clinical Applications *Edited by: R. M. Kostrzewa.*
Humana Press Inc., Totowa, NJ

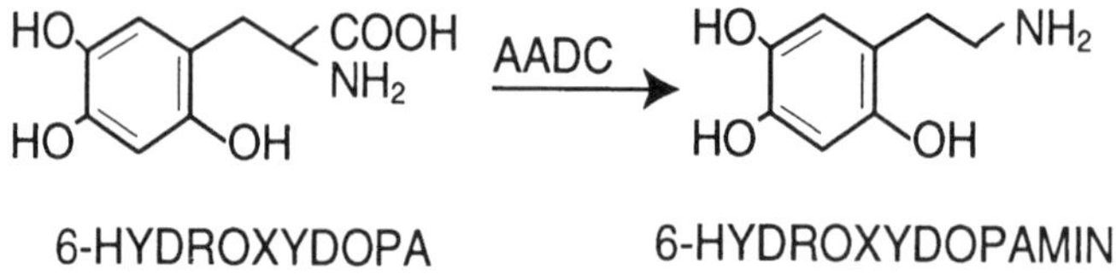

6-HYDROXYDOPA   6-HYDROXYDOPAMINE

Fig. 1. Chemical structure of 6-OHDOPA and decarboxylation to 6-OHDA.

13. Neurotoxic actions of 6-OHDOPA on CA-containing neurons have been reviewed previously in a different context *(10–12)*.

2. 6-OHDOPA AND CA-CONTAINING NEURONS IN BRAIN

2.1. Neurochemical Effects

Although the L-isomer is far more potent that the D-isomer *(7)*, overall potency of 6-OHDOPA is far less than that of 6-OHDA. A 100–150 mg/kg iv dose of 6-OHDOPA reduces mouse whole brain NE content by 30–60% and dopamine (DA) content by < 20% at 24 h. Cerebellum and neocortex are more depleted of NE (50–60%) than brainstem (20%). After 2 mo, whole-brain NE content partly recovers, but is still reduced, whereas DA content is unchanged from control *(13–16)*.

Even when administered directly into the lateral ventricles of brain (icv), 6-OHDOPA is weakly potent. The threshold dose of 6-OHDOPA for altering telencephalic or diencephalic NE is 45 µg; for cerebellum, it is 90 µg. At a 180-µg dose of 6-OHDOPA, telencephalic DA remains unaltered *(17)*. Nevertheless, as with 6-OHDA, the central effect of 6-OHDOPA is long-lived. Proposed reasons for lower efficacy of 6-OHDOPA relative to 6-OHDA include the following *(12)*:

1. Wider distribution of systemically administered 6-OHDOPA, disbursing into both central and peripheral compartments;
2. Possible greater accumulation of 6-OHDOPA in extraneuronal compartments;
3. Activation of 6-OHDOPA at a rate sufficiently slow to allow partial inactivation of generated 6-OHDA;
4. Metabolic inactivation of substantial amounts of 6-OHDOPA; and
5. Metabolic conversion of 6-OHDOPA to species lacking neurotoxic effects

2.2. Histochemical Effects

Prolonged reduction of NE in brain is reflective of destruction of noradrenergic nerve endings from innervated brain regions. This is supported by the finding that 6-OHDOPA produces retrograde accumulation of NE in preterminal axons—an effect seen with the acknowledged neurotoxin, 6-OHDA. When brain tissue of 6-OHDOPA-treated rats is prepared by the formaldehyde condensation method of Falck et al. *(18)* and observed under a fluorescence microscope, preterminal axons have brilliant green fluorescence and are greatly swollen *(13)*. From studies with 6-OHDA, this is known to be attributable to retrograde accumulation of neurotransmitter in nerves in which axoplasmic transport has been disrupted *(19,20)*.

In the late 1960s and even up to the mid-1970s, there was no reliable means of visualizing axons of CA-containing nerves. 6-OHDOPA made it possible to construct accurate neural maps of noradrenergic tracts in brain. Not only did 6-OHDOPA produce swollen axons, but 6-OHDOPA was relatively selective for noradrenergic vs dopaminergic neurons. At 6-OHDOPA doses as high as 250–400 mg/kg, DA content of mouse whole brain remains unaltered, whereas NE is reduced up to 80% *(21,22)*. For these reasons, 6-OHDOPA is particularly useful for mapping noradrenergic neuronal pathways in brain *(13,15,16,23,24)*. The entire trajectory of the dorsal bundle is able to be seen, from its origin at the locus cerulus to target brain regions. The ventral bundle either is not affected by 6-OHDOPA or is far less sensitive than the dorsal bundle to 6-OHDA actions *(23,24)*.

A relatively high focal dose of 6-OHDOPA (15 µg) in rat amygdala reduces NE and DA amygdala contents by about 30% and reduces striatal DA content by 25% at 3–5 wk. A lower dose of 6-OHDOPA (4 µg) in the amygdala produces an unexpected increase (nearly 50%) in amygdala NE and DA contents, with a concurrent 50% increase in striatal DA content *(25)*. These latter changes possibly reflect axonal accumulation of CA in axons en passant in the amygdala. After being discovered as an excitotoxin at non-NMDA receptors, 6-OHDOPA was shown to produce edematous lesions similar to that of quisqualate, when injected focally as a 150-nmol alkaline solution directly into rat substantia nigra, striatum, and frontal cortex *(8)*.

3. EVIDENCE OF 6-OHDOPA NEUROTOXICITY

Largely based on the absence of change in tyrosine hydroxylase activity in brainstem after 6-OHDOPA treatment *(26)*, it was suggested that 6-OHDOPA may act more like the synaptic granule depleter reserpine than like the neurotoxin 6-OHDA. To address this question, newborn rats were treated on the day of birth or on the first several days after birth with 6-OHDOPA (60 µg/g ip, 1–3 times at 2-d intervals), and brain samples were observed during the next 2 wk. Using argyrophilic Heimer and Fink-Heimer methods, along with electron microscopy for detecting nerve degeneration, 6-OHDOPA was indeed found to be neurotoxic. Silver particles were found in and around the locus cerulus, whereas fragmenting fibers were seen to be emanating from this noradrenergic nucleus. Shrunken perikarya displayed increased electron opacity of nuclear and cytoplasmic matrices, and had swollen mitochondria, dissociated polyribosomes, and disrupted Golgi cisternae. This profile was not seen in neighboring non-CA nuclei or in brain specimens from controls. At distal targets, degenerating fibers were evident as argyrophilic deposits and astrocytic invasion *(27,28)*. Using morphometric analysis of cresyl violet preparations, it was ultimately shown that a single injection of 6-OHDOPA (60 μg/g ip) at birth destroyed one-third of perikarya in the nucleus locus cerulus of rats. Moreover, the caudal portion of this nucleus was most susceptible, with approximately half of these perikarya being destroyed *(29)*. The reduced cell number was accompanied by a reduction of tyrosine hydroxylase activity of the nucleus locus cerulus *(30)*.

4. 6-OHDOPA AND PERIPHERAL SYMPATHETIC NEURONS

A high dose of 6-OHDOPA (100–150 mg/kg) partially depletes NE from cardiac atria and ventricles, salivary glands and spleen, but not vasa deferentia of mice. Restoration of NE in cardiac ventricles and salivary glands requires 1–2 wk *(31)*. In mouse atria that are partly depleted of endogenous NE by 6-OHDOPA, there is an associated reduction of the in vitro uptake of [^{3}H]NE and [^{3}H]metaraminol by atria *(15,16)* and a reduction in sympathetic innervation *(15,16,31)*. These findings demonstrate that 6-OHDOPA destroys sympathetic nerve fibers in adult species and that complete regeneration occurs in several weeks.

5. ONTOGENETIC EFFECT OF 6-OHDOPA

5.1. Postnatal 6-OHDOPA and CA Neurons

The ontogenetic effect of 6-OHDOPA is much different from its effect in adult species. When neonatal rats are treated with 6-OHDOPA (50 μg/g sc × 4, 2-d intervals) starting at birth or as late as the 20th day after birth, the development of NE in whole brain is impaired approx 50%, whereas that of telencephalon and hypothalamus is impaired 65 and 25%, respectively. Following a single 50-μg dose of 6-OHDOPA up to the ninth day after birth, there is prominent noradrenergic fiber sprouting and hyperinnervation of brainstem regions most proximal and rostral to the locus cerulus, as evidenced by the 50–161% elevation of brainstem NE content. When injected at 13 d after birth or later, however, 6-OHDOPA produces a long-lived reduction in brainstem NE content (< 50%). The noted alterations in brain levels of NE can be observed for at least 300 d *(32–35)*.

Of the assorted brain regions, the hippocampus is most depleted of NE after 6-OHDOPA. Typically 90% NE depletion is found in this region, even after a low dose of 6-OHDOPA *(36–38)*. The neocortex and spinal cord are moderately depleted of NE by neonatal 6-OHDOPA *(36–39)*. Retrograde accumulation of NE and noradrenergic hyperinnervation are greater in pons medulla than in midbrain *(35–37)*. Corresponding alterations in [^{3}H]NE uptake into homogenates from the cortex, hypothalamus, and pons medulla are observed *(37)*.

In cerebellum, the initial damage to noradrenergic fibers after neonatal 6-OHDOPA is followed by robust regenerative sprouting of noradrenergic fibers to an innervation density twice that of intact controls. This is evidenced by an approximate twofold elevation in NE content *(28,33,36–39)*, a twofold increase in the V_{max} for [^{3}H]NE uptake by cerebellar homogenates (synaptosome) *(38)*, increased tyrosine hydroxylase activity *(30)*, and increased number of NE fibers in cerebellum *(40–43)*. Added injections of 6-OHDOPA on days 2 and/or 4 after birth partly attenuate the sprouting response, but do not eliminate it *(28,33,36)*. However, when administered one time, starting at 3 d after birth or later, 6-OHDOPA persistently reduces adult NE content of cerebellum *(28,33,38)*. The threshold dose for neonatal 6-OHDOPA-induced alterations in NE content of spinal cord and cerebral cortex is 25 µg; for brainstem and cerebellum, it is 50 µg *(39)*. A low dose of 6-OHDOPA does not enhance the damage or improve recovery of noradrenergic nerves following the anterograde and retrograde damage of a surgical knife cut of the dorsal bundle *(44)*.

5.2. Postnatal 6-OHDOPA and Non-CA Neurons

Neonatal 6-OHDOPA treatment does not substantially alter brain levels of DA and 5-HT *(27,32,36)*. A very high dose of 6-OHDOPA (50 µg × 3, 2 d intervals from birth, sc) impairs the development of striatal DA by only 30% *(45)*.

In newborn rats treated with a moderate dose of 6-OHDOPA (100 µg sc), choline acetyltransferase (CAT) activity is elevated by about 40% in cortex and by >200% in brainstem 30–40 d later. In dorsal brainstem, a region encompassing the nucleus locus cerulus, acetylcholinesterase activity is also increased nearly 50% *(46)*. The implication of these findings is that 6-OHDOPA selectively damages noradrenergic, not dopaminergic or serotoninergic nerve endings in regions distal to their cell bodies. However, subsequent to degeneration of noradrenergic terminals in distal target regions, homotypic sprouting of noradrenergic fibers occurs in the brainstem near the nucleus of origin, the locus cerulus. Heterotypic sprouting of cholinergic fibers may accompany noradrenergic fiber sprouting.

5.3. Prenatal 6-OHDOPA

6-OHDOPA effectively crosses the placental barrier and alters the development of fetal noradrenergic neurons. When pregnant rats are treated with 6-OHDOPA (100 mg/kg iv) on the 14th day of gestation (G14) or later, there is a modest, but seemingly permanent impairment of development of NE content (20–50% depletion) in telediencephalon, cerebral cortex, hypothalamus, and "rest of brain," and as much as a two fold elevation of NE in brainstem and cerebellum. Sympathetic nerve development is not altered, as indicated by the unchanged endogenous level of NE in heart and salivary gland *(32–35)*.

Similarly, when 6-OHDOPA (50 µg/g iv × 3, 24-h intervals starting on G15) is administered to pregnant mice, offspring display a reduction in neocortical NE content, along with an elevation in pons medullary and midbrain NE content at 7 wk. When a single dose of 6-OHDOPA (60 µg/g ip) is administered to fetal mice at G17 and later, telencephalic NE is reduced, but brainstem NE is elevated in adulthood. Mice that receive 6-OHDOPA (60 µg/g ip) as early as G13 or as late as 3 d after birth have an elevation of cerebellar NE in adulthood *(30)*. These findings and comparable studies in rats illustrate that 6-OHDOPA crosses the placental barrier to alter the ontogenetic development of noradrenergic neurons.

5.4. Neonatal 6-OHDOPA and Sympathetic Neurons

Neonatal 6-OHDOPA treatment has a minor influence on sympathetic ganglia and sympathetic innervation of visceral organs *(36,39)*. At high dose, neonatal 6-OHDOPA (75 µg/g sc × 3, 2 d intervals from birth) produces a long-lived 30% reduction of NE content of rat atria and cardiac ventricle *(47)* and twofold elevation in the B_{max} of cardiac β_1-adrenoceptors with no change in the K_d; and no change in the B_{max} or K_d for cardiac β_2-adrenoceptors *(48)*. These changes are associated with a marked potentiation of isoproterenol-induced stimulation of ornithine decarboxylase activity of heart and potentiation of the positive inotropic response to NE in isolated atria *(47)*. The number of muscarinic [^{3}H]quinuclidinyl benzilate ([^{3}H]QNB) binding sites is reduced in the heart of these rats, as is the ED_{50} for the acetylcholine negative inotropic response *(49)*. Changes in the status and number of sympathetic and parasympathetic nerves and/or receptors in 6-OHDOPA-lesioned rats indicate that there is a shift away from the usual dominant parasympathetic regulation of cardiac function.

6. NERVE REGENERATION AFTER 6-OHDOPA

Because of the pronounced noradrenergic fiber regeneration and sprouting in brain subsequent to 6-OHDOPA, this neurotoxin has been used in many studies as a tool for delving into mechanisms of neural plasticity. Jaim-Etcheverry *(50)*, suspecting nerve growth factor (NGF) to be a vital neurotropin for central noradrenergic nerves as per sympathetic nerves *(51)*, studied the effects of an antiserum to NGF and found it to be ineffective in attenuating noradrenergic fiber regeneration and sprouting in the brain of 6-OHDOPA-treated rats. In effect, NGF appears to be of little if any importance for this process in brain.

However, morphine *(40–44,52)* and the opioid peptides methionine-enkephalin (met-enk), leu-enk, β-endorphin, and *d*-ala-enkephalinamide *(41)*, when coadministered with 6-OHDOPA to newborn rats, enhance recovery of NE in the cerebellum and pons medulla. These changes in NE are accompanied by increased numbers of noradrenergic fibers in the cerebellum, with particularly prominent noradrenergic hyperinnervation in and around cerebellar Purkinje cells *(40–43,52)*. This appears to be an opioid-specific effect, since the µ-receptor antagonist, naloxone, attenuates effects of morphine in pons medulla and cerebellum *(41,52)*. The findings indicate that endogenous neurotransmitters and exogenously administered analogs are able to modify recovery from damage of noradrenergic fibers from neurotoxins like 6-OHDOPA.

7. BEHAVIORAL EFFECTS OF 6-OHDOPA

7.1. 6-OHDOPA and Adult Rats

Food and water consumption are reduced in adult rats for several days after 6-OHDOPA treatment *(17)*. Shock-induced aggression develops 4 d after a 90-µg icv 6-OHDOPA dose and is evident for more than 1 wk *(53,54)*. Nociception is increased; predatory mouse killing is unaltered *(54)*. Locomotor activity in the open field is reduced for 1 wk, although latency to begin open field locomotion is elevated for a longer time, as is emotionality. The duration of decapitation convulsions is diminished after 60 µg and higher doses of 6-OHDOPA, and this effect persists for at least 70 d *(17,54)*.

7.2. 6-OHDOPA and Neonatally Treated Rats

When treated as newborns with 6-OHDOPA (60–180 µg/g ip, in 60-µg doses at 2-d intervals from birth), there is an overall increase in exploratory activity (ambulation, climbing, rearing, sniffing) during the initial 5 wk after birth. The effect is similar to that after neonatal 6-OHDA treatment, except that 6-OHDA-lesioned rats have more self-directed behaviors (eating, grooming, scratching) *(55)*. As adults, neonatal 6-OHDOPA-lesioned rats display impaired passive avoidance, less shock-induced aggression, increased general activity (i.e., ambulation, climbing, rearing, and sniffing) and more locomotor open-field activity *(55,56)*. In similarly treated rats, there is also increased apomorphine-induced (0.25–2 mg/kg ip) locomotor activity at 20 and 50 d after birth *(57)* and thyrotropin-releasing hormone- (TRH) induced behavioral arousal *(58)* that is attenuated by desipramine *(59)*. Although the antinociceptive effect of morphine is enhanced in rats lesioned neonatally with 6-OHDOPA *(60)*, such rats are minimally affected by noxious stimuli, a finding comparable to that of rats with hippocampal lesions *(61)*. In 6-OHDOPA-lesioned rats, hippocampal NE content is reduced approx 80% *(56)*. Also, rats with a neonatal 6-OHDOPA lesion display diminished conspecific odor preference in adulthood and as early as 8 d after birth *(62)*. Inhibitory avoidance learning is impaired as well *(63)*. Male copulatory behavior, food and water consumption, and thermoregulation are unimpaired in rats lesioned as neonates with 6-OHDOPA *(56)*, but their equilibrium is impaired, as indicated by increased numbers of falls from a rotating rod. Female sexual development and reproductive status are unaltered *(64)*.

In summary, damage to CA nerves by 6-OHDOPA produces measurable anatomical, biochemical, and behavioral deficits in lesioned animals.

8. SPECIFICITY OF 6-OHDOPA

8.1. DA Neurons

High-dose 6-OHDOPA (150–400 mg/kg) reduces endogenous NE content of visceral tissues and CNS without dramatically altering DA content of brain *(13,14,21,22)* or the pattern of DA innervation *(13)*. High-dose 6-OHDOPA

(50 mg/kg × 3, 2-d intervals, sc), administered to rats from birth, impedes development of striatal DA by 30% *(45)*. Therefore, it is apparent that DA neurons are susceptible to 6-OHDOPA neurotoxic effects, but DA nerves are far more insulated from such damage than noradrenergic neurons. Tuberoinfundibular DA neurons also appear to be more susceptible to 6-OHDOPA-induced damage than nigrostriatal DA neurons *(65)*.

8.2. Cholinergic Neurons

CAT, a marker enzyme for cholinergic nerves, does not appear to be altered in either mouse cardiac ventricle or whole brain after 6-OHDOPA treatment of adults *(14,31)*. In rats lesioned neonatally with 6-OHDOPA, however, CAT activity is reduced in brainstem *(46)*. Neonatal 6-OHDOPA (50 µg/g sc × 3, 2-d intervals from d 2) also enhances atropine-induced (5–20 mg/kg ip) locomotor activity between 20 and 50 d after birth and diminishes pilocarpine-induced catalepsy at 30 d and later *(66,67)*. Overall behavioral subsensitivity correlates with the observed reduction in the B_{max} for [^{3}H]QNB binding site (i.e., muscarinic sites) in striatal and mesolimbic brain regions of rats lesioned neonatally with 6-OHDOPA *(67)*. The number of [^{3}H]QNB sites in hearts from these rats is similarly reduced *(49)*.

8.3. Serotonin (5-HT) Neurons

5-HT content of mouse brain is unaltered by 6-OHDOPA, even after a dose of 400 mg/kg iv *(15,22)*. 5-HT content of rat diencephalon/mesencephalon and pons is elevated by 15–25% at 2 d after an icv 6-OHDOPA dose of 90 µg, but is unaltered in these regions and in the telencephalon and cerebellum at 14 d *(54)*.

8.4. Methemoglobinemia and 6-OHDOPA

Methemoglobinemia is one major effect that limits the dosing level of 6-OHDOPA—a limitation not evident with 6-OHDA. With high-dose 6-OHDOPA, animals frequently die after several hours to several days *(21)*. Methemoglobinemia *per se* can catalyze formation of 6-OHDOPA from tyrosine, a process facilitated by hydrogen peroxide *(68)*. It is not known if this is a factor in the delayed lethal action of 6-OHDOPA.

9. MECHANISM OF ACTION OF 6-OHDOPA ON CA NEURONS

Inhibitors of the NE transporter (e.g., *d-/l*-amphetamine, desipramine, imipramine, protriptyline, chlorpromazine) attenuate the action of 6-OHDOPA in brain and outside the brain *(14,31,69,70)*. The 6-OHDOPA-induced depletion of epinephrine in brainstem is prevented by iprindole *(70)*. Peripherally restricted decarboxylase inhibitors, such as DL-(3,4-dihydroxyphenyl)-α-hydrazino-α-methyl-propionic acid (MK-485), *N*-(DL-seryl)-*N'*-2,3,4-trihydroxybenzylhydrazine (Ro-4-4602), and carbidopa, attenuate the effects of 6-OHDOPA on sympathetic nerves and, in some instances, potentiate the effects of 6-OHDOPA on NE and DA nerves in brain *(6,7,14,21,31,69)*.

After a 100 mg/kg ip dose of 6-OHDOPA, the concentration of 6-OHDA in rat striatum is 1.9 μg/g at 15 min, a 50% decline at 1 h, and another 50% decline by 4 h. Pargyline pretreatment (100 mg/kg ip, 18 h) is associated with a doubling of 6-OHDA content after 6-OHDOPA, while carbidopa (25 mg/kg ip) is associated with a 2.4-fold elevation of striatal 6-OHDA content after 6-OHDOPA. Combined pargyline and carbidopa treatments produce a threefold elevation of striatal 6-OHDA content after 6-OHDOPA. Because reserpine pretreatment (10 mg/kg ip, 1.5 h) largely prevents accumulation of 6-OHDA, it appears that 6-OHDA is normally stored in synaptic vesicles. In this study by Evans and Cohen *(71)*, there was discordance between striatal/neocortical concentrations 6-OHDA and depletion of striatal DA/neocortical NE content at 2.5 h. Although 6-OHDA content was 8.8 times higher in striatum than in neocortex, DA content in striatum remained unaltered, whereas NE content of neocortex was reduced by 75%. As discussed by the authors, the striatum is densely innervated by DA fibers, and it is conceivable that DA fibers in striatum do not individually have a cytoplasmic concentration of 6-OHDA sufficient to produce destruction. Neocortex is sparsely innervated by NE fibers, which might individually have suprathreshold levels of 6-OHDA needed for neuronal damage. It is also possible that the time-course of 6-OHDA is different in DA vs NE neurons, and that later measures of DA and NE contents might better correlate with early concentrations of 6-OHDA after 6-OHDOPA treatment.

On the basis of the finding that the NE transport inhibitors desipramine and nomifensine elevated levels of 6-OHDA in frontal cortex by 63–248% and in striatum by 83%, 1–4 h after systemic administration of 6-OHDOPA to mice, Evans and Cohen *(72)* believe that 6-OHDOPA is accumulated by CA-containing nerves prior to decarboxylation to 6-OHDA.

The above findings, along with earlier studies already discussed, illustrate the known fact that 6-OHDOPA must be decarboxylated to 6-OHDA to produce damage to CA-containing neurons. However, there are still unanswered questions concerning the cellular sites and events associated with the molecular processes.

10. 6-OHDOPA (TOPA) EXCITOTOXICITY

More than 20 years ago, 6-OHDOPA was shown to be more excitatory to frog spinal neurons than glutamate. Because this effect was observed when 6-OHDOPA was added to the perfusion medium, but not when electrophoretically applied, it was believed that an oxidation product, not 6-OHDOPA *per se,* was responsible *(73)*. Only recently, 6-OHDOPA (150 μg) was found to be excitotoxic to chick embryo retinal preparations and to cultured hippocampal neurons. These neurotoxic effects are attenuated by the non-NMDA receptor antagonist 6-cyano-7-nitroquinoxaline-2,3-dione (CNQX), but not by the NMDA antagonists (*5R,10S*)-(+)-5-methyl-10,11-dihydro-5H-dibenzo[a,d]cyclohepten-5,10-imine (MK-801) and D-2-amino-5-phosphonovalerate (APV) *(8)*. Similar findings have been obtained with 6-OHDOPA in cultured cortical neurons from rats. An oxidation product of 6-OHDOPA, likely to be 6-OHDOPA-quinone, accounts for the excitoxic effect *(74,75)*. The abbreviation topa (trihydroxyphenylalanine) is used for 6-OHDOPA in virtually all reports on its excitotoxicity. Accordingly, in this section of the chapter, topa has been substituted for 6-OHDOPA.

Topa has high affinity for non-NMDA receptors that recognize α-amino-3-hydroxy-5-methylisoxazole-4-propionic acid (AMPA) as a ligand *(76)*. By means of autoradiography, topa was shown to effectively displace [^{3}H]AMPA binding to hippocampal subfields and is as potent as kainate in displacing [^{3}H]AMPA binding to rat striatal homogenates *(77)*. DA, DOPA, and 6-OHDA lack this effect. Also, topa does not have affinity for kainate-, PCP-, DA D_1- or D_2-receptors *(76)*. Glutathione prevents topa excitotoxicity to rat cortical neurons in culture *(78)*, by virtue of its forming glutathionyl conjugates (catecholthioethers) and thereby reducing the amount of toxic precursors *(79)*.

11. REACTIVE OXYGEN SPECIES FORMED BY 6-OHDOPA/TOPA

Oxidation of topa to quinones is nonenzymatic and occurs in nonbiologic media *(80)* in < 30 min at physiologic pH *(79)*. Likely products of topa oxidation are the cyclized, reduced leukodopachrome, leukodopaminechrome, topa-*p*-quinone, and/or its cyclized oxidized aminochromes, dopachrome or dopaminechrome *(79)*.

Formation of topa from dopa is accelerated nearly 20-fold in the presence of ascorbic acid or in the presence of both iron (Fe^{2+}) and hydrogen peroxide (H_2O_2). It has been proposed that this process is a consequence of direct oxidation of DOPA by free radicals *(80)*. Also, reactive oxygen formation and free radical generation from topa oxidation are processes that are greatly facilitated by Fe^{3+} or Cu^{2+}, using H_2O_2 as a substrate (Fenton reaction) to form highly reactive hydroxyl radical (HO˙) *(81)*. It may be significant that topa enhances release of Fe^{3+} from ferritin *(82)*. Because neuromelanin interacts with Fe, pigmented substantia nigra DA-containing neurons and pigmented locus cerulus NE-containing neurons may be particularly susceptible to neurotoxic actions of topa and like compounds. Topa quinone is a long-lived substrate, stable at physiological pH for hours *(74,83,84)*.

12. 6-OHDOPA AND MALIGNANT CELLS

Recently, topa compounds including topa quinone were found to be produced by and released from a pheochromocytoma PC12 clonal cell line to which tyrosine (30 μ*M*) was added. Topa synthesis is enhanced twofold by KCl (56 m*M*) and fivefold by a decarboxylase inhibitor, to a concentration of 5.5 pmol/10^6 cells. Topa content is reduced by GSH (100 µ*M*) *(80)*. It is notable that topa (3–30 µ*M*) is toxic to DA-containing cells in neuronal cultures derived from rat mesencephalon *(85)*. As indicated by Newcomer et al. *(80)*, topa is the first known non-NMDA neurotoxin naturally produced by nerves and may be one of the principal factors involved in neurodegenerative disorders like parkinsonism and Huntington's chorea.

Twenty-five years ago, 6-OHDA was shown to cause an immediate condensation of pigment grains, followed by long-lasting pigment dispersion in melanophores of the teleost *(86)*. Wick *(87)* later showed that L-dihydroxyphenylalanine (L-DOPA), DA, and analogs were effective antitumor agents against B-16 melanoma in vivo. 6-OHDOPA was particularly effective in killing pigmented melanotic cells, primarily by impairing DNA and RNA synthesis *(88)*. Hansson et al. *(89)* previously had shown that a tyrosinase in human melanoma cells oxygenates 2,4-L-DOPA to 6-OHDOPA, indicating that 6-OHDOPA formation is vital for the L-DOPA cytotoxic action. A tyrosinase inhibitor, 1-phenylthiourea, attenuated the effect of L-DOPA, thereby confirming the importance of tyrosinase in L-DOPA action *(90)*.

Agrup et al. *(91)* found that the sera from five of seven patients with widespread melanoma metastases contained detectable quantities of tyrosinase. Passi et al. *(92)* presented evidence that di- and tri-phenols act by generation of toxic oxygen species acting outside the cell and that enzymes scavenging reactive oxygen prevent their toxic effects. Although tyrosinase is the major enzyme involved in the L-DOPA effect on melanoma cells, appreciable amounts of L-DOPA can be nonenzymatically (spontaneously) converted to 6-OHDOPA in solution *(84)*. A lactoperoxidase also converts L-DOPA to 6-OHDOPA and *o*- and *p*-semiquinone radicals under conditions in which hydrogen peroxide is formed *(93)*.

These series of findings indicate that malignant melanoma cells are particularly susceptible to damage by the antiparkinsonian drug L-DOPA, by virtue of formation of 6-OHDOPA, and subsequent reactive oxygen.

13. 6-OHDOPA AS COFACTOR FOR AMINE OXIDASES

In this decade, topa was discovered to be a naturally occurring covalently bound cofactor at the active site of copper amine oxidase in bovine serum *(9)*, in lentil seedlings *(94)*, Gram-positive and Gram-negative bacteria *(9,95)*, and in the yeasts *Hansenula polymorpha (96)* and *Aspergillus niger (97)*. This was confirmed by mass spectrometry, UV spectroscopy, proton nuclear magnetic resonance (NMR) *(9)*, resonance Raman spectroscopy *(98)* and electron paramagnetic resonance (EPR) *(99)*. The active site peptide in this protein is homologous with a segment of cloned human kidney amiloride binding protein *(100)*. Topa is linked in the sequence -Leu-Asn-6-OHDA-Asp-Tyr- *(9)* and is believed to be derived from a tyrosyl residue, either co- or posttranslationally, by external enzymes or self-catalyzed by the prosthetic Cu^{2+} group *(93,100–102)*. In the resting state, the topa cofactor is most likely to exist as topa-*p*-quinone *(103)*.

EPR studies have established that Cu^{2+} is too distant from topa and does not interact with the topa cofactor *(99)*. Although the role of Cu^{2+} is unclear, its presence is known to enhance catalytic activity, although its absence is known to eliminate enzyme activity. Other EPR evidence indicates that a Cu^{+}-semiquinone is generated by substrates and that this principle is an intermediate that reacts with O_2 *(104,105)*. There are two active sites per enzyme dimer *(106)*. The nitrogen atom of amines is thought to bind covalently to the C-3 or C-4 carbonyl group of topa quinone, forming an amino-6-hydroxydopa semiquinone en route to conversion to an aldehyde *(9,103)*. Nucleophilic addition of the amine to topa quinone is associated with transient formation of an indole or iminoquinone intermediate *(83)*. Oxidized topa is restored by oxidation of the radical *(107,108)*.

14. CLINICAL IMPLICATIONS RELATIVE TO 6-OHDOPA

L-DOPA is the byproduct of tyrosine hydroxylation and the immediate metabolic precursor for DA synthesis in nerves. Exogenously administered L-DOPA is the most effective drug for treating Parkinson's disease, an extrapyramidal movement disorder that develops when large numbers of substantia nigra dopaminergic neurons spontaneously degenerate. L-DOPA *per se* has excitotoxic properties and in high amount will produce excitatory destruction of neurons.

L-DOPA will spontaneously and noncatalytically oxidize in solution to 6-OHDOPA *(84)*, or topa, which itself is excitotoxic to neurons with non-NMDA receptors and also neurotoxic to CA-containing neurons, including dopaminergic neurons (*see* Fig. 2). Both mechanisms may be involved in the spontaneous degeneration of dopaminergic neurons with aging. Based on this potential toxicity of L-DOPA and its derivatives, 6-OHDOPA and 6-OHDOPA quinones, antioxidant vitamins have been promoted as nutritional supplements.

Because exogenous L-DOPA is given in gram amounts for treating parkinsonism, the neurotoxic potential of L-DOPA and its derivatives 6-OHDOPA and 6-OHDOPA quinones is great. It is conceivable that L-DOPA, although producing immediate symptomatic relief of parkinsonian symptoms, may actually accelerate the progression of parkinsonism *(8)*.

The enzymatic formation of topa and topa quinone is distinctly possible, particularly in neuromelanin-containing catecholamine neurons *(84)*. However, there is still no convincing evidence for tyrosinase activity in these cells. No tyrosinase mRNA has been found in the substantia nigra or locus cerulus. However, using immunological methods, tyrosinase-like polypeptides have been found in rat brains. The absence of tyrosinase in the brain may be immaterial, since topa is known to be formed in human melanocytes in culture *(89)* and would likely be formed by melanocytes in the dermis in vivo and subsequently

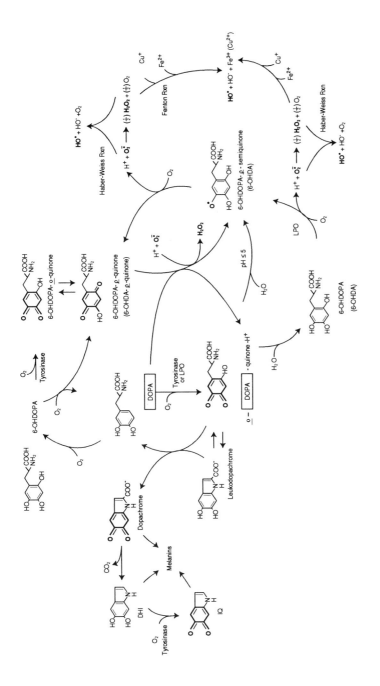

Fig. 2. In vivo formation of 6-OHDOPA and 6-OHDA from DOPA and dopamine, and schematic for generation of reactive oxygen species and melanin. Adapted from Nappi and Vass (79) and Rodriguez-Lopez et al. (81).

enter the brain. If excess topa were formed as a consequence of altered physiological or pathological states, the potential for CNS damage would be increased.

Deprenyl, the only drug reported to slow the progression of parkinsonism, is both an MAO$_B$ inhibitor and an antagonist at excitatory amino acid (EAA) receptors. This possibly could represent the mechanism of action of deprenyl. If deprenyl owed its effect to a block of EAA receptors, then it could be an important adjuvant of L-DOPA therapy. It would be of value to know if other EAA antagonists slow the progression of Parkinson's disease. This could be the avenue to newer antiparkinsonian drugs. It would also be of value to know if antioxidants *per se* have a beneficial effect in parkinsonism. These could likewise be useful for preventing degeneration of nigral neurons with aging and in preventing damage to DA neurons by drugs that may generate reactive oxygen.

Weiss et al. *(109)* have proposed that the excitotoxin β-*N*-methylamino-L-alanine (BMAA, L-BOAA) found in the chickling pea *Lathyrus sativus* is the causative agent of neurolathyrism, a motor disorder of humans in which spastic paraparesis occurs. Excitotoxins have also been proposed as causative factors in neurodegenerative disorders, such as Huntington's chorea *(110–114)*.

In ongoing studies, topa is being used as a tool in the search for neurotrophic and related factors that can rescue cells from injury by autoexcitatory toxins *(85,115)*. It is of interest that NGF synthesis was enhanced in mouse astroglia cultures by topa, which was not cytotoxic *(116)*.

6-OHDOPA is the first identifiable highly selective non-NMDA agonist that is naturally produced by nerves and could represent one of the principal factors involved in neurodegenerative disorders.

REFERENCES

1. Porter, C. C., Totaro, J. A., and Stone, C. A. (1963) Effect of 6-hydroxydopamine and some other compounds on the concentration of norepinephrine in the hearts of mice. *J. Pharmacol. Exp. Ther.* **140,** 308–316.
2. Porter, C. C., Totaro, J. A., and Burcin, A. (1965) The relationship between radioactivity and norepinephrine concentrations in the brains and hearts of mice following administration of labelled methyldopa or 6-hydroxydopamine. *J. Pharmacol. Exp. Ther.* **150,** 17–22.
3. Stone, C. A., Stavorski, J. M., Ludden, C. T., Wengler, H. C., Ross, C. A., Totaro, J. A., and Porter, C. C. (1963) Comparison of some pharmacological effects of certain 6-substituted dopamine derivatives with reserpine, guanethidine and metaraminol. *J. Pharmacol. Exp. Ther.* **142,** 147–156.
4. Tranzer, J. P. and Thoenen, H. (1967) Ultra-morphologische Veranderungen der sympatischen Nervendigunden der Katze nach Vorbehandlung mit 5- und 6-hydroxy-dopamin. *Naunyn-Schmiedebergs Arch. Pharmakol. Exp. Pathol.* **257,** 343–344.
5. Thoenen, H. and Tranzer, J. P. (1968) Chemical sympathectomy by selective destruction of adrenergic nerve endings with 6-hydroxydopamine. *Naunyn-Schmiedebergs Arch. Pharmakol. Exp. Pathol.* **261,** 271–288.

6. Ong, H. H., Creveling, C. R., and Daly, J. W. (1969) The synthesis of 2,4,5-trihydrox-yphenylalanine (6-hydroxydopa). A centrally active norepinephrine-depleting agent. *J. Med. Chem.* **12,** 458–462.

7. Berkowitz, B. A., Spector, S., Brossi, A., Focella, A., and Teitel, S. (1970) Preparation and biological properties of (–)- and (+)-6-hydroxydopa. *Experientia* **26,** 982–983.

8. Olney, J. W., Zorumski, C. F., Stewart, G. R., Price, M. T., Wang, G .J., and Labruyere, J. (1990) Excitotoxicity of L-dopa and 6-OH-dopa: implications for Parkinson's and Hunt-ington's diseases. *Exp. Neurol.* **108,** 269–272.

9. Janes, S. M., Mu, D., Wemmer, D., Smith, A. J., Kaur, S., Maltby, D., and Burlingame, A. L. (1990) A new redox cofactor in eukaryotic enzymes: 6-Hydroxydopa at the active site of bovine serum amine oxidase. *Science* **248,** 981–987.

10. Kostrzewa, R. M. and Jacobowitz, D. M. (1974) Pharmacological actions of 6-hydroxy-dopamine. *Pharmacol. Rev.* **26,** 199–288.

11. Kostrzewa, R. M. (1988) Reorganization of noradrenergic neuronal systems following neonatal chemical and surgical injury. *Prog. Brain Res.* **78,** 405–423.

12. Kostrzewa, R. M. (1989) Neurotoxins that affect central and peripheral catecholamine neurons, in *Neuromethods, vol. 12: Drugs as Tools in Neurotransmitter Research* (Boulton, A. A., Baker, G. B., and Juorio, A. V., eds.), Humana, Clifton, NJ, pp. 1–48.

13. Jacobowitz, D. and Kostrzewa, R. (1971) Selective action of 6-hydroxydopa on nora-drenergic terminals: Mapping of preterminal axons of the brain. *Life Sci.* **10,** 1329–1341.

14. Kostrzewa, R. and Jacobowitz, D. (1973) Acute effects of 6-hydroxydopa on central monoaminergic neurons. *Eur. J. Pharmacol.* **21,** 70–80.

15. Sachs, C. and Jonsson, G. (1972) Degeneration of central and peripheral noradrenaline neurons produced by 6-hydroxy-DOPA. *J. Neurochem.* **19,** 1561–1575.

16. Sachs, C. and Jonsson, G. (1972) Selective 6-hydroxy-DOPA induced degeneration of central and peripheral noradrenaline neurons. *Brain Res.* **40,** 563–568.

17. Richardson, J. S. and Jacobowitz, D. M. (1973) Depletion of brain norepinephrine by intraventricular injection of 6-hydroxydopa: A biochemical, histochemical and behavioral study in rats. *Brain Res.* **58,** 117–133.

18. Falck, B., Hillarp, N.-A., Thieme, G., and Torp, A. (1962) Fluorescence of cate-cholamines and related compounds condensed with formaldehyde. *J. Histochem. Cytochem.* **10,** 348–354.

19. Cheah, T. B., Geffen, L. B., Jarrott, B., and Ostberg, A. (1971) Action of 6-hydroxy-dopamine on lamb sympathetic ganglia, vas deferens and adrenal medulla: A combined histochemical, ultrastructural and biochemical comparison with the effects of reserpine. *Br. J. Pharmacol.* **42,** 543–557.

20. Ross, R. A., Joh, T. H., and Reis, D. J. (1975) Reversible changes in the accumulation and activities of tyrosine hydroxylase and dopamine-β-hydroxylase in neurons of nucleus locus coeruleus during the retrograde reaction. *Brain Res.* **92,** 57–72.

21. Corrodi, H., Clark, W. G., and Masuoka, D. I. (1971) The synthesis and effects of DL-6-hydroxydopa, in *6-Hydroxydopamine and Catecholamine Neurons* (Malmfors, T. and Thoenen, H., eds.), North Holland, Amsterdam, pp. 187–192.

22. Clarke, D. E., Smookler, H. H., Hadinata, J., Chi, C., and Barry, H. III (1972) Acute effects of 6-hydroxydopa and its interaction with DOPA on brain amine levels. *Life Sci.* **11,** 97–102.

23. Sachs, C., Jonsson, G., and Fuxe, K. (1973) Mapping of central noradrenaline pathways with 6-hydroxy-dopa. *Brain Res.* **63,** 249–261.

24. Tohyama, M., Maeda, T., and Shimizu, N. (1974) Detailed noradrenaline pathways of locus coeruleus neuron to the cerebral cortex with use of 6-hydroxydopa. *Brain Res.* **79,** 139–144.

25. Lenard, L. and Hahn, Z. (1982) Amygdalar noradrenergic and dopaminergic mechanisms in the regulation of hunger and thirst-motivated behavior. *Brain Res.* **233,** 115–132.

26. Thoenen, H. (1972) Chemical sympathectomy: a new tool in the investigation of the physiology and pharmacology of peripheral and central adrenergic neurons, in *Perspectives in Neuropharmacology* (Snyder, S. H., ed.), Oxford University Press, London, pp. 302–338.

27. Kostrzewa, R. M. and Harper, J. W. (1974) Effects of 6-hydroxydopa on catecholamine-containing neurons in brains of newborn rats. *Brain Res.* **69,** 174–181.

28. Kostrzewa, R. M. and Harper, J. W. (1975) Comparison of the neonatal effects of 6-hydroxydopa and 6-hydroxydopamine on growth and development of noradrenergic neurons in the central nervous system, in *Chemical Tools in Catecholamine Research,* vol. I, (Jonsson, G., Malmfors, T. and Sachs, C., eds.), North Holland, Amsterdam, pp. 181–188.

29. Clark, M. B., King, J. C., and Kostrzewa, R. M. (1979) Loss of nerve cell bodies in caudal locus coeruleus following treatment of neonates with 6-hydroxydopa. *Neurosci. Lett.* **13,** 331–336.

30. Kostrzewa, R. M., Klara, J. W., Robertson, J., and Walker, L. C. (1978) Studies on the mechanism of sprouting of noradrenergic terminals in rat and mouse cerebellum after neonatal 6-hydroxydopa. *Brain Res. Bull.* **3,** 525–531.

31. Kostrzewa, R. and Jacobowitz, D. (1972) The effect of 6-hydroxydopa on peripheral adrenergic neurons. *J. Pharmacol. Exp. Ther.* **183,** 284–297.

32. Zieher, L. M. and Jaim-Etcheverry, G. (1973) Regional differences in the long-term effect of neonatal 6-hydroxydopa treatment on rat brain noradrenaline. *Brain Res.* **60,** 199–207.

33. Jaim-Etcheverry, G. and Zieher, L. M. (1975) Alterations of the development of central adrenergic neurons produced by 6-hydroxydopa, in *Chemical Tools in Catecholamine Research,* vol. I, (Jonsson, G., Malmfors, T. and Sachs, C., eds.), North Holland, Amsterdam, pp. 173–180.

34. Zieher, L. M. and Jaim-Etcheverry, G. (1975) Different alterations in the development of the noradrenergic innervation of the cerebellum and the brain stem produced by neonatal 6-hydroxydopa. *Life. Sci.* **17,** 987–991.

35. Zieher, L. M. and Jaim-Etcheverry, G. (1975) 6-Hydroxydopa during development of central adrenergic neurons produces different long-term changes in rat brain noradrenaline. *Brain Res.* **86,** 271–281.

36. Kostrzewa, R. M. (1975) Effects of neonatal 6-hydroxydopa treatment on monamine content of rat brain and peripheral tissues. *Res. Commun. Chem. Pathol. Pharmacol.* **11,** 567–579.

37. Kostrzewa, R. M. and Garey, R. E. (1976) Effects of 6-hydroxydopa on noradrenergic neurons in developing rat brain. *J. Pharmacol. Exp. Ther.* **197,** 105–118.

38. Kostrzewa, R. M. and Garey, R. E. (1977) Sprouting of noradrenergic terminals in rat cerebellum following neonatal treatment with 6-hydroxydopa. *Brain Res.* **124,** 385–391.

39. Jaim-Etcheverry, G. and Zieher, L. M. (1977) Differential effect of various 6-hydroxy-dopa treatments on the development of central and peripheral noradrenergic neurons. *Eur. J. Pharmacol.* **45,** 105–116.

40. Harston, C. T., Morrow, A., and Kostrzewa, R. M. (1980) Enhancement of sprouting and putative regeneration of central noradrenergic fibers by morphine. *Brain Res. Bull.* **5,** 421–424.

41. Harston, C. T., Clark, M. B., Hardin, J. C., and Kostrzewa, R. M. (1981) Opiate-enhanced toxicity and noradrenergic sprouting in rats treated with 6-hydroxydopa. *Eur. J. Pharmacol.* **71,** 365–373.

42. Harston, C. T., Clark, M. B., Hardin, J. C., and Kostrzewa, R. M. (1982) Developmental localization of noradrenergic innervation to the rat cerebellum following neonatal 6-hydroxydopa and morphine treatment. *Dev. Neurosci.* **5,** 252–262.

43. Kostrzewa, R. M., Harston, C. T., Fukushima, H., and Brus, R. (1982) Noradrenergic fiber sprouting in the cerebellum. *Brain Res. Bull.* **9,** 509–517.

44. Klisans-Fuenmayor, D., Harston, C. T., and Kostrzewa, R. M. (1986) Alterations in nora-drenergic innervation of the brain following dorsal bundle lesions in neonatal rats. *Brain Res. Bull.* **16,** 47–54.

45. Nomura, Y. and Segawa, T. (1979) Striatal dopamine content reduced in developing rats treated with 6-hydroxydopa. *Jpn. J. Pharmacol.* **29,** 306–309.

46. Jaim-Etcheverry, G., Teitelman, G., and Zieher, L. M. (1975) Choline acetyltransferase activity increases in the brain stem of rats treated at birth with 6-hydroxydopa. *Brain Res.* **100,** 699–704.

47. Nomura, Y., Kajiyama, H., and Segawa, T. (1980) Hypersensitivity of cardiac beta-adrenergic receptors after neonatal treatment of rats with 6-hydroxydopa. *Eur. J. Pharmacol.* **66,** 225–232.

48. Kajiyama, H., Obara, K., Nomura, Y., and Segawa, T. (1982) The increase of cardiac beta 1-subtype of beta-adrenergic receptors in adult rats following neonatal 6-hydroxy-dopa treatment. *Eur. J. Pharmacol.* **77,** 75–77.

49. Nomura, Y., Kajiyama, H., and Segawa, T. (1979) Decrease in muscarinic cholinergic response of the rat heart following treatment with 6-hydroxydopa. *Eur. J. Pharmacol.* **60,** 323–327.

50. Jaim-Etcheverry, G., Shoemaker, W. J., Zieher, L. M., and Bloom, F. E. (1980) Anti-serum to nerve growth factor does not prevent the increase of brain stem noradrenaline after neonatal 6-hydroxydopa. *Brain Res.* **197,** 547-553.

51. Levi-Montalcini, R. and Angeletti, P. U. (1968) Nerve growth factor. *Physiol. Rev.* **48,** 534–569.

52. Kostrzewa, R. M. and Klisans-Fuenmayor, D. (1984) Development of an opioid-specific action of morphine in modifying recovery of neonatally-damaged noradrenergic fibers in rat brain. *Res. Commun. Chem. Pathol. Pharmacol.* **46,** 3–11.

53. Thoa, N. B., Eichelman, B., Richardson, J. S., and Jacobowitz, D. (1972) 6-Hydroxy-dopa depletion of brain norepinephrine and the function of aggressive behavior. *Science* **178,** 75–77.

54. Richardson, J. S., Cowan, N., Hartman, R., and Jacobowitz, D. M. (1974) On the behavioral and neurochemical actions of 6-hydroxydopa and 5,6-dihydroxytryptamine in rats. *Res. Commun. Chem. Pathol. Pharmacol.* **8,** 29–44.

55. Morgan, D. N., McLean, J. H., and Kostrzewa, R. M. (1979) Effects of 6-hydroxydopamine and 6-hydroxydopa on development of behavior. *Pharmacol. Biochem. Behav.* **11,** 309–312.

56. McLean, J. H., Kostrzewa, R. M., and May, J. G. (1976) Behavioral and biochemical effects of neonatal treatment of rats with 6-hydroxydopa. *Pharmacol. Biochem. Behav.* **4,** 601–607

57. Nomura, Y. and Segawa, T. (1978) Apomorphine-induced locomotor stimulation in developing rats treated with 6-hydroxydopa. *Eur. J. Pharmacol.* **50,** 153–156.

58. Nomura, Y. and Oki, K. (1980) TRH-induced behavioral arousal in developing rats pretreated with 6-hydroxydopa. *Pharmacol. Biochem. Behav.* **12,** 925–930.

59. Oki, K., Nomura, Y., and Segawa, T. (1982) Involvement of central noradrenergic system in thyrotropin-releasing hormone-induced behavioral excitement in 6-OHDOPA-treated, infant rats. *J. Pharmacobiodyn.* **5,** 716–719.

60. Slater, P. and Blundell, C. (1978) The effects of a permanent and selective depletion of brain catecholamines on the antinociceptive action of morphine. *Naunyn-Schmiedebergs Arch. Pharmacol.* **305,** 227–232.

61. Kimura, T. (1957) Electromyographic studies of effects of chlorpromazine on nociceptively and proprioceptively induced reflex pattern of rabbits hindlimb. *Jpn. J. Pharmacol.* **6,** 162–174.

62. Cornwell-Jones, C. A. and Bollers, H. R. (1983) Neonatal 6-hydroxydopa alters conspecific odor investigation by male rats. *Brain Res.* **268,** 291–294.

63. Cornwell-Jones, C. A., Decker, M. W., Chang, J. W., Cole, B., Goltz, K. M., Tran, T., and McGaugh, J. L. (1989) Neonatal 6-hydroxydopa, but not DSP-4, elevates brainstem monoamines and impairs inhibitory avoidance learning in developing rats. *Brain Res.* **493,** 258–268.

64. McLean, J. H., Glasser, R. S., Kostrzewa, R. M., and May, J. G. (1980) Effects of neonatal 6-hydroxydopa on behavior in female rats. *Pharmacol. Biochem. Behav.* **13,** 863–868.

65. Lin, J. Y., Mai, L. M., and Pan, J. T. (1993) Effects of systemic administration of 6-hydroxydopamine, 6-hydroxydopa and 1-methyl-4-phenyl-1,2,3,6-tetrahydropyridine (MPTP) on tuberoinfundibular dopaminergic neurons in the rat. *Brain Res.* **624,** 126–130.

66. Nomura, Y. and Segawa, T. (1978) Muscarinic hyposensitivity in the developing rat pretreated with 6-hydroxydopa. *Eur. J. Pharmacol.* **50,** 431–435.

67. Nomura, Y., Kajiyama, H., Nakata, Y., and Segawa, T. (1979) Muscarinic cholinergic binding in striatal and mesolimbic areas of the rat: Reduction by 6-hydroxydopa. *Eur. J. Pharmacol.* **58,** 125–131.

68. Agrup, G., Hansson, C., Rorsman, H., Rosengren, E., and Tegner, E. (1983) Methaemoglobin-catalysed formation of dopa and 6-OH-dopa from tyrosine. *Acta. Derm. Venereol.* **63,** 152–155.

69. Jonsson, G. and Sachs, C. (1973) Pharmacological modifications of the 6-hydroxyDOPA induced degeneration of central noradrenaline neurons. *Biochem. Pharmacol.* **22,** 1709–1716.

70. VonVoigtlander, P. F. and Losey, E. G. (1978) 6-Hydroxydopa depletes both brain epinephrine and norepinephrine: Interactions with antidepressants. *Life Sci.* **23,** 147–150.

71. Evans, J. M. and Cohen, G. (1989) Studies on the formation of 6-hydroxydopamine in mouse brain after administration of 2,4,5-trihydroxyphenylalanine (6-hydroxyDOPA). *J. Neurochem.* **52,** 1461–1467.

72. Evans, J. and Cohen, G. (1993) Catecholamine uptake inhibitors elevate 6-hydroxydopamine in brain after administration of 6-hydroxydopa. *Eur. J. Pharmacol.* **232,** 241–245.

73. Biscoe, T. J., Evans, R. H., Headley, P. M., Martin, M. R., and Watkins, J. (1976) Structure activity relations of excitatory amino acids on frog and rat spinal neurones. *Br. J. Pharmacol.* **58,** 373–382.

74. Rosenberg, P. A., Loring, R., Xie, Y., Zaleskas, V., and Aizenman, E. (1991) 2,4,5-trihydroxyphenylalanine in solution forms a non-*N*-methyl-D-aspartate glutamatergic agonist and neurotoxin. *Proc. Natl. Acad. Sci. USA* **88,** 4865–4869.

75. Aizenman, E., White, W. F., Loring, R. H., and Rosenberg, P. A. (1990) A 3,4-dihydroxyphenylalanine oxidation product is a non-*N*-methyl-D-aspartate glutamatergic agonist in rat cortical neurons. *Neurosci. Lett.* **116,** 168–171.

76. Cha, J.-H. J., Dure, J. L. S., IV, Sakurai, S. Y., Penney, J. B., and Young, A. B. (1991) 2,4,5-Trihydroxyphenylalanine (6-hydroxy-DOPA) displaces [^{3}H]AMPA binding in rat striatum. *Neurosci. Lett.* **132,** 55–58.

77. Kunig-G., Hartmann, J., Niedermeyer, G., Deckert, J., Ransmayr, G., Heinsen, H., Beckmann, H., and Riederer, P. (1994) Excitotoxins L-beta-oxalyl-amino-alanine (L-BOAA) and 3,4,6-trihydroxyphenylalanine (6-OH-DOPA) inhibit [3H] alpha-amino-3-hydroxy-5-methyl-4-isoxazole-propionic acid (AMPA) binding in human hippocampus. *Neurosci. Lett.* **169,** 219–222.

78. Aizenman, E., Boeckman, F. A., and Rosenberg, P. A. (1992) Glutathione prevents 2,4,5-trihydroxyphenylalanine excitotoxicity by maintaining it in a reduced, non-active form. *Neurosci. Lett.* **144,** 233–236.

79. Nappi, A. J. and Vass, E. (1994) The effects of glutathione and ascorbic acid on the oxidations of 6-hydroxydopa and 6-hydroxydopamine. *Biochem. Biophys. Acta* **1201,** 498–504.

80. Newcomer, T.A., Rosenberg, P.A., and Aizenman, E. (1995) Iron-mediated oxidation of 3,4-dihydroxyphenylalanine to an excitotoxin. *J. Neurochem.* **64,** 1742–1748.

81. Rodriguez-Lopez, J. N., Banon-Arnao, M., Martinez-Ortiz, F., Tudela, J., Acosta, M., Varon, R., and Garcia-Canovas, F. (1992) Catalytic oxidation of 2,4,5-trihydroxyphenylalanine by tyrosinase: identification and evolution of intermediates. *Biochim. Biophys. Acta* **1160,** 221–228.

82. Lode, H. N., Bruchelt, G., Rieth, A. G., and Niethammer, D. (1990) Release of iron from ferritin by 6-hydroxydopamine under aerobic and anaerobic conditions. *Free Radical Res. Commun.* **11,** 153–158.

83. Kano, K., Mori, T., Uno, B., Goto, M., and Ikeda, T. (1993) Characterization of topa quinone cofactor. *Biochim. Biophys. Acta* **1157,** 324–331.

84. Newcomer, T. A., Palmer, A. M., Rosenberg, P. A., and Aizenman, E. (1993) Nonenzymatic conversion of 3,4-dihydroxyphenylalanine to 2,4,5-trihydroxyphenylalanine and 2,4,5-trihydroxyphenylalanine quinone in physiological solutions. *J. Neurochem.* **61,** 911–920.

85. Skaper, S. D., Negro, A., Facci, L., and Dal-Toso, R. (1993) Brain-derived neurotrophic factor selectively rescues mesencephalic dopaminergic neurons from 2,4,5-trihydroxyphenylalanine-induced injury. *J. Neurosci. Res.* **34,** 478–487.

86. Follenius, E. (1971) Action de la 6 OH dopamine sur les melanophores et sur l'adaptation chromatique chez le poisson teleosteen *Gasterosteus aculeatus. Comptes Rendes Acad. Sci.* **272,** 733–736.

87. Wick, M. M. (1978) Dopamine: a novel antitumor agent active against B-16 melanoma in vivo. *J. Invest. Dermatol.* **71,** 163,164.

88. Wick, M. M., Byers, L., and Ratliff, J. (1979) Selective toxicity of 6-hydroxydopa for melanoma cells. *J. Invest. Dermatol.* **72,** 67–69.

89. Hansson, C., Rorsman, H., Rosengren, E., and Wittbjer, A. (1985) Production of 6-hydroxydopa by human tyrosinase. *Acta Derm. Venereol.* **65,** 154–157.

90. Morrison, M. E., Yagi, M. J., and Cohen, G. (1985) In vitro studies of 2,4-dihydroxyphenylalanine, a prodrug targeted against malignant melanoma cells. *Proc. Natl. Acad. Sci. USA* **82,** 2960–2964.

91. Agrup, P., Carstam, R., Wittbjer, A., Rorsman, H., and Rosengren, E. (1989) Tyrosinase activity in serum from patients with malignant melanoma. *Acta Derm. Venereol.* **69,** 120–124.

92. Passi, S., Picardo, M., and Nazzaro-Porro, M. (1987) Comparative cytotoxicity of phenols in vitro. *Biochem. J.* **245,** 537–542.

93. Metodiewa, D., Reszka, K., and Dunford, H. B. (1989) Oxidation of the substituted catechols dihydroxyphenylalanine methyl ester and trihydroxyphenylalanine by lactoperoxidase and its compounds. *Arch. Biochem. Biophys.* **274,** 601–608.

94. Rossi, A., Petruzzelli, R., and Agro, A. F. (1992) cDNA-derived amino-acid sequence of lentil seedlings' amine oxidase. *FEBS Lett.* **301,** 253–257.

95. Cooper, R. A., Knowles, P. F., Brown, D. E., McGuirl, M. A., and Dooley, D. M. (1992) Evidence for copper and 3,4,6-trihydroxyphenylalanine quinone cofactors in an amine oxidase from the gram-negative bacterium Escherichia coli K-12. *Biochem. J.* **288,** 337–340.

96. Plastino, J. and Klinman, J. P. (1995) Limited proteolysis of Hansenula polymorpha yeast amine oxidase: Isolation of a C-terminal fragment containing both a copper and quinocofactor. *FEBS Lett.* **371,** 276–278.

97. Schilling, B. and Lerch, K. (1995) Amine oxidases from *Aspergillus niger:* Identification of a novel flavin-dependent enzyme. *Biochim. Biophys. Acta* **1233,** 529–537.

98. Brown, D. E., McGuirl, M. A., Dooley, D. M., Janes, S. M., Mu, D., and Klinman, J. P. (1991) The organic functional group in cooper-containing amine oxidases. Resonance Raman spectra are consistent with the presence of topa quinone (6-hydroxydopa quinone) in the active site. *J. Biol. Chem.* **266,** 4049–4051.

99. Greenaway, F. T., O'Gara, C. Y., Marchena, J. M., Poku, J. W., Urtiaga, J. G., and Zou, Y. (1991) EPR studies of spin-labeled bovine plasma amine oxidase: The nature of the substrate-binding site. *Arch. Biochem. Biophys.* **285,** 291–296.

100. Mu, D., Medzihradszky, K. F., Adams, G. W., Mayer, P., Hines, W. M., Burlingame, A. L., Smith, A. J., Cai, D., and Klinman, J. P. (1994) Primary structures for a mammalian cellular and serum copper amine oxidase. *J. Biol. Chem.* **269,** 9926–9932.

101. Mu, D., Janes, S. M., Smith, A. J., Brown, D. E., Dooley, D. M., and Klinman, J. P. (1992) Tyrosine codon corresponds to *topa* quinone at the active site of copper amine oxidases. *J. Biol. Chem.* **267,** 7979–7982.

102. McIntire, W. S. (1994) Quinoproteins. *FASEB J.* **8,** 513–521.

103. Pedersen, J. Z., el-Sherbini, S., Finazzi-Agro, A., and Rotilio, G. (1992) A substrate-cofactor free radical intermediate in the reaction mechanism of copper amine oxidase. *Biochemistry* **31,** 8–12.

104. Dooley, D. M., McGuirl, M. A., Brown, D. E., Turowski, P. N., McIntire, W. S., and Knowles, P. F. A Cu(I)-semiquinone state in substrate-reduced amine oxidases. *Nature* **349,** 262–264.

105. Matsuzaki, R., Suzuki, S., Yamaguchi, K., Fukui, T., and Tanizawa, K. (1995) Spectroscopic studies on the mechanism of the topa quinone generation in bacterial monoamine oxidase. *Biochemistry* **34,** 4524–4530.

106. Janes, S. M. and Klinman, J. P. (1991) An investigation of bovine serum amine oxidase active site stoichiometry: Evidence for an aminotransferase mechanism involving two carbonyl cofactors per enzyme dimer. *Biochemistry* **30,** 4599–4605.

107. Bellelli, A., Finazzi Agro, A., Floris, G., and Brunori, M. (1991) On the mechanism and rate of substrate oxidation by amine oxidase from lentil seedlings. *J. Biol. Chem.* **266,** 20,654–20,657.

108. Dooley, D. M., McIntire, W. S., McGuirl, M. A., Cote, C. E., and Bates, J. L. (1990) Characterization of the active site of *Arthrobacter P1* methylamine oxidase: evidence for copper–quinone interactions. *J. Am. Chem. Soc.* **112,** 2782–2789.

109. Weiss, J. H., Koh, J. Y., and Choi, D. W. (1989) Neurotoxicity of beta-*N*-methylamino-L-alanine (BMAA) and beta-*N*-oxalylamino-L-alanine (BOAA) on cultured cortical neurons. *Brain Res.* **497,** 64–71.

110. Calne, D. B., McGeer, E., Eisen, A., and Spencer, P. (1986) Alzheimer's disease, Parkinson's disease and motoneurone disease: abiotrophic interaction between ageing and environment. *Lancet* **ii,** 1067–1070.

111. Kurland, L. T. (1988) Amyotrophic lateral sclerosis and Parkinson's disease complex on Guam linked to an environmental neurotoxin. *Trends Neurosci.* **11,** 51–54.

112. Meldrum, B. and Garthwaite, J. (1991) Excitatory amino acid neurotoxicity and neurodegenerative disease, in *The Pharmacology of Excitatory Amino Acids* (Lodge, D. and Collingridge, G. L., eds.), A TiPS Special Report, pp. 54–61.

113. Meldrum, B. and Garthwaite, J. (1991) Excitatory amino acid neurotoxicity and neurodegenerative disease. *Trends Pharmacol. Sci.* **11,** 379–387

114. Spencer, P. S., Ludolph, A., Dwivedi, M. P., Roy, D. N., Hugon, J., and Schaumburg, H. H. (1986) Lathyrism: evidence for role of the neuroexcitatory amino acid BOAA. *Lancet* **ii,** 1066–1067.

115. Skaper, S. D., Facci, L., Sciavo, N., Vantini, G., Moroni, F., Dal Toso, R., and Leon, A. (1992) Characterization of 2,4,5-trihydroxyphenylalanine neurotoxicity *in vitro* and protective effects of ganglioside GM1: Implications for Parkinson's disease. *J. Pharmacol. Exp. Ther.* **263,** 1440–1446.

116. Murase, K., Hattori, A., Kohno, M., and Hayashi, K. (1993) Stimulation of nerve growth factor synthesis/secretion in mouse astroglial cells by coenzymes. *Biochem. Mol. Biol. Int.* **30,** 615–621.

2-Chloroethylamines (DSP4 and Xylamine)
Toxic Actions on Noradrenergic Neurons

Guillermo Jaim-Etcheverry

1. CHARACTERIZATION OF 2-CHLOROETHYLAMINES (2-CEAS)

In the early 1970s, the group of Svante Ross in Sweden investigated a series of compounds chemically related to the adrenergic neuron blocker bretylium in an attempt to identify those that could cross the blood–brain barrier and, once in the brain, cyclize to form quaternary ammonium derivatives. Although some of these tertiary haloalkylamines had the predicted neuron-blocking activity, one of these compounds, N-(2-chloroethly)-N-ethyl-2-bromobenzylamine (DSP4), whose chemical structure is shown in Fig. 1, lacked this property, but produced a long-lasting inhibition of norepinephrine (NE) uptake by brain slices when injected to rodents (1–3). Xylamine, a closely related molecule, which was originally described by Kreuger and Cook (4) and is also shown in Fig. 1, was extensively investigated by Arthur Cho and his group in Los Angeles and found to have essentially similar properties to those of DSP4 (5–8).

Apart from irreversibly blocking NE uptake in neurons containing NE, these 2-CEA compounds markedly deplete endogenous NE both in the peripheral and the central nervous system (3,9–11). Although this reduction of NE is transitory in the periphery, it is long-lasting in the CNS.

Since the beginning of the 1980s, DSP4 has been mainly used to study the influence of the reduced function of noradrenergic neurons on behavioral parameters as well as the biochemistry and pharmacology of the adrenergic synapse. Xylamine, on the other hand, has been mostly used to explore the molecular interaction of these compounds with the catecholamine transport system of adrenergic terminals.

The major aspects of the action of 2-CEA compounds have been extensively reviewed (12,13), and therefore, only the essential aspects of their biological activity, mechanism of action, and major applications will be discussed.

From: Highly Selective Neurotoxins: Basic and Clinical Applications *Edited by: R. M. Kostrzewa.*
Humana Press Inc., Totowa, NJ

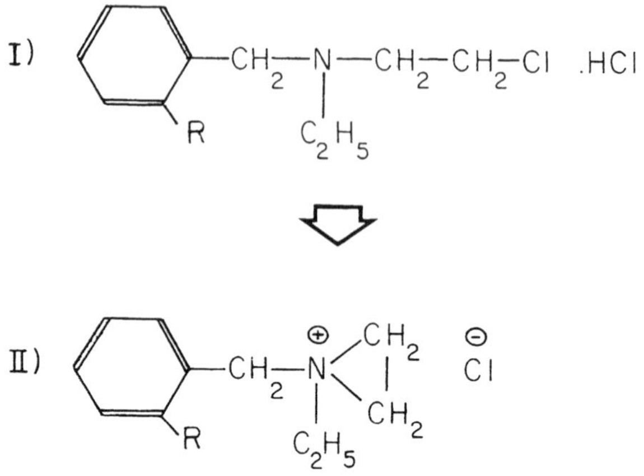

Fig. 1. Chemical structure of the 2-CEAs DSP4 and xylamine. The parent compounds (I) cyclize in solution to the corresponding quaternary aziridinium ion (II), which is the active toxic compound. This in turn may be hydrolyzed to yield a 2-hydroxyethylamine, and a dimeric piperazinium compound is also formed between the parent amine and its aziridinium ion (reproduced from ref. *12*).

2. BIOLOGICAL EFFECTS OF 2-CEAS

As already mentioned, the most remarkable actions produced by these 2-CEA compounds, either in vitro or after injection, are the strong inhibition of NE uptake and the depletion of endogenous NE in both peripheral and central noradrenergic neurons. The characteristics of these actions have led to the conclusion that these effects result from a toxic action on the neurons.

2.1. Impairment of NE Uptake

When injected systemically into rodents, 2-CEA compounds markedly inhibit NE uptake by central and peripheral noradrenergic neurons. This effect is dose-related and appears very rapidly, i.e., 1 h after injection. The inhibition of NE uptake by cortical synaptosomes has an ED_{50} of 20 mg/kg ip, but maximal effects are achieved with 100 mg/kg ip. This uptake-inhibiting effect, which is also demonstrable in vitro, is blocked by pretreating the animal with NE uptake blockers, such as desipramine (DMI) or with amphetamine, but is unaffected by reserpine *(3,9–11)*.

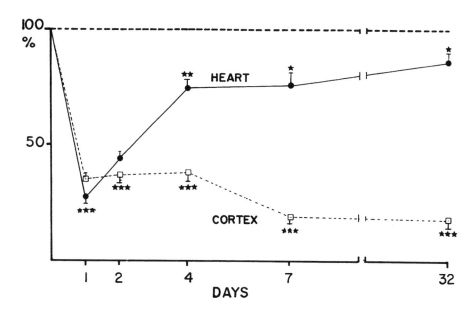

Fig. 2. Content of endogenous NE of the heart and the cerebral cortex of adult rats killed at different times after a single injection of DSP4 (50 mg/kg ip). The results are expressed as percentages of untreated controls ± SE. *$P < 0.05$; **$P < 0.01$; ***$P < 0.001$ (reproduced from ref. *12*).

2.2. Depletion of Endogenous NE

As was also mentioned, apart from inhibiting NE uptake by central and peripheral noradrenergic neurons, these compounds produce when injected a rapid depletion of endogenous NE in these neurons. Both their capacity to take up NE and their endogenous stores gradually recover in the periphery, whereas in the central nervous systems, their actions are long-lasting, as shown in Fig. 2 *(3,9–11)*. Although NE levels are diminished for at least 8 mo in certain brain areas *(3)*, in some of them there is a tendency to recovery with time *(14,15)*.

The administration of DSP4 increases turnover rate in the brain regions that show the greatest depletion of NE, such as in the cortex and the hippocampus, i.e., those areas innervated by neurons in the locus ceruleus (LC) *(15,16)*. Exposure of brain slices in vitro to DSP4 produces a rapid and marked depletion of endogenous NE levels, most probably as a result of the continued stimulation of NE release *(17)*. Although after in vivo administration DSP4 does not modify catecholamines in the adrenal medulla, the compound interacts with adrenomedullary cells in culture: it inhibits NE reuptake, stimulates NE release, and produces morphological changes in these cells *(18)*.

As will be discussed below, DSP4 and xylamine act as precursors of an active aziridinium compound. This is confirmed by the finding that similar effects on peripheral noradrenergic neurons are produced by the injection of the aziridinium derivative of DSP4 or of xylamine, although no changes are observed in brain NE when the ion is injected, because it does not cross the blood–brain barrier. However, if this barrier is bypassed by injecting the aziridinium derivative directly into the brain, the effect on noradrenergic neurons is similar to that produced by the parent compound *(19)*.

2.3. Evidence for Neurotoxicity of 2-CEAs

The biochemical effects of 2-CEA compounds on noradrenergic neurons have suggested the possibility that they act as neurotoxins. This has been much debated and the real existence of a true degenerative action of 2-CEA compounds on these neurons has been questioned (for references, *see 20*).

However, the existence of such a neurotoxic effect is supported by the finding of a marked decrease in the activity of another biochemical marker of noradrenergic neurons: the enzyme dopamine-β-hydroxylase (DβH), which synthesizes NE. The activity of this enzyme decreases in the brain and the heart after a single injection of DSP4 (50 mg/kg ip), an effect that is also prevented by pretreatment with NE uptake blockers, such as DMI. Although in the heart enzyme activity decreases after a lag period of 2–4 d and recovers after 1–2 wk, in the brain, DβH was significantly reduced 8 mo after treatment *(3)*.

Several morphological studies have been performed to characterize this postulated neurodegenerative action of 2-CEA compounds. After DSP4 injection, the histochemical fluorescence for catecholamines disappears very rapidly in the terminal arborizations of peripheral sympathetic neurons, such as those in the iris and the atria, but NE accumulates proximally, as is characteristically observed in neuronal degenerative processes. As is the case for NE levels, NE uptake, and DβH activity, this fluorescence recovers with time. Fluorescent nerve terminals are also lost from brain areas receiving NE innervation, and NE accumulates in preterminal swollen axons. Although in many regions of the brain the lost fluorescent terminals do not seem to recover, in others, reinnervation seems to take place, confirming the results of biochemical studies *(10,11,15)*.

It has been suggested that DSP4 does not destroy NE terminals, but may produce instead an intraneuronal lesion that leads to the accumulation of TH and the depletion of NE within the terminal fields *(21)*. However, it has been possible to demonstrate that there are two different phases in the response of central NE axons to DSP4 administration: an acute phase characterized by the precipitous loss of transmitter and a neurodegenerative phase in which DβH is lost and the NE axons are structurally disintegrated. Confirming these toxic effects, degenerating NE axons have been observed a few days after treatment, and glial fibrillary acidic protein immunoreactivity greatly increases throughout the brain with a peak 7 d after DSP4 injection *(22)*.

As is the case with other neurotoxic compounds, the projections originating from neurons in the LC are those most affected by DSP4 *(23)*. This selectivity of DSP4 toward LC neurons seems to be related to the significant difference found in the affinity of DSP4 for the NE uptake carrier in synaptosomes obtained from areas innervated by these neurons in comparison with that found in synaptosomes of, for example, the hypothalamus *(24)*.

Although the terminals of neurons in the LC appear to be the target of DSP4 action, the cell bodies of these neurons seem to be spared as also are those in peripheral ganglia *(25)*. However, although the cell bodies and preterminal axons in the LC are not markedly affected immediately after a single injection of DSP4, there is a slow and progressive loss of noradrenergic cell bodies in the area: 30% of LC neurons disappear within 2 mo (roughly 20% are eliminated in 2 wk), and an additional 30% are lost 1 yr after injection. This time-course suggests that the cell bodies are lost as a result of a retrograde degeneration following the destruction of their terminal arborizations, but surviving LC neurons exhibit a vigorous regenerative response, which leads to a progressive restoration of the NE innervation pattern in the forebrain. The pattern of this compensatory reinnervation seems to be controlled by the nature of the target area because it is region-specific: the forebrain is markedly reinnervated while this process is very limited in the brainstem, the cerebellum, or the spinal cord (for discussion, *see 20*).

2.4. 2-CEAs and Developing Noradrenergic Neurons

Similarly to what happens with other neurotoxic compounds that affect the NE-containing neurons, such as 6-OHDA or its precursor amino acid 6-OHDOPA, DSP4 injected to newborn rodents alters the subsequent development of noradrenergic neurons (for description of these effects, *see 9,10,26*). DSP4 injected during the first days after birth produces biochemical and morphological changes that are compatible with an NE hyperinnervation of the brainstem and of the cerebellum accompanied by a destruction of the terminal fields in the cerebral cortex and the spinal cord. These changes, which are dose-dependent and persistent, can be prevented by pretreatment with DMI or with pargyline.

The fact that DSP4 crosses the placenta makes possible the study of its interaction with noradrenergic neurons at the prenatal stage. The changes are similar to those produced by injecting DSP4 at birth, and in this case, the timing of the injection is critical in determining the nature and extension of the long-term changes.

Although 6-OHDA given at birth destroys numerous NE-containing cell bodies in peripheral sympathetic ganglia and produces a partial, but permanent peripheral sympathectomy, DSP4 lacks this effect probably because of a lower neurotoxic potency. Thus, 2-CEA compounds interact with NE-containing neurons at the level of their uptake system because their effects are preventable by NE uptake blockers. This initial action triggers a series of changes in these neurons, which apparently lead to the destruction of their terminal fields. This process seems to be reversible only in the periphery most probably because of differences between the regenerative capacity of peripheral and central noradrenergic neurons. The ability of DSP4 or xylamine to reduce central NE by a single ip injection has converted these compounds into useful tools for studying the contribution made by central noradrenergic systems to complex behavioral phenomena as well as to endocrine regulatory mechanisms.

3. SPECIFICITY AND POTENCY

In most of the studies in which DSP4 has been used as a tool to study the effects of the impairment of central noradrenergic neurons on behavioral or neurochemical parameters, the compound has been injected in doses that range between 50 and 100 mg/kg ip. It has also been injected intracerebrally, intraventricularly, and intrathecally. For in vivo studies with xylamine, mostly doses in the range of 10–50 mg/kg were used.

DSP4 does not modify the endogenous levels of brain adrenaline, striatal dopamine (DA) or cortical acetylcholine, GABA, glutamic acid, glycine, or aspartic acid. It slightly depletes DA in the hippocampus and the cerebellum, but this effect seems to be related to the presence of DA as a precursor in noradrenergic neurons (for references, *see 12,13*). When exposed to synaptosomal preparations, unlike what happens in the intact animal, xylamine exhibits selectivity toward DA terminals in the striatum *(27)*.

In the rat brain, neurons containing serotonin are affected by DSP4 as shown by a 20–30% reduction of serotonin levels and uptake. This effect, which is not observed in mice, is greater in adult rats treated at birth with DSP4 and can be prevented by pretreating the animals with the specific serotonin uptake blocker zimelidine without interfering with the actions of DSP4 or noradrenergic neurons *(26)*.

Since it is chemically related to phenoxybenzamine, DSP4 blocks α-adrenergic receptors with a 10 times lower potency. This effect is reversible and much less sustained than the inhibition of NE uptake, because it disappears 6 h after injecting the compound *(1,3)*.

DSP4 apparently interacts with opiate receptors. The binding of 2-CEA compounds competes with naloxone for opiate binding sites and, like opiates, DSP4 reduces leuteinizing hormone secretion in the rat, an effect prevented by naloxone, but not by DMI *(28)*.

4. MOLECULAR MECHANISMS OF ACTION OF 2-CEAS

2-CEA compounds are alkylating agents that form covalent bonds with electrophilic centers on or near their site of action. The effects of these one-arm mustards are irreversible and involve an initial ring closure of the neutral amine to the corresponding aziridinium ion (Fig. 1). 2-CEA compounds are very rapidly transformed in aqueous solutions: DSP4 has a half-life of 7 min at 37°C and pH 7.4 *(1)*. The major species derived from the parent compounds are the aziridinium ion, its hydrolysis product, and a piperazinium dimer formed between the parent nitrogen mustard and the aziridinium ion (*see* description in *12,29*). This aziridinium derivative is the active compound that strongly reacts with nucleophilic functions, such as thiols and amines. The binding of the aziridinium ion to a cellular site determines the formation of a covalent bond with a neighboring nucleophilic function, and this disrupts the normal activity of the binding site.

These compounds, which show specificity for the NE transport system, exert their action on adrenergic synapses through a mechanism that has been studied both in vivo and in vitro. In summary, 2-CEA compounds bind selectively to the uptake system for NE and irreversibly inhibit the transporter, i.e., they require a functional transporter to produce their effects. This is why their actions are prevented by blocking this transport system with, for example, DMI. The characteristics of the molecular interactions between 2-CEA compounds and the NE transport system have been reviewed in detail *(12,13)*. In vitro studies carried out with radiolabeled xylamine or, more properly, with its stable aziridinium derivative, essentially show that the compound labels a 55-kDa peptide that seems to be associated with NE uptake.

The precise mechanism by which these changes in the adrenergic synapse produced by 2-CEA compounds trigger the depletion of endogenous NE is not definitively established apart from the known fact that the aziridinium ion is the active species. Pretreatment with inhibitors of monoamine oxidase (MAO) activity, particularly that of MAO-B, block the NE depletion from rat brain produced by DSP4 *(30,31)*. DSP4 has also been shown to have the capacity to inhibit MAO activity *(32)*. Since DSP4 is not inactivated by MAO, the protective effect of an MAO-B inhibitor, such as *l*-deprenyl, might be the result of an interference with the interaction of DSP4 with the NE uptake site. Recent studies using deprenyl and several other MAO-B inhibitors suggest that the mechanisms underlying their neuroprotective effects may be more complex than anticipated *(33–36)*. There is also some evidence that the depleting actions of 2-CEA compounds may involve mitochondria, which, in turn, via the production of ATP, may interfere with the storage capacity of the neuron.

5. APPLICATIONS OF 2-CEA IN RESEARCH

2-CEA compounds produce marked changes in the dynamics of adrenergic synapses leading to the reduction of intraneuronal levels of NE and of the amount released into the synaptic cleft. These actions together with several properties of these compounds make them particularly suitable for chemically lesioning noradrenergic neurons: their ability to cross the blood–brain barrier when injected systemically in adult rodents, their marked selectivity for noradrenergic neurons, and their good general tolerance. The capacity of DSP4 to cross the placenta in pregnant animals has also proven to be of great usefulness to study the ontogenesis of noradrenergic neurons.

Moreover, as a result of the alterations in the content and release of NE at the adrenergic synapse, postsynaptic receptors undergo adaptive changes that have been the subject of many investigations (for references, *see 13*). Thus, animals treated with either DSP4 or xylamine constitute interesting models to analyze the mechanisms and significance of the up- and downregulation of adrenergic receptors. These actions may be relevant to study the basic mechanisms underlying mental diseases, such as depression, in which these processes might be involved.

The characteristics of the complex response of brain adrenergic innervation to 2-CEA compounds provide an interesting model to study the biochemical and morphological phenomena that accompany the regenerative responses in the central nervous system both in newborn and in adult animals.

Because of their affinity for the NE uptake transporter to which they are irreversibly bound, 2-CEA compounds and xylamine, in particular, have been useful in the characterization of the molecular mechanisms by which NE is taken up by adrenergic neurons, a process that has a great biological and therapeutic significance.

From the analysis of the effects of 2-CEA compounds, it can be concluded that this family of site-directed alkylating agents has a great pharmacological potential for manipulating the function of neurons, and the more precise characterization of their actions will most probably lead not only to the development of new compounds, but also to useful therapeutic applications.

REFERENCES

1. Ross, S. B., Johansson, J. G., Lindborg, B., and Dahlbom, R. (1973) Cyclizing compounds. Tertiary *N*-(2-bromobenzyl)-*N*-haloalkylamines with adrenergic blocking action. *Acta Pharm. Suec.* **10,** 29–42.
2. Ross, S. B. and Renyi, A. L. (1976) On the long-lasting inhibitory effect of *N*-(2-chloroethyl)-*N*-ethyl-2-bromobenzylamine (DSP-4) on the active uptake of noradrenaline. *J. Pharm. Pharmacol.* **28,** 458,459.
3. Ross, S. B. (1976) Long-term effects of *N*-2-chloroethyl-*N*-ethyl-2-bromobenzyl-amine hydrochloride on noradrenergic neurones in the rat brain and heart. *Br. J. Pharmacol.* **58,** 521–527.
4. Kreuger, C. A. and Cook, D. A. (1975) Synthesis and adrenergic blocking properties of some alkylating analogs of bretylium. *Arch. Int. Pharmacodyn. Ther.* **218,** 96–115.
5. Kammerer, R. C., Amiri, B., and Cho, A. K. (1979) Inhibition of uptake of catecholamines by benzylamine derivatives. *J. Med. Chem.* **22,** 352–355.

6. Cho, A. K., Ranson, R. W., Fischer, J. B., and Kammerer, R. C. (1980) The effects of xylamine, a nitrogen mustard, on [³H] norepinephrine accumulation in rabbit aorta. *J. Pharmacol. Exp. Ther.* **214**, 324–327.

7. Fischer, J. B. and Cho, A. K. (1982) Inhibition of [³H] norepinephrine uptake in organ cultured rat superior cervical ganglia by xylamine. *J. Pharmacol. Exp. Ther.* **220**, 115–119.

8. Fischer, J. B., Waggaman, L. A., Ransom, R. W., and Cho, A. K. (1983) Xylamine, an irreversible inhibitor of norepinephrine uptake is transported by the same uptake mechanism in cultured rat superior cervical ganglia. *J. Pharmacol. Exp. Ther.* **226**, 650–655.

9. Jaim-Etcheverry, G. and Zieher, L. M. (1980) DSP-4: a novel compound with neurotoxic effects on noradrenergic neurons of adult and developing rats. *Brain Res.* **188**, 513–523.

10. Jonsson, G., Hallman, H., Ponzio, F., and Ross, S. (1981) DSP4 (*N*-(2-chloroethyl)-*N*-ethyl-2-bromobenzylamine)—a useful denervation tool for central and peripheral noradrenaline neurons. *Eur. J. Pharmacol.* **72**, 173–188.

11. Dudley, M. W., Butcher, L. L., Kammerer, R. C., and Cho, A. K. (1981) The actions of xylamine on central noradrenergic neurons. *J. Pharmacol. Exp. Ther.* **217**, 834–840.

12. Jaim-Etcheverry and Zieher, L. M. (1983) 2-chloroethylamines: new chemical tools for the study of the noradrenergic neurons. *Trends Pharmacol. Sci.* **4**, 473–475.

13. Dudley, M. W., Howard, B. D., and Cho, A. K. (1990) The interaction of the beta-haloethyl benzylamines, xylamine, and DSP-4 with catecholaminergic neurons. *Ann. Rev. Pharmacol. Toxicol.* **30**, 387–403.

14. Wolfman, C., Abó, V., Calvo, D., Medina, J., Dajas, F., and Silveira, R. (1994) Recovery of central noradrenergic neurons one year after the administration of the neurotoxin DSP4. *Neurochem. Int.* **25**, 395–400.

15. Hallman, H., Sundstrom, E., and Jonsson, G. (1984) Effects of the noradrenaline neurotoxin DSP4 on monoamine neurons and their transmitter turnover in rat CNS. *J. Neural Transm.* **60**, 89–102.

16. Logue, M. P., Growden, J. H., Coviella, I. L., and Wurtman, R. J. (1985) Differential effects of DSP4 administration on regional brain norepinephrine turnover in rats. *Life Sci.* **37**, 403–409.

17. Landa, M. E., Rubio, M. C., and Jaim-Etcheverry, G. (1984) The neurotoxic compound *N*-(2-chloroethyl)-*N*-ethyl-2-bromobenzylamine hydrochloride (DSP4) depletes endogenous norepinephrine and enhances release of [³H] norepinephrine from rat cortical slices. *J. Pharmacol. Exp. Ther.* **231**, 131–136.

18. Boksa, P., Aitken, D., and Meaney, M. (1989) Effects of catecholaminergic neurotoxin *N*-(2-chloroethyl)-*N*-ethyl-2-bromobenzylamine (DSP-4) on adrenal chromaffin cells in culture. *Biochem. Pharmacol.* **38**, 1491–1498.

19. Zieher, L. M. and Jaim-Etcheverry, G. (1980) Neurotoxicity of *N*-(2-chloroethyl)-*N*-ethyl-2-bromobenzylamine hydrochloride (DSP 4) on noradrenergic neurons is mimicked by its cyclic aziridinium derivative. *Eur. J. Pharmacol.* **65**, 249–256.

20. Fritschy, J. M. and Grzanna, R. (1992) Restoration of ascending noradrenergic projections by residual locus coeruleus neurons: compensatory response to neurotoxin-induced cell death in the adult rat brain. *J. Comp. Neurol.* **321**, 421–441.

21. Booze, R. M., Hall, J. A., Cress, N. M., Miller, S. D., and Davis, J. N. (1988) DSP-4 treatment produces abnormal tyrosine hydroxylase immunoreactivity in rat hippocampus. *Exp. Neurol.* **101**, 75–86.

22. Fritschy, J. M., Geffard, M., and Grzanna, R. (1990) The response of noradrenergic axons to systemically administered DSP-4 in the rat: an immunohistochemical study using antibodies to noradrenaline and dopamine-β-hydroxylase. *J. Chem. Neuroanat.* **3,** 309–323.

23. Grzanna, R. Berger, U., Fritschy, J. M., and Geffard, M. (1989) The acute action of DSP-4 on central norepinephrine axons: biochemical and immunohistochemical evidence for differential effects. *J. Histochem. Cytochem.* **37,** 1435–1442.

24. Zaczek, R., Fritschy, J. M., Culp, S., De Souza, E. B., and Grzanna, R. (1990) Differential effects of DSP-4 on noradrenaline axons in cerebral cortex and hypothalamus may reflect heterogeneity of noradrenaline uptake sites. *Brain Res.* **522,** 308–314.

25. Fritschy, J. M. and Grzanna, R. (1989) Immunohistochemical analysis of the neurotoxic effects of DSP-4 identifies two populations of noradrenergic axon terminals. *Neuroscience* **30,** 181–197.

26. Jonsson, G., Hallman, H., and Sundstrom, E. (1982) Effects of the noradrenaline neurotoxin DSP4 on the postnatal development of cental noradrenaline neurons in the rat. *Neuroscience* **7,** 2895–2908.

27. Liang, P., Chang, S. S., and Cheng, J. T. (1993) Inhibitory effect of xylamine on the uptake of [³H] dopamine into isolated striatal synaptosomes of the rats. *Neurosci. Lett.* **153,** 181–184.

28. Wilkinson, M., Jacobson, W., and Wilkinson, D. A. (1985) The noradrenergic neurotoxins DSP4 and xylamine bind to opiate receptors. *Brain Res. Bull.* **14,** 493–495.

29. Ranson, R. W., Kammerer, R. C., and Cho, A. K. (1982) Chemical transformations of xylamine (*N*-2′-chloroethyl-*N*-ethyl-2-methylbenzylamine) in solution. Pharmacological activity of the species derived from this irreversible norepinephrine uptake inhibitor. *Mol. Pharmacol.* **21,** 380–386.

30. Hallman, H. and Jonsson, G. (1984) Pharmacological modifications of the neurotoxic action of the noradrenaline neurotoxin DSP-4 on central noradrenaline neurons. *Eur. J Pharmacol.* **103,** 269–278.

31. Gibson, C. J. (1987) Inhibition of MAO B, but no MAO A, blocks DSP-4 toxicity on central NE neurons. *Eur. J. Pharmacol.* **141,** 135–138.

32. Lyles, G. A. and Callingham, B. A. (1981) The effects of DSP-4 on monoamine oxidase activities in tissues of the rat. *J. Pharm. Pharmacol.* **33,** 632–638.

33. Finnegan, K. T., Skratt, J. S., Irwin, I., DeLanney, L. E., and Langston, J. W. (1990) Protection against DSP-4-induced neurotoxicity by deprenyl is not related to its inhibition by MAO-B. *Eur. J. Pharmacol.* **184,** 119–126.

34. Finnegan, K. T. (1993) Neurotoxins and monoamine oxidase inhibition. *Mov. Disord.* **8 (Suppl. 1),** S14–S19.

35. Yu, P. H., Davis, B. A., Fang, J., and Boulton, A. A. (1994) Neuroprotective effects of some monoamine oxidase-B inhibitors against DSP-4-induced noradrenaline depletion in the mouse hippocampus. *J. Neurochem.* **63,** 1820–1828.

36. Zhang, X., Zuo, D. M., and Yu, P. H. (1995) Neuroprotection by *R*(−)-deprenyl and *N*-2-hexyl-*N*-methylpropargylamine on DSP-4, a neurotoxin, induced degeneration of noradrenergic neurons in rat locus coeruleus. *Neurosci. Lett.* **186,** 45–48.

Joyce E. Royland and J. William Langston

1. INTRODUCTION AND BACKGROUND

1-Methyl-4-phenyl-1,2,3,6-tetrahydropyridine (MPTP) is one of the most selective neurotoxins. However, this is not the sole reason it has gained widespread recognition. Rather, the spectacular way that its biologic effects were first recognized, which was a direct result of the clandestine manufacture of illicit drugs *(1)*, thrust this previously obscure compound squarely into the limelight. Although in retrospect at least one earlier case had been reported *(2)*, a direct association between exposure to MPTP and central nervous system damage was not made until the early 1980s, when a group of young heroin abusers in northern California mysteriously developed parkinsonism after using a new "designer drug" known as China White. The cause of their clinical syndrome was eventually traced back to a contaminant of the "synthetic heroin," which proved to be MPTP *(3)*.

The physical findings in these young addicts, which included bradykinesia, resting tremor, and rigidity *(4–6)*, were strikingly similar to those seen in idiopathic Parkinson's disease, a neurodegenerative disorder typically seen in the elderly. Their shuffling gait, loss of postural control, shaking hands, and facial "mask" gave them an appearance that was nearly indistinguishable from classical Parkinson's disease *(5)*. At first there was some controversy regarding whether or not a true parkinsonian rest tremor was present in these individuals, but electrophysiologic studies demonstrated a tremor pattern that was identical to that seen in the idiopathic disease *(7)*. Even the less obvious symptoms of Parkinson's disease, such as seborrhea, hypophonia, and micrographia, were seen in these young addicts.

Once MPTP was recognized as the offending agent, a second cohort of addicts was identified in which the exposure levels were much lower. On motor assessment scales, these individuals scored in a range that was very similar to patients with early Parkinson's disease *(8)*. Interestingly, positron emission tomography (PET) scans in this group revealed ^{18}F-labeled L-dihydroxypheny-

From: Highly Selective Neurotoxins: Basic and Clinical Applications *Edited by: R. M. Kostrzewa.*
Humana Press Inc., Totowa, NJ

lalanine (L-DOPA) uptake to be decreased compared to controls, but still greater than in Parkinson's disease *(9,10)* and more severely affected MPTP patients *(11)*. Examination of cognitive abilities showed both mildly and severely affected MPTP-exposed individuals to have a pattern of subtle defects similar to that seen in nondemented Parkinson's patients *(12,13)*. Thus, it is now well accepted that MPTP can induce the full range of symptoms seen throughout the course of Parkinson's disease, and that it may be possible to use this model to test the usefulness of PET scans and cognitive examinations to aid in diagnosis of early asymptomatic disease.

In addition to manifesting all the classical symptoms of Parkinson's disease, MPTP patients responded to the standard antiparkinsonian therapy in a manner that was highly analogous to that which is seen in the idiopathic disease. Even more remarkable was the fact that they also suffered virtually all of the typical side effects of therapy, including end-of-dose wearing off, the "on/off" fluctuations, dyskinesias, and hallucinations *(4)*. These observations were among the first to indicate that MPTP had tremendous potential as a model for drug development and for the study of the side effects of therapy as well.

Information on the human pathology of MPTP is limited. In the one patient who died with MPTP-induced parkinsonism, changes in the substantia nigra were very similar to that seen with advanced Parkinson's disease *(2,3)*. Neuromelanin, a pigment found in high concentrations within dopamine (DA) neurons in the substantia nigra (SN), was found extracellularly and in microglia cells, an indication of DA neuronal cell death. Another major postmortem diagnostic feature of Parkinson's disease is the presence of eosinophilic intraneuronal inclusions known as Lewy bodies. A Lewy body-like structure was described in this patient, but classical Lewy bodies were not present.

The discovery of a simple molecule that so faithfully replicated most, if not all of the features of a major neurodegenerative disease in humans quickly led to an explosion of basic research. This chapter will review what has been learned about MPTP over the ensuing years since the discovery of its biologic effects, the implications of these findings, and some possibilities for future research. We need to point out, however, that the factors that make MPTP such a valuable model for Parkinson's disease and many of the broader questions relating to neurodgenerative disease make it difficult to cover the entire literature on this subject in a single review. Thus, only selected areas will be covered here, with an emphasis on more recent discoveries, or those areas that are either controversial or deemed most promising for new research. Those facets of MPTP neurotoxicity not covered (or only briefly reviewed) will doubtless provide material for future chapters.

2. ANIMAL MODELS AND SPECIES VARIABILITY

Although self-administration of MPTP had tragic consequences for a group of young drug abusers, it also gave birth to the first good model of Parkinson's disease. That the toxic effects of MPTP were recognized first in humans has provided a rare opportunity to compare observations in humans with those in experimental animals. That MPTP so closely mimics the symptoms and pathology of Parkinson's disease makes correlating experimental models to the human condition that much easier. Although MPTP toxicity has been used to mimic parkinsonism in species as diverse as cows and goldfish, by far the two most frequently used models are rodents and primates. Rodents are easy to house and relatively inexpensive, making them an obvious choice for experimental use. However, primates provide the only complete behavioral model for the disease. This section focuses on the differences in response to MPTP between rodents and primates, and the problems or benefits of working with each model.

2.1. Rodent Studies

Rodents treated with MPTP differ substantially from primates in a number of ways. For example, except under stressed conditions *(14)*, they typically do not exhibit behavioral deficits reminiscent of Parkinson's disease. Furthermore, the neurochemical deficits induced by MPTP tend to recover with time *(15)*. One of the most intriguing (and at times vexing) differences between these species relates to sensitivity to MPTP. In contrast to primates, rodents tolerate very high doses of MPTP (up to 10 times higher than monkeys on a body weight basis). Also, it is much more difficult to induce neuronal degeneration in rodents using systemically administered MPTP *(16–20)*. Among rodents, mice (particularly C57BL/6 mice) appear to be more sensitive to systemically administered MPTP than rats *(21,22)*. MPTP has been shown to damage the striatal DA terminals *(15)* and to decrease tyrosine hydroxylase immunoreactivity in the striatum *(23–25)* of mice. However, inducing neuronal degeneration in mice has proven to be somewhat more challenging. Although the majority of earlier studies failed to demonstrate neuronal death after MPTP, except in older animals *(26)*, more recent studies using very sensitive counting techniques have shown a mild degree of neuronal depletion in the SN of MPTP-treated mice *(27)*.

2.2. Primate Studies

In contrast to rodents, primates are exquisitely sensitive to the neurodegenerative effects of MPTP and also provide a faithful behavioral model of Parkinson's disease. As such, this model is invaluable for screening new drugs, determining new ways to evaluate disease progression, and for research into fetal or cloned cell transplants.

In monkeys, there are two phases of MPTP toxicity. The first appears to be an acute, short-term, pharmacologic effect lasting a few hours *(28)*. This is followed by a chronic syndrome consisting of bradykinesia, postural abnormalities, hypophonia, and freezing episodes *(29–34)*. With the possible exception of resting tremor, which has been described in only the African green monkey *(35)*, all the classical signs of Parkinson's disease are manifested in these animals. MPTP-treated monkeys also respond to commonly used medications in Parkinson's disease, including L-DOPA and DA agonists *(29,30)* in a manner similar to patients with Parkinson's disease, including most if not all of the side effects of long-term treatment with L-DOPA *(36,37)*. Deficits in perceptual motor activity *(38–40)* even suggest cognitive impairment similar to that seen in MPTP-exposed humans and patients with Parkinson's disease.

There is some question concerning whether the putamen is more affected than caudate in MPTP-lesioned monkeys, as it is an idiopathic Parkinson's disease *(41)*. Earlier reports found the caudate to have greater DA depletion *(42,43)*, whereas later reports found the reverse *(28,44)*. A recent study by Wüllner et al. *(45)* in *macaca fascicularis* monkeys, in which DA transporter (DAT) levels were monitored by PET, supported a pattern of putamen-to-caudate sensitivity that was similar to idiopathic Parkinson's disease. Why there has been so much disparity between investigators in regard to this important point is unclear, but factors, such as species differences or the use of different dosing regimens, may play a role. In any case, the question remains an important one because of its relevance to Parkinson's disease.

One pathological consequence of Parkinson's disease that has not been unequivocally demonstrated in MPTP animal models is the presence of Lewy bodies. Lewy bodies are intraneuronal inclusions seen in several neurodegenerative diseases. Several classes of proteins have been identified in Lewy bodies, including epitopes of cytoskeletal proteins, microtubule proteins, and ubiquitin, a protein believed to play a role in nonlysosomal degradation of short-lived and aberrant proteins *(46–50)*. Although typical Lewy bodies have not been seen in MPTP-treated animals, round eosinophilic inclusions have been found in old monkeys treated with the compound *(51,52)*. These inclusions are distributed in a manner similar to Lewy bodies in the parkinsonian brain, and it has been suggested that they represent an immature form of the Lewy body, which, if given time, might develop into the classic form.

3. UPTAKE, STORAGE, DISTRIBUTION, AND ELIMINATION

Systemically administered MPTP is initially distributed widely throughout the body, after which it is rapidly eliminated from all organs except the eye *(53–55)*. Although it crosses the blood–brain barrier *(56)*, MPTP itself does not appear to be toxic *(57)*; rather, the compound appears to require biotransformation into a toxic metabolite via a two-step process involving monoamine oxidase (MAO) *(58–60)*. MAO mediates the first step, conversion of MPTP to the 2,3-dihydropyridinium intermediate (MPDP+) *(61)*. MPDP+ is then thought to oxidize spontaneously, forming the 1-methyl-4-phenylpyridinium ion (MPP+) *(58)*. The oxidation of MPTP to MPTP+ can occur by both the A and B isoforms of MAO, but MPTP is a better substrate for MAO B *(60)*, and MAO B is generally thought to be the main mediator of MPTP metabolism.

Many studies support the hypothesis that MPP+ is the primary toxic metabolite of MPTP *(53,54)*. MPP+ is toxic in tissue culture *(62–64)* and produces long-term DA depletion after intraventricular injection in rodents *(65,66)*. Furthermore, blocking the biotransformation of MPTP to MPP+ by inhibiting MAO B is very effective in preventing toxicity after systemic administration of MPTP *(21,31,32,67)*.

MPP+ is transformed and accumulated in the periphery as well as in the central nervous system (CNS). However, MPP+ formed in the periphery does not contribute to brain toxicity, since it cannot readily pass the blood–brain barrier because of its charged configuration. This observation may provide one of the explanations concerning why rodents are less susceptible to CNS toxicity, since MAO B is found in much higher levels in the cerebrocapillary endothelium of rats than in primates *(68)*. By converting MPTP to its charged metabolite at the cerebrocapillary endothelium, which constitutes the blood–brain barrier, ingress of MPTP/MPP+ into the CNS would be effectively blocked.

One observation that has puzzled investigators from the beginning is that MPP+ concentrations in the adrenal medulla are 5–10 times greater than in other tissues *(55)*, yet this organ is relatively resistant to the effects of MPTP both in vivo and in vitro. Studies on cultured adrenal chromaffin cells have shown active uptake and storage of MPP+ by chromaffin vesicles *(69,70)*. These investigators suggested that sequesting MPP+ into cytoplasmic vesicles may actually be protective. In support of this hypothesis is the observation that reserpine, which disrupts vesicular storage, potentiates MPTP-induced DA depletion in the striatum *(71)*.

Since MPP+ cannot readily enter the CNS, most if not all of this toxic metabolite must be generated from MPTP that has already crossed the blood–brain barrier. Originally it was suggested that the conversion of MPTP to MPP+ took place within the target neuron—i.e., DA neurons of the nigrostriatal system *(54)*. However, this hypothesis was challenged by the observation that MAO B is present in astrocytes and serotonergic (5HT) neurons *(72)*, but is absent from DA neurons *(73)*. Much to everyone's surprise, it was also found that MPP+, but not MPTP, is actively taken up by the DAT *(74,75)*.

It is now generally agreed that astrocytes are the primary site for the generation of MPP+ *(76–78)*. Recently, this issue has been studied in detail using primary cultures of mouse astrocytes *(64)*. It was found that both MPDP+ and MPP+ can cross the astrocyte cell membrane into the extracellular space in a manner independent of cytotoxicity, and that MPP+ can be formed extracellularly, presumably by autoxidation of MPDP+.

Once in the extracellular space, MPP+ is available for uptake into DA neurons by the DA transporter, at concentrations as low as the nanomolar range *(75)*. It was thought that this mechanism was entirely responsible for the accumulation of MPP+ into DA neurons. However, it has been shown that at high enough concentrations, MPP+ is taken up even if the uptake system is blocked by specific uptake blockers, such as mazindol and nomifensine *(79)*. One explanation for this observation is that MPP+, in spite of the fact that it is charged, appears capable of crossing cell membranes along a concentration gradient. This may be possible because of charge delocalization throughout its pyridinium ring *(80)*. This carrier-independent uptake could function at levels of MPP+ known to accumulate in the extracellular space in MPTP-treated mice and monkeys *(55,81)*.

Elimination of MPTP from the body probably takes place through a number of pathways. Peripheral metabolism of MPTP by vascular and liver MAOs to the water-soluble MPP+ almost certainly allows for its clearance by the kidneys *(82)*. In brain, alternate biotransformation pathways form other metabolites of MPTP, including an *N*-oxide *(83)*, a demethylated tetrahydropyridine, and three pyridones (1-methyl-4-phenyl-5,6-dihydro-2-1H-pyridinone, 1-methyl-4-phenyl-2-1H-pyridinone, and 1-methyl-4-phenyl-2-piperidinone) *(84)*. These compounds could potentially act as detoxification pathways by reducing the available MPTP for conversion to MPP+. Evidence for this possibility comes from tissue-culture studies utilizing astrocytes in which a flavin-containing monooxygenase oxide-like enzyme has been identified. This enzyme catalyzes MPTP to MPTP *N*-oxide, resulting in *N*-oxide accumulation in the media of MPTP-treated astrocytes. If this excretion of *N*-oxide from astrocytes occurs in vivo, it would provide a mechanism whereby the *N*-oxide could be released from astrocytes and presumably eliminated from the brain *(85)*.

4. TARGETING OF THE NIGROSTRIATAL SYSTEM

Although the oxidation of MPTP to MPP+ is necessary for the development of nigrostriatal toxicity, this transformation alone does not explain the selectivity of the damage within the brain. A key player on the road to selectivity is probably the DAT system. Active uptake of MPP+ by the DAT almost certainly serves to target DA neurons by providing a conduit for accumulation of MPP+, which far exceeds that of other neuronal populations. Lower-affinity noradrenergic or serotoninergic monoamine uptake systems might take up MPP+ when extracellular levels are at their highest after an acute MPTP dose, but they are less likely to accumulate toxic levels with low doses or once extracellular levels drop. Compared to other cell types, such as astrocytes, which lose MPP+ by diffusion to the extracellular space, thereby preventing toxic levels from building up *(85)*, DA neurons may actually recycle any lost MPP+ through the re-uptake mechanism, preventing its removal.

It also seems possible that the uptake system may be responsible, at least in part, for the differences in species susceptibility to MPTP. For example, it has been shown that reuptake blockers that prevent MPTP-induced damage in mouse striatum are much less effective in monkey striatum *(86–91)*. This could be explained by species differences in uptake capacity, but may also simply be owing to the relatively short half-life of the uptake blockers combined with the prolonged MPP+ half-life in monkeys.

MPTP even appears to target specific areas within the nigrostriatal system. For example, a primary target of MPTP neurotoxicity in the brain is the terminal field of the nigrostriatal DA system, where DA depletion of 90% is readily achieved, particularly in monkeys *(29,33,43)*. Although there is near-universal agreement that nerve terminals are more subject to MPTP-induced damage, at least in rodents, studies in monkeys have suggested that DA depletion in SN may be affected first, followed by the caudate and putamen *(28)*. This raises the possibility that DA cell bodies may be the initial site of MPTP action in monkeys, presumably after its entry into cell bodies via dendritic DATs. Evidence in support of this hypothesis is provided by studies in which chemical ablation of glia in the SN prevent MPTP-induced cell loss *(92)*. On the other hand, ^{3}H-MPTP binds with high affinity to the caudate nucleus *(93)*. Also, Santiago et al. *(94)* using in vivo microdialysis to monitor MPP+-invoked DA release as a measure of damage, found that terminals in the striatum were more sensitive than cell bodies/dendrites in the SN. Until further studies are done, it is likely that the exact site of primary damage in monkeys will remain controversial. Such information might provide clues concerning the primary site of damage in Parkinson's disease as well.

Evidence of damage to other brain areas (similar to those affected in idiopathic parkinsonism) with MPTP has been reported by several groups. In the monkey, these include the ventral tegmental region, locus cerulus, and retrorubral area *(51,95–97)*. At least one study has suggested that the site of MPTP-induced damage appears to be dose-dependent, with larger doses affecting more sites. This could be a reflection of the different sensitivities of these areas to MPTP, possibly because of their respective proportions of neurons that project to the striatum *(97)*. This idea is supported by evidence that nigropallidal DA projections are relatively spared compared to nigrostriatal projections in MPTP-treated monkeys *(98)*. Other studies point to age as an important factor in governing whether or not other anatomical areas are affected *(23,51)*.

Neuromelanin has been frequently implicated as another key factor in determining the selectivity of MPTP. The presence of neuromelanin was first recognized over two centuries ago *(99)*, but its function, if any, is still a mystery. Composed of a heterogeneous mixture of protein, lipofuscin, and melanin pigment, it is derived from the polymerization of autoxidized DA products and other catecholamines *(100)*, and accumulates throughout life in the catecholaminergic nuclei of the midbrain in an age- and species-dependent manner *(101)*. A loss of neuromelanin is characteristic of Parkinson's disease and is associated with a loss of pigment-bearing neurons in the SN *(102,103)*. This observation, plus the fact that the most heavily pigmented cells are the first to degenerate in Parkinson's disease, has led to the hypothesis that neuromelanin plays an active role in disease progression *(104)*.

However, both protective and destructive capabilities have been variously attributed to this enigmatic substance *(105)*. Those arguing for a protective role point out that neuromelanin can transduce potentially harmful molecular energy into innocuous forms, aid in removing reactive free radicals and maintaining cellular redox states, and bind potentially harmful compounds *(106–110)*. On the other hand, neuromelanin can interact with reactive metal ions, resulting in the generation of reactive oxygen species (ROS) *(111,112)*, participate in production of cytotoxic quinones *(67,100)*, and act as a "sink" for the slow release of toxic xenobiotics.

Several areas of research have linked MPTP to neuromelanin. Differential species sensitivity to MPTP correlates with the presence (monkeys) or absence (rodents) of this compound *(113,114)*, and it has been reported that neurons containing neuromelanin demonstrate a greater vulnerability to MPTP *(115)*. It has been shown that MPTP and MPP+ can bind with high affinity to both neuromelanin and synthetic melanin *(116–118)*, and that chloroquine, which competes with MPP+ for neuromelanin binding, reduces the toxicity of MPTP in the monkey *(119)*. It is also known that synthetic melanin can oxidize MPDP+ to its toxic metabolite, MPP+ *(120,121)*. In view of these observations, Melamed and colleagues *(122)* designed an experiment in which they were able partially to overcome the natural resistance to MPTP toxicity in rats by injecting synthetic neuromelanin directly into the striatum. However, neuromelamin cannot be the entire answer to either neuronal selectivity or differences in species susceptibility, since species that do not have neuromelanin, such as the mouse, still exhibit some degree of sensitivity to MPTP. Therefore, it remains to be determined if neuromelanin plays a key role or simply a supporting role in the selectivity and/or sensitivity to MPTP.

5. NEUROPHARMACOLOGY AND NEUROCHEMISTRY

5.1. MPTP and DA Release

One of the early observations on the mechanism of action of MPTP, beyond its ability to cause nigrostriatal dopaminergic neurodegeneration, was its ability to cause significant release of DA from fibers in the striatum *(123–127)*. Most of these studies demonstrated a calcium dependence on the release, implying a synaptosomal mechanism. In fact, it has been reported that MPTP toxicity is attenuated in the SN of mice and monkeys by calcium channel blockers *(128)*. On the other hand, at least two groups have reported DA release to be calcium-independent, suggesting some nonsynaptic mechanism *(125,129)*. In support of a nonsynaptic mechanism, a recent study by Wilson et al. *(130)*, using mouse striatal slices, found that blocking neostriatal transmission did not prevent DA depletion; rather, the depletion still occurred in the absence of calcium. Also, in contrast to other studies that reported MPTP/MPP+ to enhance deploraization-evoked DA release *(126,131)*, these investigators showed that MPP+ irreversibly inhibits synaptic transmission *(130,132)*. Thus, there are still major questions remaining in regard to the mechanism by which MPP+ induces DA release, and even as to the exact role of calcium.

5.2. MPTP and DA Transporters

As noted earlier, MPP+ appears to be a high-affinity substrate for the DAT, a phenomenon that is probably key to its selectivity for the nigrostriatal DA system. The importance of the DAT was confirmed quite early when it was shown that DA reuptake blockers prevented MPTP or MPP+ toxicity *(86,133)*, and that treatment with MPP+ or its analogs blocked ^{3}H-DA uptake into striatal synaptosomes *(134,135)*. Recently, it was shown that MPTP/MPP+ also has a direct effect on the DAT. MPTP is known to release DA. This event may actually be damaging to the transporter, since DA itself is a potential source for ROS, including hydrogen peroxide (H_2O_2) and quinones. In addition, MPTP is suspected to be involved in the production of ROS through its effect on electron transport. There is substantial evidence that oxidative stress adversely affects the transporter system. For example, Berman et al. *(136)* found that treating nigral synaptosomes with ascorbate in the presence of ferric iron, a combination that is known to promote lipid peroxidation, inhibited DAT uptake by 45%. They also found synaptosomal treatment with xanthine + xanthine oxidase, as a free radical generating system, decreased DAT uptake 80%. Using a similar lipid peroxidation model, Ramassamy et al. *(137)* found similar uptake inhibition, but also determined that preloaded synaptosomes retained the ability to release DA when stimulated with K+. They hypothesized that the DAT complex was damaged under these conditions, but not vesicular release. However, in a study utilizing striatal and cortical tissue slices, Drukarch et al. *(138)* found H_2O_2 treatment to induce a nonexocytotic release of DA and norepinephrine by interfering with vesicular storage. They suggested that the high sensitivity of the dopamingergic system to oxidative stress is more a function of transporter sensitivity than the presence of DA. Regardless of the mechanism, it appears that MPTP/MPP+ has a complex set of effects on the DAT, many of which are yet to be clearly understood.

Distinct from cell membrane transporters are the intraneuronal vesicular transporters, a family of carriers that accumulate cytoplasmic amine neurotransmitters into storage vesicles. The energy required for amine accumulation by the vesicular monoamine transporter (VMT) is an ATP-driven H+ pump that exchanges H+ for cytoplasmic amines *(139)*. In the brain, MPP+ is an inhibitor of the VMT, interacting at a site separate from that of reserpine, the classical synaptic vesicle uptake blocker *(140,141)*. In the periphery, vesicular transporters have been shown to suppress MPP+ toxicity by sequestering it from the cytoplasm where it has access to mitochondria and other cell components *(142)*. Recently, a gene from PC12 cells was cloned and found to confer MPP+ resistance in Chinese hamster ovary cells by transporting the MPP+ from the cytoplasm to intracellular vesicles *(143)*. The amino acid sequence predicted for by this gene is highly homologous to the amino acid sequence determined for an isolated bovine adrenal VMT, confirming that the gene that conferred resistance did indeed code for a VMT *(139)*. Thus, it appears that peripheral VMTs are important in attenuating toxicity of xenobiotics, such as MPTP, as part of the body's defense system to sequester and remove such toxic compounds. These fascinating observations are likely to lead to insights into the sequence of events that occur after MPP+ reaches the intracellular space, some of which may be protective.

5.3. MPTP and Norepinephrine

Although the primary target of MPTP is the nigrostriatal dopaminergic system, a number of reports suggest that it may affect other brain catecholamines as well. Decreases in cortical norepinephrine levels by MPTP treatment were reported quite early by Sundstrom and Jonsson *(144)*. That these decreases could be blocked by MAO B inhibition and norepinephrine uptake blockers pointed to MPP^+ as a general substrate for the catecholaminergic uptake system. In a recent study, Pifl and coworkers *(145)* directly compared the uptake of MPP^+ by DA and norepinephrine transporters. Using COS-7 (monkey kidney) or SK-N-MC (human neuroblastoma) cells transiently transfected with either cloned human DA or norepinephrine transporters, they found that the affinity for MPP^+ was actually greater for the norepinephrine transporter. This observation would not predict selectivity for DA neurons. The data showed that the DAT took up MPP^+ at a rate approximately equal to that for DA, but that the norepinephrine transporters took MPP^+ up at only half the rate for norepinephrine. Thus, higher levels of MPP^+ would probably be needed in the cortex to compete effectively with endogenous levels of norepinephrine. Also, the turnover rate for MPP^+, which reflects how many molecules of substrate can be transported by one transport molecule per unit time, was five times higher in the DA clone. Together these data predict that DA neurons are more likely to accumulate toxic amounts of MPP^+ in vivo, and are certainly supportive of the hypothesis that the DAT is in large part responsible for the selectivity of MPTP for DA neurons.

Other studies suggest that the noradrenergic system is involved in the sequence of events that lead to MPTP toxicity. For example, lesioning of the noradrenergic locus ceruleus (LC) both exacerbates MPTP-induced striatal DA loss and SN cell loss in mice and monkeys, and impairs the recovery that normally occurs after MPTP treatment *(146–148)*. Similarly, inhibiting noradrenergic α_2-adrenoceptors with yohimbine has been shown to enhance MPTP toxicity in mice, whereas treatment with clonidine, an α_2-agonist, prevented loss of striatal DA and tyrosine hydroxylase activity *(149)*. Interestingly, MPP^+ levels were not diminished by clonidine treatment, but rather its clearance was delayed. Although the exact mechanisms by which the noradrenergic system is interacting with MPTP are not yet worked out, further studies in this area seem more than warranted, particularly since degeneration of the noradrenergic LC is a common feature of the neuronal degeneration, which occurs in Parkinson's disease.

5.4. MPTP and Serotonin

The very earliest behavioral observations in rats were strongly indicative of a serotoninergic syndrome, with symptoms, such as tremor, rigidity, salivation, retropulsion, hypertonus, and "Straub tail" *(17)*. Similar behavioral effects in monkeys immediately after MPTP dosing also suggest a serotonergic effect, at least in the acute stages of toxicity *(28)*. These early behavioral observations led investigators to test for an effect of MPTP on 5HT. It was found that MPTP treatment increased 5HT levels in the midbrain of rats and monkeys *(150)*. In vivo microdialysis experiments demonstrated that striatal MPP+ application caused a dose-dependent increase in extracellular 5HT and a decrease in its metabolite, 5-hydroxyindole acetic acid (5HIAA), suggesting increased release and/or decreased metabolism contributed to higher levels *(151)*. In contrast, a study by Pérez-Otaño et al. *(152)*, which employed chronic MPTP dosing in marmosets, showed a profound and long-lasting depletion of striatal and extrastriatal 5HT.

An interesting study by Mitra et al. *(153)* was specifically designed to explore both the serotonergic and dopaminergic phases of MPTP toxicity based on behavior and on monoamine levels in several areas of the brain, including the SN and caudate putamen. Using Balb/c mice, they found an initial excitatory motor activity phase lasting approx 90 min followed by an inhibitory phase that lasted for an additional 5–7 h after MPTP treatment. The excitatory phase correlated with increased 5HT levels and decreased utilization as indicated by a decreased 5HIAA/5HT ratio. The excitatory phase could be blocked by methysergide, a nonspecific 5HT blocker, without affecting the inhibitory phase. The inhibitory phase correlated with a decreased DA level and increased turnover as indicated by increased homovanillic acid/DA ratios. The bradykinesia and akinesia of the inhibitory phase could be blocked with the DA agonist, apomorphine, but only if the agonist was given 90 min after MPTP treatment. These studies clearly show that MPTP has at least an acute effect on the 5HT system, and it seems likely that this system may be responsible for the acute syndrome induced by MPTP in rodents as suggested by the very early behavioral observations noted above.

5HT neurons projecting from midbrain raphe nuclei are known to modulate DA activity *(154)*. Recent studies have shown that 5HT causes a dose-dependent release of striatal DA, an effect that appears to be owing, at least in part, to activation of striatal 5HT receptors *(155,156)*. Other studies have shown that 5HT displaces DA from storage vesicles and thus indirectly induces DA outflow *(157,158)*. Thus, 5HT neuronal participation in MPTP toxicity could reasonably be expected. However, destruction of striatal 5HT fibers with 5,7-dihydroxytryptamine *(159)* or blocking 5HT uptake with fluoxetine *(160)* have been shown to have no effect on the DA-depleting effects of MPTP, so the serotonergic system may not play a direct role in regard to the toxic effects of MPTP.

In addition to mediating at least some of the acute behavioral effects of MPTP, 5HT neurons may participate in recovery from MPTP toxicity. In a recent study by Gaspar and coworkers *(161)* involving long-term consequences of MPTP-induced hemiparkinsonism in cebus monkeys, immunocytochemical data indicated that 5HT fiber sprouting was acting as a compensatory mechanism following DA neuronal loss. The study of serotonergic participation in MPTP toxicity has highlighted this system's interaction with the nigrostriatal pathway, and opened up some interesting prospects for further work on modulation and compensatory mechanisms between these neuronal systems.

6. MECHANISMS OF NEUROTOXICITY

Numerous mechanisms of neurotoxicity have been proposed over the years since MPTP's biologic effects were first recognized. In this section, we review these various theories, and review the experimental evidence for and against each. Interestingly, these theories in many ways represent a microcosm of those that have been put forward as the cause of Parkinson's disease.

6.1. Free Radicals and Oxidative Stress

6.1.1. The Paraquat Hypothesis

Soon after its discovery, the similarity in structure between MPP+ and the well-known herbicide, paraquat, was noted. The mechanism of action of paraquat involves redox cycling, during which ROS are formed *(162)*, and it was suggested that MPP+ might undergo redox cycling in a similar manner. This theory was at first greeted with great enthusiasm. However, Frank et al. *(163)* soon pointed out that the reduction potential for redox recycling for MPP+ is too high to occur easily under normal biological conditions. Furthermore, *in situ* studies *(163,164)* as well as experiments using oxidatively stressed hepatocytes (a model for paraquat toxicity) *(165,166)* failed to demonstrate a paraquat-like action for MPTP. Thus, the "redox recycling" hypothesis was largely abandoned, but considerable interest has continued in the possible role for free radicals in MPTP toxicity.

6.1.2. The Role of DA

Perhaps the most frequently cited alternative "free radical hypothesis" relates to the effects of MPTP on endogenous DA. As noted earlier, MPTP is known to cause SN cell loss and striatal DA efflux *(167,168)* with increased DA turnover *(28)*. In the normal degradative pathway of DA by MAO, one molecule of H_2O_2 is produced for every molecule of DA consumed. H_2O_2 can cause lipid peroxidation and/or can be further metabolized to other ROS. In situations where DA neurons are reduced in number, as in Parkinson's disease or after MPTP treatment, remaining DA neurons are likely to compensate by producing more DA. This in turn would lead to greater production of H_2O_2, resulting in more free radicals and greater cell death, setting up a self-perpetuating cycle.

That DA itself can induce oxidative stress has been supported by experiments showing that enhanced DA release increases markers of oxidative stress, such as oxidized glutathione (GSH), and that MAO inhibition partially blocks this increase *(169,170)*. In addition to H_2O_2 produced during normal metabolism, a number of other ROS can be generated spontaneously from DA *(171*, for review, *see 172)*. Interestingly, it has been reported that MPTP and MPTP-like compounds potentiate DA autoxidation in vitro *(173)*. In the presence of oxygen,

DA is rapidly converted to an orthoquinone derivative plus superoxide or peroxide *(174,175)*. All these products are unstable and potentially cytotoxic. The orthoquinone derivatives rapidly form covalent bonds with sulfhydryl compounds, such as GSH and cysteine, and with amino acids; such binding could thereby alter enzyme activities, membrane permeability, or transport systems *(175)*. In support of this possibility, Hastings et al. *(176)* have shown that intrastriatal injections of DA increase levels of free and protein-bound cysteinyl catechols by 7- to 20-fold above control levels, and are associated with neuronal loss and gliosis. This damage was greatly reduced by simultaneous injection of ascorbic acid or GSH, indicating an oxidative process.

This "DA hypothesis" of oxidative stress in the nigrostriatal system is often invoked as a potential cause for the degenerative process that occurs in idiopathic Parkinson's disease as well. This of course is troubling for clinicians because the mainstay of treatment for the disease, L-DOPA, is a precursor to DA and might be expected to exacerbate this process (putting coals on the fire, as it were), if indeed it is operative in Parkinson's disease. L-DOPA treatment has been shown to cause a severalfold increase in concentration of DA and its metabolites in the nigrostriatal system *(177)*, which, in theory at least, should increase the risk for reactive species production and consequent oxidative damage. However, definitive evidence that this happens has not been readily forthcoming. In fact, several groups have actually found chronic L-DOPA treatment ameliorates age-induced decreases in motor function and increases the life-span in mice and rats *(178,179)*. On the other hand, treating Parkinson's patients with MAO inhibitors or pergolide, a DA agonist, has been shown to delay the need for L-DOPA *(180–185)* and possibly to have neuroprotective effects *(186–188)*. Since either of these treatments would decrease the turnover of endogenous DA, they may work by reducing the potential for oxidative stress.

6.1.3. Iron as a Provocateur

Another factor in favor of free radical production playing a role in MPTP toxicity is the presence of high levels of iron in the SN *(189,190)*, since iron is known to participate in oxygen-radical-mediated reactions *(191)*. MPTP induces iron accumulation in the SN of monkeys when injected into the striatum (caudate or putamen). This is thought not to be a direct response of MPTP, but rather to be resulting from enhanced transferrin-mediated uptake of iron in response to neuronal injury, where it is thought to play a role in neuronal recovery and repair *(192)*. In MPTP-treated mice, both striatal transferrin receptors and mazindol binding (a DAT blocker) decreased initially and then increased with apparent recovery. The time-course of recovery indicates that transferrin receptors recover first, suggesting a compensatory mechanism in iron transport prior to terminal recovery *(193)*.

The idea of a possible role for transition metals in MPTP toxicity is not new. In 1985, Poirier et al. *(194)* related the presence of transition metals in the SN to its selective vulnerability to MPTP. Iron has been shown to bioactivate the conversion of MPTP to MPP+ in glial cultures, even in the presence of MAO A and B inhibition. This nonenzymatic conversion has been hypothesized to occur via generation of the superoxide radical from oxidation of ferrous iron in the presence of oxygen. The superoxide radical formed would react with MPTP to produce MPDP+, followed by the rapid nonenzymatic oxidation of MPDP+ to MPP+, a process facilitated by the ability of glial cells to keep iron in its more reactive reduced state *(195)*.

Not surprisingly, iron has also been implicated in the pathology of Parkinson's disease through its ability to act as a catalyst in the auto-oxidation of DA *(196)*. Iron is known to accumulate in the brains of parkinsonian patients *(197–199)* and in aging rodents and humans *(200)*. This accumulation has been shown to target the SN in PD both from postmortem specimens *(201,202)* and from in vivo studies *(203)*. One pathway for iron accumulation would be the binding of ferrotransferrin to a specific high-affinity receptor found on glial and neuronal cell surfaces *(204)*. An autoradiographic study in postmortem brains showed ^{125}I-ferrotransferrin binding sites to be increased in the caudate and putamen, but not the SN of PD patients. An inverse relationship between iron binding density and iron content was found, suggesting uptake by the dopaminergic terminals in the striatum and translocation to the cell bodies in the SN. Parkinson's patients with the greatest level of disability were found to have the greatest level of binding. On the other hand, Mash et al. *(193)* found parkinsonians to have reduced transferrin binding in the putamen. In contrast to the data in human parkinsonian brain, MPTP treatment tended to decrease binding in these same areas in monkey brains, probably owing to the large loss of DA neurons *(205)*.

As is obvious from the preceding discussion, the data on iron are extensive, but it is still not clear whether its presence in the nigrostriatal pathway is causative or secondary to pathological changes. There is no clinical evidence that iron intoxication or metabolic disorders of iron metabolism are associated with a parkinsonian syndrome. On the other hand, there are recent reports of regional differences in levels of the iron transport proteins, transferrin and lactotransferrin *(206,207)*, and the iron storage protein, ferritin *(208)* in parkinsonian vs normal brains, suggesting that disturbances in iron regulation play a role in disease pathogenesis. That the greatest iron accumulation in PD is seen in patients with the longest duration or greatest severity of symptoms suggests that the presence of iron could as easily be a response to neurodegeneration as be the initiator of cell death. Iron is known to localize primarily within nonneuronal cells in the SN *(209)* and increased levels may reflect infiltration by glial cells into damaged areas. However, iron is reported to be increased in SN neurons and is found in Lewy bodies *(199)*. That the acute damage of MPTP results in iron accumulation and changes in iron metabolism, which reflect changes seen in idiopathic Parkinson's disease, lends support for the possible role of transition metals in disease progression.

6.1.4. The Role of Endogenous Antioxidant Systems

Given the potential for radical formation both in normal metabolism and toxic insult, cells have evolved a number of systems to protect against such compounds. These include the enzyme superoxide dismutase (SOD), which reduces superoxide anion to H_2O_2 and oxygen, and the enzymes GSH peroxidase and catalase, which catalyze the reduction of H_2O_2 to water. Maintaining high levels of reduced GSH is critical to cellular detoxification of ROS *(210)*. It has even been suggested that GSH depletion, rather than the level of ROS, is the critical component of oxidative stress *(211)*.

MPTP and MPP+ have been shown to increase ROS both in vitro *(212)* and in vivo *(213)*. MPP+ is also known to inhibit electron transport at Complex I, thus generating ROS *(214,215)*. If free radicals play a role in MPTP toxicity, then one would expect endogenous protective mechanisms, such as the enzymatic scavenging of radicals, to be affected. This is shown to be the case with SOD, the formation of which is enhanced by MPTP *(216)*. Also Przedborski et al. *(217)* found transgenic mice that overproduce copper/zinc-SOD to be MPTP-resistant. That free radical scavenging is not completely successful was demonstrated in a study by Rojas and Rios *(218)*. They found increased striatal lipid peroxidation, a common consequence of excessive H_2O_2 generation, after intracerebroventricular MPP+ treatment. Furthermore, work by Desole and coworkers *(219,220)* implicated xanthine oxidase-mediated oxidative stress in MPTP toxicity.

What are the implications for oxidative stress in Parkinson's disease? One factor may be that the levels of SOD, which functions as an oxygen radical scavenger, are high in the normal SN relative to other brain areas, such as the cortex and cerebellum *(221)*. Also, there is evidence that the pigmented cells of the SN that are thought to be most vulnerable to the disease process *(222)* express high levels of this enzyme *(223)*. However, there seems to be no compensatory increase in systems involved in removing the H_2O_2 generated by SOD. The level of GSH peroxidase in the SN *(224)* is unchanged, and levels of reduced GSH are much lower in SN as well *(225,226)*. These data suggest that the SN has diminished ability to handle oxidative stress.

In Parkinson's disease the situation may be even worse. Both cytosolic and mitochondrial levels of nigral and striatal SOD are increased over control brains *(227,228)*. This would suggest an increased need and/or capacity to remove oxygen radicals. Since no mutations have been found in the genes for either SOD or catalase in Parkinson's patients *(229)*, increases in levels would have to be in response to local environment. In parkinsonism, levels of catalase and GSH peroxidase are decreased or unchanged *(224,230)*, and GSH content is significantly decreased *(225,226,231)*. Since GSH is lost to the cell only when it acts as a conjugating agent in removing free radicals, decreases in its levels could reflect its consumption during the process of free radical scavenging. Thus, so far there is only indirect evidence that oxidative stress plays a role in idiopathic Parkinson's disease.

6.1.5. Neuroprotection with Antioxidants: Do They Work?

Studies to date have fallen short of definitively proving that oxygen stress and free radical formation play a role in MPTP toxicity. Similarly, studies undertaken to protect against toxicity by the use of antioxidants have provided evidence both in favor of *(226,232,233)* and against *(226,234)* such a role. However, in many of the studies favoring a protective role, other factors have frequently been found to play a part. For example, where GSH was thought to be decreased by MPTP treatment, it was found that increased efflux rather than oxidation was occurring *(235)*, or where ascorbic acid was thought to be acting as an antioxidant, an effect on MPP+ uptake was found *(236)*. Also, where diethyldithiocarbamate (DDC), a known inhibitor of SOD, was thought to enhance MPTP toxicity by potentiating the oxidative stress induced, it was found that DDC also increased the MAO-mediated conversion of MPTP to MPP+ and to decrease MPP+ elimination from the brain *(237)*.

6.1.6. Conclusion

Although there has been an enormous amount of work on the "free radical hypothesis" of both MPTP and in regard to Parkinson's disease, definitive answers are still not available for either. However, evidence is still accumulating, and the possibility exists that free radicals and oxidative stress may play a role in site-specific, localized damage in both conditions. In any case, research in this area is likely to continue well into the foreseeable future.

6.2. The Role of Mitochondria

MPTP was first reported to interact with the mitochondrial electron transport chain in 1985, when it was found that MPP+ inhibited NADH-linked oxidation *(238,239)*. In a series of experiments over the next few years, it was discovered that MPP+ inhibits Complex I of the respiratory chain by binding at the same site as rotenone between NADH dehydrogenase and coenzyme Q *(240)*. The mitochondrial hypothesis quickly gained eminence as a key aspect of MPTP toxicity, since the consequences of inhibiting the electron transport system include energy crisis with severe energy impairment with depletion of ATP, loss of cellular membrane potential, and ultimately, if the energy crisis is severe enough, cell death.

For this mechanism to be operative in vivo, however, MPP+ must not only be taken up into SN neurons via the DAT, but it must also gain entrance into mitochondria *(241)*. It has now been shown that MPP+ is indeed accumulated in mitochondria, reaching 40-fold the concentration found outside the mitochondrial membrane, and that that process is largely driven by the mitochondrial membrane potential *(242)*. Furthermore, there is now a large body of evidence that this sequestration has important functional consequences. For example, MPTP and its metabolite MPP+ have been found to deplete ATP in hepatocytes *(243)*, in isolated mouse brain mitochondria *(244)*, in a neuroblastoma/glial cell line *(245)*, and in astrocytes *(85)*. Functional proof of loss of oxidative metabolism with MPTP was demonstrated in a study by Di Monte et al. *(246)* in which they were able to circumvent MPTP toxicity in cultured hepatocytes by providing fructose as an alternate energy source. In a follow-up astrocyte study, ATP depletion was found to precede MPP+-induced cell mortality *(85)*.

These studies have subsequently been extended to the in vivo setting. Chan and colleagues *(247,248)*, after the use of focal microwave irradiation to kill brain degradative enzymes, have now clearly demonstrated in vivo decreases in striatal and mesencephalic ATP levels in mice within hours after systemic MPTP treatment. These decreases appear to be selective for the nigrostriatal system, since levels in the frontal cortex or cerebellum were not decreased. The decreases in ATP levels found were small (15–20%), probably reflecting dilution by nondopaminergic cell types, and largely recovered by 24 h posttreatment. In a similar study that employed direct intrastriatal injection of MPP+ into rats and a freeze-clamp method to stop enzymatic ATP degradation, Storey et al. *(249)* also found ATP to be decreased. In their study, however, ATP levels did not recover in 24 h, but continued to decline up to the last time-point at 48 h. Differences in route of drug administration and tissue sampling may have contributed to the differences noted between these two investigations. The importance of energy state in MPTP neurotoxicity was further demonstrated in work done by Chan and colleagues *(250)* in which reduction of endogenous energy supplies by pretreatment with 2-deoxyglucose, a competitive inhibitor of cellular glucose, enhanced MPTP-induced ATP loss.

An interesting observation, which has received little attention, is the fact that MPP+ also inhibits the α-ketoglutarate dehydrogenase complex of the tricarboxylic acid cycle *(251)*, an enzyme complex that is also reduced in Parkinson's disease *(252)*. Inhibition of the α-ketoglutarate dehydrogenase complex would affect complex II of the respiratory chain by decreasing availability of the substrate, succinate, which is generated from α-ketoglutarate in the tricarboxylic acid cycle. Thus, MPP+ has the potential to have a double effect on energy metabolism.

It was not long before the connection between energy depletion and MPTP led investigators to search for a similar impairment in Parkinson's disease. To date, impaired mitochondrial ATP has been reported in SN, striatum, muscle, and platelets of parkinsonian patients *(253–267, for review see 260)*. There currently appears to be agreement that the SN and platelets are affected, but other tissues, such as muscle, remain the subject of controversy. Also, although complex I of the respiratory chain is the primary target of MPTP, other electron transport complexes are also reported to be decreased. Variations in results might be attributable to the use of homogenates vs mitochondrial isolation preps, time postmortem, or assay. There has also been a question concerning whether drug treatment in the PD patients may have affected enzyme activities. A variety of neuroleptic drugs *(268)* as well as L-DOPA *(269)* have been shown to have an inhibitory effect on complex I. Overall, the evidence suggests that this is not the case, since Benecke et al. *(264)* found no abnormalities in Complex I in

L-DOPA-treated patients who had other forms of parkinsonism, and Haas et al. *(267)* found decreases in electron transport chain enzymes in early untreated parkinsonian patients.

The cause of the mitochondrial defects in Parkinson's disease remains unclear. They could be owing to either a genetic defect or some toxic response to endogenous or exogenous factors. A number of laboratories have searched for a mutation in the mitochondrial genome (mtDNA), but nothing has been found that would definitively point to a defect in energy metabolism or to a difference between control and brains from patients who had Parkinson's disease *(262,270–275)*. Perhaps the most interesting and elegant study done to date to explore this question further comes from Swerdlow and coworkers *(276)*. These investigators inserted mitochondria isolated from platelets of PD patients into a clonal line of human neuroblastoma cells from which endogenous mtDNA had been removed. In these cells, termed "cybrids," they found a 20% decrease in complex I activity, increased oxygen radical production, and increased sensitivity to MPP+-induced cell death. These data indicate that the respiratory deficit is genetic and that it arises from the mitochondria. That such a genetic component has not been identified in tissues from parkinsonians could be owing to differences between the cell line and normal cells. For example, lack of heterogeneity in mitochondria in a cloned line would make a defect easier to identify. Alternatively, locating possibly multiple small defects in the mitochondrial DNA of parkinsonians would be extremely difficult and time-consuming. Ikebe et al. *(274)* recently sequenced the entire mitochondrial genome in five parkinsonian patients. They were unable to find any point mutations common to all the patients, but did find mutations in each of the patients, which they suggested may contribute to decreased function. Somewhat confounding these results is the fact that mitochondrial deletions are also found in control brains *(270)*.

6.3. Cytosolic Calcium

Maintenance of low cytoplasmic free calcium (Ca^{2+}) levels is of critical importance to cell function. Elevated cytoplasmic free Ca^{2+} can activate a number of different degradative enzymes, such as phospholipases, which can degrade cell membranes *(277)*, and certain proteases, which can degrade the proteins of cell membranes and cytoskeleton. The consequences of this activation can be disrupted cell function, axonal transport, cell membranes, and eventual cell death. Normally, extracellular Ca^{2+} levels are approx 10,000 times as high as cytoplasmic levels. This low cytoplasmic level is maintained by a number of pathways, including a Na^+/Ca^{2+} exchange, active extrusion of Ca^{2+} from the cell by a Ca^{2+}-ATPase-mediated transport, as well as sequestering of intracellular Ca^{2+} by endoplasmic reticulum and mitochondria. This section discusses what is known (and not known) about the relationship between MPTP and Ca^{2+} homeostasis.

In a model of hepatocyte toxicity *(278)*, MPTP and MPP+ caused a rapid depletion of the mitochondrial Ca^{2+} followed by a sustained increase in the cytoplasmic free Ca^{2+} levels. The increase in cytoplasmic free Ca^{2+} was probably owing in part to the inhibition by MPTP of the plasma membrane Ca^{2+}-ATPase and loss of the cells' ability to extrude Ca^{2+}, since no change was noted in Ca^{2+} influx. The MPTP- and MPP+-mediated release of Ca^{2+} from the mitochondrial pool was probably owing in large part to MPP+'s inhibition of electron transport at complex I and the resulting ATP depletion. This ATP depletion would also decrease the energy-dependent Ca^{2+}-ATPase activity and thus contribute to the increase in cytosolic free Ca^{2+} levels. That both MPTP-induced cell death and the increase in cytosolic free Ca^{2+} were protected against by inhibition of the MAO B-mediated conversion of MPTP to MPP+ supports the thesis of a role for Ca^{2+} in cell toxicity.

Another mechanism by which MPTP/MPP+ could influence Ca^{2+} homeostasis is through disruption of membrane integrity. Sun and coworkers *(279)* reported that hydroxyl radical formation from autoxidation of DA after MPP+-induced release resulted in lipid peroxidation, membrane permeability changes, and increased Ca^{2+} influx. However, MPTP has been found to cause an increase in intracellular Ca^{2+} within 5 s and to have a disorganizing effect on fibroblast cytoskeleton, suggesting a direct effect of MPTP on membranes *(280)*. It is also of interest that these fibroblasts do not show the MPTP-induced decrease of ATP that is typically found in brain and liver. Thus, Ca^{2+} influx independent of mitochondrial dysfunction may also be occurring, at least in fibroblasts *(281)*.

Changes in Ca^{2+} homeostasis have not been measured in Parkinson's disease. This is not surprising in view of the difficulty in accurately measuring in vivo changes in Ca^{2+} levels in a subset of cells. Also, changes are likely to be transient, with a short-term change setting off a chain of events that may lead to long-term consequences on cell survival. One piece of evidence for Ca^{2+} participating in parkinsonism is that nigral neurons that contain the Ca^{2+} binding protein, calbindin, seem to be preserved relative to calbindin-negative cells. This suggests that if excess free Ca^{2+} can be complexed, the neuron has a better chance of survival *(282–284)*. A similar protection has been found against MPTP-induced nigral neurodegeneration in nonhuman primates *(283)*.

7. CURRENT CONTROVERSIES AND FUTURE RESEARCH DIRECTIONS

Although a vast number of papers have been published on the toxicity and mechanisms of action of MPTP, many facets of its action are still undecided. The previous sections have touched on these in the course of a more general discussion. However, in some areas, there are still more questions than answers, and in others, new research vistas are just opening up. For these reasons, we have chosen to organize the last section of this article around these areas.

7.1. MPTP and Excitatory Amino Acids

Excitatory amino acids (EAAs) have frequently been implicated in the process of neurodegeneration *(285–287)*, so it was logical to explore the possibility that they might be a participant in the sequence of events leading to neuronal degeneration after MPTP exposure. Indeed, an early report by Sonsalla and colleagues *(285)* found that blocking the glutamatergic NMDA receptor with MK-801 (dizocilpine) partially prevented MPTP-induced striatal DA depletion in mice, but had no effect on tyrosine hydroxylase activity, nor was there any difference between MK-801 stereoisomers. In a follow-up experiment, the effects of intranigral injections of MK-801 to rats were investigated, but no protection against DA depletion or tyrosine hydroxylase activity changes was found, even after repeated administration of MK-801 *(288)*. These investigators concluded that NMDA-related excitotoxicity does not play a role in MPTP-induced neurodegeneration. In support of these findings, an in vivo microdialysis study in which MPP+ was infused into the striatum did produce release of the EAAs, glutamate, and aspartate, but only at 10 times the concentration required to release DA *(289)*. It was found that even though MK-801 did block the EAA release, it had no effect on MPP+-induced DA release, again suggesting that NMDA was unlikely to play a major role in MPTP toxicity.

This story became more complicated in 1991, when Turski et al. *(290)* reported that systemic dosing or direct intranigral infusion of a variety of EAA antagonists over time did in fact protect against MPP+-induced cell loss in the SN of rats. That same year, Zuddas et al. *(291)* reported that multiple systemic injections of MK-801 to monkeys provided partial protection against MPTP-induced striatal DA depletion and nigral cell loss, as well as completely preventing behavioral parkinsonism induced by MPTP. Furthermore, decortication, which removes striatal glutaminergic innervation, was found to protect partially against striatal damage *(249)* or to protect completely against MPP+-induced cell loss *(292)*, again suggesting a role for NMDA receptors in MPTP neurotoxicity. Other studies over the last few years have been conflicting, with some favoring such a role *(293–295)* and others failing to do so *(296–298)*.

It does seem likely that there is at least some type of relationship between MPTP and EAAs. For example, it has been shown that MPTP, at high enough concentrations, causes EAA release *(289)*. In support of an effect on glutamate and aspartate release, Chan et al. *(299)* reported that treating mice with MPTP selectively decreased tissue levels of these two amino acids. This effect, however, was short-lived. This same group also found only transient protection by MK-801 against MPTP toxicity *(300)*. In addition to the early, but temporary (8-h) protection afforded by MK-801 against MPTP-induced DA depletion in the striatum, a longer-lasting (1 wk), but still temporary protection against DA depletion in the ventral mesencephalon was seen. However, there was no protection against cell loss in the SN *(27)*. These investigators concluded that MK-801 provided a short-term "pharmacologic" protection against MPTP-induced DA depletion, but no long-term protection.

Further support for a pharmacological role for MK-801 is presented in a study by Clark and Reuben *(301)* on MPTP- and MPP+-induced DA release from rat striatal synaptosomes. As might be expected if NMDA receptor activation plays a role in this release, MK-801 inhibited MPTP- and MPP+-induced DA release in a dose-dependent manner. However, changing other factors that should have also blocked the activation of the NMDA receptors (e.g., concentrations of Mg^{2+} or Ca^{2+} in the incubation media) had no effect. The inhibition seen was NMDA-specific, since other NMDA antagonists did not inhibit MPTP-induced DA release, nor was NMDA, the specific receptor agonist, able to induce DA release as would be the case if the DA depletion were EAA-mediated. All these data point to a role for MK-801 in inhibiting MPTP-induced DA release that is independent of its role as an NMDA antagonist. Clark and Reuben concluded that MK-801 inhibited DA uptake in striatal synaptosomes at least in part by inhibition of the DA transporter. Thus the protective effect of MK-801 reported by some may have been via an ability to inhibit MPP+ uptake by the dopaminergic neurons.

There are several other interesting leads regarding the relationship between EAAs and MPTP. For example, at least one study has suggested that energy impairment may be involved. In this investigation, low levels of MPP+ sufficient to block complex I, but not to cause extensive neuronal damage, exacerbated glutamate toxicity *(295)*, a phenomenon that was blocked by MK-801. Oxidative stress may also play a role, since MK-801 has been shown to block the DDC exacerbation of MPTP toxicity *(302)*. DDC is known to inhibit the endogenous oxygen radical scavenging system, SOD. The interactions of these exogenous and endogenous systems are likely to provide insights into possible disease mechanisms and provide one of the more important areas for future research.

7.2. MPTP and Aging

The concept of aging has not always been explored as a critical factor in neurotoxicity. However, when one takes into account that many neuropathological conditions are of a chronic long-term nature, and that some are almost exclusively of a late onset (e.g., Alzheimer's and Parkinson's), then the importance of aging, at least when it comes to the modeling and study of these diseases, becomes obvious. Certainly, this has proven to be the case for MPTP, especially in regard to its use as a model for Parkinson's disease, an age-related disorder.

Most of the work done on MPTP and aging has been done in mice and rats. The first work done in young (6–8 wk) and mature (8–12 mo) *(26)* or young (6 wk) and aged mice (21 mo) *(23)* established a shift to the left in the dose–response curve for MPTP in older animals. Older rats (12 mo) were found to be more than twice as sensitive as neonates (7–10 d) to MPTP-induced DA depletion and SN cell loss *(303)*. Recent more-detailed studies have shown that the pattern is biphasic, with the greatest change in sensitivity occurring between young (1–2 mo) and mature mice (8–10 mo), with little additional increase in sensitivity between mature and middle-aged mice (16 mo), and with aged mice (24 mo) actually being somewhat less sensitive than the middle-aged group *(304)*.

Other areas of the brain appear to become more sensitive to the effects of MPTP with aging as well. However, in contrast to the SN and striatum, which demonstrate sensitivity at all ages, other brain areas appear to be sensitive only in aged animals. Gupta and coworkers *(23)* reported decreases in catecholamine fluorescence in the ventral tegmental and LC areas only in aged mice. Also, a study by Date and colleagues *(305)* found a similar effect between young and mature mice (12 mo) in the ventral tegmental area. In addition, they found decreases in norepinephrine and 5HT only in the older animals. These studies are of special interest, because they may more closely mimic the idiopathic disease, since the mesolimbic dopaminergic pathway as well as noradrenergic and serotoninergic systems have been implicated in Parkinson's disease *(306)*.

Biochemical studies carried out to determine the basis for the increased sensitivity in older mice revealed no changes in the catecholamine uptake system *(26)*, in the elimination of MPP+ from the brain *(66)*, or even in the sensitivity of the striatum to intraventricular injections of MPP+ *(66)*. However, the possibility that greater levels of MAO in the brains of aging mice *(307)* might contribute to increased conversion of MPTP to its toxic form, MPP+, was confirmed nearly a decade ago *(303,308)*. Also, in fact, MAO B activity *(304)* and levels of striatal MPP+ *(309)* closely reflected the biphasic pattern of MPTP sensitivity found, suggesting that pharmacokinetics rather than pharmacodynamics underlie the changes in susceptibility that occur during the aging process with MPTP, at least in mice.

Recently, aging studies have been extended to primates, since they are closer to humans phylogenetically. Squirrel monkeys, for example, show a normal age-related decline in nigrostriatal DA, similar to what is thought to occur in humans *(310)*. Several nonsystematic studies have reported that older nonhuman primates are more sensitive to the effects of MPTP *(311)*, with a greater likelihood of LC damage *(312)*. Also, the appearance of eosinophilic inclusion bodies, possibly an immature form of the Lewy bodies characteristic of Parkinson's disease, occurs predominantly in older animals *(51,52)*. More recently a carefully controlled systematic study of this question showed unequivocally that primates are increasingly more sensitive to the effects of MPTP throughout their life-span. Somewhat surprisingly, however, there was no indication of an age-related increase in MAO, such as exists in mice *(313)*.

Unfortunately, the data on human levels of monoamines in aging are not clear-cut. Some authors report an increase in MAO B, but not MAO A *(314–317)*. Others report the reverse *(318,319)*, and some authors report both to be increased *(320)* or neither to be changed *(321)*. The variability in the above studies is probably owing in large part to the use of postmortem tissues and the lack of controls for tissue handling, patient treatment, and/or time between death and sampling. These factors would not come into play in the controlled primate studies. In addition, recent in vivo PET studies in humans found no differences in monoamine levels between young and older individuals *(321,323)*. These data make it unlikely that MAOs play a significant role in MPTP or parkinsonian syndromes in primates when it comes to age-related changes in toxicity, raising the intriguing question: If increasing sensitivity is not the result of increasing levels of MAO B, then what is it the result of? The answer will have to await future investigations.

An interesting sideline to the MPTP aging study is the relative efficiency of neuronal recovery mechanisms between young and old mice. Initially it was shown that, although MPTP-induced DA depletion recovers in about 5 mo in young mice, no recovery was found in older mice in that same time frame *(324)*. If, however, mice were examined after a longer time (16–24 mo), both age groups exhibited complete recovery of striatal DA levels *(309)*. These data show that although the systems for repair or induction of compensatory mechanisms are still available in older mice, they function at a lower level than in young mice. This could have consequences on the ability of animals to handle insult later in life.

Another factor that may contribute to sensitivity in aging animals is a decrease in mitochondrial electron transport *(325)*. Since energy impairment is an important factor in MPTP toxicity as well as being implicated in Parkinson's disease, decreased mitochondrial function with aging could become an important factor in cell sensitivity. In a study investigating the effects of MPP+ on activity of respiratory chain complexes in brain mitochondria of mice from three different age groups, Desai et al. *(326)* found electron transport deficits after MPP+ to be greater in the brains of old vs young mice. Thus, there exists not only the potential for more MPP+ to be produced in the nigrostriatal system of mice, but the potential for it to have a greater effect on energy metabolism.

The possibility of increased sensitivity in the mitochondrial respiratory chain has not been examined in the primate. Also, little work has been done on the relationship of repair or compensatory mechanisms to aging in nonprimate species and none at all in primates. So much remains to be done, but the exciting aspect is that we now appear to have a tool to investigate directly factors that make the nigrostriatal system more sensitive to neurodegeneration during the aging process.

7.3. MPTP and Nitric Oxide (NO)

Recently, NO has been implicated in the toxicity of MPTP. Neuronal NO is synthesized from L-arginine by the enzyme, NO synthase (NOS) *(327,328)* in a Ca^{2+}-calmodulin-dependent process *(329)*. In 1995, pretreatment with 7-nitroindazole, a potent inhibitor of the neuronal NOS *(330)*, was shown to protect completely against MPTP-induced depletion of DA and its metabolites, 3,4-dihydroxyphenylacetic acid and homovanillic acid, in mouse striatum *(331)*. Shortly thereafter, Przedborski et al. *(332)*, using tyrosine hydroxylase immunoreactivity and cell counts, were able to show 7-nitroindazole also protected against MPTP-induced dopaminergic cell death in mouse SN. A role for NO was also supported by experiments carried out in mice lacking the gene for neuronal NOS. These neuronal NOS knockout mice were found to be resistant to the effects of MPTP. The work with 7-nitroindazole and MPTP has been repeated in baboons, with much the same results for striatal DA and SN cell counts *(333)*.

Inhibition of complex I has long been put forth as a major factor in MPTP toxicity. NO or its reactive metabolite, peroxynitrite, has also been shown to interact with the mitochondrial respiratory chain. NO reversibly inhibits complex IV (cytochrome oxidase) *(334,335)* and peroxynitrite irreversibly inhibits complexes, I, II, and III *(336,337)*. A current theory of NO-induced toxicity is that increased NO levels inhibit complex IV, which leads to an increase in oxygen radical production. NO then reacts with those oxygen radicals to form peroxynitrite *(338)*, which then irreversibly blocks electron transport, resulting in ATP depletion, membrane depolarization, and cell death. The role MPTP plays in blocking complex I may help initiate the irreversible peroxynitrite electron transport blockade, or MPTP-induced increase in cytoplasmic Ca^{2+} may initiate No synthesis and a similar cascade of events.

These experiments are exciting in that they add a new dimension to MPTP-induced neurodegeneration and possibly neurodegeneration in general. A note of caution, however, on interpretation of the 7-nitroindazole experiments needs to be taken. Recent observations suggest 7-nitroindazole may play a role in inhibition of MPP^+ formation as well as a role in NOS inhibition *(338a,338b)*.

7.4. MPTP and Apoptosis

Naturally occurring cell death is common in the developing nervous system. This programmed cell death or "apoptosis" can occur in up to 50% of the neurons formed during embryogenesis *(339)*. However, the occurrence of apoptosis in neurodegeneration, if it occurs at all, is likely to be a much rarer phenomenon. Most neurodegenerative diseases are slowly progressive, chronic conditions in which only a few cells at any one time would be likely to exhibit the characteristic morphology. This contrasts sharply with an agent, such as MPTP, causing an acute toxicological insult, where far more cells may be affected at a given time. Even in this setting, the window for apoptosis to occur may only be a few hours, making detection difficult. Nonetheless, the search for apoptosis in MPTP toxicity has elicited a fair amount of interest.

There are good reasons to expect that MPTP would be capable of inducing apoptosis, given its known effects on the electron transport chain and ATP production. For example, depletion of ATP can cause activation of endonucleases via increased intracellular calcium, which can in turn damage DNA and cause the apoptotic process *(340,341)*. Furthermore, declining ATP production may lead to mitochondrial membrane depolarization and loss of mitochondrial Ca^{2+} sequestering capabilities, a sequence of events that has been shown to precipitate neuronal cell death by apoptosis. However, the literature on MPTP and apoptosis is still in its nascent stage. MPP^+ has been reported to induce apoptosis in cultured mesencephalic *(342)* and cerebellar granule cells *(343)*, but in vivo studies have been conflicting. Jackson-Lewis and colleagues *(344)* failed to detect any apoptotic cells in either the SN or ventral tegmental area in mice after systemic administration of MPTP. However, a similar study done at the Karolinska Institutet showed a significant number of midbrain apoptotic cells *(345)*.

There is also conflicting evidence regarding apoptosis in Parkinson's disease, with at least one group finding apoptotic cells *(346)*, but another failing to find any evidence of such a process *(347)*. Refinement of detection techniques and standardization of the parameters used to identify apoptosis may be necessary before the question of whether or not it occurs in either MPTP-induced parkinsonism or the idiopathic disease can be answered. If additional studies clearly indicate a role for apoptosis in either of these two conditions, then investigating the mechanisms by which it is occurring becomes of great interest. MPTP would be an invaluable experimental tool to study this process.

7.5. MPTP and Trophic Factors

MPTP has found many niches in the world of neuroscience research, but one of the most interesting and potentially important of these relates to its use as a model in which to study neuroprotection. At the current time, growth factors are at the forefront in the race to find new neuroprotective agents for Parkinson's disease, and typically the MPTP model has been one of the first places that investigators have chosen to test the mettle of new neuroprotective agents.

7.5.1. Neurotrophins

The first growth factor to be well characterized, nerve growth factor (NGF) *(348)*, is often considered to be the prototypic neurotrophic factor. Although NGF has been shown in in vitro studies to have no effect on differentiation or survival of dopaminergic neurons in mesencephalic culture *(349–352)*, there has been a report of protection against MPTP-induced striatal DA depletion after intraventricular injection of NGF in mice *(353)*. However, the absence of receptors for NGF on DA neurons would seem to make a direct effect unlikely. As more was learned about the specific actions of NGF on sympathetic neurons, striatal and basal forebrain cholinergic neurons, and certain sensory neurons, the search for trophic factors that would affect other neuronal populations began to intensify. The search led eventually to the discovery of a family of NGF-like factors, which collectively are called neurotrophins, and includes NGF, brain-derived neurotrophic factor (BDNF), and neurotrophins 3 (NT-3) and 4/5 (NT-4/5) *(354)*.

Although the structures of the neurotrophins are similar, each possesses distinctive neuronal specificities. BDNF is often referred to as the neurotrophic factor for the dopaminergic system *(350,351)*. In contrast to NGF, BDNF does promote survival and differentiation of dopaminergic neurons in culture *(350,351,355)*. It has also been shown to protect against dopaminergic neuronal damage induced by both MPP+ and 6-hydroxydopamine treatment of mesencephalic cultures *(351,352)*.

Once it became apparent that BDNF could protect dopaminergic neurons from specific neurotoxins in vitro, studies were quickly extended to experimental animals. In vivo protection was described in an elegant study by Frim et al. *(356)*, in which immortalized rat fibroblasts genetically engineered to produce BDNF were implanted near the SN 7 d before infusing MPP+ into the striatum of rats. They found a marked increase in SN dopaminergic neuronal survival after MPTP treatment in the rats implanted with BDNF-producing fibroblasts vs those implanted with normal fibroblasts. The experiments were then extended to nonhuman primates in a study carried out in Japanese green monkeys *(357)*. In this study, experimental animals received BDNF continuously for 10 d into the cisterna magna via an osmotic minipump; in control animals, the vehicle carrier was infused. MPTP was given systemically on days 1 and 4 to induce a parkinsonian syndrome, and analyses of behavioral parameters as well as pathological damage were made. BDNF completely protected against the parkinsonism induced by MPTP in the first week after treatment, and significantly reduced behavioral deficits normally seen at 2 wk postdosing. Histological examination of the SN at the end of the 2-wk behavioral study similarly showed partial protection against the neurodegenerative effects of MPTP in the BDNF-treated group.

Following the reports of BDNF action in promoting dopaminergic neuronal survival, NT-3 and NT-4/5 were found to have similar effects *(358,359)*. In a comprehensive evaluation of the neurotrophin gene family, Kirschner and colleagues *(360)* compared the capabilities of NGF, BDNF, NT-3, and NT-5 against MPP+ toxicity. Systemic administration of the neurotrophins immediately prior to intracerebral infusion of MPP+ into neonatal rat brains showed that NGF, BDNF, and NT-5, but not NT-3 significantly protected against MPP+-induced striatal lesions (BDNF > NT-5 > NGF).

One might have predicted the failure of NT-3 to protect based on what is known about neurotrophin receptors. The activity of the individual neurotrophins is related to the location and number of their high-affinity receptors. Neurotrophins bind to tyrosine protein kinase receptors when they are mediating their trophic effects. Three different receptor subtypes, named trkA, trkB, and trkC, have generally overlapping specificities for the various members of the neurotrophin family. However, NT-3 binds only to the trkC receptor, which is not found in the striatum, and thus NT-3 could not exert a protective effect on this tissue.

One possible mechanism by which neurotrophins may be protecting against MPTP toxicity could be related to their ability to act as antioxidants, since they have been shown to increase the activity of enzymes that remove ROS. For example, NGF increases the activity of catalase and GSH peroxidase in PC12 cells *(361,362)* and that of SOD and GSH reductase in rats *(363)*. Pretreatment with BDNF completely blocks the 6-hydroxydopamine-induced increase in oxidized GSH and increases the activity of the protective enzyme, GSH reductase, 100% in a neuroblastoma cell line *(352)*. Also, Kirschner et al. *(360)* have shown that both NGF and BDNF reduce the MPP+-induced hydroxyl radical formation in neonatal rat brains. Thus, it seems quite possible that the protective effects of neurotrophins could be the result of their ability to reduce oxidative stress.

Finally, in their study on BDNF gene expression, Hung and Lee *(364)* found the mesolimbic dopaminergic pathway to express more than twofold higher basal levels of BDNF mRNA than the nigrostriatal pathway, presumably conferring the resistance long recognized between the two systems. Although chronic MPTP treatment did increase SN BDNF message levels, they were still less than basal levels in the ventral tegmental area. This suggests that neurons possess mechanisms for protection or repair of systems that are probably being overwhelmed by MPTP.

7.5.2. Mitogenic Growth Factors

Another family of growth factors that have been shown to ameliorate MPTP toxicity are the mitogenic growth factors, which include acidic (aFGF) and basic fibroblast growth factor (bFGF), endothelial growth factor (EGF), and insulin-like growth factor (IGF). bFGF has been shown to enhance survival of dopaminergic neurons in mesencephalic culture in most studies *(349,365,366)*, but Beck and colleagues *(355)*, in a later morphometric analysis, reported a reduction in neurite outgrowth after bFGF treatment. It has also been reported that the increase in DA uptake activity induced by either bFGF or EGF in mesencephalic cultures is dependent on glial cell proliferation, since removing the proliferating glial cells eliminates the increase *(349,367)*. However, Unsicker et al. *(368)*, using single neuronal cultures, did find bFGF to enhance cell survival. Thus, the effect of the mitogenic factors appears to include both an indirect action via glial cells as well as a direct action on neurons.

When it comes to neuroprotection and MPTP, the results of investigations using this neurotrophic family have been mixed. Both aFGF and bFGF have been reported to enhance DA neuronal recovery after MPTP in mice *(369,370)*, in experiments where multiple injections during drug delivery caused significant mechanical damage to the brain. In a later study where care was taken to reduce brain trauma, DA recovery was not seen *(371)*. Jin and Iacovitti *(372)* have put forth an interesting hypothesis concerning how trauma during dosing may be related to neuroprotection. They found unilateral striatal infusion of aFGF compounds to MPTP-lesioned mice increased tyrosine hydroxylase activity and DOPAC levels 20–35% without affecting DA levels, suggesting that FGF enhanced DA synthesis and turnover. The recovery they found was greatest when animals were dosed with a muscle-derived differentiation factor, which included both aFGF and its activators *(373)*. Accordingly, it has been suggested that the mitogenic growth factors aFGF and bFGF require the simultaneous presence of endogenous activators for their protective effects. Endogenous compounds that may be provided by glial cells in the above culture studies or by infiltrating glial cells in response to inflammation owing to trauma in the early in vivo MPTP studies, serve to activate the mitogenic growth factors.

Few studies have been done to date on the mechanisms by which mitogenic growth factors may function. One mechanistic study was carried out by Kirschner and colleagues *(360)*, in their study on neurotrophic protection. They found bFGF to attenuate MPP+-induced hydroxyl radical formation in a manner similar to that seen with the neurotropins *(see above)*. In addition, they found bFGF to attenuate increases in striatal lactate concentrations that result from mitochondrial dysfunction after MPP+ treatment. They interpreted these data to mean that bFGF was protecting against impairment of oxidative metabolism, possibly via an effect on oxygen radical scavenging or maintenance of Ca^{2+}-buffering capacity. These studies indicate how much remains to be discovered about the pathways these mitogenic factors utilize and the roles they play in neuroprotection.

7.5.3. Glial Cell Line-Derived Neurotrophic Factor (GDNF)

As indicated by the mitogenic growth factor studies, neurons are not the only cell type involved in neurotrophic actions in the CNS. In fact, glia probably represent a much greater potential source for such endogenous compounds. The best-characterized glial trophic factor is GDNF. Several studies have examined whether or not GDNF can also protect against neurotoxicity. The data support a role for GDNF in repair as well as in protection. For example, the apomorphine-induced rotational behavior that occurs after 6-hydroxydopamine striatal lesioning was reduced by treatment with GDNF given 4 wk later. In addition, nigral, but not striatal DA recovered to control levels over time *(374)*. In an in vivo MPTP study in mice, GDNF given by intracerebral injection protected against DA depletion, striatal terminal damage, and nigral cell loss if given prior to systemic MPTP treatment. If given after MPTP, GDNF significantly restored terminal densities and DA levels *(375)*. A similar role for recovery was found in an in vitro mesencephalic dopaminergic neuronal culture model. Hou et al. *(376)* reported that although GDNF did not protect against MPP+ as measured by DA uptake and tyrosine hydroxylase staining, after MPP+ was removed, GDNF protected DA neurons from the further cell death and stimulated regrowth of DA fibers that were damaged.

The area of neurotrophic factors in neurodegeneration is a swiftly growing field. These endogenous compounds have vast potential for therapeutic intervention in trauma or disease if they can be identified and targeted to the necessary systems. Neurotrophic factors are already being tested in studies to improve fetal transplant survival with an eye toward a cure for Parkinson's disease *(377)*. In a fascinating series of studies, it has been suggested that recovery of MPTP-induced parkinsonism is owing as much to the trauma of the fetal transplant surgery as to the implant itself (for review, *see 378*), at least in the short-term. Wang and colleagues *(378)* have shown that implantation of pellets containing interleukin-1 into the brains of rats can stimulate behavioral and neuropathological recovery after 6-hydroxydopamine lesioning. The hypothesis is that the interleukin-1 induces an inflammation response that causes reactive astrocytes to produce neurotrophic factors and consequent neuronal sprouting and recovery. They suggest a similar sequence of events is partially responsible for the recovery found in experimental transplant work. The developing field of neurotrophic factors represents a myriad of areas for research in the neurosciences. As a model of neurotoxicity, MPTP will continue to be a valuable tool for ascertaining the role of growth factors in neuroprotection and recovery and for developing therapeutic uses.

8. SUMMARY

In this chapter, we have tried to give a brief historical background, to include what is new in MPTP biochemistry and physiological action, to report on some current hypotheses for mechanism(s) of toxicity, and to suggest some areas of future work. The value of MPTP in Parkinson's research cannot be overestimated. The discovery that MAO B is important for its conversion to MPP+ led to the testing of MAO inhibitors as neuroprotective agents in Parkinson's disease. Likewise, the discovery of a role for MPTP in energy impairment led to the successful search for a similar impairment in parkinsonism. The continuing search for MPTP mechanism of toxicity will hopefully provide new research directions, not only for Parkinson's, but for other neurodegenerative diseases as well. Overall, the MPTP story is a graphic demonstration of just how valuable a highly selective neurotoxin can be for scientific investigation, particularly when it targets the very same population of cells that are primarily affected in a major human disease.

REFERENCES

1. Lewin, R. (1984) Trail of ironies to Parkinson's disease. *Science* **224,** 1083–1085.
2. Davis, G. C., Williams, A. C., Markey, S. P., Ebert, M. H., Caine, E. D., Reichert, C. M., and Kopin, I. J. (1979) Chronic parkinsonism secondary to intravenous injection of meperidine analogues. *Psychiatry Res.* **1,** 249–254.
3. Langston, J. W., Ballard, P. A., Tetrud, J. W., and Irwin, I. (1983) Chronic parkinsonism in humans due to a product of meperidine analog synthesis. *Science* **219,** 979–980.
4. Langston, J. W. and Ballard, P. A. (1984) Parkinsonism induced by 1-methyl-4-phenyl-1,2,3,6-tetrahydropyridine (MPTP): Implications for treatment and the pathogenesis of Parkinson's disease. *Can. J. Neurol. Sci.* **11***,* 160–165.
5. Ballard, P. A., Tetrud, J. W., and Langston, J. W. (1985) Permanent human parkinsonism due to 1-methyl-4-phenyl-1,2,3,6-tetrahydropyridine (MPTP): Seven cases. *Neurology* **35,** 949–956.
6. Ruttenber, A. J., Garbe, P. L., Kalter, H. D., Castro, K. G., Tetrud, J. W., Porter, P., Irwin, I., and Langston, J. W. (1986) Meperidine analog exposure in California narcotics abusers: Initial epidemiologic findings, in *MPTP: A Neurotoxin Producing a Parkinsonian Syndrome* (Markey, S. P., Castagnoli, N., Jr., Trevor, A. J., and Kopin, I. J., eds.), Academic, New York, pp. 339–353.
7. Tetrud, J. W. and Langston, J. W. (1992) Tremor in MPTP-induced parkinsonism. *Neurology* **42,** 407–410.

8. Tetrud, J. W., Langston, J. W., Garbe, P. L., and Ruttenber, A. J. (1989) Mild parkinsonism in persons exposed to 1-methyl-4-phenyl-1,2,3,6-tetrahydropyridine (MPTP). *Neurology* **39,** 1483–1487.

9. Calne, D. B., Langston, J. W., Martin, W. R. W., Stoessl, A. J., Ruth, T. J., Adam, M. J., Pate, B. D., and Schulzer, M. (1985) Positron emission tomography after MPTP: observations relating to the cause of Parkinson's disease. *Nature* **317,** 246–248.

10. Martin, W. R. W., Palmer, M. R., Patlak, C. S., and Calne, D. B. (1989) Nigrostriatal function in humans studied with positron emission tomography. *Ann. Neurol.* **26,** 535–542.

11. Widner, H., Tetrud, J. W., Rehncrona, S., Brundin, P., Bjorklund, A., Lindvall, O., and Langston, J. W. (1993) Fifteen months follow-up on bilateral embryonic mesencephalic grafts in two cases of severe MPTP-induced parkinsonism, in *Advances in Neurology*, vol. 60, *Parkinson's Disease: From Basic Research to Treatment. Proceedings of the 10th International Symposium on Parkinson's Disease* (Narabayashi, H., Nagatsu, T., Yanagisawa, N., and Mizuno, Y., eds.), Raven, New York, pp. 729–733.

12. Stern, Y. and Langston, J. W. (1985) Intellectual changes in patients with MPTP-induced parkinsonism. *Neurology* **35,** 1506–1509.

13. Stern, Y., Tetrud, J. W., Martin, W. R. W., Kutner, S. J., and Langston, J. W. (1990) Cognitive change following MPTP exposure. *Neurology* **40,** 261–264.

14. Weihmuller, F. B., Hadjiconstantinou, M., and Bruno, J. P. (1988) Acute stress or neuroleptics elicit sensorimotor deficits in MPTP-treated mice. *Neurosci. Lett.* **85,** 137–142.

15. Ricaurte, G. A., Langston, J. W., DeLanney, L. E., Irwin, I., Peroutka, S. J., and Forno, L. S. (1986) Fate of nigrostriatal neurons in young mature mice given 1-methyl-4-phenyl-1,2,3,6-tetrahydropyridine: A neurochemical and morphological reassessment. *Brain Res.* **376,** 117–124.

16. Boyce, S., Kelly, E., Reavill, C., Jenner, P., and Marsden, C. D. (1984) Repeated administration of N-methyl-4-phenyl-1,2,5,6-tetrahydropyridine to rats is not toxic to striatal dopamine neurons. *Biochem. Pharmacol.* **33,** 1747–1752.

17. Chiueh, C. C., Markey, S. P., Burns, R. S., Johannessen, J. N., Pert, A., and Kopin, I. J. (1984) Neurochemical and behavioral effects of systemic and intranigral administration of *N*-methyl-4-pheynl-1,2,3,6-tetrahydropyridine in the rat. *Eur. J. Pharmacol.* **100,** 189–194.

18. Sahgal, A., Andrews, J. S., Biggins, J. A., Candy, J. M., Edwardson, J. A., Keith, A. B., Turner, J. D., and Wright, C. (1984) N-methyl-4-phenyl-1,2,3,6-tetrahydropyridine (MPTP) affects locomotor activity without producing a nigrostriatal lesion in the rat. *Neurosci. Lett.* **48,** 179–184.

19. Donnan, G. A., Kaczmarczyk, S. J., Solopotias, T., Rowe, P. J., Kalnins, R. M., Vajda, F. J. E., and Mendelsohn, F. A. O. (1986) The neurochemical and clinical effects of 1-methyl-4-phenyl-1,2,3,6-tetrahydropyridine in small animals. *Clin. Exp. Neurol.* **22,** 155–164.

20. Jarvis, M. F. and Wagner, G. C. (1990) 1-Methyl-4-phenyl-1,2,3,6-tetrahydropyridine-induced neurotoxicity in the rat: Characterization and age-dependent effects. *Synapse* **5,** 104–112.

21. Heikkila, R. E., Manzino, L., Cabbat, F. S., and Duvoisin, R. C. (1984) Protection against the dopaminergic neurotoxicity of 1-methyl-4-pheynl-1,2,5,6-tetrahydropyridine by monoamine oxidase inhibitors. *Nature* **311,** 467–469.

22. Sonsalla, P. K. and Heikkila, R. E. (1988) Neurotoxic effects of 1-methyl-4-phenyl-1,2,3,6-tetrahydropyridine (MPTP) and methamphetamine in several strains of mice. *Prog. Neuropsychopharmacol. Biol. Psychiatry* **12,** 345–354.

23. Gupta, M., Gupta, B. K., Thomas, R., Bruemmer, V., Sladek, J. R., Jr., and Felten, D. L. (1986) Aged mice are more sensitive to 1-methyl-4-phenyl-1,2,3,6-tetrahydropyridine than young adults. *Neurosci. Lett.* **70,** 326–331.

24. Mori, S., Fujitake, J., Kuno, S., and Sano, Y. (1988) Immunohistochemical evaluation of the neurotoxic effects of 1-methyl-4-phenyl-1,2,3,6-tetrahydropyridine (MPTP) on dopaminergic nigrostriatal neurons of young adult mice using dopamine and tyrosine hydroxylase antibodies. *Neurosci. Lett.* **90,** 57–62.

25. Seniuk, N. A., Tatton, W. G., and Greenwood, C. E. (1990) Dose-dependent destruction of the coeruleus-cortical and nigral-striatal projections by MPTP. *Brain Res.* **527,** 7–20.

26. Ricaurte, G. A., Irwin, I., Forno, L. S., DeLanney, L. E., Langston, E. B., and Langston, J. W. (1987) Aging and MPTP-induced degeneration of dopaminergic neurons in the substantia nigra. *Brain Res.* **403,** 43–51.

27. Chan, P., Di Monte, D. A., Langston, J. W., and Janson, A. M. (1997) (+)MK-801 does not prevent MPTP-induced loss of nigral neurons in mice. *J. Pharmacol. Exp. Ther.* **280,** 439–446.

28. Irwin, I., DeLanney, L. E., Forno, L. S., Finnegan, K. T., Di Monte, D. A., and Langston, J. W. (1990) The evolution of nigrostriatal neurochemical changes in the MPTP-treated squirrel monkey. *Brain Res.* **531,** 242–252.

29. Burns, R. S., Chiueh, C. C., Markey, S. P., Ebert, M. H., Jacobowitz, D. M., and Kopin, I. J. (1983) A primate model of parkinsonism: selective destruction of dopaminergic neurons in the pars compacta of the substantia nigra by *N*-methyl-4-phenyl-1,2,3,6-tetrahydropyridine. *Proc. Natl. Acad. Sci. USA* **80,** 4546–4550.

30. Langston, J. W., Forno, L. S., Rebert, C. S., and Irwin, I. (1984) Selective nigral toxicity after systemic administration of 1-methyl-4-phenyl-1,2,5,6-tetrahydropyridine (MPTP) in the squirrel monkey. *Brain Res.* **292,** 390–394.

31. Langston, J. W., Irwin, I., Langston, E. B., and Forno, L. S. (1984) Pargyline prevents MPTP-induced parkinsonism in primates. *Science* **225,** 1480–1482.

32. Cohen, G., Pasik, P., Cohen, B., Leist, A., Mytilineou, C., and Yahr, M. D. (1985) Pargyline and deprenyl prevent the neurotoxicity of 1-methyl-4-phenyl-1,2,3,6-tetrahydropyridine (MPTP) in monkeys. *Eur. J. Pharmacol.* **106,** 209–210.

33. Jenner, P., Rose, S. P., Nomoto, M., and Marsden, C. D. (1986) MPTP-induced parkinsonism in the common marmoset: behavioral and biochemical effects. *Adv. Neurol.* **45,** 183–186.

34. Di Monte, D. A. and Langston, J. W. (1993) MPTP-induced parkinsonism in nonhuman primates, in *Current Concepts in Parkinson's Disease Research* (Schneider, J. S. and Gupta, M., eds.), Hogrefe and Huber Publishers, Toronto, pp. 159–179.

35. Tetrud, J. W., Langston, J. W., Redmond, D. E., Jr., Roth, R. H., Sladek, J. R., Jr., and Angel, R. W. (1986) MPTP-induced tremor in human and non-human primates [Abstract]. *Neurology* **36(Suppl. 1),** 308.

36. Clarke, C. E., Boyce, S., Robertson, R. G., Sambrook, M. A., and Crossman, A. R. (1989) Drug-induced dyskinesia in primates rendered hemiparkinsonian by intracarotid administration of 1-methyl-4-phenyl-1,2, 3,6-tetrahydropyridine (MPTP). *J. Neurol. Sci.* **90,** 307–314.

37. Schneider, J. S. (1989) Levodopa-induced dyskinesias in Parkinsonian monkeys: Relationship to extent of nigrostriatal damage. *Pharmacol. Biochem. Behavior* **34**, 193–196.

38. Schneider, J. S., Unguez, G. A., Yuwiler, A., Berg, S. C., and Markham, C. H. (1988) Deficits in operant behaviour in monkeys treated with *N*-methyl-4-phenyl-1,2,3,6-tetrahydropyridine (MPTP). *Brain* **111**, 1265–1285.

39. Schneider, J. S. (1990) Chronic exposure to low doses of MPTP. II. Neurochemical and pathological consequences in cognitively-impaired, motor asymptomatic monkeys. *Brain Res.* **534**, 25–36.

40. Schneider, J. S. and Kovelowski, C. J., II (1990) Chronic exposure to low doses of MPTP. I. Cognitive deficits in motor asymptomatic monkeys. *Brain Res.* **519**, 122–128.

41. Kish, S. J., Shannak, K., and Hornykiewicz, O. (1988) Uneven pattern of dopamine loss in the striatum of patients with idiopathic Parkinson's disease. *N. Engl. J. Med.* **318(14)**, 876–880.

42. Elsworth, J. D., Deutch, A. Y., Redmond, D. E., Jr., Sladek, J. R., Jr., and Roght, R. H. (1987) Differential responsiveness to 1-methyl-4-phenyl-1,2,3,6-tetrahydropyridine toxicity in subregions of the primate substantia nigra and striatum. *Life Sci.* **40**, 193–202.

43. Pifl, C., Schingnitz, G., and Hornykiewicz, O. (1988) The neurotoxin MPTP does not reproduce in the rhesus monkey the interregional pattern of striatal dopamine loss typical of human idiopathic Parkinson's disease. *Neurosci. Lett.* **92**, 228–233.

44. Moratalla, R., Quinn, B., DeLanney, L. E., Irwin, I., Langston, J. W., and Graybiel, A. M. (1992) Differential vulnerability of primate caudate-putamen and striosome-matrix dopamine systems to the neurotoxic effects of 1-methyl-4-phenyl-1,2,3,6-tetrahydropyridine. *Proc. Natl. Acad. Sci. USA* **89**, 3859–3863.

45. Wüllner, U., Pakzaban, P., Brownell, A.-L., Hantraye, P., Burns, L., Shoup, T., Elmaleh, D., et al. (1994) dopamine terminal loss and onset of motor symptoms in MPTP-treated monkeys. A positron emission tomography study with ^{11}C-CFT. *Exp. Neurol.* **126**, 305–309.

46. Goldman, J. E., Yen, S. H., Chiu, F. C., and Peress, N. S. (1983) Lewy bodies of Parkinson's disease contain neurofilament antigens. *Science* **221**, 1082–1084.

47. Forno, L. S., Sternberger, L. A., Sternberger, N. H., Strefling, A. M., Swanson, K., and Eng, L. F. (1986) Reaction of Lewy bodies with antibodies to phosphorylated and non-phosphorylated neurofilaments. *Neurosci. Lett.* **64**, 253–258.

48. Galloway, P. G., Grundke-Iqbal, I., Iqbal, K., and Perry, G. (1988) Lewy bodies contain epitopes both shared and distinct from Alzheimer neurofibrillary tangles. *J. Neuropathol. Exp. Neurol.* **47**, 654–663.

49. Love, S., Saitoh, T., Saitoh, T., Quijada, S., Cole, G. M., and Terry, R. D. (1988) Alz-50, ubiquitin, and tau reactivity of neurofibrillary tangles, Pick bodies and Lewy bodies. *J. Neuropathol. Exp. Neurol.* **47**, 393–405.

50. Bancher, C., Lassmann, H., Budka, H., Jellinger, K. A., Grundke-Iqbal, I., Iqbal, K., Wiche, G., Seitelberger, F., and Wisniewski, H. M. (1989) An antigenic profile of Lewy bodies: Immunocytochemical indication for protein phosphorylation and ubiquitination. *J. Neuropathol. Exp. Neurol.* **48**, 81–93.

51. Forno, L. S., Langston, J. W., DeLanney, L. E., Irwin, I., and Ricaurte, G. A. (1986) Locus ceruleus lesions and eosinophilic inclusions in MPTP-treated monkeys. *Ann. Neurol.* **20**, 449–455.

52. Forno, L. S., Langston, J. W., DeLanney, L. E., and Irwin, I. (1988) An electron microscopic study of MPTP-induced inclusion bodies in an old monkey. *Brain Res.* **448,** 150–157.

53. Langston, J. W., Irwin, I., Langston, E. B., and Forno, L. S. (1984) 1-Methyl-4-phenylpyridinium ion (MPP+): Identification of a metabolite of MPTP, a toxin selective to the substantia nigra. *Neurosci. Lett.* **48,** 87–92.

54. Markey, S. P., Johannessen, J. N., Chiueh, C. C., Burns, R. S., and Herkenham, M. A. (1984) Intraneuronal generation of a pyridinium metabolite may cause drug-induced parkinsonism. *Nature* **311,** 464–467.

55. Irwin, I. and Langston, J. W. (1985) Selective accumulation of MPP+ in the substantia nigra: A key to neurotoxicity? *Life Sci.* **36,** 207–212.

56. Riachi, N. J., Harik, S. I., Kalaria, R. N., and Sayre, L. M. (1988) On the mechanisms underlying 1-methyl-4-phenyl-1,2,3,6-tetrahydropyridine neurotoxicity. II. Susceptibility among mammalian species correlates with the toxin's metabolic patterns in brain microvessels and liver. *J. Pharmacol. Exp. Ther.* **244(2),** 443–448.

57. Sayre, L. M., Arora, P. K., Iacofano, L. A., and Harik, S. I. (1986) Comparative toxicity of MPTP, MPP+ and 3,3-dimethyl-MPDP+ to dopaminergic neurons of the rat substantia nigra. *Eur. J. Pharmacol.* **124,** 171–174.

58. Castagnoli, N., Jr., Chiba, K., and Trevor, A. J. (1985) Potential bioactivation pathways for the neurotoxin 1-methyl-4-phenyl-1,2,3,6-tetrahydropyridine (MPTP). *Life Sci.* **36(3),** 225–230.

59. Chiba, K., Peterson, L. A., Castagnoli, K. P., Trevor, A. J., and Castagnoli, N., Jr. (1985) Studies on the molecular mechanism of bioactivation of the selective nigrostriatal toxin 1-methyl-4-phenyl-1,2,3,6-tetrahydropyridine. *Drug Metab. Dispos.* **13,** 342–347.

60. Singer, T. P., Salach, J. I., Castagnoli, N., Jr., and Trevor, A. J. (1986) Interactions of the neurotoxic amine 1-methyl-4-phenyl-1,2,3,6-tetrahydropyridine with monoamine oxidases. *Biochem. J.* **235,** 785–789.

61. Chiba, K., Trevor, A. J., and Castagnoli, N., Jr. (1984) Metabolism of the neurotoxic tertiary amine, MPTP, by brain monoamine oxidase. *Biochem. Biophys. Res. Commun.* **120,** 574–578.

62. Mytilineou, C. and Cohen, G. (1984) 1-Methyl-4-phenyl-1,2,3,6-tetrahydropyridine destroys dopamine neurons in explants of rat embryo mesencephalon. *Science* **225,** 529–531.

63. Sanchez-Ramos, J. R., Barrett, J. N., Goldstein, M., Weiner, W. J., and Hefti, F. (1986) 1-Methyl-4-phenylpyridinium (MPP+) but not 1-methyl-4-phenyl-1,2,3,6-tetrahydropyridine (MPTP) selectively destroys dopaminergic neurons in cultures of dissociated rat mesencephalic neurons. *Neurosci. Lett.* **72,** 215–220.

64. Di Monte, D. A., Wu, E. Y., Irwin, I., DeLanney, L. E., and Langston, J. W. (1992) Production and disposition of 1-methyl-4-phenylpyridinium in primary cultures of mouse astrocytes. *Glia* **5,** 48–55.

65. Bradbury, A. J., Costall, B., Domeney, A. M., Jenner, P., Kelly, M. E., Marsden, C. D., and Naylor, R. J. (1986) 1-Methyl-4-phenylpyridine is neurotoxic to the nigrostriatal dopamine pathway. *Nature* **319,** 56–57.

66. Irwin, I., Ricaurte, G. A., DeLanney, L. E., and Langston, J. W. (1988) The sensitivity of nigrostriatal dopamine neurons to MPP+ does not increase with age. *Neurosci. Lett.* **87,** 51–56.

67. Fuller, R. W. and Hemrick-Luecke, S. K. (1985) Influence of selective, reversible inhibitors of monoamine oxidase on the prolonged depletion of striatal dopamine by 1-methyl-4-phenyl-1,2,3,6-tetrahydropyridine in mice. *Life Sci.* **37,** 1089–1096.

68. Kalaria, R. N., Mitchell, M. J., and Harik, S. I. (1987) Correlation of 1-methyl-4-phenyl-1,2,3,6-tetrahydropyridine neurotoxicity with blood-brain barrier monoamine oxidase activity. *Proc. Natl. Acad. Sci. USA* **84,** 3521–3525.

69. Reinhard, J. F., Jr., Diliberto, E. J., Jr., Viveros, O. H., and Daniels, A. J. (1987) Subcellular compartmentalization of 1-methyl-4-phenylpyridinium with catecholamines in adrenal medullary chromaffin vesicles may explain the lack of toxicity to adrenal chromaffin cells. *Proc. Natl. Acad. Sci. USA* **84,** 8160–8164.

70. Reinhard, J. F., Jr., Carmichael, S. W., and Daniels, A. J. (1990) Mechanisms of toxicity and cellular resistance to 1-methyl-4-phenyl-1,2,3,6-tetrahydropyridine and 1-methyl-4-phenylpyridinium in adrenomedullary chromaffin cell cultures. *J. Neurochem.* **55,** 311–320.

71. Reinhard, J. F., Jr., Daniels, A. J., and Viveros, O. H. (1988) Potentiation by reserpine and tetrabenazine of brain catecholamine depletions by MPTP (i-methyl-4-phenyl-1,2,3,6-tetrahydropyridine) in the mouse: Evidence for subcellular sequestration as basis for cellular resistance to the toxicant. *Neurosci. Lett.* **90,** 349–353.

72. Levitt, P., Pintar, J. E., and Breakefield, X. O. (1982) Immunocytochemical demonstration of monoamine oxidase B in brain astrocytes and serotonergic neurons. *Proc. Natl. Acad. Sci. USA* **79,** 6385–6389.

73. Westlund, K. N., Denney, R. M., Kochersperger, L. M., Rose, R. M., and Abell, C. W. (1985) Distinct monoamine oxidase A and B populations in primate brain. *Science* **230,** 181–183.

74. Chiba, K., Trevor, A. J., and Castagnoli, N., Jr. (1985) Active uptake of MPP+, a metabolite of MPTP, by brain synaptosomes. *Biochem. Biophys. Res. Commun.* **128,** 1229–1232.

75. Javitch, J. A., D'Amato, R. J., Strittmatter, S. M., and Snyder, S. H. (1985) Parkinsonism-inducing neurotoxin, *N*-methyl-4-phenyl-1,2,3,6-tetrahydropyridine: Uptake of the metabolite *N*-methyl-4-phenylpyridine by dopamine neurons explains selective toxicity. *Proc. Natl. Acad. Sci. USA* **82,** 2173–2177.

76. Ransom, B. R., Kunis, D. M., Irwin, I., and Langston, J. W. (1987) Astrocytes convert the parkinsonism-inducing neurotoxin, MPTP, to its active metabolite, MPP+. *Neurosci. Lett.* **75,** 323–328.

77. Schinelli, S., Zuddas, A., Kopin, I. J., Barker, J. L., and Di Porzio, U. (1988) 1-Methyl-4-phenyl-1,2,3,6-tetrahydropyridine metabolism and 1-methyl-4-phenylpyridinium uptake in dissociated cell cultures from the embryonic mesencephalon. *J. Neurochem.* **50,** 1900–1907.

78. Di Monte, D. A., Wu, E. Y., Irwin, I., DeLanney, L. E., and Langston, J. W. (1991) Biotransformation of 1-methyl-4-phenyl-1,2,3,6-tetrahydropyridine in primary cultures of mouse astrocytes. *J. Pharmacol. Exp. Ther.* **258,** 594–600.

79. Scotcher, K. P., Irwin, I., DeLanney, L. E., Langston, J. W., and Di Monte, D. A. (1991) Mechanism of accumulation of the 1-methyl-4-phenylpyridinium species into mouse brain synaptosomes. *J. Neurochem.* **56,** 1602–1607.

80. Reinhard, J. F., Daniels, A. J., and Painter, G. R. (1990) Carrier-independent entry of 1-methyl-4-phenylpyridinium (MPP+) into adrenal chromaffin cells as a consequence of charge delocalization. *Biochem. Biophys. Res. Commun.* **168,** 1143–1148.

81. Irwin, I., Langston, J. W., and DeLanney, L. E. (1987) 4-Phenylpyridine (4PP) and MPTP: The relationship between striatal MPP+ concentrations and neurotoxicity. *Life Sci.* **40,** 731–740.

82. Riachi, N. J., Arora, P. K., Sayre, L. M., and Harik, S. I. (1988) Potent neurotoxic fluorinated 1-methyl-4-phenyl-1,2,3,6-tetrahydropyridine analogs as potential probes in models of Parkinson's disease. *J. Neurochem.* **50,** 1319–1321.

83. Cashman, J. R. and Ziegler, D. M. (1986) Contribution of *N*-oxygenation to the metabolism of MPTP (1-methyl-4-phenyl-1,2,3,6-tetrahydropyridine) by various liver preparations. *Mol. Pharmacol.* **29,** 163–167.

84. Arora, P. K., Riachi, N. J., Harik, S. I., and Sayre, L. M. (1988) Chemical oxidation of 1-methyl-4-phenyl-1,2,3,6-tetrahydropyridine (MPTP) and its in vivo metabolism in rat brain and liver. *Biochem. Biophys. Res. Commun.* **152,** 1339–1347.

85. Di Monte, D. A., Wu. E. Y., and Langston, J. W. (1992) Role of astrocytes in MPTP metabolism and toxicity. *Ann. NY Acad. Sci.* **648,** 219–228.

86. Melamed, E., Rosenthal, J., Globus, M., Cohen, O., and Uzzan, A. (1985) Suppression of MPTP-induced dopaminergic neurotoxicity in mice by nomifensine and L-dopa. *Brain Res.* **342,** 401–404.

87. Ricaurte, G. A., Langston, J. W., DeLanney, L. E., Irwin, I., and Brooks, J. D. (1985) Dopamine uptake blockers protect against the dopamine depleting effects of 1-methyl-4-phenyl-1,2,3,6-tetrahydropyridine (MPTP) in the mouse striatum. *Neurosci. Lett.* **59,** 259–264.

88. Sundstrom, E., Goldstein, M., and Jonsson, G. (1986) Uptake inhibition protects nigrostriatal dopamine neurons from the neurotoxicity of 1-methyl-4-phenylpyridine (MPP+) in mice. *Eur. J. Pharmacol.* **131,** 289–292.

89. Mayer, R. A., Kindt, M. V., and Heikkila, R. E. (1986) Prevention of MPTP-induced neurotoxicity by dopamine uptake inhibitors, in *MPTP; A Neurotoxin Producing a Parkinsonian Syndrome* (Markey, S. P., Castagnoli, N., Jr., Trevor, A. J., and Kopin, I. J., eds.), Academic, Orlando, FL, pp. 585–590.

90. Schultz, W., Scarnati, E., Sundstrom, E., Tsutsumi, T., and Jonsson, G. (1986) The catecholamine uptake blocker nomifensine protects against MPTP-induced parkinsonism in monkeys. *Exp. Brain Res.* **63,** 216–220.

91. Schultz, W., Scarnati, E., Sundstrom, E., and Romo, R. (1989) Protection against 1-methyl-4-phenyl-1,2,3,6-tetrahydropyridine-induced Parkinsonism by the catecholamine uptake inhibitor nomifensine: Behavioral analysis in monkeys with partial striatal dopamine depletions. *Neuroscience* **31,** 219–230.

92. Takada, M., Li, Z. K., and Hattori, T. (1990) Astroglial ablation prevents MPTP-induced nigrostriatal neuronal death. *Brain Res.* **509,** 55–61.

93. Javitch, J. A., Uhl, G. R., and Snyder, S. H. (1984) Parkinsonism-inducing neurotoxin, *N*-methyl-4-phenyl-1,2,3,6-tetrahydropyridine: characterization and localization of receptor binding sites in rat and human brain. *Proc. Natl. Acad. Sci. USA* **81,** 4591–4595.

94. Santiago, M., Machado, A., and Cano, J. (1996) Nigral and striatal comparative study of the neurotoxic action of 1-methyl-4-phenylpyridinium ion: involvement of dopamine uptake system. *J. Neurochem.* **66,** 1182–1190.

95. Mitchell, I. J., Cross, A. J., Sambrook, M. A., and Crossman, A. R. (1986) *N*-methyl-4-phenyl-1,2,3,6-tetrahydropyridine-induced parkinsonism in the monkey: neurochemical pathology and regional brain metabolism. *J. Neural Transm.* **XX,** 41–46.

96. Schneider, J. S., Yuwiler, A., and Markham, C. H. (1987) Selective loss of subpopulations of ventral mesencephalic dopaminergic neurons in the monkey following exposure to MPTP. *Brain Res.* **411,** 144–150.

97. German, D. C., Durach, A., Askari, S., Speciale, S. G., and Bowden, D. M. (1988) 1-Methyl-4-phenyl-1,2,3,6-tetrahydropyridine-induced parkinsonian syndrome in macaca fascicularis: Which midbrain dopaminergic neurons are lost? *Neuroscience* **24,** 161–174.

98. Parent, A., Lavoie, B., Smith, Y., and Bedard, P. J. (1990) The dopaminergic nigropallidal projection in primates: Distinct cellular origin and relative sparing in MPTP-treated monkeys. *Adv. Neurol.* **53,** 111–116.

99. D'Azur, V. (1786) Traite d'anatomie et de physiologie, cited by Marsden, C. D.: brain melanin in *Pigments in Pathology,* 1969, (Wolman, M., ed.), Academic, New York, pp. 395–420.

100. Graham, D. G. (1978) Oxidative pathways for catecholamines in the genesis of neuromelanin and cytotoxic quinones. *Mol. Pharmacol.* **14,** 633–643.

101. Marsden, C. D. (1983) Neuromelanin and Parkinson's disease. *J. Neural Transm.* **Suppl. 19,** 121–141.

102. Forno, L. S. (1966) Pathology of parkinsonism—a preliminary report of 24 cases. *J. Neurosurg.* **24,** 266–271.

103. Forno, L. S. and Alvord, E. C., Jr. (1974) Depigmentation in the nerve cells of the substantia nigra and locus coeruleus in parkinsonism. *Adv. Neurol.* **5,** 195–202.

104. Mann, D. M. A. and Yates, P. O. (1983) Possible role of neuromelanin in the pathogenesis of Parkinson's disease. *Mech. Age Dev.* **21,** 193–203.

105. Lindquist, N. G., Larsson, B. S., and Lyden-Sokolowski, A. (1987) Neuromelanin and its possible protective and destructive properties. *Pigment Cell Res.* **1,** 133–136.

106. Commoner, B., Townsend, J., and Pake, G. E. (1954) Free radicals in biological materials. *Nature* **174,** 689–691.

107. McGinness, J. and Proctor, P. (1973) The importance of the fact that melanin is black. *J. Theor. Biol.* **39,** 677–680.

108. Gan, E. V., Haberman, H. F., and Menon, I. A. (1976) Electron transfer properties of melanin. *Arch. Biochem. Biophys.* **173,** 666–672.

109. Lindquist, N. G. (1972) Accumulation in vitro of ^{35}S-chlorpromazine in the neuromelanin of human substantia nigra and locus coeruleus. *Arch. Int. Pharmacodyn. Ther.* **200,** 190–195.

110. Salazar, M., Sokoloski, T. D., and Patil, P. N. (1978) Binding of dopaminergic drugs by the neuromelanin of the substantia nigra, synthetic melanins and melanin granules. *Fed. Proc.* **37,** 2403–2407.

111. Swartz, H. M., Sarna, T., and Zecca, L. (1992) Modulation by neuromelanin of the availability and reactivity of metal ions. *Ann. Neurol.* **32 Suppl.,** S69–S75.

112. Youdim, M. B. H., Ben-Shachar, D., and Riederer, P. (1994) The enigma of neuromelanin in Parkinson's disease substantia nigra. *J. Neural Transm.* **Suppl. 43,** 113–122.

113. Marsden, C. D. (1969) Brain melanin, in *Pigments in Pathology* (Wolman, M., ed.), Academic, New York, pp. 395–420.

114. Barden, H. and Levine, S. (1983) Histochemical observations on rodent brain neuromelanin. *Brain Res. Bull.* **10,** 847–851.

115. Herrero, M. T., Hirsch, E. C., Kastner, A., Ruberg, M., Luquin, M. R., Laguna, J., Javoy-Agid, F., Obeso, J. A., and Agid, Y. (1993) Does neuromelanin contribute to the vulnerability of catecholaminergic neurons in monkeys intoxicated with MPTP. *Neuroscience* **56,** 499–511.

116. D'Amato, R. J., Lipman, Z. P., and Snyder, S. H. (1986) Selectivity of the parkinsonian neurotoxin MPTP: toxic metabolite MPP+ binds to neuromelanin. *Science* **231,** 987–989.

117. D'Amato, R. J., Benham, D. F., and Snyder, S. H. (1987) Characterization of the binding of *N*-methyl-4-phenylpyridine, the toxic metabolite of the parkinsonian neurotoxin *N*-methyl-4-phenyl-1,2,3,6-tetrahydropyridine, to neuromelanin. *J. Neurochem.* **48,** 653–658.

118. Lindquist, N. G., Lyden-Sokolowski, A., and Larsson, B. S. (1986) Accumulation of a parkinsonism-inducing neurotoxin in melanin-bearing neurons: Autoradiographic studies on 3H-MPTP. *Acta Phamacol. Toxicol.* **59,** 161–164.

119. D'Amato, R. J., Alexander, G. M., Schwartzman, R. J., Kitt, C. A., Price, D. L., and Snyder, S. H. (1987) Evidence for neuromelanin involvement in MPTP-induced neurotoxicity. *Nature* **327,** 324–326.

120. Wu, E. Y., Chiba, K., Trevor, A. J., and Castagnoli, N., Jr. (1986) Interactions of the 1-methyl-4-phenyl-2,3-dihydropyridinium species with synthetic dopamine-melanin. *Life Sci.* **39,** 1695–1700.

121. Korytowski, W., Felix, C. C., and Kalyanaraman, B. (1987) Mechanism of oxidation of 1-methyl-4-phenyl-2,3,-dihydropyridinium (MPDP+). *Biochem. Biophys. Res. Commun.* **144(2),** 692–698.

122. Melamed, E., Soffer, D., Rosenthal, J., Pikarsky, E., and Reches, A. (1987) Effect of intrastriatal and intranigral administration of synthetic neuromelanin on the dopaminergic neurotoxicity of MPTP in rodents. *Neurosci. Lett.* **83,** 41–46.

123. Pileblad, E., Nissbrandt, H., and Carlsson, A. (1984) Biochemical and functional evidence for a marked dopamine releasing action of *N*-methyl-4-phenyl-1,2,3,6-tetrahydropyridine (NMPTP) in mouse brain. *J. Neural Transm.* **60,** 199–203.

124. Chang, G. D. and Ramirez, V. D. (1986) The mechanism of action of MPTP and MPP+ on endogenous dopamine release from the rat corpus striatum superfused in vitro. *Brain Res.* **368,** 134–140.

125. Sirinathsinghji, D. J. S., Heavens, R. P., and McBride, C. S. (1988) Dopamine-releasing action of 1-methyl-4-phenyl-1,2,3,6-tetrahydropyridine (MPTP) and 1-methyl-4-phenylpyridine (MPP+) in the neostriatum of the rat as demonstrated in vivo by the push-pull perfusion technique: Dependence on sodium but not calcium ions. *Brain Res.* **443,** 101–116.

126. Huang, S.-J. and Chiueh, C. C. (1990) Calcium-dependent potentiation of potassium depolarization-evoked release of dopamine from nigrostriatal (A9) terminals by picomoles of MPP+. *FASEB J.* **4,** A605–A608.

127. Lang, K., Huang, S.-J., Miyake, H., and Chiueh, C. C. (1990) Calcium ion fluxes mediate sustained release of endogenous striatal dopamine by toxic doses of MPP+ in vivo. *FASEB J.* **4,** A605.

128. Kupsch, A., Gerlack, M., Pupeter, S. C., et al. (1995) Pretreatment with nimodipine prevents MPTP-induced neurotoxicity at the nigral but not at the striatal level in mice. *Neuroreport* **6,** 621–625.

129. Schmidt, C. J., Matsuda, L. A., and Gibb, J. W. (1984) In vitro release of tritiated monoamines from rat CNS tissue by the neurotoxic compound 1-methyl-4-phenyl-1,2,3,6-tetrahydropyridine. *Eur. J. Pharmacol.* **103**, 255–260.

130. Wilson, J. A., Lau, Y. S., Gleeson, J. G., and Wilson, J. S. (1991) The action of MPTP on synaptic transmission is affected by changes in Ca^{2+} concentrations. *Brain Res.* **541**, 342–346.

131. Chiueh, C. C. and Huang, S.-J. (1991) MPP+ enhances potassium evoked striatal release through a Ω-conotoxin-insensitive, tetrodotoxin- and nimodipine-sensitive calcium dependent mechanism. *Ann. NY Acad. Sci.* **635**, 393–396.

132. Wilson, J. A., Wilson, J. S., and Weight, F. F. (1986) MPTP causes a non-reversible depression of synaptic transmission in mouse neostriatal brain slice. *Brain Res.* **368**, 357–360.

133. Mayer, R. A., Kindt, M. V., and Heikkila, R. E. (1986) Prevention of the nigrostriatal toxicity of 1-methyl-4-phenyl-1,2,3,6-tetrahydropyridine by inhibitors of 3,4-dihydroxyphenylethylamine transport. *J. Neurochem.* **47**, 1073–1079.

134. Johnson, E. A., Wu, E. Y., Rollema, H., Booth, R. G., Trevor, A. J., and Castagnoli, N., Jr. (1989) 1-Methyl-4-phenylpyridinium (MPP+) analogs: In vivo neurotoxicity and inhibition of striatal synaptosomal dopamine uptake. *Eur. J. Pharmacol.* **166**, 65–74.

135. Athwal, N. S. S., Ramsden, D. B., Simpson, M., and Williams, A. C. (1996) Inhibition of dopamine uptake into PC-12 cells by analogues of 1-methyl-4-phenyl-1,2,3,6-tetrahydropyridine (MPTP). *Parkinsonism and Related Disorders* **2(1)**, 1–6.

136. Berman, S. B., Zigmond, M. J., and Hastings, T. G. (1996) Dopamine transport function: modulation by oxidative stress, in *Neurodegenerative Diseases 1995: Molecular and Cellular Mechanisms and Therapeutic Advances—XVth Washington International Symposium, Washington, DC* (Fiskum, G., ed.), Plenum, New York, p. 136.

137. Ramassamy, C., Girbe, F., Christen, Y., and Cosentin, J. (1995) Peroxidation of synaptosomes alters the dopamine uptake complex but spares the exocytotic release of dopamine. *Neurodegeneration* **4(2)**, 155–160.

138. Drukarch, B., Schepens, H. T. W. J., Langeveld, C. H., and Stoof, J. C. (1996) The vesicular storage properties of the nigro-striatal dopaminergic projection are involved in its high sensitivity to reactive oxygen species, in *Neurodegenerative Diseases 1995: Molecular and Cellular Mechanisms and Therapeutic Advances, XVth Washington International Symposium, Washington, DC* (Fiskum, G., ed.), Plenum, New York, p. 67.

139. Stern-Bach, Y., Keen, J. N., Bejerano, M., Steiner-Mordoch, S., Wallach, M., Findlay, J. B. C., and Schuldiner, S. (1992) Homology of a vesicular amine transporter to a gene conferring resistance to 1-methyl-4-phenylpyridinium. *Proc. Natl. Acad. Sci. USA* **89**, 9730–9733.

140. Vaccari, A., Del Zompo, M., Melis, F., Gessa, G. L., and Rossetti, Z. L. (1991) Interaction of 1-methyl-4-phenylpyridinium ion and tyramine with a site putatively involved in the striatal vesicular release of dopamine. *Br. J. Pharmacol.* **104**, 573–574.

141. Schuldiner, S., Steiner-Mordoch, S., Yelin, R., Wall, S. C., and Rudnick, G. (1993) Amphetamine derivatives interact with both plasma membrane and secretory vesicle biogenic amine transporters. *Mol. Pharmacol.* **44**, 1227–1231.

142. Lesch, K. P., Heils, A., and Riederer, P. (1996) The role of neurotransporters in excitotoxicity, neuronal cell death, and other neurodegenerative processes. *J. Mol. Med.* **74**, 365–378.

143. Liu, Y., Roghani, A., and Edwards, R. H. (1992) Gene transfer of a reserpine-sensitive mechanism of resistance to *N*-methyl-4-phenylpyridinium. *Proc. Natl. Acad. Sci. USA* **89,** 9074–9078.

144. Sundstrom, E. and Jonsson, G. (1985) Pharmacologoical interferance with the toxic action of 1-methyl-4-phenyl-1, 2, 3, 6-tetrahydropyridine (MPTP) on central catecholamine neurons in the mouse. *Eur. J. Pharmacol.* **110,** 293–299.

145. Pifl, C., Hornykiewicz, O., Giros, B., and Caron, M. G. (1996) Catecholamine transporters and 1-methyl-4-phenyl-1,2,3,6-tetrahydropyridine neurotoxicity: studies comparing the cloned human noradrenaline and human dopamine transporter. *J. Pharmacol. Exp. Ther.* **277(3),** 1437–1443.

146. Mavridis, M., Degryse, A.-D., Lategan, A. J., Marien, M. R., and Colpaert, F. C. (1991) Effects of locus coeruleus lesions on Parkinsonian signs, striatal dopamine and substantia nigra cell loss after 1-methyl-4-phenyl-1,2,3,6-tetrahydropyridine in monkeys: A possible role for the locus coeruleus in the progression of Parkinson's disease. *Neuroscience* **41,** 507–523.

147. Marien, M., Briley, M., and Colpaert, F. (1993) Noradrenaline depletion exacerbates MPTP-induced striatal dopamine loss in mice. *Eur. J. Pharmacol.* **236,** 487–489.

148. Bing, G., Zhang, Y., Watanabe, Y., McEwen, B. S., and Stone, E. A. (1994) Locus coeruleus lesions potentiate neurotoxic effects of MPTP in dopaminergic neurons of the substantia nigra. *Brain Res.* **668,** 261–265.

149. Fornai, F., Alessandri, M. G., Fascetti, F., Vaglini, F., and Corsini, G. U. (1995) Clonidine suppresses 1-methyl-4-phenyl-1,2,3,6-tetrahydropyridine-induced reductions of striatal dopamine and tyrosine hydroxylase activity in mice. *J. Neurochem.* **65,** 704–709.

150. Chiueh, C. C. (1988) Dopamine in the extrapyramidal motor function: A study based upon the MPTP-induced primate model of parkinsonism. *Ann. NY Acad. Sci.* **515,** 226–238.

151. Miyake, H. and Chiueh, C. C. (1989) Effects of MPP+ on the release of serotonin and 5-hydroxyindoleacetic acid from rat striatum in vivo. *Eur. J. Pharmacol.* **166,** 49–55.

152. Pérez-Otaño, I., Herrero, M. T., Oset, C., de Ceballos, M. L., Luquin, M. R., Obeso, J. A., and Del R'o, J. (1991) Extensive loss of brain dopamine and serotonin induced by chronic administration of MPTP in the marmoset. *Brain Res.* **567,** 127–132.

153. Mitra, N., Mohanakumar, K. P., and Ganguly, D. K. (1992) Dissociation of serotoninergic and dopaminergic components in acute effects of 1-methyl-4-phenyl-1,2,3,6-tetrahydropyridine in mice. *Brain Res. Bull.* **28,** 355–364.

154. Soubrie, P., Reisine, T. D., and Glowinski, J. (1984) Functional aspects of serotonin transmission in the basal ganglia: a review and in vivo approach using the push-pull cannula technique. *J. Neurosci.* **13,** 605–625.

155. Benloucif, S. and Galloway, M. P. (1991) Facilitation of dopamine release in vivo by serotonin agonists: pharmacological characterization. *Eur. J. Pharmacol.* **200,** 1–8.

156. Bonhomme, N., de Deuwaerdere, P., Le Moal, M., and Spampinato, U. (1995) Evidence for 5-HT$_4$ receptor subtype involvement in the enhancement of striatal dopamine release induced by serotonin: a microdialysis study in the halothane-anesthetized rats. *Neuropharmacology* **34,** 269–279.

157. Andrews, D. W., Patrick, R. L., and Burchas, J. (1978) The effects of 5-hydroxytryptophan and 5-hydroxytryptamine on dopamine synthesis and release in rat brain striatal synaptosomes. *J. Neurochem.* **30,** 465–470.

158. de Deuwaerdere, P., Bonhomme, N., Lucas, G., Le Moal, M., and Spampinato, U. (1996) Serotonin enhances striatal dopamine outflow in vivo through dopamine uptake sites. *J. Neurochem.* **66,** 210–215.

159. Melamed, E., Pikarski, E., Goldberg, A., Rosenthal, J., Uzzan, A., and Conforti, N. (1986) Effect of serotonergic, corticostriatal and kainic acid lesions on the dopaminergic neurotoxicity of 1-methyl-4-phenyl-1,2,3,6-tetrahydropyridine (MPTP) in mice. *Brain Res.* **399,** 178–180.

160. Melamed, E., Rosenthal, J., Cohen, O., Globus, M., and Uzzan, A. (1985) Dopamine but not norepinephrine or serotonin uptake inhibitors protect mice against neurotoxicity of MPTP. *Eur. J. Pharmacol.* **116,** 179–181.

161. Gaspar, P., Febvret, A., and Colombo, J. (1993) Serotonergic sprouting in primate MTP-induced hemiparkinsonism. *Exp. Brain Res.* **96,** 100–106.

162. Bus, J. S. and Gibson, J. E. (1984) Paraquat: model for oxidant-initiated toxicity. *Environ. Health Perspect.* **55,** 37–46.

163. Frank, D. M., Arora, P. K., Blumer, J. L., and Sayre, L. M. (1987) Model study on the bioreduction of paraquat, MPP+, and analogs. Evidence against a "redox cycling" mechanism in MPTP neurotoxicity. *Biochem. Biophys. Res. Commun.* **147,** 1095–1104.

164. Linkous, C. A., Schaich, K. M., Forman, A., and Borg, D. C. (1988) An electrochemical study of the neurotoxin 1-methyl-4-phenyl-1,2, 3,6-tetrahydropyridine and its oxidation products. *Bioelectrochem. Bioenerget.* **19,** 477–490.

165. Di Monte, D. A., Sandy, M. S., Ekstrom, G., and Smith, M. T. (1986) Comparative studies on the mechanisms of paraquat and 1-methyl-4-phenylpyridine (MPP+) cytotoxicity. *Biochem. Biophys. Res. Commun.* **137,** 303–309.

166. Ekstrom, G., Di Monte, D. A., Sandy, M. S., and Smith, M. T. (1987) Comparative toxicity and antioxidant activity of 1-methyl-4-phenyl-1,2,3,6-tetrahydropyridine and its moanamine oxidase B-generated metabolites in isolated hepatocytes and liver microsomes. *Arch. Biochem. Biophys.* **255,** 14–18.

167. Santiago, M., Rollema, H., De Vries, J. B., and Westerink, B. H. C. (1991) Acute effects of intranigral application of MPP+ on nigral and bilateral striatal release of dopamine simultaneously recorded by microdialysis. *Brain Res.* **538,** 226–230.

168. Matsubara, K., Idzu, T., Kobayashi, Y., Gonda, T., Okunishi, H., and Kimura, K. (1996) Differences in dopamine efflux induced by MPP+ and b-carbolinium in the striatum of conscious rats. *Eur. J. Pharmacol.* **315,** 145–151.

169. Spina, M. B. and Cohen, G. (1989) Dopamine turnover and glutathione oxidation: Implications for Parkinson disease. *Proc. Natl. Acad. Sci. USA* **86,** 1398–1400.

170. Werner, P. and Cohen, G. (1991) Intramitochondrial formation of oxidized glutathione during the oxidation of benzylamine by monoamine oxidase. *FEBS Lett.* **280(1),** 44–46.

171. Chiueh, C. C., Miyake, H., and Peng, M.-T. (1993) Role of dopamine, autoxidation, hydroxyl radical generation, and calcium overload in underlying mechanisms involved in MPTP-induced parkinsonism, in *Advances in Neurology*, vol. 60, *Parkinson's Disease: From Basic Research to Treatment. Proceedings of the 10th International Symposium on Parkinson's Disease* (Narabayashi, H., Nagatsu, T., Yanagisawa, N., and Mizuno, Y., eds.), Raven, New York, pp. 251–258.

172. Irwin, I. and Langston, J. W. (1995) Endogenous toxins as potential etiologic agents in Parkinson's disease, in *Etiology of Parkinson's Disease* (Ellenberg, J. H., Koller, W. C., and Langston, J. W., eds.), Marcel Dekker, New York, pp.153–201.

173. Barbeau, A., Poirier, J., Dallaire, L., Rucinska, E., Buu, N. T., and Donaldson, J. (1986) Studies on MPTP, MPP+ and paraquat in frogs and in vitro, in *MPTP: A Neurotoxin Producing A Parkinsonian Syndrome* (Markey, S. P., Castagnoli, N., Jr., Trevor, A. J., and Kopin, I. J., eds.), Academic, Orlando, FL, pp. 85–103.

174. Tse, D. C. S., McCreery, R., and Adams, R. N. (1976) Potential oxidative pathways of brain catecholamines. *J. Med. Chem.* **19(1)**, 37–40.

175. Graham, D. G., Tiffany, S. M., Bell, W. R., Jr., and Gutknecht, W. F. (1978) Autoxidation versus covalent binding of quinones as the mechanism of toxicity of dopamine, 6-hydroxydopamine, and related compounds toward C1300 neuroblastoma cells in vitro. *Mol. Pharmacol.* **14**, 644–653.

176. Hastings, T. G., Lewis, D. A., and Zigmond, M. J. (1996) Role of oxidation in the neurotoxic effects of intrastriatal dopamine injections. *Proc. Natl. Acad. Sci. USA* **93**, 1956–1961.

177. Hornykiewicz, O. (1974) The mechanisms of action of L-dopa in Parkinson's disease. *Life Sci.* **15**, 1249–1259.

178. Hefti, F., Melamed, E., Bhawan, J., and Wurtman, R. J. (1981) Long-term administration of L-dopa does not damage dopaminergic neurons in the mouse. *Neurology* **31**, 1194–1195.

179. Perry, T. L., Yong, V. W., Ito, M., Foulks, J. G., Wall, R. A., Godin, D. V., and Clavier, R. M. (1984) Nigrostriatal dopaminergic neurons remain undamaged in rats given high doses of L-dopa and carbidopa chronically. *J. Neurochem.* **43**, 990–993.

180. Tetrud, J. W. and Langston, J. W. (1989) The effect of deprenyl (selegiline) on the natural history of Parkinson's disease. *Science* **245**, 519–522.

181. Parkinson Study Group (1989) Effect of deprenyl on the progression of disabililty in early Parkinson's disease. *N. Engl. J. Med.* **321**, 1364–1371.

182. Myllyla, V. V., Sotaniemi, K. A., Vuorinen, J. A., and Heinonen, E. H. (1992) Selegiline as initial treatment in de novo parkinsonian patients. *Neurology* **42**, 339–343.

183. Lichter, D., Kurlan, R., Miller, C., and Shoulson, I. (1988) Does pergolide slow the progression of Parkinson's disease? A 7-year follow-up study [Abstract]. *Neurology* **38**, 122.

184. Zimmerman, T. and Sage, J. I. (1989) Long-term pergolide treatment and progression of Parkinson's disease [Abstract]. *Neurology* **39**, 200.

185. Zimmerman, T. and Sage, J. I. (1991) Comparison of combination pergolide and levodopa to levodopa alone after 63 months of treatment. *Clin. Neuropharm.* **14**, 165–169.

186. Knoll, J. (1988) Extension of life span of rats by long-term (−) deprenyl treatment. *Mt. Sinai J. Med.* **55(1)**, 67–74.

187. Felten, D. L., Felten, S. Y., Fuller, R. W., Romano, T. D., Smalstig, E. B., Wong, D. T., and Clemens, J. A. (1992) Chronic dietary pergolide preserves nigrostriatal neuronal integrity in aged-Fischer-344 rats. *Neurobiol. Aging* **13**, 339–351.

188. Felten, D. L., Felten, S. Y., Steece-Collier, K., Date, I., and Clemens, J. A. (1992) Age-related decline in the dopaminergic nigrostriatal system: The oxidative hypothesis and protective strategies. *Ann. Neurol.* **32 Suppl.**, S133–S136.

189. Sofic, E., Riederer, P., Heinsen, H., Beckmann, H., Reynolds, G. P., Hebenstreit, G., and Youdim, M. B. H. (1988) Increased iron (III) and total iron content in post mortem substantia nigra of parkinsonian brain. *J. Neural Transm.* **74**, 199–205.

190. Sofic, E., Paulus, W., Jellinger, K. A., Riederer, P., and Youdim, M. B. H. (1991) Selective increase of iron in substantia nigra zona compacta of Parkinsonian brains. *J. Neurochem.* **56**, 978–982.

191. Halliwell, B. and Gutteridge, J. M. C. (1985) The importance of free radicals and catalytic metal ions in human diseases. *Mol. Aspects Med.* **8**, 89–193.

192. Mochizuki, H., Imai, H., Endo, K., Yokomizo, K., Murata, Y., Hattori, N., and Mizuno, Y. (1994) Iron accumulation in the substantia nigra of 1-methyl-4-phenyl-1,2,3,6-tetrahydropyridine (MPTP)-induced hemiparkinsonian monkeys. *Neurosci. Lett.* **168**, 251–253.

193. Mash, D. C., Pablo, J., Buck, B. E., Sanchez-Ramos, J. R., and Weiner, W. J. (1991) Distribution and number of transferrin receptors in Parkinson's disease and in MPTP-treated mice. *Exp. Neurol.* **114**, 73–81.

194. Poirier, J., Donaldson, J., and Barbeau, A. (1985) The specific vulnerability of the substantia nigra to MPTP is related to the presence of transition metals. *Biochem. Biophys. Res. Commun.* **128**, 25–33.

195. Di Monte, D. A., Schipper, H. M., Hetts, S., and Langston, J. W. (1995) Iron-mediated bioactivation of MPTP in glial cultures. *Glia* **15**, 203–206.

196. Jenner, P., Dexter, D. T., Schapira, A. H. V., and Marsden, C. D. (1990) Free radical involvement and altered iron metabolism as a cause of Parkison's disease, in *The Assessment of Therapy of Parkinsonism* (Marsden, C. D. and Fahn, S., eds.), Parthenon, New Jersey, pp. 17–30.

197. Earle, K. M. (1968) Studies on Parkinson's disease including X-ray fluorescent spectroscopy of formalin fixed brain tissue. *J. Neuropathol. Exp. Neurol.* **27(1)**, 1–14.

198. Dexter, D. T., Carter, C. J., Wells, F. R., Agid, F. J., Agid, Y., Lees, A. J., Jenner, P., and Marsden, C. D. (1989) Basal lipid peroxidation in substantia nigra is increased in Parkinson's disease. *J. Neurochem.* **52**, 381–389.

199. Hirsch, E. C., Brandel, J.-P., Galle, P., Agid, F. J., and Agid, Y. (1991) Iron and aluminum increase in the substantia nigra of patients with Parkinson's disease: An x-ray microanalysis. *J. Neurochem.* **56**, 446–451.

200. Schipper, H. M. (1991) Gomori-positive astrocytes: biological properties and implications for neurologic and neuroendocrine disorders. *Glia* **4**, 365–377.

201. Dexter, D. T., Wells, F. R., Agid, F. J., Agid, Y., Lees, A. J., Jenner, P., and Marsden, C. D. (1987) Increased nigral iron content in postmortem parkinsonian brain. *Lancet* **2**, 1219–1220.

202. Riederer, P., Sofic, E., Rausch, W. D., Schmidt, B., Reynolds, G. P., Jellinger, K. A., and Youdim, M. B. H. (1989) Transition metals, ferritin, glutathione, and ascorbic acid in Parkinsonian brains. *J. Neurochem.* **52**, 515–520.

203. Ryvlin, P., Broussolle, E., Piollet, H., Viallet, F., Khalfallah, Y., and Chazon, G. (1995) Magnetic resonance imaging evidence of decreased putamenal iron content in idiopathic Parkinson's disease. *Arch. Neurol.* **52(6)**, 583–588.

204. Aisen, P. (1992) Entry of iron into cells: a new role for the transferrin receptor in modulating iron release from transferrin. *Ann. Neurol.* **32**, s62–s68.

205. Faucheux, B. A., Herrero, M. T., Villares, J., Levy, R., Javoy-Agid, F., Obeso, J. A., Hauw, J. J., Agid, Y., and Hirsch, E. C. (1995) Autoradiographic localization and

density of [125I]ferrotransferrin binding sites in the basal ganglia of control subjects, patients with Parkinson's disease and MPTP-lesioned monkeys. *Brain Res.* **691**, 115–124.

206. Connor, J. R., Snyder, B. S., Arosio, P., Loeffler, D. A., and LeWitt, P. A. (1995) A quantitative analysis of isoferritins in select regions of aged, parkinsonian, and Alzheimer's diseased brains. *J. Neurochem.* **65**, 717–724.

207. Leveugle, B., Faucheux, B. A., Bouras, C., Nillesse, N., Spik, G., Hirsch, E. C., Agid, Y., and Hof, P. R. (1996) Cellular distribution of the iron-binding protein lactotransferring in the mesencephalon of Parkinson's disease cases. *Acta Neuropathol.* **91**, 566–572.

208. Loeffler, D. A., Connor, J. R., Juneau, P. L., Snyder, B. S., Kanaley, L., DeMaggio, A. J., Nguyen, H., Brickman, C. M., and LeWitt, P. A. (1995) Transferrin and iron in normal, Alzheimer's disease, and Parkinson's disease brain regions. *J. Neurochem.* **65**, 710–716.

209. Hill, J. M. (1989) Comments on "Putative biological mechanisms of the effect of iron deficiency on brain chemistry and behavior." *Am. J. Clin. Nutr.* **50**, 616–617.

210. Di Monte, D. A., Chan, P., and Sandy, M. S. (1992) Glutathione in Parkinson's disease: A link between oxidative stress and mitochondrial damage. *Ann. Neurol.* **32 Suppl.**, S111–S115.

211. Starke, P. E. and Farber, J. L. (1985) Ferric iron and superoxide ions are required for the killing of cultured hepatocytes by hydrogen peroxide. Evidence for the participation of hydroxyl radicals formed by an iron-catalyzed Haber-Weiss reaction. *J. Biol. Chem.* **260**, 10,099–10,104.

212. Rossetti, Z. L., Sotgiu, A., Sharp, D. E., Hadjiconstantinou, M., and Neff, N. H. (1988) 1-Methyl-4-phenyl-1,2,3,6-tetrahydropyridine (MPTP) and free radicals in vitro. *Biochem. Pharmacol.* **37(23)**, 4573–4574.

213. Wu, R.-M., Chiueh, C. C., Pert, A., and Murphy, D. L. (1993) Apparent antioxidant effect of 1-deprenyl on hydroxyl radical formation and nigral injury elicited by MPP+ in vivo. *Eur. J. Pharmacol.* **243**, 241–247.

214. Hasegawa, E., Takeshige, K., Oishi, T., Murai, Y., and Minakami, S. (1990) 1-Methyl-4-phenylpyridinium (MPP+) induces NADH-dependent superoxide formation and enhances NADH-dependent lipid peroxidation in bovine heart submitochondrial particles. *Biochem. Biophys. Res. Commun.* **170**, 1049–1055.

215. Cleeter, M. W. J., Cooper, J. M., and Schapira, A. H. V. (1992) Irreversible inhibition of mitochondrial complex I by 1-methyl-4-phenylpyridinium: Evidence for free radical involvement. *J. Neurochem.* **58**, 786–789.

216. Poirier, J. and Barbeau, A. (1985) A catalyst function for MPTP in superoxide formation. *Biochem. Biophys. Res. Commun.* **131(3)**, 1284–1289.

217. Przedborski, S., Kostic, V., Jackson-Lewis, V., Naini, A. B., Simonetti, S., Fahn, S., Carlson, E., Epstein, C. J., and Cadet, J. L. (1992) Transgenic mice with increased Cu/Zn-superoxide dismutase activity are resistant to *N*-methyl-4-phenyl-1,2,3,6-tetrahydropyridine-induced neurotoxicity. *J. Neurosci.* **12**, 1658–1667.

218. Rojas, P. and Rios, C. (1993) Increased striatal lipid peroxidation after intracerebroventricular MPP+ administration to mice. *Pharmacol Toxicol.* **72**, 364–368.

219. Desole, M. S., Miele, M., Esposito, G., Fresu, L. G., Migheli, R., Zangani, D., Sircana, S., Grella, G., and Miele, E. (1995) Neuronal antioxidant system and MPTP-induced

oxidative stress in the striatum and brain stem of the rat. *Pharmacol. Biochem. Behav.* **51(4)**, 581–592.

220. Desole, M. S., Esposito, G., Fresu, L., Migheli, R., Sircana, S., Delogu, R., Miele, M., and Miele, E. (1996) Further investigation of allopurinol effects of MPTP-induced oxidative stress in the striatum and brain stem of the rat. *Pharmacol. Biochem. Behav.* **54(2)**, 377–383.

221. Marttila, R. J., Lorentz, H., and Rinne, U. K. (1988) Oxygen toxicity protecting enzymes in Parkinson's disease: Increase of superoxide dismutase-like activity in the substantia nigra and basal nucleus. *J. Neurol. Sci.* **86**, 321–331.

222. Hirsch, E. C., Graybiel, A. M., and Agid, Y. (1988) Melanized dopaminergic neurons are differentially susceptible to degeneration in Parkinson's disease. *Nature* **334**, 345–348.

223. Ceballos, I., Lafon, M., Javoy-agid, F., et al. (1990) Superoxide dismutase and Parkinson's disease. *Lancet* **335**, 1035–1036.

224. Kish, S. J., Morito, C., and Hornykiewicz, O. (1985) Glutathione peroxidase activity in Parkinson's disease brain. *Neurosci. Lett.* **58**, 343–346.

225. Perry, T. L., Godin, D. V., and Hansen, S. (1982) Parkinson's disease: A disorder due to nigral glutathione deficiency? *Neurosci. Lett.* **33**, 305–310.

226. Perry, T. L., Yong, V. W., Clavier, R. M., Jones, K., Wright, J. M., Foulks, J. G., and Wall, R. A. (1985) Partial protection from the dopaminergic neurotoxin *N*-methyl-4-phenyl-1,2,3,6-tetrahydropyridine by four different antioxidants in the mouse. *Neurosci. Lett.* **60**, 109–114.

227. Saggu, H., Cooksey, J., Dexter, D. T., Wells, F. R., Lees, A. J., Jenner, P., and Marsden, C. D. (1989) A selective increase in particulate superoxide dismutase activity in parkinsonian substantia nigra. *J. Neurochem.* **53**, 692–697.

228. Poirier, J., Dea, D., Baccichet, A., and Thiffault, C. (1994) Superoxide dismutase expression in Parkinson's disease. *Ann. NY Acad. Sci.* **738**, 116–120.

229. Parboosingh, J. S., Rousseau, M., Rogan, F., Amit, Z., Chertkow, H., Johnson, W. G., Manganaro, F., Schipper, H. N., Curran, T. J., Stoessl, A. J., and Rouleau, G. A. (1995) Absence of mutations in superoxide dismutase and catalase genes in patients with Parkinson's disease. *Arch. Neurol.* **52**, 1160–1163.

230. Poirier, J. and Thiffault, C. (1993) Are free radicals involved in the pathogenesis of idiopathic Parkinson's disease? *Eur. Neurol.* **33(Suppl 1)**, 38–43.

231. Perry, T. L. and Yong, V. W. (1986) Idiopathic Parkinson's disease, progressive supranuclear palsy and glutathione metabolism in the substantia nigra. *Neurosci. Lett.* **67**, 269–274.

232. Sershen, H., Reith, M. E. A., Hashim, A., and Lajtha, A. (1985) Protection against 1-methyl-4-phenyl-1,2,3,6-tetrahydropyridine neurotoxicity by the antioxidant ascorbic acid. *Neuropharmacology* **24(12)**, 1257–1259.

233. Wagner, G. C., Jarvis, M. F., and Carelli, R. M. (1985) Ascorbic acid reduces the dopamine depletion induced by MPTP. *Neuropharmacology* **24(12)**, 1261–1262.

234. Martinovits, G., Melamed, E., Cohen, O., Rosenthal, J., and Uzzan, A. (1986) Systemic administration of antioxidants does not protect mice against the dopaminergic neurotoxicity of 1-methyl-4-phenyl-1,2,3,6-tetrahydropyridine (MPTP). *Neurosci. Lett.* **69**, 192–197.

235. Di Monte, D. A., Sandy, M. S., and Smith, M. T. (1987) Increased efflux rather than oxidation is the mechanism of glutathione depletion by 1-methyl-4-phenyl-1,2,3,6-tetrahydropyridine (MPTP). *Biochem. Biophys. Res. Commun.* **148,** 153–160.

236. Sershen, H., Debler, E. A., and Lajtha, A. (1987) Effect of ascorbic acid on the synaptosomal uptake of [^{3}H]MPP+, [^{3}H]dopamine, and [^{14}C]GABA. *J Neurosci Res,* **17,** 298–301.

237. Irwin, I., Wu, E. Y., DeLanney, L. E., Trevor, A. J., and Langston, J. W. (1987) The effect of diethyldithiocarbamate (DDC) on the biodisposition of MPTP: An explanation for enhanced neurotoxicity. *Eur. J. of Pharmacol.* **141,** 209–217.

238. Nicklas, W. J., Vyas, I., and Heikkila, R. E. (1985) Inhibition of NADH-linked oxidation in brain mitochondria by 1-methyl-4-phenyl-1,2,3,6-tetrahydropyridine. *Life Sci.* **36,** 2503–2508.

239. Poirier, J. and Barbeau, A. (1985) 1-Methyl-4-phenyl-pyridinium-induced inhibition of nicotinamide adenosine dinucleotide cytochrome c reductase. *Neurosci. Lett.* **62,** 7–11.

240. Ramsay, R. R., Krueger, M. J., Youngster, S. K., Gluck, M. R., Casida, J. E., and Singer, T. P. (1991) Interaction of 1-methyl-4-phenylpyridinium ion (MPP+) and its analogs with the rotenone/piericidin binding site of NADH dehydrogenase. *J. Neurochem.* **56,** 1184–1190.

241. Di Monte, D. A. (1991) Mitochondrial DNA and Parkinson's disease. *Neurology* **41 Suppl. 2,** 38–42.

242. Wu, E. Y., Smith, M. T., Bellomo, G., and Di Monte, D. A. (1990) Relationships between the mitochondrial transmembrane potential, ATP concentration, and cytotoxicity in isolated rat hepatocytes. *Arch. Biochem. Biophys.* **282(2),** 358–362.

243. Di Monte, D. A., Jewell, S. A., Ekstrom, G., Sandy, M. S., and Smith, M. T. (1986) 1-methyl-4-phenyl-1,2,3,6-tetrahydropyridine (MPTP) and 1-methyl-4-phenylpyridine (MPP+) cause rapid ATP depletion in isolated hepatocytes. *Biochem. Biophys. Res. Commun.* **137(1),** 310–315.

244. Mizuno, Y., Suzuki, K., Sone, N., and Saitoh, T. (1987) Inhibition of ATP synthesis by 1-methyl-4-phenylpyridinium ion (MPP+) in isolated mitochondria from mouse brains. *Neurosci. Lett.* **81,** 204–208.

245. Kutty, R. K., Santostasi, G., Horng, J., and Krishna, G. (1991) MPTP-induced ATP depletion and cell death in neuroblastoma x glioma hybrid NG 108–15 cells: Protection by glucose and sensitization by tetraphenylborate. *Toxicol. Appl. Pharmacol.* **107,** 377–388.

246. Di Monte, D. A., Sandy, M. S., Blank, L., and Smith, M. T. (1988) Fructose prevents 1-methyl-4-phenyl-1,2,3,6-tetrahydropyridine (MPTP)-induced ATP depletion and toxicity in isolated hepatocytes. *Biochem. Biophys. Res. Commun.* **153,** 734–740.

247. Chan, P., DeLanney, L. E., Irwin, I., Langston, J. W., and Di Monte, D. A. (1991) Rapid ATP loss caused by 1-methyl-4-phenyl-1,2,3,6-tetrahydropyridine in mouse brain. *J. Neurochem.* **57,** 348–351.

248. Chan, P., DeLanney, L. E., Irwin, I., Langston, J. W., and Di Monte, D. A. (1992) MPTP-induced ATP loss in mouse brain. *Ann. NY Acad. Sci.* **648,** 306–308.

249. Storey, E., Hyman, B. T., Jenkins, B., Brouillet, E., Miller, J. M., Rosen, B. R., and Beal, M. F. (1992) 1-Methyl-4-phenylpyridinium produces excitotoxic lesions in rat striatum as a result of impairment of oxidative metabolism. *J. Neurochem.* **58,** 1975–1978.

250. Chan, P., Langston, J. W., Irwin, I., DeLanney, L. E., and Di Monte, D. A. (1993) 2-Deoxyglucose enhances 1-methyl-4-phenyl-1,2,3,6-tetrahydropyridine-induced ATP loss in the mouse brain. *J. Neurochem.* **61,** 610–616.

251. Mizuno, Y., Saitoh, T., and Sone, N. (1987) Inhibition of mitochondrial alpha-ketoglutarate dehydrogenase by 1-methyl-4-phenylpyridinium ion. *Biochem. Biophys. Res. Commun.* **143(3),** 971–976.

252. Mizuno, Y., Matuda, S., Yoshino, H., Mori, H., Hattori, N., and Ikebe, S. (1994) An immunohistochemical study on alpha-ketoglutarate dehydrogenase complex in Parkinson's disease. *Ann. Neurol.* **35,** 204–210.

253. Hornsby, P. J. (1989) Parkinson's disease, vitamin E, and mitochondrial energy metabolism. *Arch. Neurol.* **46,** 840–841.

254. Bindoff, L. A., Birch-Machin, M. A., Cartlidge, N. E. F., Parker, W. D., Jr., and Turnbull, D. M. (1991) Respiratory chain abnormalities in skeletal muscle from patients with Parkinson's disease. *J. Neurol. Sci.* **104,** 203–208.

255. Hattori, K., Tanaka, M., Sugiyama, S., Obayashi, T., Ito, T., Satake, T., Hanaki, Y., Asai, J., Nagano, M., and Ozawa, T. (1991) Age-dependent increase in deleted mitochondrial DNA in the human heart: Possible contributory factor to presbycardia. *Am. Heart J.* **121,** 1735–1742.

256. Boyson, S. J. (1991) Parkinson's disease and the electron transport chain. *Ann. Neurol.* **30,** 330–331.

257. Parker, W. D., Jr. (1991) Preclinical detection of Parkinson's disease: Biochemical approaches. *Neurology* **41 Suppl. 2,** 34–36.

258. Krige, D., Carroll, M. T., Cooper, J. M., Marsden, C. D., and Schapira, A. H. V. (1992) Platelet mitochondrial function in Parkinson's disease. *Ann. Neurol.* **32,** 782–788.

259. Nakagawa-Hattori, Y., Yoshino, H., Kondo, T., Mizuno, Y., and Horai, S. (1992) Is Parkinson's disease a mitochondrial disorder? *J. Neurol. Sci.* **107,** 29–33.

260. Schapira, A. H. V., Mann, V. M., Cooper, J. M., Drige, D., Jenner, P. J., and Marsden, C. D. (1992) Mitochondrial function in Parkinson's disease. *Ann. Neruol.* **32,** S116–S124.

261. Shoffner, J. M. and Wallace, D. C. (1992) Heart disease and mitochondrial DNA mutations. *Heart Dis. Stroke* **1,** 235–241.

262. Wallace, D. C., Shoffner, J. M., Watts, R. L., Juncos, J. L., and Torroni, A. (1992) Mitochondrial oxidative phosphorylation defects in Parkinson's disease. *Ann. Neurol.* **32,** 113,114.

263. Yoshino, H., Nakagawa-Hattori, Y., Kondo, T., and Mizuno, Y. (1992) Mitochondrial complex I and II activities of lymphocytes and platelets in Parkinson's disease. *J. Neural Transm. Park. Dis. Dement. Sect.* **4,** 27–34.

264. Benecke, R., Strümper, P., and Weiss, H. (1993) Electron transfer complexes I and IV of platelets are abnormal in Parkinson's disease but normal in Parkinson-plus syndromes. *Brain* **116,** 1451–1463.

265. Cardellach, F., Marti, M. J., Fernandez-Sola, J., Marin, C., Hoek, J. B., Tolosa, E., and Urbano-Marquez, A. (1993) Mitochondrial respiratory chain activity in skeletal muscle from patients with Parkinson's disease. *Neurology* **43,** 2258–2262.

266. Blin, O., Desnuelle, C., Rascol, O., Borg, M., Peyro Saint Paul, H., Azulay, J. P., Bill, F., Figarella, D., Coulom, F., Pellissier, J. F., Montastruc, J. L., Chatel, M., and Serratrice, G. (1994) Mitochondrial respiratory failure in skeletal muscle from patients with Parkinson's disease and multiple system atrophy. *J. Neurol. Sci.* **125,** 95–101.

267. Haas, R. H., Nasirian, F., Nakano, K., Ward, D., Pay, M., Hill, R., and Shults, C. W. (1995) Low platelet mitochondrial Complex I and Complex II/III activity in early untreated Parkinson's disease. *Ann. Neurol.* **37,** 714–722.

268. Burkhardt, C., Kelly, J. P., Lim, Y. H., Filley, C. M., and Parker, W. D., Jr. (1993) Neuroleptic medications inhibit complex I of the electron transport chain. *Ann. Neurol.* **33,** 512–517.

269. Przedborski, S., Jackson-Lewis, V., Muthane, U., Jiang, H., Ferreira, M., Naini, A. B., and Fahn, S. (1993) Chronic levodopa administration alters cerebral mitochondrial respiratory chain activity. *Ann. Neurol.* **34,** 715–723.

270. Ikebe, S., Tanaka, M., Ohno, K., Sato, W., Hattori, K., Kondo, T., Mizuno, Y., and Ozawa, T. (1990) Increase of deleted mitochondrial DNA in the striatum in Parkinson's disease and senescence. *Biochem. Biophys. Res. Commun.* **170,** 1044–1048.

271. Lestienne, P., Nelson, J., Riederer, P., Jellinger, K. A., and Reichmann, H. (1990) Normal mitochondrial genome in brain from patients with Parkinson's disease and complex I defect. *J. Neurochem.* **55,** 1810–1812.

272. Cortopassi, G. A. and Arnheim, N. (1990) Detection of a specific mitochondrial DNA deletion in tissues of older humans. *Nucleic Acids Res.* **18(23),** 6927–6933.

273. Ozawa, T., Tanaka, M., Ino, H., Ohno, K., Sano, T., Wada, Y., Yoneda, M., Tanno, Y., Miyatake, T., Tanaka, T., Itoyama, S., Ikebe, S., Hattori, N., and Mizuno, Y. (1991) Distinct clustering of point mutations in mitochondrial DNA among patients with mitochondrial encephalomyopathies and with Parkinson's disease. *Biochem. Biophys. Res. Commun.* **176,** 938–946.

274. Ikebe, S., Tanaka, M., and Ozawa, T. (1995) Point mutations of mitochondrial genome in Parkinson's disease. *Mol. Brain Res.* **28,** 281–295.

275. Lücking, C. B., Kosel, S., Mehraein, P., and Graeber, M. B. (1995) Absence of the mitochondrial A7237T mutation in Parkinson's disease. *Biochem. Biophys. Res. Commun.* **211(2),** 700–704.

276. Swerdlow, R. H., Parks, J. K., Miller, S. W., Tuttle, J. B., Trimmer, P. A., Sheehan, J. P, Bennett, J. P., Jr., Davis, R. E., and Parker, W. D., Jr. (1996) Origin and functional consequences of the complex I defect in Parkinson's disease. *Ann. Neurol.* **40,** 663–671.

277. Chien, K. R., Pfau, R. G., and Farber, J. L. (1979) Ischemic myocardial cell injury. Prevention by chlorpromazine of an accelerated phospholipid degradation and associated membrane dysfunction. *Am. J. Pathol.* **97,** 505–530.

278. Kass, G. E. N., Wright, J. M., Nicotera, P., and Orrenius, S. (1988) The mechanism of 1-methyl-4-phenyl-1,2,3,6-tetrahydropyridine toxicity: role of intracellular calcium. *Arch. Biochem. Biophys.* **260,** 789–797.

279. Sun, C. J., Johannessen, J. N., Gessner, W., Namura, I., Singhaniyom, W., Brossi, A., and Chiueh, C. C. (1988) Neurotoxic damage to the nigrostriatal system in rats following intranigral administration of MPDP+ and MPP+. *J. Neural Transm.* **74,** 75–86.

280. Cappelletti, G., Brambilla, E., Maci, R., and Camatini, M. (1991) N-methyl-4-phenyl-1,2,3,6-tetrahydropyridine (MPTP) induced cytoskeletal alterations on "Swiss 3T3" mouse fibroblasts. *Neurosci. Lett.* **129,** 149–152.

281. Urani, C., Brambilla, E., Santagostino, A., and Camatini, M. (1994) 1-methyl-4-phenyl-1,2,3,6-tetrahydropyridine (MPTP) affects the actin cytoskeleton and calcium level of Swiss 3T3 mouse fibroblasts. *Toxicology* **91,** 117–126.

282. Yamada, T., McGeer, P. L., Baimbridge, K. G., and McGeer, E. G. (1990) Relative sparing in Parkinson's disease of substantia nigra dopamine neurons containing calbindin-D$_{28K}$. *Brain Res.* **526,** 303–307.

283. German, D. C., Manaye, K. F., Sonsalla, P. K., and Brooks, B. A. (1992) Midbrain dopaminergic cell loss in Parkinson's disease and MPTP-induced Parkinsonism: Sparing of calbindin-D$_{28k}$-containing cells. *Ann. NY Acad. Sci.* **648,** 42–62.

284. Hirsch, E. C., Mouatt, A., Thomasset, M., Javoy-Agid, F., Agid, Y., and Graybiel, A. M. (1992) Expression of calbindin D28K-like immunoreactivity in catecholaminergic cell groups of the human midbrain: normal distribution and distribution in Parkinson's disease. *Neurodegeneration* **1,** 83–93.

285. Sonsalla, P. K., Nicklas, W. J., and Heikkila, R. E. (1989) Role for excitatory amino acids in methamphetamine-induced nigrostriatal dopaminergic toxicity. *Science* **243,** 398–400.

286. Finnegan, K. T., Skratt, J. J., Irwin, I., and Langston, J. W. (1990) The *N*-methyl-D-aspartate (NMDA) receptor antagonist, dextrorphan, prevents the neurotoxic effects of 3,4-methylenedioxymethamphetamine (MDMA) in rats. *Neurosci. Lett.* **105,** 300–306.

287. Whetsell, W. O., Jr. (1997) Current concepts of excitotoxicity. *J. Neuropathol. Exp. Neurol.* **55,** 1–13.

288. Sonsalla, P. K., Zeevalk, G. D., Manzino, L., Giovanni, A., and Nicklas, W. J. (1992) MK-801 fails to protect against the dopaminergic neuropathology produced by systemic 1-methyl-4-phenyl-1,2,3,6-tetrahydropyridine in mice or intranigral 1-methyl-4-phenylpyridinium in rats. *J. Neurochem.* **58,** 1979–1982.

289. Carboni, S., Melis, F., Pani, L., Hadjiconstantinou, M., and Rossetti, Z. L. (1990) The non-competitive NMDA-receptor antagonist MK-801 prevents the massive release of glutamate and aspartate from rat striatum induced by 1-methyl-4-phenylpyridinium (MPP+). *Neurosci. Lett.* **117,** 129–133.

290. Turski, L., Bressler, K., Rettig, K.-J., Lŝschmann, P.-A., and Wachtel, H. (1991) Protection of substantia nigra from MPP+ neurotoxicity by *N*-methyl-D-aspartate antagonists. *Nature* **349,** 414–418.

291. Zuddas, A., Vaglini, F., Fornai, F., Fascetti, F., and Corsini, G. U. (1991) MK-801 prevents MPTP-induced nigrostriatal neuronal death in monkeys and mice [Abstract]. *Soc. Neurosci.* **17(1),** 716.

292. Srivastava, R., Brouillet, E., Beal, M. F., Storey, E., and Hyman, B. T. (1993) Blockade of 1-methyl-4-phenylpyridinium ion (MPP+) nigral toxicity in the rat by prior decortication or MK-801 treatment: A stereological estimate of neuronal loss. *Neurobiol. Aging* **14,** 295–301.

293. Tabatabaei, A., Perry, T. L., Hansen, S., and Krieger, C. (1992) Partial protective effect of MK-801 on MPTP-induced reduction of striatal dopamine in mice. *Neurosci. Lett.* **141,** 192–194.

294. Brouillet, E. and Beal, M. F. (1993) NMDA antagonists partially protect against MPTP induced neurotoxicity in mice. *Neuroreport* **4,** 387–390.

295. Sawada, H., Shimohama, S., Tamura, Y., Kawamura, T., Akaike, A., and Kimura, J. (1996) Methylphenylpyridinium ion (MPP+) enhances glutamate-induced cytotoxicity against dopaminergic neurons in cultured rat mesencephalon. *J. Neurosci. Res.* **43,** 55–62.

296. Kupsch, A., Lsschmann, P.-A., Sauer, H., Arnold, G., Renner, P., Pufal, D., Burg, M., Wachtel, H., Ten Bruggencate, G., and Oertel, W. H. (1992) Do NMDA receptor antagonists protect against MPTP–toxicity? Biochemical and immunocytochemical analyses in black mice. *Brain Res.* **592,** 74–83.

297. Michel, P. P. and Agid, Y. (1992) The glutamate antagonist, MK-801, does not prevent dopaminergic cell death induced by the 1-methyl-4-phenylpyridinium ion (MPP+) in rat dissociated mesencephalic cultures. *Brain Res.* **597,** 233–240.

298. Finiels-Marlier, F., Marini, A. M., Williams, P., and Paul, S. M. (1993) The N-methyl-D-aspartate antagonist MK-801 fails to protect dopaminergic neurons from 1-methyl-4-phenylpyridinium toxicity in vitro. *J. Neurochem.* **60,** 1968–1971.

299. Chan, P., Di Monte, D. A., and Langston, J. W. (1994) Effects of 1-methyl-4-phenyl-1,2,3,6-tetrahydropyridine (MPTP) on levels of glutamate and aspartate in the mouse brain. *Brain Res.* **647,** 249–254.

300. Chan, P., Langston, J. W., and Di Monte, D. A. (1993) MK-801 temporarily prevents MPTP-induced acute dopamine depletion and MPP+ elimination in the mouse striatum. *J. Pharmacol. Exp. Ther.* **267,** 1515–1520.

301. Clarke, P. B. S. and Reuben, M. (1995) Inhibition by dizocilpine (MK-801) of striatal dopamine release induced by MPTP and MPP+: possible action at the dopamine transporter. *Br. J. Pharmacol.* **114,** 315–322.

302. Vaglini, F., Fascetti, F., Fornai, F., Maggio, R., and Corsini, G. U. (1994) (+)MK-801 prevents the DDC-induced enhancement of MPTP toxicity in mice. *Brain Res.* **668,** 194–203.

303. Jarvis, M. F. and Wagner, G. C. (1985) Age-dependent effects of 1-methyl-4-phenyl-1,2,3,6-tetrahydropyridine (MPTP). *Neuropharmacology* **24,** 581–583.

304. Irwin, I., Finnegan, K. T., DeLanney, L. E., Di Monte, D. A., and Langston, J. W. (1992) The relationships between aging, monoamine oxidase, striatal dopamine and the effects of MPTP in C57BL/6 mice: A critical reassessment. *Brain Res.* **572,** 224–231.

305. Date, I., Felten, D. L., and Felten, S. Y. (1990) Long-term effect of MPTP in the mouse brain in relation to aging: Neurochemical and immunocytochemical analysis. *Brain Res.* **519,** 266–276.

306. Hornykiewicz, O. (1982) Imbalance of brain monoamines and clinical disorders, in *Progress in Brain Research*, vol. 55, *Chemical Transmission in the Brain: The Role of Amines, Amino Acids and Peptides* (Buijs, R. M., Pevet, P., and Swaab, D. F., eds.), Elsevier Biomedical, Amsterdam, pp. 419–429.

307. Benedetti, M. S. and Keane, P. E. (1980) Differential changes in monoamine oxidase A and B activity in the aging rat brain. *J. Neurochem.* **35,** 1026–1032.

308. Langston, J. W., Irwin, I., and DeLanney, L. E. (1987) The biotransformation of MPTP and disposition of MPP+: The effects of aging. *Life Sci.* **40,** 749–754.

309. Irwin, I., Delanney, L. E., and Langston, J. W. (1992) Studies in the C57BL/6 mouse, in *Advances in Neurology*, vol. 60, *Parkinson's Disease: From Basic Research to Treatment. Proceedings of the 10th International Symposium on Parkinson's Disease* (Narabayashi, H., Nagatsu, T., Yanagisawa, N., and Mizuno, Y., eds.), (Raven, New York, pp. 197–206.

310. Irwin, I., DeLanney, L. E., McNeill, T. H., Chan, P., Forno, L. S., Murphy, G. M., Jr., Di Monte, D. A., Sandy, M. S., and Langston, J. W. (1994) Aging and the nigrostriatal dopamine system: a non-human primate study. *Neurodegeneration* **3,** 251–265.

311. Rose, S. P., Nomoto, M., Jackson, E. A., Gibb, W. R. G., Jaehnig, P., Jenner, P., and Marsden, C. D. (1993) Age-related effects of 1-methyl-4-phenyl-1,2,3,6-tetrahydropyridine treatment of common marmosets. _Eur. J. Pharmacol._ **230,** 177–185.

312. Langston, J. W., Irwin, I., Forno, L. S., and DeLanney, L. E. (1987) Parkinson's disease, aging, and MPTP: clinical and experimental observations, in _Recent Developments in Parkinson's Disease_ (Fahn, S., Marsden, C. D., Goldstein, M., and Calne, D. B., eds.), MacMillan Healthcare Information, Florham Park, NJ, pp. 59–74.

313. Irwin, I., DeLanney, L. E., Chan, P., Sandy, M. S., Di Monte, D. A., and Langston, J. W. (1997) Nigrostriatal monoamine oxidase A and B in aging squirrel monkeys and C57BL/6 mice. _Neurobiol. Aging,_ in press.

314. Fowler, C. J., Wilberg, A., Oreland, L., Marcusson, J., and Winblad, B. (1980) The effect of age on the activity and molecular properties of human brain monoamine oxidase. _J. Neural Transm._ **49,** 1–20.

315. Gottfries, C. G. (1987) Pharmacology of mental aging and dementia disorders. _Clin. Neuropharm._ **10(4),** 313–329.

316. Gottfries, C. G. (1990) Neurochemical aspects on aging and diseases with cognitive impairment. _J. Neurosci. Res._ **27,** 541–547.

317. Robinson, D. S., Davis, J. M., Nies, A., Ravaris, C. L., and Sylwester, D. (1971) Relation of sex and aging to monoamine oxidase activity of human Brain plasma, and platelets. _Arch. Gen. Psychiatry_ **24,** 536–539.

318. Grote, S. S., Moses, S. G., Robins, E., Hudgens, R. W., and Croninger, A. B. (1974) A stuy of selected catecholamine metabolizing enzymes: a comparison of depressive suicides and alcoholic suicides with controls. _J. Neurochem._ **23,** 791–802.

319. Shih, J. C. (1975) Multiple forms of monoamine oxidase and aging, in: _Aging,_ vol. 1, (Brody, H., Harman, D., and Ordy, J. M., eds.), Raven, New York, pp. 191–198.

320. Cote, L. J. and Kremzner, L. T. (1983) Biochemical changes in normal aging in human brain, in _The Dementias_ (Mayeux, R. and Rosen, W. G., eds.), Raven, New York, pp. 19–30.

321. Samorajski, T. and Rolsten, C. (1973) Age and regional differences in the chemical composition of brains of mice, monkeys and humans, in _Progress in Brain Research,_ vol. 40, (Ford, D. H., eds.), Elsevier Science Publishing, Amsterdam, pp. 253–265.

322. Fowler, J. S., MacGregor, R. R., Wolf, A. P., Arnett, C. D., Dewey, S. L., Schlyer, D., Christman, D., Logan, J., Smith, M., Sachs, H., Aquilonius, S. M., Bjurling, P., Halldin, C., Hartvig, P., Leenders, K. L., Lundquist, H., Oreland, L., Stalnacke, C. G., and Langstrom, B. (1987) Mapping human brain monoamine oxidase A and B with 11C-labeled suicide inactivators and PET. _Science_ **235,** 481–485.

323. Fowler, J. S., Volkow, N. D., Wang, G. J., Pappas, N., Logan, J., MacGregor, R., Alexoff, D., Shea, C., Schlyer, D., Wolf, A. P., Warner, D., Zezulkova, I., and Cilento, R. (1996) Inhibition of monoamine oxidase in the brains of smokers. _Nature_ **379,** 733–736.

324. Ricaurte, G. A., DeLanney, L. E., Irwin, I., and Langston, J. W. (1987) Older dopaminergic neurons do not recover from the effects of MPTP. _Neuropharmacology_ **26(1),** 97–99.

325. Shigenaga, M. K., Hagen, T. M., and Ames, B. N. (1994) Oxidative damage and mitochondrial decay in aging. _Proc. Natl. Acad. Sci._ **91,** 10,771–10,778.

326. Desai, V. G., Feuers, R. J., Hart, R. W., and Ali, S. F. (1996) MPP+-induced neurotoxicity in mouse is age-dependent: evidenced by the selective inhibition of complexes of elctron transport. *Brain Res.* **715,** 1–8.

327. Stuehr, D. J., Kwon, N. S., Nathan, C. F., and Griffith, O. W. (1991) Nw-hydroxy-L-arginine is an intermedite in the biosynthesis of nitric oxide from L-arginine. *J. Biol. Chem.* **266,** 6259–6263.

328. Marletta, M. A. (1993) Nitric oxide synthase structure and mechanism. *J. Biol. Chem.* **268(17),** 12,231–12,234.

329. Bredt, D. S. and Snyder, S. H. (1990) Isolation ofnitric oxide synthetase, a calmodulin-requiring enzyme. *Proc. Natl. Acad. Sci. USA* **87,** 682–685.

330. Babbedge, R. C., Bland-Ward, P. A., Hart, S. L., and Moore, P. K. (1993) Inhibition of rat cerebellar nitric oxide synthase by 7-nitro indazole and related substituted indazoles. *Br. J. Pharmacol.* **110,** 225–228.

331. Schulz, J. B., Matthews, R. T., Muqit, M. M. K., Browne, S. E., and Beal, M. F. (1995) Inhibition of neuronal nitric oxide synthase by 7-nitroindazole protects against MPTP-induced neurotoxicity in mice. *J. Neurochem.* **64,** 936–939.

332. Przedborski, S., Jackson-Lewis, V., Yokoyama, R., Shibata, T., Dawson, V. L., and Dawson, T. M. (1996) Role of neuronal nitric oxide in 1-methyl-4-phenyl-1,2,3,6-tetrahydropyridine (MPTP)-induced dopaminergic neurotoxicity. *Proc. Natl. Acad. Sci. USA* **93,** 4565–4571.

333. Hantraye, P., Brouillet, E., Ferrante, R. J., Palfi, S., Dolan, R., Matthews, R. T., and Beal, M. F. (1996) inhibition of neuronal nitric oxide synthase prevents MPTP-induced parkinsonism in baboons. *Nature Med.* **2,** 1017–1021.

334. Cassina, A. and Radi, R. (1996) Differential inhibitory action of nitric oxide and peroxynitrite onb mitochondrial electron transport. *Arch. Biochem. Biophys.* **328,** 309–316.

335. Lizasoain, I., Moro, M. A., Knowles, R. G., Darley-Usmar, V., and Moncada, S. (1996) Nitric oxide and peroxynitrite exert distinct effects on mitochondrial respiration which are differentially blocked by glutathione or glucose. *Biochem. J.* **314,** 877–880.

336. Stadler, J., Billiar, T. R., Curran, R. D., Stuehr, D. J., Ochoa, J. B., and Simmons, R. L. (1991) Effect of exogenous and endogenous nitric oxide on mitochondrial respiration of rat hepatocytes. *Am. J. Physiol.* **260,** C910–C916.

337. Radi, R., Rodriguez, M., Castro, L., and Telleri, R. (1994) Inhibition of mitochondrial electron transport by peroxynitrite. *Arch. Biochem. Biophys.* **308(1),** 89–95.

338. Huie, R. E. and Padmaja, S. (1993) The reaction of NO with superoxide. *Free Radical Res. Commun.* **18,** 195–199.

338a. Castagnoli, K., Palmer, S., Anderson, A., Bueters, T., and Castagnoli, N., Jr. (1997) The neuronal nitric oxide synthase inhibitor 7-nitroindazole also inhibits the monoamine oxidase-B catalyzed oxidation of 1-methyl-4-phenyl-1,2,3,6-tetrahydropyridine. *Chem. Res. Toxicol.* **10,** 364–368.

338b. Di Monte, D. A., Royland, J. E., Anderson, A., Castagnoli, K., Castagnoli, N., Jr., and Langston, J. W. (1997) Inhibition of monoamine oxidase contributes to the protective effect of 7-nitroindazole against MPTP neurotoxicity. *J. Neurochem.,* in press.

339. Oppenheim, R. W. (1991) Cell death during development of the nervous system. *Ann. Rev. Neurosci.* **14,** 453–501.

340. McDonald, H. R. and Lees, R. K. (1990) Programmed cell death of autoreactive thymocytes. *Nature* **343**, 642–644.

341. Mochizuki, H., Goto, K., Mori, H., and Mizuno, Y. (1996) Histochemical detection of apoptosis in Parkinson's disease. *J. Neurol. Sci.* **137**, 120–123.

342. Mochizuki, H., Nakamura, N., Nishi, K., and Mizuno, Y. (1994) Apoptosis is induced by 1-methyl-4-phenylpyridinium ion (MPP+) in ventral mesencephalic-striatal co-culture in rat. *Neurosci. Lett.* **170**, 191–194.

343. Dipasquale, B., Marini, A. M., and Youle, R. J. (1991) Apoptosis and DNA degradation induced by 1-methyl-4-phenylpyridinium in neurons. *Biochem. Biophys. Res. Commun.* **181**, 1442–1448.

344. Jackson-Lewis, V., Jakowec, M. W., Burke, R. E., and Przedborski, S. (1995) Time course and morphology of dopaminergic neuronal death caused by the neurotoxin 1-methyl-4-phenyl-1,2,3,6-tetrahydropyridine. *Neurodegeneration* **4**, 257–269.

345. Janson, A. M. (1996) Neuronal cell death by apoptosis *in vivo* in the 1-methyl-4-phenyl-1,2,3,6-tetrahydropyridine (MPTP) mouse model of Parkinson's disease (PD). *Soc. Neurosci. Abstracts* **22**, 721.

346. Mochizuki, H., Goto, K., Mori, H., and Mizuno, Y., (1996) Histochemical detection of apoptosis in Parkinson's disease. *J. Neurol. Sci.* **137**, 120–123.

347. Dragunow, M., Faull, L. M., Lawlor, P., Beilharz, E. J., Singleton, K., Walker, E. B., and Mee, E. (1995) In situ evidence for DNA fragmentation in Huntington's disease striatum amd Alzheimer's disease temporal lobe. *Neuroreport* **6**, 1053–1057.

348. Levi-Montalci, R. (1987) The nerve growth factor 35 years later. *Science* **237**, 1154–1162.

349. Knusel, B., Michel, P. P., Schwaber, J. S., and Hefti, F. (1990) Selective and nonselective stimulation of central cholinergic and dopaminergic development in vitro by nerve growth factor, basic fibroblast growth factor, epidermal growth factor, insulin and the insulin-like growth factors I and II. *J. Neurosci.* **10(2)**, 558–570.

350. Knusel, B., Winslow, W., Rosenthal, A., Burton., L. E., Seid, D. D., Nickolics, K., and Hefti, F. (1991) Promotion of central cholinergic and dopaminergic neuron differentiation by brain derived neurotrophic factor but not neurotrophin-3. *Proc. Natl. Acad. Sci. USA* **88**, 961–965.

351. Hyman, C., Hofer, M., Barde, Y. A., Juhasz, M., Yancopoulos, G. D., Squinto, S. P., and Lindsay, R. M. (1991) BDNF is a neurotrophic factor for dopaminergic neurons of the substantia nigra. *Nature* **350**, 230–232.

352. Spina, M. B., Squinto, S. P., Miller, J., Lindsay, R. M., and Hyman, C. (1992) Brain-derived neurotrophic factor protects dopamine neurons against 6-hydroxydopamine and *N*-methyl-4-phenylpyridinium ion toxicity: Involvement of the glutathione system. *J. Neurochem.* **59**, 99–106.

353. Garcia, E., Rios, C., and Sotelo, J. (1992) Ventricular injection of nerve growth factor increases dopamine content in the striata of MPTP-treated mice. *Neurochem. Res.* **17**, 979–982.

354. Glass, D. J. and Yancopoulos, G. D. (1993) The neurotrophins and their receptors. *Trends Cell. Biol.* **3**, 262–268.

355. Beck, K. D., Knusel, B., and Hefti, F. (1993) The nature of the trophic action of brain-derived neurotrophic factor, des(1-3)-insulin-like growth factor-1, and basic fibroblast

growth factor on mesencephalic dopaminergic neurons developing in culture. *Neuroscience* **52(4)**, 855–866.

356. Frim, D. M., Uhler, T. A., Galpern, W. R., Beal, M. F., Breakefield, X. O., and Isacson, O. (1994) Implanted fibroblasts genetically engineered to produce brain-derived neurotrophic factor prevent 1-methyl-4-phenylpyridinium toxicity to dopaminergic neurons in the rat. *Proc. Natl. Acad. Sci. USA* **91**, 5104–5108.

357. Tsukahara, T., Takeda, M., Shimohama, S., Ohara, O., and Hashimoto, N. (1995) Effects of brain-derived neurotrophic factor on 1-methyl-4-phenyl-1,2,3,6-tetrahydropyridine-induce parkinsonsim in monkeys. *Neurosurgery* **37(4)**, 733–741.

358. Hyman, C., Juhasz, M., Jackson, C., Radziejewski, C., and Linsay, R. M. (1991) Effects of BDNF and NT-3 on dopaminergic and GABAergic neurons of the ventral mesencephalon. *Soc. Neurosci. Abstracts* **17**, 908.

359. Hyman, C., Juhasz, M., Jackson, C., Wright, P., Ip, N. Y., and Lindsay, R. M. (1994) Overlapping and distinct actions of the neurotrophins, BDNF, NT-3, and NT-4/5, on cultured dopaminergic and GABAergic neurons of the ventral mesencephalon. *J. Neurosci.* **14**, 335–347.

360. Kirschner, P. B., Jenkins, B. G., Schulz, J. B., Finkelstein, S. P., Matthews, R. T., Rosen, B. R., and Beal, M. F. (1996) NGF, BDNF and NT-5, but not NT-3 protect against MPP+ toxicity and oxidative stress in neonatal animals. *Brain Res.* **713**, 178–185.

361. Jackson, G. R., Apffel, L., Werrbach-Perez, K., and Perez-Polo, J. R. (1990) Role of nerve growth factor in oxidant-antioxidant balance and neuronal injury. I. Stimulation of hydrogen peroxide resistance. *J. Neurosci. Res.* **25**, 360–368.

362. Pan, Z. and Perez-Polo, J. R. (1993) Role of nerve growth factor in oxidant homeostasis: glutathione metabolism. *J. Neurochem.* **61**, 1713–1721.

363. Nistico, G., Cirolo, M. R., Fishkin, K., Iannone, M., De Martino, A., and Rotilio, G. (1992) NGF restores decrease in catalase activity and increases superoxide dismutase and glutathione peroxidase activity in the brain of aged rats. *Free Radical Biol. Med.* **12**, 177–181.

364. Hung, H. C. and Lee, E. H. Y. (1996) The mesolimbic dopaminergic pathway is more resistant than the nigrostriatal dopaminergic pathway to MPTP and MPP+ toxicity: role of BDNF gene expression. *Mol. Brain Res.* **41**, 16–26.

365. Ferrari, G., Minozzi, M. C., Toffano, G., Leon, A., and Skaper, S. D. (1989) Basic fibroblast growth factor promotes the survival and development of mesencephalic neurons in culture. *Dev. Biol.* **133**, 140–147.

366. Mayer, E., Dunnett, S. B., Pellitteri, R., and Fawcett, J. W. (1993) Basic fibroblast growth factor promotes the survival of embryonic ventral mesencephalic dopaminergic neurons—I. Effects *in vitro*. *Neuroscience* **56**, 379–388.

367. Engele, J. and Bohn, M. C. (1991) The neurotrophic effects of fibroblast growth factors on dopaminergic neurons *in vitro* are mediated by mesencephalic glia. *J. Neurosci.* **11**, 3070–3078.

368. Unsicker, K., Reichert-Preibsch, H., and Wewetzer, K. (1992) Stimulation of neuron-survival by basic FGF and CNTF is a direct effect and not mediated by non-neuronal cells: evidence from single cells cultures. *Dev. Brain Res.* **65**, 285–288.

369. Date, I., Notter, M. F. D., Felten, S. Y., and Felten, D. L. (1990) MPTP-treated young mice but not aging mice show partial recovery of the nigrostriatal dopaminergic system

by stereotaxic injection of acidic fibroblast growth factor (aFGF). *Brain Res.* **526,** 156–160.

370. Otto, D. and Unsicker, K. (1990) Basic FGF reverses chemical and morphological deficits in the nigrostriatal system of MPTP-treated mice. *J. Neurosci.* **10,** 1912–1921.

371. Farris, T. W., DiStefano, L., and Schneider, J. S. (1994) Intranigral infusion of CNTF, but not bFGF, EGF or TGFb1, restores striatal DOPAC but not dopamine levels in MPTP-treated mice. *Soc. Neurosci. Abstracts* **20,** 1646.

372. Jin, B. K. and Iacovitti, L. (1996) Dopamine differentiation factors increase striatal dopaminergic function on 1-methyl-4-phenyl-1,2,3,6-tetrahydropyridine-lesioned mice. *J. Neurosci. Res.* **43,** 331–334.

373. Du, X. Y., Stull, N. D., and Iacovitti, L. (1994) Novel expression of the tyrosine hydroxylase gene requires both acidic fibroblast growth factor and an activator. *J. Neurosci.* **14,** 7688–7694.

374. Hoffer, B. J., Hoffmann, A., Bowenkamp, K., Huettl, P., Hudson, J., Martin, D., Lin, L. F. H., and Gerhardt, G. A. (1994) Glial cell line-derived neurotrophic factor reverses toxin-induced injury to midbrain dopaminergic neurons in vivo. *Neurosci. Lett.* **182,** 107–111.

375. Tomac, A., Lindqvist, E., Lin, L. F. H., Ogren, S. O., Young, D., Hoffer, B. J., and Olson, L. (1995) Protection and repair of the nigrostriatal dopaminergic system by GDNF in vivo. *Nature* **373,** 335–339.

376. Hou, J. G. G., Lin, L. F. H., and Mytilineou, C. (1996) Glial cell line-derived neurotrophic factor exerts neurotrophic effects on dopaminergic neurons in itro and promotes their survival and regrowth after damage by 1-methyl-4-phenylpyridinium. *J. Neurochem.* **66,** 74–82.

377. Zeng, B. Y., Jenner, P., and Marsden, C. D. (1996) Altered motor function and graft survival produced by basic fibroblast growth factor in rats with 6-OHDA lesions and fetal ventral mesencephalic grafts are associated with glial proliferation. *Exp. Neurol.* **139,** 214–226.

378. Wang, J., Bankiewicz, K. S., Plunkett, R. J., and Oldfield, E. H. (1994) Intrastriatal implantation of interleukin-1. Reduction of parkinsonism in rats by enhancing neuronal sprouting from residual dopaminergic neurons in the ventral tegmental area of the midbrain. *J. Neurosurg.* **80,** 484–490.

7

Haloperidol-Derived Pyridinium Metabolites

Structural and Toxicological Relationships to MPP+-Like Neurotoxins

Cornelis J. Van der Schyf, Etsuko Usuki, Susan M. Pond, and Neal Castagnoli, Jr.

1. INTRODUCTION

Schizophrenia is a chronic disorder that usually develops before the age of 35 yr and affects about 1% of the population *(1)*. The introduction of neuroleptics, drugs that are characterized by their antagonist properties at the dopamine (DA) D2-receptor, has allowed most schizophrenic patients to be treated on an outpatient basis. These drugs remain the cornerstone in the management of these patients. The two principal classes of "typical" neuroleptics are the phenothiazines and the butyrophenones. Haloperidol (HP, **1**; *see* Scheme 1), the subject of this chapter, is the prototypic butyrophenone-based neuroleptic and one of the most frequently prescribed antipsychotic agents. A new class of "atypical" neuroleptics, best represented by the drug clozapine, is thought to mediate an antipsychotic effect through selective interactions with DA D4- and serotonin (specifically 5HT2A) receptors *(2,3)* but may exhibit low affinity also for the DA D2-receptor *(3)*.

Although the "typical" neuroleptics have revolutionized the treatment of schizophrenia, they cause a wide range of adverse side effects with the most troublesome being extrapyramidal syndromes, including tardive dyskinesia (TD), which develops relatively late in the course of drug treatment. The pathogenesis of this condition remains poorly understood *(4)*. We will review some of the past and current hypotheses relating to TD, and examine the potential contribution that neurotoxic pyridinium metabolites derived from HP may make to the pathogenesis of this syndrome.

2. TARDIVE DYSKINESIA (TD)

TD is regarded as one of the most serious side effects of neuroleptic drug treatment in part because of its potential irreversibility *(5)*. The symptoms,

From: Highly Selective Neurotoxins: Basic and Clinical Applications *Edited by: R. M. Kostrzewa.*
Humana Press Inc., Totowa, NJ

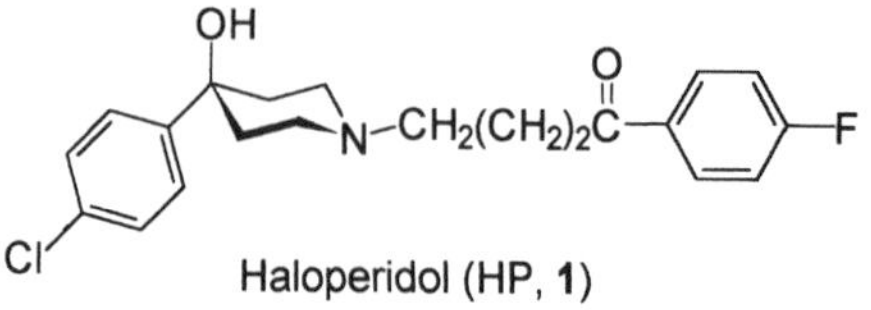

Haloperidol (HP, **1**)

Scheme 1

abnormal involuntary movements that often involve the orofacial musculature and less frequently the trunk and limbs, usually become evident after 12 or more months of neuroleptic therapy. Although the overall prevalence of TD is estimated to be about 15–20% *(3,6,7)* and increases with age and duration of therapy *(8)*, an estimation of the prevalence of true TD requires knowledge of the frequency of spontaneous dyskinesia, a syndrome that is clinically similar to TD *(9)*, but that is not induced by neuroleptics *(8)*. TD also may be related to another extrapyramidal side effect known as drug-induced parkinsonism (DIP).

It is generally accepted that TD is a progressive syndrome with gradual worsening of symptoms during long-term treatment *(8)*. In a prospective study of 50 patients on depot neuroleptics, the incidence of TD and DIP increased over the 7.1 yr the patients were monitored *(10)*. In this study, DIP was found to increase as TD worsened *(10)*. TD usually increases or becomes manifest for the first time after a reduction in drug dosage or after the withdrawal of long-term neuroleptic treatment. Consequently, neuroleptics appear to mask the symptoms of the syndrome they induce.

In some patients, TD improves and may disappear if neuroleptic treatment is discontinued and not resumed for a number of years. In other cases, however, especially in elderly patients, the syndrome persists even though drug therapy has been stopped. The prevalence of irreversible TD, defined as a condition lasting for more than 5 yr after drug withdrawal, may be lower than once believed *(8)*. Marsden *(11)* has estimated that 60% of TD cases resolve within 5 yr. This implies that about 6–8% of all patients receiving long-term neuroleptics develop irreversible dyskinesia *(8)*.

The longer the treatment and the higher the dose, the greater the risk of TD *(6)*. It is not yet possible to differentiate between various traditional neuroleptics with respect to TD-inducing capacity. The potent antidopaminergic neuroleptic drugs, such as HP, however, have a stronger TD-masking and parkinsonism-evoking effect as well as a greater tendency to lead to the development of TD compared to weaker antidopaminergic drugs, such as thioridazine *(8,12)*.

2.1. The Etiology of TD

2.1.1. Neuropathology

Some postmortem studies in human brain have suggested an association between TD and striatal damage *(13)*, but it is generally accepted that conventional neuropathologic examination of the human brain does not reveal any gross lesion in patients with TD. Animal models have been used to reduce the number of confounding variables that pertain in human studies, such as those of the disease, the drugs administered, their dose and duration, age, gender, and intercurrent illnesses, and to optimize the preparation of the brain for neuropathologic and ultrastructural examination. Such studies have revealed subtle abnormalities *(14–21)*. Several other theories attempt to propose a viable mechanism for the origins of drug-induced TD.

2.1.2. DA Receptor Upregulation

The development of TD is generally thought to be a consequence of DA receptor hypersensitivity *(22)*. Neuroleptic drugs block postsynaptic DA receptors, which causes an increase in acetylcholine turnover that may be linked to neuroleptic-induced parkinsonism. Chronic DA receptor blockade is thought to evoke a receptor hypersensitivity that can be demonstrated behaviorally as an increased sensitivity to DA agonists. The hypersensitivity is considered to be partly related to an upregulation of DA receptors in response to the drug-mediated receptor blockade *(8)*.

Several problems with this theory have been identified. Although it has been reported that DA agonists may sometimes increase TD, in many cases, they fail to do so *(5)*. Furthermore, animals rapidly develop neuroleptic-induced DA hypersensitivity, often after a single injection, whereas TD is generally observed in patients only after several months or even years of drug treatment *(8)*. This hypersensitivity also is not permanent, and normal receptor sensitivity and density usually recover within a few weeks of termination of neuroleptic administration *(23)*. Furthermore, no increase in the number of DA receptors has been found in the brains of TD patients in postmortem studies compared to control brains *(8)*. On the contrary, a decrease has been found in the number of D2-receptors in a postmortem study of the brains of TD patients who had been treated with neuroleptics, but had been off the drug for at least 1 yr before death *(24)*. The upregulation theory also does not explain the coexistence of parkinsonism and TD in many patients *(5)*.

2.1.3. The γ-Aminobutyric Acid (GABA) Hypothesis

A relationship between TD and reduced activity in a subgroup of GABA neurons in the striatum may exist. Animal studies have shown that more than one type of GABA neuron projects from the striatum. GABA neurons projecting from the anterior striatum to the lateral segment of the globus pallidus are inhibited by DA, whereas GABA neurons projecting from the posterior part of the striatum to the reticular zone of the substantia nigra and the medial-segment of the globus pallidus are excited by DA *(25)*. Neuroleptic drugs excite the former type of GABA neurons and inhibit the latter. An increase in GABA activity in the lateral segment of the globus pallidus is limited to parkinsonism-like symptoms, whereas the injection of a GABA antagonist into the medial segment of the globus pallidus and the reticular zone of the substantia nigra gives rise to hyperkinetic movements. This may imply that neuroleptic drugs acting in the anterior part of the striatum can give rise to parkinsonism-like symptoms and, when acting in the posterior part of the striatum, may give rise to hyperkinetic movements. Such dual functions offer a possible explanation to account for the coexistence of parkinsonism and TD *(26,27)*.

A study performed in 1984 by Gunne and colleagues using dyskinetic monkeys supports this hypothesis. Their study revealed a decrease of glutamate decarboxylase (GAD, a GABA-synthesizing enzyme) activity in the substantia nigra as well as in the medial segment of the globus pallidus and in the nucleus subthalamicus in the postmortem brains of these animals *(28)*. There appears to be a link between decreased GABA function in certain striatofugal neurons and TD development *(8)*. It is not yet known whether these biochemical changes also imply neuron loss. Long-term neuroleptic treatment may reduce GABA turnover in neurons running from the most posterior part of the striatum to the medial segment of the globus pallidus and to the reticular zone in the substantia nigra. This reduction in GABA turnover together with GABA receptor hypersensitivity may be one of the fundamental biochemical changes underlying TD development *(8)*. However, many questions involving the therapeutic implications of the GABA hypothesis remain unanswered. Particularly confounding in this regard are data suggesting that neither GABA receptor agonists nor GABA transaminase inhibitors display any specific anti-TD effect *(29)*.

2.1.4. Cholinergic Cell Loss

Cell damage or loss of the large cholinergic interneurons situated in the striatum may also explain the permanence of TD *(5,17)*. More selective histochemical techniques will be required to verify this hypothesis, because the proposed cell loss has not been studied thoroughly *(5)*. These populations of cells contribute only 2–3% of the cells found in the striatum *(30)*, so this mechanism would result in only a small proportion of cell loss. Damage to these cells could result from chronic overactivity owing to D2-receptor blockade *(5)*. These striatal cholinergic interneurons are inhibited by DA, acting through D2-receptors *(31)*. The chronic blockade of D2-receptors would result in a prolonged state of overactivity resulting from the disinhibition of these cholinergic neurons *(5)*.

Some clinical evidence supports this proposed mechanism for the induction of dyskinetic syndromes *(5)*. Reynolds et al. *(24)* reported a decrease in D2-receptors in the striatum of neuroleptic-free TD patients. This would be in agreement with the hypothesis of cholinergic cell loss, because a major proportion of striatal D2-receptors are situated on cholinergic neurons *(5)*. Also, the atypical neuroleptic drug clozapine, which is thought to owe its therapeutic efficacy to D1 blockade *(3,32)*, appears to carry a minor risk of TD development compared to HP *(8)*.

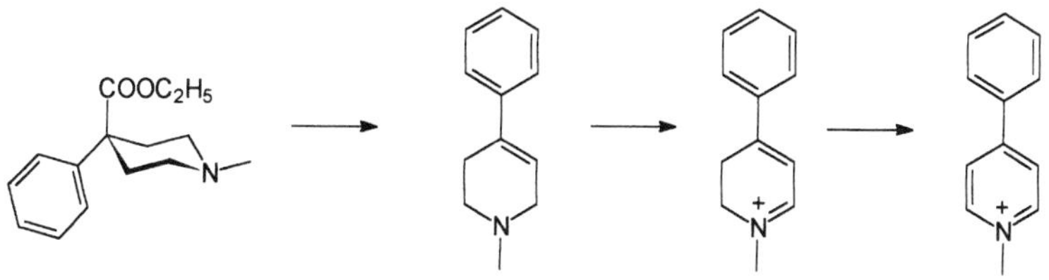

Fig. 1. Conversion of the propionoxypiperidine derivative MPPP (**2**) to neurotoxic pyridinium metabolite **5** via MPTP (**3**) and the dihydropyridinium intermediate (**4**).

3. THE 1-METHYL-4-PHENYL-1,2,3,6-TETRAHYDROPYRIDINE (MPTP) MODEL FOR PARKINSON'S DISEASE

The discovery that the neurotoxin MPTP ([**3**] Fig. 1) causes irreversible parkinsonism in humans has provided, for the first time, a model for one of the most prevalent neurodegenerative disorders in humans, Parkinson's disease *(33)*. Administration of MPTP causes the selective destruction of the DA-ergic neurons of the pars compacta of the substantia nigra (SN) *(33,34)*. Similar clinical symptoms and lesions were observed in monkeys *(35,36)* and mice *(37,38)*. The rat, however, appears to be resistant to MPTP *(39)*. The MPTP-induced syndrome in monkeys is thought to be a useful animal model for Parkinson's disease *(36)*.

This chemical is not new, its synthesis having been reported by Ziering and colleagues in 1947 *(40)*. Clues regarding its toxicological properties were overlooked for many years *(41)*. MPTP's biological effects were first reported in the scientific literature in the 1950s during the screening for new analgesics when aspects of its general toxicity were discovered *(42)*.

The first published record of MPTP's neurotoxic effects described a 23-yr-old drug abuser who, after self-administration of a substance thought to be 1-methyl-4-propionoxy-4-phenylpiperidine (MPPP **2**), developed a parkinsonian syndrome *(34)*. The patient responded positively to treatment with antiparkinsonian agents. Discontinuance of drug therapy, however, resulted in the reappearance of the parkinsonism, indicating the irreversibility of the syndrome. Analyses of the apparatus found in the home laboratory revealed traces of MPPP and MPTP. These compounds were tested in laboratory rats, but failed to produce the parkinsonian symptoms.

Interest in the neurotoxic effects of MPTP was renewed in 1983 when Langston and Ballard reported their observations on four patients, aged 26–42 yr, all of whom developed severe parkinsonism within days of using an illicit

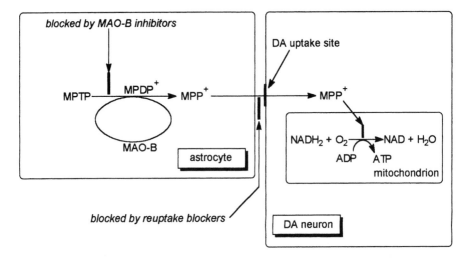

Fig. 2. Processes involved in MPTP neurotoxicity. *See text* for a detailed description.

"street heroin" drug, which was found to contain large quantities of MPTP *(33).* Subsequent studies have demonstrated that MPTP is biotransformed through a two-electron oxidation catalyzed by the flavoenzyme monoamine oxidase B (MAO-B) to yield the 1-methyl-4-phenyl-2,3-dihydropyridinium intermediate MPDP+ (**4**), which undergoes a second two-electron oxidation to form the 1-methyl-4-phenylpyridinium species MPP+ (**5**) *(43,44).* The MAO-B (and MAO-A) substrate properties of 1,4-disubstituted tetrahydropyridine derivatives, such as MPTP, were unexpected because no other cyclic amines have been found to display such activity.

MAO-B inhibitors were found to protect both mice *(45)* and monkeys *(46)* against the neurodegenerative effects of MPTP, suggesting that the brain MAO-B-generated pyridinium metabolite MPP+ is the ultimate neurotoxin *(43).* This suspicion was confirmed by evidence that MPP+ is transported by the DA transporter and localizes in the terminals of the nigrostriatal neurons *(47,48).* This relatively selective uptake of MPP+ may partly explain the fact that MPTP targets dopaminergic nigrostriatal neurons *(49)* (Fig. 2). Following nerve terminal uptake, MPP+ is concentrated inside the inner mitochondrial membrane *(50),* where it inhibits complex I of the electron transport chain *(51).* Because pyruvate and glutamate are the main metabolic and energy-producing substrates for brain mitochondria *(52),* the inhibition of the oxidation of these substances in the brain and the resulting depletion of adenosine triphosphate (ATP) *(53)* could account for the inability of neurons to perform functions critical for their survival *(54).*

4. STUDIES ON HP AND ITS BIOTRANSFORMATION TO POTENTIALLY NEUROTOXIC PYRIDINIUM METABOLITES

HP, a classical D2-receptor antagonist *(55,56)*, has been used for more than 25 years to treat psychotic disorders, such as schizophrenia *(57)*, and remains one of the most frequently prescribed neuroleptic drugs *(58)*. HP also is used to treat hyperexitable children and to control the symptoms of Tourette's syndrome *(59)*. As already discussed, the therapeutic usefulness of this drug is limited by the acute extrapyramidal side effects, such as dystonia, akathisia, and parkinsonism, as well as late-onset disorders, such as TD.

Animal studies have revealed subtle abnormalities in the brain ultrastructure after treatment with neuroleptics *(14–21)*. Recently, Roberts and coworkers *(4)* treated rats with HP for 6 mo and compared the differences in ultrastructure of the striatum between rats that developed a low or high number of vacuous chewing movements (VCMs), which are used as an indicator of TD. The ultrastructural abnormalities observed were either more severe or present only in rats with a high VCM rate. Of particular interest was the observation that the number of mitochondria in the high VCM group was reduced, whereas those that remained had hypertrophied. These abnormal features had not returned to normal 4 wk after withdrawal of the drug. These findings are provocative because MPTP's neurotoxicity is thought to be mediated by a mitochondrial lesion caused by the pyridinium metabolite MPP$^+$. It may be reasonable, therefore, to speculate that the irreversible dyskinesias observed in patients following chronic treatment with HP may involve damage to striatal mitochondria caused by an HP-derived pyridinium metabolite. This proposal also is consistent with the report that several MPTP analogs are biotransformed to neurotoxic pyridinium metabolites *(49,60)*. In addition, a variety of pyridinium and related quaternized azaheteroaromatic systems display MPP$^+$ behavior when delivered intracerebrally by the microdialysis technique *(61–63)*.

The structural features present in HP also may predispose this compound to an oxidative pathway leading to a pyridinium metabolite. Ring α-carbon oxidation of piperidine derivatives (**6**) bearing a leaving group (Lv) at C-4 generates the corresponding iminium metabolites **7** that, via the enamines **8**, will undergo spontaneous conversion to the dihydropyridinium intermediates **9** (Fig. 3). Dihydropyridinium compounds, such as **9**, undergo spontaneous autoxidation to form the pyridinium products **10** *(64,65)*. HP (**1**) and its principal circulating metabolite, reduced HP (RHP, **15**), are candidates for such reaction sequences (**1** → **11** → **12** → **13** → **14**, in the case of HP, and **15** → **16** → **17** → **18** → **19**, in the case of RHP). Alternatively, simple dehydration of **1** and **15** generates the corresponding tetrahydropyridine derivatives **20** and **21**, respectively, which, following ring α-carbon oxidation, would yield the same dihydropyridinium intermediates **13** and **18**.

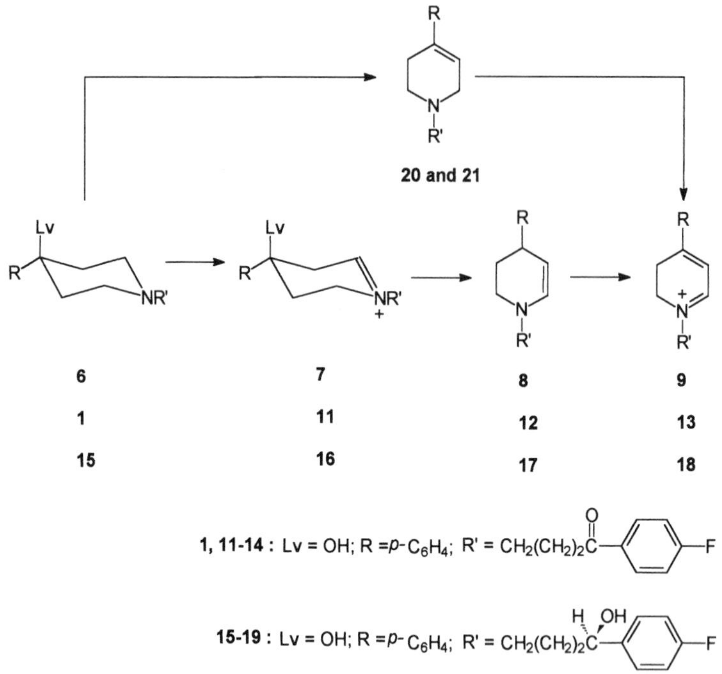

Fig. 3. Chemical structures of compounds discussed in the text, also illustrating the proposed catalytic pathway for the cytochrome P450-catalyzed (P450 3A4) oxidation of HP (**1**) to HPP+ (**14**).

In view of these considerations, the metabolic fate of HP (**1**) has been examined in rodents *(66,67)*, baboons *(68)*, and humans *(69,70)*, and that of its dehydration product HPTP (4-(4-chlorophenyl)-1,4-(4-fluorophenyl)-4-oxobutyl-1,2, 3,6-tetrahydropyridine, **20**) in rodents *(66,67)* and baboons *(68)* employing high-performance liquid chromatography–mass spectrometry/mass spectrometry (LC-MS/MS) and LC-fluorescence techniques designed to detect pyridinium metabolites. The results of the studies in rodents and baboons have established that both HP and HPTP are biotransformed to the HP pyridinium metabolite HPP+ (**14**), as is HP in humans. Unlike the corresponding conversion of MPTP to the neurotoxic pyridinium species MPP+, the oxidative biotransformation of HP to HPP+ is not mediated by MAO-B *(65)*. Neither HP nor HPTP is a substrate for purified bovine liver MAO-B, although both are biotransformed to the pyridinium product by rat and human liver microsomal preparations *(65)*. Recent findings *(71–73)* have led to the identification of cytochrome P450 3A4 as the principal catalyst responsible for these oxidative conversions.

More recent studies have documented the conversion of HP to a second pyridinium metabolite, the 4-(4-chlorophenyl)-1,4-(4-fluorophenyl)-4-hydroxybutyl-

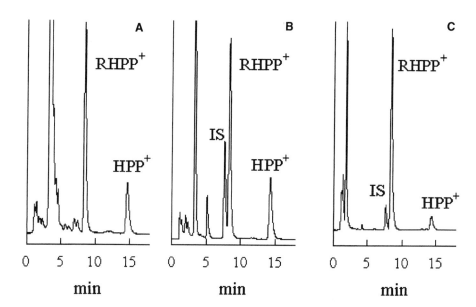

Fig. 4. HPLC chromatograms showing HPP+ and RHPP+ in urine extracts obtained from HP- (**A**) and HPTP- (**B**) treated baboons and HP-treated humans (**C**). IS = internal standard.

pyridinium species RHPP+ (**19**) in the human *(69,70,73)*. Both HP and HPTP also are biotransformed in mice *(67)* and baboons *(68)* to RHPP+. Figure 4 displays the HPLC chromatograms of urine extracts obtained from HP- and HPTP-treated baboons and HP-treated humans showing the presence of HPP+ and RHPP+. RHPP+ is derived from RHP (**15**), the major circulating metabolite of HP that is generated by the reduced nicotinamide adenine dinucleotide phosphate (NADPH)-dependent ketone reductase *(74)*, and from HPP+, via the ketone reductases *(72)*.

These findings have prompted an assessment of the neurotoxic properties of HPP+ and RHPP+. HPP+ is toxic to dopaminergic and serotonergic neurons present in cultures of rat embryonic mesencephalic tissue *(75)* and to cultured neuroblastoma cells *(76)*. Microdialysis studies in the rat indicate that HPP+ is about 10% as effective as MPP+ in causing the irreversible depletion of DA in striatal nerve terminals *(66)*. The serotonergic toxicities of HPP+ and MPP+ appear to be comparable *(77)*. Similar to MPP+, HPP+ is also reported to be a substrate for the brain synaptosomal DA transporter *(75,76)*. Finally and, perhaps most importantly, HPP+ is an even more potent inhibitor of the mitochondrial electron transport chain than is MPP+ *(77)*. Data on the neurotoxic potential of RHPP+ remain to be determined. All of these data are consistent with the proposal that HPP+ could cause a biochemical lesion that would contribute to the morphological changes seen in the striatal mitochondria of HP-treated rats and therefore to the TD associated with chronic HP treatment in humans (66,73,77).

5. SUMMARY AND CONCLUSIONS

The arguments summarized above together with the evidence that HP is metabolized in humans to HPP+ and RHPP+ raise the possibility that these pyridinium metabolites may play a role in TD observed in patients treated with HP. At this time, however, there is little histopathological evidence to suggest that HP itself is neurotoxic (14,18). Furthermore, the fact that the structurally unrelated phenothiazine neuroleptics also cause TD would indicate that if HPP+ and/or RHPP+ does contribute to the mediation of these effects, then these pyridinium compounds should interact with a system that is also compromised by the phenothiazines and possibly by MPP+. One possible common target would be the neuronal mitochondrial respiratory chain, since an abundance of evidence supports the proposal that the key event responsible for the neuronal degeneration observed in MPTP-treated animals is the inhibition of complex I in nigrostriatal neurons (51,78). Burckhardt and coworkers (79) recently reported that both HP and chlorpromazine (a phenothiazine neuroleptic drug) inhibit complex I in rat brain mitochondria, HP at concentrations comparing favorably with values reported for the inhibition of mouse liver mitochondria by MPP+ (77,80). Furthermore, several authors have recently provided compelling evidence of an association between Parkinson's disease and a mitochondrial defect (81–84). Support for this thesis comes also from our recent discovery of dicarboxylic acids associated with defects in mitochondrial respiration in the urine of HPTP-treated baboons (85).

The fact that HPP+ and RHPP+ are well-documented human metabolites of HP supports the possibility that these pyridinium neurotoxins may have access to the central nervous system in HP-treated patients. Recent evidence in fact suggests that both HPP+ and RHPP+ sequester in the brain of HP-treated patients examined at postmortem (86). It is important to recognize that TD most commonly appears only after very long-term exposure to neuroleptics (6). What is needed to make more credible the proposed in vivo neurotoxicity of HPP+ and RHPP+ is a well-documented central nervous system lesion in the HP- or HPTP-treated animal. We have very recently reported that long-term treatment with HPTP influences DA ligand binding in the basal ganglia of baboons, as measured in vivo by single-photon emission computed tomography (SPECT) (87). These data suggest that HPTP treatment exerts significant effects on the central nervous system that may contribute to some of the neurological disorders, including TD, that are seen in patients undergoing chronic HP therapy.

REFERENCES

1. Hafner, H. and Anderheiden, W. (1997) Epidemiology of schizophrenia. *Can J. Psychiatry* **42,** 139–151.

2. Melzer, H. Y. and Nash, J. F. (1991) Effects of antipsychotic drugs on serotonin receptors. *Pharmacol. Rev.* **43,** 587–604.

3. Seeman, P., Corbett, R., and Van Tol, H. H. M. (1997) Atypical neuroleptics have low affinity for dopamine D-2 receptors or are selective for D-4 receptors. *Neuropsychopharmacology* **16,** 93–110.

4. Roberts, R. C., Gaither, L. A., Gao, X. M., Kashyap, S. M., and Tamminga, C. A. (1995) Ultrastructural correlates of haloperidol-induced oral dyskinesias in rat striatum. *Synapse* **20,** 234–243.

5. Miller, R. and Chouinard, G. (1993) Loss of striatal cholinergic neurons as a basis for tardive dyskinesia and L-Dopa-induced dyskinesias, neuroleptic-induced supersensitivity psychosis and refractory schizophrenia. *Biol. Psychiatry* **34,** 713–738.

6. Saltz, B. L., Woerner, M. G., Kane, J. M., Lieberman, J. A., Alvir, J. M., Bergmann, K. J., Blank, K., Koblenzer, J., and Kahaner, K. (1991) Prospective study of tardive dyskinesia incidence in the elderly. *J. Am. Med. Assoc.* **266,** 2402–2406.

7. Casey, D. E. and Hasen, T. E. (1984) Spontaneous dyskinesia, in *Neuropsychiatric Movement Disorders* (Jeste, D. V. and Waytt, R. J., eds.), America Psychiatry Press, Washington, DC, pp. 68–95.

8. Gerlach, J. and Casey, D. E. (1988) Tardive dyskinesia. *Acta Psychiatry Scand.* **77,** 369–378.

9. Khot, V. and Wyatt, R. J. (1991) Not all that moves is tardive dyskinesia. *Am. J. Psychiatry* **148,** 661–666.

10. Elliot, K. J., Lewis, S., El-Mallakh, R. S., Looney, S. W., Caudill, R., and Bacani-Oropilla, T. (1994) The role of parkinsonism and antiparkinsonism in the subsequent development of tardive dyskinesia. *Ann. Clin. Psychiatry* **6,** 197–203.

11. Marsden, C. D. (1985) Is tardive dyskinesia a unique disorder? in *Dyskinesia Research and Treatment* (Casey, D. E., Chase, T., Christensen, A. V., and Gerlach, J., eds.), (*Psychopharmacology* **Suppl. 2**) Springer, Berlin, pp. 64–71.

12. Gerlach, J. and Simmelsgaard, H. (1978) Tardive dyskinesia during and following treatment with haloperidol, haloperidol + biperiden, thioridazine, and clozapine. *Psychopharmacologia* **59,** 105–112.

13. Christensen, E., Moller, J. E., and Faurbye, A. (1970) Neuropathologic investigation of 28 brains from patients with dyskinesias. *Acta Psychiatrica Scand.* **46,** 14–23.

14. Benes, F. M., Paskewich, P. A., and Domesick, V. B. (1983) Haloperidol-induced plasticity of axon terminals in rat substantia nigra. *Science* **221,** 969–971.

15. Benes, F. M., Paskewich, P. A., Davidson, J., and Domesick, V. B. (1985) The effects of haloperidol on synaptic patterns in the rat striatum. *Brain Res.* **329,** 265–274.

16. Dom, R. (1967) Local glial reaction of the CNS of albino-rats in response to the administration of a neuroleptic drug (butyrophenone). *Acta Neurol. Psychiatrica Belg.* **67,** 755–762.

17. Mahadik, S. P., Laev, H., Korenovsk, A., and Karpiak, S. E. (1988) Haloperidol alters rat CNS cholinergic system: enzymatic and morphological analyses. *Biol. Psychiatry* **24,** 199–217.

18. Meshul, C. K. and Casey, D. E. (1989) Regional, reversible ultrastructural changes in rat brain with chronic neuroleptic treatment. *Brain Res.* **489,** 338–346.

19. Meshul, C. K. and Tan, S. E. (1994) Haloperidol-induced morphological alterations are associated with changes in calcium/calmodulin kinase II activity and glutamate immunoreactivity. *Synapse* **18,** 205–217.

20. Meshul, C. K., Buckman, J. F., Allen, C., Riggan, J. P., and Feller, D. J. (1996) Activation of corticostriatal pathway leads to similar morphological changes observed following haloperidol treatment. *Synapse* **22**, 350–361.

21. Meshul, C. K., Stallbaumer, R. K., Taylor, B., and Janowsky, A. (1994) Haloperidol-induced morphological changes in striatum are associated with glutamate synapses. *Brain Res.* **648**, 181–195.

22. Klawans, H. L., Coetz, C. G., and Perlik, S. (1980) Tardive dyskinesia: review and update. *Am. J. Psychiatry* **137**, 900–908.

23. Tarsy, D. and Baldessarini, R. J. (1974) Behavioural supersensitivity to apomorphine following treatment with drugs which interfere with synaptic function of catecholamines. *Neuropharmacology* **13**, 927–940.

24. Reynolds, G. P., Brown, J. E., McCall, J. E., and McKay, A. V. P. (1992) Dopamine receptor abnormalities in the striatum and pallidum in tardive dyskinesia: a post mortem study. *J. Neural Transm.* **87**, 225–230.

25. Scheel-Krüger, J. and Arnt, J. (1985) New aspects on the role of dopamine, acetylcholine, and GABA in the development of tardive dyskinesia, in *Dyskinesia Research and Treatment* (Casey, D. E., Chase, T., Christensen, A. V., and Gerlach, J., eds.) (*Psychopharmacology* **Suppl. 2**) Springer, Berlin, pp. 46–71.

26. Casey, D. E., Chase, T., Christensen, A. V., and Gerlach, J. (1985) *Dyskinesia Research and Treatment.* (*Psychopharmacology* **Suppl. 2**) Springer, Berlin, p. 230.

27. Richardson, M. A., Pass, R., and Craig, T. J. (1986) The coexistence of parkinsonism-like symptoms and tardive dyskinesia, in *Biological Psychiatry* (Shagass, C., Josiassen, R. C., Bridger, W. H., Weiss, K. L., Stoff, D., and Simpson, G. M., eds.), Elsevier, New York, pp. 1211–1213.

28. Gunne, L. M, Häggeström, J. E., and Sjöquist, B. (1984) Association with persistent neuroleptic-induced dyskinesias of regional changes in brain GABA synthesis. *Nature* **309**, 347–349.

29. Casey, D. E., Povlsen, U. J., Meidahl, B., and Gerlach, J. (1985) Neuroleptic induced-tardive dyskinesia and parkinsonism: changes during several years of continuing treatment. *Psychopharmacol. Bull.* **22**, 250–253.

30. DiFiglia, M. (1987) Synaptic organization of cholinergic neurons in the monkey striatum. *J. Comp. Neurol.* **255**, 245–258.

31. Stoff, J, C., Verheijden, P. M. F. H., and Leysen, J. E. (1987) Stimulation of D-2 receptors in rat nucleus accumbens slices inhibits dopamine and acetylcholine release, but not cyclic AMP formation. *Brain Res.* **423**, 364–368.

32. Ellenbroek, B. A., Artz, M. T., and Cools A. R. (1991) The involvement of dopamine D1 and D2 receptors in the effects of the classical neuroleptic haloperidol and the atypical neuroleptic clozapine. *Eur. J. Pharmacol.* **196**, 103–108.

33. Langston, J. W., Ballard, P., Tetrud, J. W., and Irwin, I. (1983) Chronic Parkinsonism in humans due to a product of meperidine-analog synthesis. *Science* **219**, 979,980.

34. Davis, G. C., Williams, A. C., Markey, S. P., Ebert, M. H., Caine, E. D., Reichert, C. M., and Kopin, I. J. (1979) Chronic parkinsonism secondary to intravenous injection of meperidine analogues. *Psychiatry Res.* **1**, 249–254.

35. Burns, R. S., Chiueh, C. C., Markey, S. P., Ebert, M. H., Jacobowitz, D. M., and Kopin, I. J. (1983) A primate model of parkinsonism: selective destruction of dopaminergic

neurons in the pars compacta of the substantia nigra by *N*-methyl-4-phenyl-1,2,3,6-tetrahydropyridine. *Proc. Natl. Acad. Sci. USA* **80**, 4546–4550.

36. Langston, J. W., Forno, L. S., Rebert, C. S., Rebert, C. S., and Irwin, I. (1984) Selective nigral toxicity after systematic administration of 1-methyl-4-phenyl-1,2,5,6-tetrahydropyrine (MPTP) in the squirrel monkey. *Brain Res.* **292**, 390–394.

37. Heikkila, R. E., Hess, A., and Duvoisin, R. C. (1984) Dopaminergic neurotoxicity of 1-methyl-4-phenyl-1,2,5,6-tetrahydropyridine in mice. *Science* **224**, 1451–1453.

38. Hallman, H., Lange, J., Olson, L., Stromberg, I., and Jonsson, G. (1985) Neurochemical and histochemical characterization of neurotoxic effects of 1-methyl-4-phenyl-1,2,3,6-tetrahydropyridine on brain catecholamine neurons in the mouse. *J. Neurochem.* **44**, 117–127.

39. Chiueh, C. C., Markey, S. P., Burns, R. S., Johannessen, J. N., Pert, A., and Kopin, I. J. (1984) Neurochemical and behavioral effects of systemic and intranigral administration of *N*-methyl-4-phenyl-1,2,3,6-tetrahydropyridine in the rat. *Eur. J. Pharmacol.* **100**, 189–191.

40. Ziering, A., Berger, L., Heineman, S. D., and Lee, J. (1947) Piperidine derivatives. III. 4-Arylpiperidines. *J. Org. Chem.* **12**, 894–903.

41. Markey, S. P. and Schmuff, N. R. (1986) The pharmacology of the parkinsonian syndrome producing neurotoxin MPTP (1-methyl-4-phenyl-1,2,3,6-tetrahydropyridine) and structurally related compounds. *Med. Res. Rev.* **6**, 389–429.

42. Buchi, M., Prost, H., Eichenberger, H., and Lieberherr, R. (1952) Synthese und analgetische wirkung einiger 1-methyl-4-phenyl-piperidin-(4)-alkylsulfone. *Helv. Chim. Acta* **35**, 1527–1536.

43. Chiba, K., Trevor, A., and Castagnoli, N., Jr. (1984) Metabolism of the neurotoxic tertiary amine, MPTP, by brain monoamine oxidase. *Biochem. Biophys. Res. Commun.* **120**, 574–578.

44. Castagnoli, N., Jr., Chiba, K., and Trevor, A. J. (1985) Potential bioactivation pathways for the neurotoxin 1-Methyl-4-phenyl-1,2,3,6-tetrahydropyridine (MPTP). *Life Sci.* **36**, 225–230.

45. Heikkila, R. E., Manzino, L., Cabbat, F. S., and Duvoisin, R. C. (1984) Protection against the dopaminergic neurotoxicity of 1-methyl-4-phenyl-1,2,5,6-tetrahydropyridine by monoamine oxidase inhibitors. *Nature* **311**, 467–469.

46. Langston, J. W., Irwin, L., Langston, E. B., and Forno, L. S. (1984) Pargyline prevents MPTP-induced Parkinsonism in primates. *Science* **225**, 1480–1482.

47. Javitch, J. A., D'Amato, R. J., Strittmatter, S. M., and Snyder, S. H. (1985) Parkinsonian-inducing neurotoxin, *N*-methyl-4-phenyl-1,2,3,6-tetrahydropyridine: uptake of the metabolite *N*-methyl-4-phenylpyridine by dopamine neurons explains selective toxicity. *Proc. Natl. Acad. Sci. USA* **82**, 2173–2177.

48. Chiba, K., Trevor, A., and Castagnoli, N., Jr. (1985) Active uptake of MPP+, a metabolite of MPTP by brain synaptosomes. *Biochem. Biophys. Res. Commun.* **128**, 1228–1232.

49. Youngster, S. K., Sonsalla, P. K., Sieber, B. A., and Heikkila, R. E. (1989) Structure–activity study of the mechanism of 1-methyl-4-phenyl-1,2,3,6-tetrahydropyridine (MPTP)-induced neurotoxcicity. I. Evaluation of the biological activity of MPTP analogs. *J. Pharm. Exp. Ther.* **249**, 820–828.

50. Sayre, L. M. (1989) Biochemical mechanism of action of the dopaminergic neurotoxin 1-methyl-4-phenyl-1,2,3,6-tetrahydropyridine (MPTP). *Toxicol. Lett.* **48**, 121–149.

51. Nicklas, W. J., Vyas, I., and Heikkila, R. E. (1985) Inhibition of NADH-linked oxidation in brain mitochondria by 1-methyl-4-phenyl-1,2,5,5-tetrahydropyridine. *Life Sci.* **36,** 2503–2508.

52. Clark, J. B. and Nicklas, W. J. (1970) The metabolism of rat brain mitochondria: preparation and characterization. *J. Biol. Chem.* **245,** 4724–4731.

53. Di Monte, D., Jewell, S. A., Ekström, G., Sandy, M. S., and Smith, M. T. (1986) 1-Methyl-4-phenyl-1,2,3,6-tetrahydropyridine (MPTP) and 1-methyl-4-phenylpyridine (MPP+) cause rapid ATP depletion in isolated hepatocytes. *Biochem. Biophys. Res. Commun.* **137,** 310–315.

54. Gerlach, M., Riederer, P., Przuntek, and Youdim, M. B. H. (1991) MPTP mechanisms of neurotoxicity and their implications for Parkinson's disease. *Eur. J. Pharmacol.* **208,** 273–286.

55. Hyttel, J. and Christensen, A. V. (1983) Biochemical and pharmacological differentiation of neuroleptic effect on dopamine D-1 and D-2 receptors. *J. Neural Transm. Suppl.* **18,** 157–164.

56. Graybiel, A. M., Besson, M. J., and Weber, E. (1989) Neuroleptic sensitive binding sites in the nigrostriatal system: evidence for differential distribution of sigma sites in the substantia nigra, pars compacta of the cat. *J. Neurosci.* **9,** 326–338.

57. Osborne, P. G., O'Conner, W. T., Beck, O., and Ungerstedt, U. (1994) Acute versus chronic haloperidol: relationship between tolerance to catalepsy and striatal and accumbens dopamine, GABA and acetylcholine release. *Brain Res.* **634,** 20–30.

58. Wysowski, D. K. and Baum, C. (1989) Antipsychotic drug use in the United States. *Arch. Gen. Psychiatry* **46,** 929–932.

59. Baldessarini, R. J. (1991) Drugs and the treatment of psychiatric disorders, in *The Pharmacological Basis of Therapeutics* (Goodman, A. G., Rall, T. W., Nies, A. S., and Taylor, P., eds.), Pergamon, New York, pp. 383–435.

60. Johnson, E. A., Wu, E. Y., Rollema, H., Booth, R. G., Trevor, A., and Castagnoli, N., Jr. (1989) 1-Methyl-4-phenylpyridinium (MPP+) analogs: *in vivo* neurotoxicity and inhibition of striatal synaptosomal dopamine uptake. *Eur. J. Pharmacol.* **166,** 65–74.

61. Rollema, H., Booth, R. G., and Castagnoli, N., Jr. (1988) *In vivo* dopaminergic neurotoxicity of the 2-β-methylcarbolinium ion, a potential endogenous MPP+ analog. *Eur. J. Pharmacol.* **153,** 131–134.

62. Booth, G., Castagnoli, N., Jr., and Rollema, H. (1989) Intracerebral microdialysis neurotoxicity studies of quinoline and isoquinoline derivatives related to MPTP/MPP+. *Neurosci. Lett.* **100,** 306–312.

63. Nicklas, W. J., Saporito, M., Basma, A., Geller, H. M., and Heikkila, R. E. (1992) Mitochondrial mechanisms of neurotoxicity. *Ann. NY Acad. Sci.* **648,** 28–36.

64. Wu, E., Shinka, T., Caldera-Munoz, P., Yoshizumi, H., Trevor, A., and Castagnoli, N., Jr. (1988) Metabolic studies on the nigrostriatal toxin MPTP and its MAO B generated dihydropyridinium metabolite MPDP+. *Chem. Res. Toxicol.* **1,** 186–194.

65. Subramanyam, B., Woolf, T., Castagnoli, N., Jr. (1991) Studies on the *in vitro* conversion of haloperidol to a potentially neurotoxic pyridinium metabolite. *Chem. Res. Toxicol.* **4,** 123–128.

66. Subramanyam, B., Rollema, H., Woolf, T., and Castagnoli, N., Jr. (1990) Identification of a potentially neurotoxic pyridinium metabolite of haloperidol in rats. *Biochem. Biophys. Res. Commun.* **166,** 238–244.

67. Van der Schyf, C. J., Castagnoli, K., Usuki, E., Fouda, H. G., Rimoldi, J. M., and Castagnoli, N., Jr. (1994) Metabolic studies on haloperidol and its tetrahydropyridine analog in C57BL/6 mice. *Chem. Res. Toxicol.* **7,** 281–285.

68. Avent, K. M., Usuki, E., Eyles, D. W., Keeve, R., Van der Schyf, C. J., Castagnoli, N., Jr., and Pond, S. M. (1996) Haloperidol and its tetrahydropyridine derivative (HPTP) are metabolized to potentially neurotoxic pyridinium species in the baboon. *Life Sci.* **59,** 1473–1482.

69. Subramanyam, B., Pond, S. M., Eyles, D. W., Whiteford, H. A., Fouda, H. G., and Castagnoli, N., Jr. (1991) Identification of potentially neurotoxic pyridinium metabolite in the urine of schizophrenic patients treated with haloperidol. *Biochem. Biophy. Res. Commun.* **181,** 573–578.

70. Eyles, D. W., McLennan, H. R., Jones, A., McGrath, J. J., Stedman, T. J., and Pond, S. M. (1994) Quantitative analysis of two pyridinium metabolites of haloperidol patients with schizophrenia. *Clin. Pharmacol. Ther.* **56,** 512–520.

71. Igarashi, K., Kasuya, F., Fukui, M., Usuki, E., and Castagnoli, N., Jr. (1995) Studies on the metabolism of haloperidol (HP): the role of CYP3A in the production of the neurotoxic pyridinium metabolite HPP+ found in rat brain following IP administration of HP. *Life Sci.* **57,** 2439–2446.

72. Usuki, E., Pearce, R., Parkinson, A., and Castagnoli, N., Jr. (1996) Studies on the conversion of haloperidol and its tetrahydropyridine dehydration product to potentially neurotoxic pyridinium metabolites by human liver microsomes. *Chem. Res. Toxicol.* **9,** 800–806.

73. Eyles, D. W., McGrath, J. J., and Pond, S. M. (1996) Formation of pyridinium species of haloperidol in human liver and brain. *Psychopharmacology (Berl.)* **125,** 214–219.

74. Eyles, D. W. and Pond, S. M. (1992) Stereospecific reduction of haloperidol in human tissues. *Biochem. Pharmacol.* **44,** 867–871.

75. Bloomquist, J., King, E., Wright, A., Mytilineou, C., Kimura, K., Castagnoli, K., and Castagnoli, N., Jr. (1994) 1-Methyl-4-phenylpyridinium-like neurotoxicity of a pyridinium metabolite derived from haloperidol: cell culture and neurotransmitter uptake studies. *J. Pharmacol. Exp. Ther.* **270,** 822–830.

76. Fang, J. and Yu, P. H. (1995) Effect of haloperidol and its metabolites on dopamine and noradrenaline uptake in rat brain slices. *Psychopharmacology (Berl.)* **121,** 379–384.

77. Rollema, H., Skolnik, M., D'Engelbronner, J., Igarashi, K., Usuki, E., and Castagnoli, N., Jr. (1994) MPP+-like neurotoxicity of a pyridinium metabolite derived from haloperidol: *in vivo* microdialysis and *in vitro* mitochondrial studies. *J. Pharmacol. Exp. Ther.* **268,** 380–387.

78. Vyas, I., Heikkila, R. E., and Nicklas, W. J. (1986) Studies on the neurotoxicity of 1-methyl-4-phenyl-1,2,3,6-tetrahydropyridine: inhibition of NAD-linked substrate oxidation by its metabolite, 1-methyl-4-phenylpyridinium. *J. Neurochem.* **46,** 1501–1507.

79. Burckhardt, C., Kelly, J. P., Lim, Y. H., Filley, C. M., and Parker, W. D. (1993) Neuroleptic medications inhibit complex 1 of the electron transport chain. *Ann. Neurol.* **33,** 512–517.

80. Saporito, M. S., Heikkila, R. E., Youngster, S. K., Nicklas, W. J., and Geller, H. M. (1992) Dopaminergic neurotoxicity of 1-methyl-4-phenylpyridinium analogs in cultured neurons: relationship to the dopamine uptake system and inhibition of mitochondrial respiration. *J. Pharmacol. Exp. Ther.* **260,** 1400–1409.

81. Parker, W. D., Jr., Boyson, S. J., and Parks, J. K. (1989) Abnormalities of the electron transport chain in idiopathic Parkinson's disease. *Ann. Neurol.* **26,** 719–723.

82. Mizuno, Y., Ohta, S., Tanaka, M., Takamiya, S., Suzuki, K., Sato, T., Oya, H., Ozawa, T., and Kagawa, Y. (1989) Deficiencies in complex I subunits of the respiratory chain in Parkinson's disease. *Biochem. Biophys. Res. Commun.* **163,** 1450–1455.

83. Schapira, A. H., Cooper, J. M., Dexter, D., Clark, J. B., Jenner, P., and Marsden, C. D. (1990) Mitochondrial complex I deficiency in Parkinson's disease. *J. Neurochem.* **54,** 823–827.

84. Schoffner, J. M., Watts, R. L., Juncos, J. L., Torroni, A., and Wallace, D. C. (1991) Mitochondrial oxidative phosphorylation defects in Parkinson's disease. *Ann. Neurol.* **30,** 332–339.

85. Mienie, L. J., Bergh, J. J., Van Staden, E., Steyn, S. J., Pond, S. M., Castagnoli, N., Jr., and Van der Schyf, C. J. (1997) Metabolic defects caused by treatment with the tetrahydropyridine analog of haloperidol (HPTP), in baboons. *Life Sci.* **61,** 265–272.

86. Eyles, D. W., Avent, K. M., Stedman, T. J., and Pond, S. M. (1997) Two pyridinium metabolites of haloperidol are present in the brain of patients at post-mortem. *Life Sci.* **60,** 529–534.

87. Van der Schyf, C. J., Dormehl, I. C., Oliver, D. W., Hugo, N., Keeve, R., Müller-Gärtner, H. W., Pond, S., and Castagnoli, N., Jr. (1996) Long-term treatment with the tetrahydropyridine analog (HPTP) of haloperidol influences dopamine ligand binding in baboon brain: an ^{123}I-Iodobenzamide (IBZM) SPECT study. *Mol. Brain Res.* **43,** 251–258.

8

The Neurotoxic Effects of Continuous Cocaine and Amphetamine in Habenula

Gaylord Ellison

1. STIMULANT PSYCHOSES

The experiments to be described have grown out of attempts to develop animal models of psychosis, and especially of schizophrenia. Consequently, they represent a different approach to neurotoxicology than most studies in this area. One might be drawn toward toxicological studies of brain for a variety of reasons. One would be the attempt to determine what agents in the environment have toxic effects; this line of research would eventually focus on the structural classes of such toxic agents. Also, others might be interested in discovering the various mechanisms underlying cell death, or damage to particular neuronal elements; this leads to the investigation of local microenvironments in the neuron and what disrupts them. However, the studies to be described in this chapter have been guided by an attempt to mimic human psychotic states in animals in an effort to develop animal models of schizophrenia, and consequently, the toxicological implications of this research were not planned—the research dictated this direction. As such, these studies represent an interesting intersection of psychopathology, neuropharmacology, and neurotoxicology. They lead to the suggestion that certain neural pathways in brain may be the "weak links" in psychopathological disorders. Because they are based in part on observations from human drug addicts, the "weak-link pathways" they discover will be only a subset of those in brain—those with a reward component. However, compared to the other strategies, they have strong implications for the psychiatric clinic and for the neural foundations of psychopathology.

There are, in humans, certain drug-induced states that can be indistinguishable to the untrained observer from endogenous psychoses, such as occur in acute schizophrenic episodes. In order to develop animal models of these states, one attempts to mimic them by inducing similar drug-induced states in animals

From: Highly Selective Neurotoxins: Basic and Clinical Applications *Edited by: R. M. Kostrzewa.*
Humana Press Inc., Totowa, NJ

and, in controlled studies, thereby clarify the altered neural mechanisms that underlie these abnormal states.

It is generally recognized that there are two principal drug models of psychosis in humans: the stimulant-induced psychoses and phencyclidine-induced psychosis (Table 1). The stimulant psychoses are observed following chronic amphetamine or cocaine abuse. We have previously reviewed the extensive literature indicating the emergence of a paranoid-like psychosis in chronic amphetamine and cocaine addicts, the chief symptoms of which are motor stereotypies, paranoid delusions, sensory hallucinations (including parasitosis, or the delusion of bugs or snakes on the skin), and a loosening of associations *(1–3)*. This literature on amphetamine abuse has also been reviewed by Connell in 1958 *(4)*, Bell in 1965 *(5)*, and Ellinwood in 1967 *(6)*, and for cocaine abuse by Siegal in 1977 *(7)*, Lesko et al. in 1982 *(8)*, Gawin in 1986 *(9)*, and Manschreck et al. in 1988 *(10)*. A particularly interesting feature of this psychosis is the pronounced parasitosis, or the delusion of bugs or snakes on the skin *(11–13)*. The parasitotic grooming that develops in animals given stimulants will be discussed later.

In order to induce a model stimulant psychosis in animals, it is of paramount importance not only to give the proper drugs, but also to do so in the proper drug regimen. The development of "speed runs" appears to be a key factor for the induction of stimulant psychoses. It was recognized long ago *(4)* that a true amphetamine psychosis only appears in chronic addicts, that most amphetamine addicts eventually come to self-administer amphetamine every few hours (in binges lasting 3–5 d or even more), and that toward the end of these binges, they reliably develop paranoid delusions and hallucinations *(14)*. In fact, every controlled study eliciting an overt amphetamine psychosis in humans has involved continuous, low-dose administration of the drug every hour for days *(15,16)*. As the number of binges increases, the paranoia appears earlier and earlier during the binge, and can eventually appear on initial drug exposure *(17* as discussed in *3)*. There is a similarly extensive literature from cocaine addict populations of "runs," which eventually lead to paranoia. Satel et al. (1992) found that every one of their subjects who had experienced cocaine-induced paranoia did so while on a "binge" of from 6 h to 5 d in duration *(18)*.

Many years ago, in an effort to mimic amphetamine "speed runs" in animals, we developed a slow-release silicone pellet containing amphetamine base (releasing 20 mg over a 5-d period), and found that rats and nonhuman primates implanted with this pellet showed stages of behavioral alterations that were similar in sequence to those that had been reported in the controlled studies in humans, although the precise behaviors elicited were much more complex in the higher organisms. In rats, continuous amphetamine administration initially resulted in a period during which a sensitization to motor stereotypies elicited by amphetamine developed *(19)*, followed by a "late stage" (3–5 d after pellet implantation) in which the motor stereotypies decreased and certain distinctive

Table 1
Two Drug Models of Psychosis

Stimulant psychoses
    Produced by chronic amphetamine or cocaine abuse
    Well-documented in addicts who develop "speed runs"
    Chief symptoms are stereotypies, paranoid delusions, parasitosis and other
        sensory hallucinations, and loosening of associations
    Evidence of persisting alterations in nervous system ("Reactivation")
Phencyclidine and ketamine psychosis
    Produced by NMDA antagonists (phencyclidine, ketamine)
    "Binging" intake pattern develops in addicts
    Chief symptoms are flat effect, depersonalization, body image distortion,
        amnesia, catatonia, thought disturbances
    Evidence of persisting memory deficits

"late-stage" behaviors emerged, including limb flicks, wet-dog shakes, spontaneous startle responses, and abnormal social behaviors *(20)*. A similar progression, but with even more distinctive and varied late-stage behaviors occurs in monkeys *(1,21)*. Many of these behaviors have been called "hallucinogen-like" because they are normally induced by hallucinogens, whereas they are suppressed by acute injections of amphetamine. Another distinctive late-stage behavior is excited parasitotic grooming episodes. In monkeys, this is expressed as rapid, slapping hand movements directed at the skin surface moving from limb to limb *(21)*; in rats this is expressed as a change from the normal body washing and grooming sequence to a body-biting sequence similar to that of a dog inflicted with fleas *(22)*. There are close similarities between the amphetamine- and cocaine-induced parasitotic effects in humans and in these animal studies *(23)*.

2. NEUROTOXIC EFFECTS OF CONTINUOUS D-AMPHETAMINE IN CAUDATE

These late-stage behaviors induced by continuous amphetamine have a number of distinct neurochemical correlates in brain. Amphetamine continuously administered for 5 d induces alterations, including a downregulation of dopamine D2-receptors in striatum *(24)* and a progressive shift of heightened glucose metabolism away from striatal and toward mesolimbic structures *(1)*. However, one of the most striking effects of continuously administered amphetamine is its well-documented neurotoxic effects on dopamine terminals in caudate. Studies of catecholamine fluorescence in animals administered continuous amphetamine *(25–27)* reveal the appearance of swollen, distinct axons with multiple enlarged varicosities and stump-like endings; similar observations were made using silver stains for degenerating axons *(28)*.

The unique capability of continuous amphetamine administration to induce degeneration of dopamine terminals in the caudate nucleus has been validated using a variety of techniques, as reviewed elsewhere in this book. The amphetamine can be delivered by slow-release silicone pellets, minipumps, very frequent injections, or as substantial and frequent doses of methamphetamine, which has a slower rate of clearance and is considerably more potent at releasing dopamine *(29–31)*. Furthermore, Fuller and Hemrick-Luecke in 1980 found that an amphetamine injection administered in combination with drugs that slow its metabolism becomes neurotoxic to caudate dopamine terminals *(32)*. The amphetamine- or methamphetamine-induced damage to dopamine endings can be prevented by pretreatments or concurrent administration of drugs, such as a tyrosine hydroxylase inhibitor *(33)*, dopamine uptake inhibitors *(34,35)*, and noncompetitive antagonists of *N*-methyl-D-aspartate (NMDA) *(36,37)*. Studies of methamphetamine-induced neurotoxicity, which have been reviewed by Seiden and Ricaurte in 1987 *(38)*, typically employ doses that are comparatively higher than those using *d*-amphetamine, and these doses also induce damage to serotonin cells, are lethal to some of the animals, and induce widespread neuronal degeneration in a variety of other structures *(39)*. Hyperthermia contributes extensively to this degeneration, and this extremely vigorous methamphetamine regimen is probably a poor way to investigate selective neurotoxicity and discover "weak links."

One of the most interesting aspects of this neurotoxic effect is that only continuous amphetamine induces it. If the exact same amount of d-amphetamine (20 mg over 5 d, or about 12 mg/kg/d) is given as five daily injections, once each day, there is no neurotoxicity observed. This was initially a rather surprising finding, for the peak brain levels achieved after such large single injections are enormously greater than those when the drug is administered around-the-clock. However, it now appears that for a number of pharmacological agents, prolonged plasma levels are more crucial for producing neurotoxicity than are larger, but more transient plasma levels. Apparently neuronal systems have developed more effective ways to cope with sudden and brief insults than with progressive, more prolonged ones. If neurons are given huge amounts of drug, but a period to recover, they do not die, but if not allowed a recovery period, toxicological effects are immense.

3. NEUROTOXIC EFFECTS OF CONTINUOUS STIMULANTS IN FASCICULUS RETROFLEXUS (FR)

We recently attempted to determine if these findings with amphetamine could be generalized to cocaine psychosis. Cocaine, like amphetamine, also potentiates dopamine at the receptor, is a sympathomimetic, and leads to "speed runs" in chronic addicts who, in some cases, develop a paranoid psychosis similar in many aspects to that induced by amphetamine. The question for the dopamine model of psychosis that grew out of the amphetamine literature was whether continuous cocaine would also have a neurotoxic effect on dopamine terminals in caudate, i.e., if this was an anatomical correlate of the paranoia. Since continuous cocaine cannot be reliably administered via osmotic minipumps because of local vasoconstrictive and necrosis-inducing properties, an alternative drug delivery system was needed. Consequently, we developed a cocaine silicone pellet with a release rate of 103 mg cocaine base over 5 d and found it to induce behavioral stages similar to those caused by continuous amphetamine (initial hyperactivity, the evolution of stereotypies, a crash stage, and finally late-stage behaviors, including limb flicks, wet-dog shakes, and parasitotic grooming episodes *(40)*. We then looked for persisting alterations in dopamine receptors produced by continuous cocaine, as would be expected following DA terminal damage in striatum. We found no such changes at 14 d following continuous cocaine administration, although a parallel group that had received continuous amphetamine showed large changes in striatal dopamine D1- and D2-receptors *(41)*. However, the rats that had received continuous cocaine did show persisting alterations in ACh and GABA receptors in caudate, perhaps indicating that continuous cocaine had produced a somewhat different kind of neurotoxicity in caudate, possibly one postsynaptic to dopamine receptors. Therefore, in collaborative studies with Robert Switzer of Neuroscience Associates, this issue was investigated using silver-stain studies to assess neural degeneration *(42,43)*. By using minimally toxic doses and then searching for selective degeneration in brain, one can search for the weak links in neuronal circuitry induced by continuous stimulants with the assumption that these "weak" or "especially vulnerable" pathways, when overdriven by incessant stimulant-induced activity, eventually degenerate, leaving the brain in a persistently altered state.

Rats were given continuous amphetamine, continuous cocaine, or no drugs for 5 d, and then their brains were removed and examined for degeneration at various times following cessation of drug administration *(44)*. The entire brain from the olfactory nucleus to the mesencephalon was screened. The continuous amphetamine animals were found to evidence quite substantial degeneration in caudate. However, there was essentially no degeneration observed in caudate in the cocaine animals, but a very distinctive pattern of extensive degeneration was observed after either continuous amphetamine or cocaine in brain regions, which had been total unexpected: the lateral habenula (LHb) and fasciculus retroflexus (FR). Many of these long degenerating axons, when observed after several days had elapsed following pellet removal, showed classical anatomical signs of disintegration, such as axons beginning to fragment and the appearance of corkscrew or stump-like endings. These degenerating axons were almost exclusively in the mantle (as opposed to the core) of FR. Figure 1 shows this dramatic degeneration of FR after 5 d of continuous cocaine in a saggital section.

These results, coupled with the existing literature, have implications for models of stimulant-produced psychosis and paranoia. It is clear that amphetamine and cocaine are similar in that they are both strong stimulants with potent actions in potentiating dopamine, and both lead to a pattern of drug intake in addicts in which the drug is taken repeatedly over prolonged periods. For both

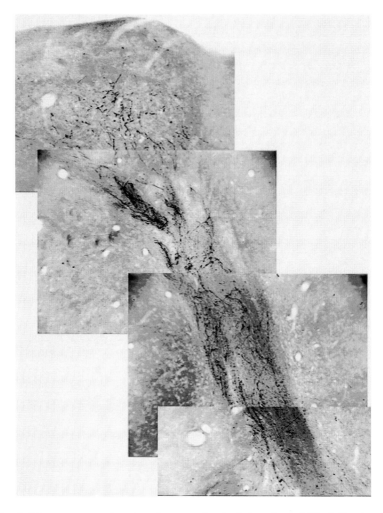

Fig. 1. Photomontage showing degeneration in habenula and FR following 5 d of continuous cocaine. At the top of the figure is LHb; the more ventral three sections follow FR. Because FR moves slightly laterally as it projects more ventrally, the bottom two sections are from a section slightly more lateral than the top two. Multiple long, darkly stained axons and swollen varicosities can be traced throughout FR.

drugs, these "runs" or binges produce a progressive dysphoria and paranoia followed by a rebound depression on drug discontinuation. Furthermore, when given continuously to animals, both drugs eventually induce comparable late-stage behaviors. However, these two drugs are markedly different in their persisting effects in caudate. Continuous amphetamine has neurotoxic effects on dopamine terminals and dopamine receptors in caudate; continuous cocaine does

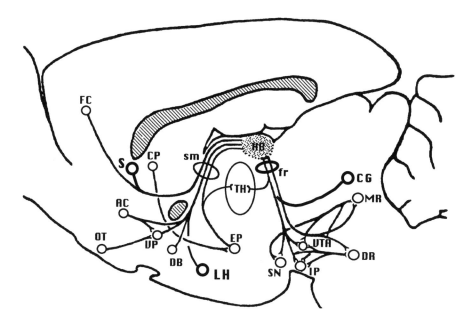

Fig. 2. Schematic representation of some of the chief inputs and outputs of the habenular complex. Abbreviations: FC, frontal cortex; OT, olfactory tubercle; AC, nucleus accumbens; CP, caudate-putamen; DB, nucleus of the diagonal band; VP, ventral pallidum; sm, stria medullaris thalami; EP, entopeduncular nucleus; fr, fasciculus retroflexus; TH, thalamic nuclei, including dorsalmedial, ventral anterior, and ventral lateral; HB habenula; SN substantia nigra; VTA ventral tegmental area; IP interpeduncular nucleus; MR medial raphe nucleus; DR dorsal raphe nucleus. Major descending pathways as shown entering sm, passing through or synapsing in habenula, and descending in fr to a variety of mesencephalic structures. Collaterals from EP and HB to thalamus are also shown.

not. Continuous cocaine produces persisting alterations in GABA and ACh receptors, whereas continuous amphetamine does not. However, the two drugs are quite similar in their ability to induce degeneration of axons in LHb extending ventrally into FR. A logical conclusion would be that it is the neurotoxic alterations in LHb and FR that play a critical role in mediating the paranoid psychosis that follows the continuous use of these stimulants and the persistently altered paranoid reactions to the drug that develop in chronic addicts.

4. THE HABENULA, FR, AND THE ANATOMY OF PARANOIA

These recent findings suggest a need to re-evaluate the role of the LHb and FR in the mediation of dopamine-related circuitry. Figure 2 illustrates the principal connections of the habenula as described in the classical anatomical studies by Herkenham and Nauta *(45,46)* and others. The inputs consist predominantly of pathways traveling in stria medullaris terminating in either the medial or lateral habenular nuclei, with two subdivisions: a medial "septal-limbic" and a lateral "pallidal-limbic" one. The principal input for medial habenula is cholinergic fibers arising from the septal area (nearly every septal cell projects to the medial habenula), but there are also projections from nucleus accumbens and the diagonal band of Broca. The major input to LHb is GABA fibers from the medial (or internal) globus pallidus (in primates) or its homolog in rat, the entopeduncular nucleus, but there are also inputs from limbic forebrain, including the lateral hypothalamus, diagonal band of Broca, substantia innominata, lateral preoptic area, nucleus accumbens, frontal cortex, and the suprachiasmatic nucleus. Both nuclei also receive less extensive ascending afferents from the central gray and medial raphe, and the LHb receives dopaminergic inputs from the substantia nigra and ventral tegmental area.

The principal efferent fibers from the medial habenula, including cholinergic, glutaminergic, and substance P fibers, travel in the core of FR to the interpeduncular nucleus, the ventral tegmental area, raphe nuclei, and substantia nigra. The LHb has more varied outputs, with axons traveling principally in the periphery or mantle region of the FR sending projections to several thalamic (mediodorsal and ventromedial) and hypothalamic (lateral, septal, and preoptic) nuclei. However, the principal efferents from LHb are to midbrain nuclei, such as the dorsal and medial raphe nuclei (constituting one of the major inputs to raphe), to ventral tegmental area and substantia nigra pars compacta, and also to central gray.

Sutherland in 1982 described some of the functional roles of what he termed this "dorsal diencephalic conduction system" *(47)*. It has anatomical and functional connections to modulate important functions, such as sensory gating through the thalamus, pain gating through the central gray and raphe, and mediation of motor stereotypies and reward mechanisms through the substantia nigra and the ventral tegmental area. Lesions of habenula produce a wide variety of behavioral alterations, including alterations in self-stimulation, pain inhibition, avoidance learning, and sexual and maternal behaviors *(3)*.

Studies of glucose utilization have consistently shown the habenula to be highly sensitive to dopamine agonists and antagonists, and in fact is the most sensitive region in brain to DA agonists, such as cocaine *(48)*. The dorsal diencephalic system has major and predominantly inhibitory connections onto dopamine-containing cells. The descending control of monoamine and other mesencephalic cells carried in FR appears to consist largely of inhibitory influences. Sasaki et al. in 1990 found that they could markedly attenuate methamphetamine-induced inhibition of substantia nigra cells by making lesions of the habenula, of the entopeduncular nucleus, or transections of the stria medullaris *(49)*. These studies support an important role of the dorsal diencephalic conduction system in inhibiting dopamine cell bodies and in mediating part of the negative feedback from limbic and striatal dopamine receptors onto dopamine cell bodies. These are ideal connections for the mediation of psychosis on both anatomical and functional grounds. The descending influences from dopamine-rich and limbic structures are quite unique in brain in that striatal and limbic inputs directly converge. In addition, this circuitry apparently mediates a major part of the descending control over serotonin cells of the raphe complex (in fact, they represent the chief input in all of brain to raphe). An implication of this is that owing to the amphetamine- or cocaine-induced degeneration of the FR fibers, the higher brain areas might no longer be able to regulate dopaminergic and serotoninergic cell firing, and especially to inhibit these midbrain cells.

5. DO THE FIBERS IN FR THAT DEGENERATE AFTER COCAINE BINGES CARRY NEGATIVE FEEDBACK FROM DOPAMINE-RICH REGIONS ONTO DA CELL BODIES?

There is additional evidence that the LHb and FR mediate part of the negative feedback from dopamine-rich regions onto dopamine cell bodies. Lesions of either stria medullaris, LHb, or FR increase dopamine turnover in prefrontal cortex, nucleus accumbens, and striatum *(50,51)*, and electrical stimulation of the habenula inhibits dopamine-containing cells in substantia nigra and ventral tegmental area *(52)*. Several recent observations from this laboratory clarify some of the long-lasting effects of continuous cocaine administration and also provide indirect evidence consistent with the hypothesis that the degenerating axons carry part of the dopamine-mediated negative feedback. We have found that there are long-lasting sequelae of 5 d of continuous treatment with the cocaine pellet that suggest correlates of the neurotoxicity observed in brain. Cocaine pellet-pretreated rats, when tested several weeks following pellet explant, act frightened in open-field tests. At the beginning of the test, they initially "freeze," remaining immobile for prolonged periods *(41)*, and when tested over prolonged periods in novel environments, they remain hyperactive far longer than the controls (Fig. 3). This suggests a lack of habituation to novel sensory stimulation in these animals. Both of these observations are highly consistent with increased DA turnover after lesions of LHb. We recently attempted to test the hypothesis that the axons that degenerate in FR and LHb following continuous cocaine mediate part of the negative feedback from DA receptors onto DA cell bodies using microdialysis techniques *(53)*. Rats were pretreated with either cocaine or control pellets for 5 d, and then 14 d later, microdialysis probes were lowered into the caudate nucleus. Baseline dopamine and GABA levels were not significantly different in the two groups. However, when the animals were perfused locally with the D1 agonist SKF38393, the controls showed a large decrease in striatal dopamine overflow and dopaminergic metabolites compared to the cocaine animals (Fig. 4). Because D1-receptors are largely postsynaptic in caudate, where dopamine release is governed largely by presynaptic mechanisms, this result suggests a deficiency in the negative feedback pathways extending from caudate onto substantia nigra and ventral tegmental area cell bodies, or locally within striatum.

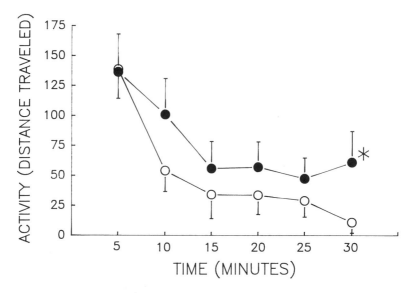

Fig. 3. When tested in a novel environment, several weeks after pellet removal, cocaine animals remain hyperactive longer than the controls. * Means different from controls, $p < .05$. O—O Control, ●—● cocaine.

It has also been reported that FR lesions in rats lead to decreased spontaneous alternation *(54)*. Similarly, we found that cocaine pellet-pretreated rats show evidence of long-lasting deficits in spontaneous alteration. Thus, animals treated with the cocaine pellet and then given a recovery period show a number of behavioral and biochemical alterations similar to those of animals following lesions of LHb or FR.

6. REPEATED COCAINE BOUTS: PROGRESSIVE EFFECTS ON BEHAVIOR AND TOXICITY

We recently conducted a study that did not work out as had been predicted, but that yielded enormously provocative results. The original experimental design represented an initial attempt to determine if there is any regeneration of the degenerating fibers in FR following the cocaine pellet. Four groups were prepared. A "single-pellet exposure group" was implanted with cocaine pellets and sacrificed 6 d later, 1 d after pellet removal. The amount of degeneration in LHb and FR in this group was compared with that in a second group ("pellet, 10-d recovery, second pellet"), which were rats implanted with a cocaine pellet for 5 d, then given a 10-d recovery period, and finally implanted with a second cocaine pellet for 5 d and sacrificed 1 d later. We hypothesized that little further degeneration would be observed in this group, since the tracts in these animals had recently degenerated and minimal recovery time had been given. A third group ("pellet, 3-mo recovery, second pellet") was implanted with a 5-d pellet,

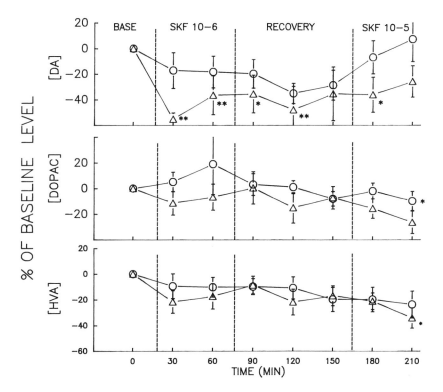

Fig 4. Dopamine and Dopac overflow in caudate, and activity levels of the rats, at several weeks following explantation of cocaine or control pellets. SKF38393 was perfused through the probe during the times shown. △—△ Control, ○—○ cocaine.

given a 3-mo drug-free recovery period, and then implanted with a second 5-d pellet and sacrificed a day later. It was hypothesized that if any regeneration had occurred, this group would now again show more degeneration in FR. A fourth group, "14-d injection," was given 14 daily injections of cocaine, a 10-d drug-free period, and then implanted with the cocaine pellet for 5 d and sacrificed a day later. This type of intermittent cocaine drug regimen has quite different effects on behavior than does continuous cocaine, so comparisons of degeneration in this group with that in the other groups were also of interest.

The actual results were quite different from our expectations. Compared to the single-pellet exposure group, there was as much, and in fact slightly more degeneration in the LHb and FR in the 10-d recovery group (Fig. 5). Thus, rather than providing evidence for any regeneration, this result seems to imply that a single cocaine pellet exposure causes degeneration in only a proportion of the vulnerable fibers, since a second pellet administered relatively shortly thereafter (the "pellet, 10-d recovery, second pellet" group) induces appreciable fur-

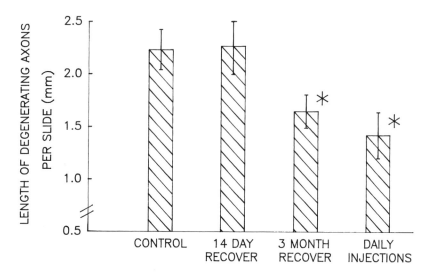

Fig. 5. Total amount of degeneration (sum of all axon lengths) in from one slide unilaterally. A blind observer sketched degenerating fibers using camera lucida and the resulting ink traces were quantified using NIMH Image. * Means significantly less than control, $P < .05$. "CONTROL": sham pellet, 14 d recover, coc. pellet. "14 DAY": coc. pellet, 14 d recover, coc. pellet. "3 MONTH": coc. pellet, 3 mo recover, coc. pellet. "DAILY INJECT": 14 coc. inj.; 14 d recover, coc. pellet.

ther degeneration. This is an important finding, since it indicates that repeated cocaine bouts spaced a week or two apart appear to be extraordinarily neurotoxic. These results imply that a "binge" induced by a single cocaine pellet does not induce degeneration in all the susceptible fibers, but leaves some of these fibers in a weakened and vulnerable state. It became clear that prior to this study, we had never really observed animals with the full extent of cocaine-induced degeneration.

A second unexpected finding was that pretreatment by spaced daily injections (14 daily injections, each of 10 mg/kg cocaine) actually produces an appreciable tolerance to neurotoxic effects induced by the drug given continuously, even though they showed a marked potentiation of stereotyped behaviors induced by the subsequent pellet. We could conclude this because in this study, we also measured behavior during the pellet exposure, videotaping the animals automatically every 2 h throughout the 5 d of the cocaine pellet exposures. Figure 6 shows that as reported previously, rats implanted with the cocaine pellet go through stages of behavior, from initial exploratory behavior, best measured by cage crossings, to motor stereotypies, and finally to "late-stage" behaviors, including what appears to be parasitotic grooming (i.e., a shift from the rat's normal grooming sequence to one involving biting or nibbling at various skin areas similar to a dog infested with fleas). The results of this experiment revealed substantially

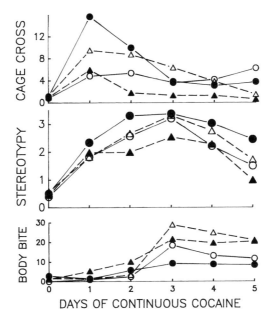

DAYS OF CONTINUOUS COCAINE

Fig. 6. Activity (cage crossings), motor stereotypy ratings, duration of body biting, and so forth, in the four groups. When rats are pretreated with daily injections of cocaine and then implanted with a cocaine pellet, they show more initial hyperactivity and then motor stereotypies than any other group, but less "late-stage" behaviors. Pretreatment before cocaine pellet: ○—○ none, ●—● 14 d inj., △– –△ pellet—2 wk, ▲– –▲ pellet—3 mo.

heightened behavioral alterations in both "reimplant" groups—both heightened stereotypies and then increased "late-stage" behaviors—i.e., the behavior was highly correlated with the amount of degeneration observed. Thus, on implantation with the second pellet, both the 10-d and the 3-mo animals showed even more late-stage behavior than with their first pellet exposure. Figure 6, on the left, shows the total duration of "body biting" in the four experimental groups during the first ("cocaine") or second pellet exposure (other two groups). This shows the potentiation of the distinctive parasitotic-like behavior, especially in the rats given a 3-mo period between implantation of the first and second pellet. Although pretreatments with cocaine injections or cocaine pellets (unless they are rapidly spaced) generally induce tolerance to neurotoxic effects induced by the drug, there is a clear lack of correlation between some behavioral indices, such as motor stereotypies, and neurotoxic effects.

In addition to replicating the stages of continuous stimulants—initial hyperactivity, stereotypy, "late stage" of limb flicks, wet-dog shakes, parasitotic grooming—these findings add a new twist to the abundant literature on the tolerance and sensitization induced by continuous and intermittent stimulants.

Although pretreatment with intermittent injections led to heightened hyperactivity and motor stereotypies, but lessened "late-stage" behaviors induced by a
subsequent cocaine pellet, the pellet pretreatment led to lessened stereotypies,
but heightened "late-stage" behaviors as indicated by parasitotic grooming.
Clearly, the persisting effects of these different drug regimens are much more
complex than previously imagined.

It thus appears that repeated bouts of cocaine exposure in rats may produce
progressively summating alterations in brain and behavior, and that we have
never really observed the fully developed "late-stage" hallucinatory syndrome of
behavior or investigated the full ramifications of how extensive the correlated
alterations in brain can be. However a highly recurrent theme in studies of both
amphetamine and cocaine addicts *(18)* is how paranoia and parasitosis progressively evolve in the confirmed addict, eventually reaching the point where the
initial drug intake can induce them. The cocaine addicts studied by Satel et al.
(18), who showed the full syndrome of binge-limited paranoia, had been addicts
for over 2 yr and had consumed an enormous estimated quantity of cocaine
each (1.34 + 1.7 kg). The repeated pellet preparation may develop into an extraordinarily interesting paradigm not only for the study of chronic cocaine abuse,
but also for more general models of sensory hallucinations, such as parasitosis
and of paranoia. These findings may have therapeutic implications as well as
general scientific ones, because the progressive development of parasitosis and
paranoia is often cited by addicts as critical for their seeking treatment. This
"repeated binge" preparation should prove perfect for the study of metabolic
and other regional brain changes correlated with "late-stage" behaviors.

7. WHERE ARE THE CELL BODIES THAT GIVE RISE TO FR DEGENERATION?

There are two distinct possibilities for where the cell bodies that give rise to
the degenerating axons following continuous amphetamine or cocaine are
located. They could be in LHb, projecting ventrally through FR, but they could
also be in midbrain cell groups. The dopaminergic cells of the substantia nigra
or ventral tegmental area give rise to ascending dopamine axons terminating in
habenula, and the raphe nuclei also project to habenula, as does the central gray.

Three lines of evidence point to the cell bodies in LHb as the source. The
first is related to the fact that the degenerating axons are quite highly concentrated in the mantle of FR. When the anterograde tracer PHAL is injected into
LHb *(55)*, the pattern of staining observed mirrors almost exactly that seen in
the degenerating fibers: a high concentration of descending fibers in the mantle
of FR, with some fibers then entering thalamic nuclei, but the majority terminating in such regions as the ventral tegmental area. The ascending fibers such
as from substantia nigra and ventral tegmental area projecting to LHb are not so
rigidly confined to FR. Some of the degenerating axons in FR following cocaine

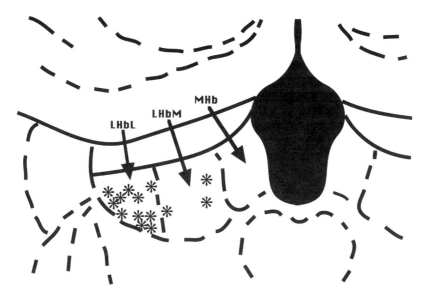

Fig. 7. Location of silver-impregnated cell bodies following repeated cocaine bouts. These degenerating cells are palely stained, but are concentrated in LHb in the same regions as C-*fos*-stained cells following acute cocaine injections.

show a morphological pattern identical to those described by Araki et al. *(55)* in normal animals labeled with injections into LHb.

A second line of evidence comes from studies in our laboratory by Giri Sulur. Rats were injected with PHAL in LHb using the Araki et al. *(55)* protocol, then given 7 d for anterograde transport to occur, and then implanted with either amphetamine or cocaine pellets for 5 d. When the animals were sacrificed 2 d after pellet removal, PHAL-stained fibers were observed in FR that had the distinctive characteristics of degenerating fibers (fragmented axons, corkscrew-shaped axons, and "end stumps"). This finding means that at least some of the degenerating axons have cell bodies in LHb.

The third line of evidence comes from the study described previously of animals given repeated cocaine bouts. When these animals were stained for degeneration, only a few stained cell bodies were observed (principally in the repeated pellet groups), but of these, most were concentrated in the most lateral part of the LHb, with a few in the more medial portion of LHb (Fig. 7). Furthermore, cell counts from cresyl violet sections from these same animals indicated a decreased number of cells in LHb in the animals given repeated cocaine bouts compared to the controls. When considered altogether, these data support the hypothesis that most, if not all, of the degenerating axons are from cells in LHb.

8. WHAT ARE THE MECHANISMS
OF THIS NEUROTOXICITY?

In most cases of neurotoxicity induced by drugs of abuse, the neurotoxic effects are observed in brain regions where glucose metabolism is markedly heightened by the drug. Examples are the neurotoxicity produced in caudate by continuous amphetamine *(3)* and the toxicity in several limbic regions produced by NMDA antagonists, such as phencyclidine *(56)*.

The neurotoxic effects of continuous cocaine and amphetamine in LHb and FR are unusual in that they are so strongly correlated with a decrease in glucose metabolism as measured by 2DG uptake in the affected structures. An immense number of studies of glucose utilization have consistently shown that although virtually all dopamine agonists increase glucose metabolism in dopamine-rich regions, such as caudate nucleus, nucleus accumbens, substantia nigra, and ventral tegmental area, they markedly decrease glucose metabolism in the habenula *(3)*. Indeed, in some studies, glucose metabolism in the habenula is the most sensitive region in all of the brain to low doses of dopamine agonists, such as cocaine. Another characteristic of the toxicity in LHb is that the conditions of drug administration sufficient to induce this effect must be continuous and extremely prolonged, on the order of many days. This was dramatically validated when it was found that very high doses of methamphetamine given for 8–10 h, although producing extraordinary degeneration in caudate-putamen and many other brain regions, were relatively ineffective in producing degeneration in LHb and FR *(39)*.

Recent data suggest a reason for this, as well as a proposed mechanism for the toxicity. Glucose metabolism, as reflected by 2DG uptake, typically reflects the summed activity in terminals whereas c-*fos* induction represents the heightened activity in cell bodies *(57)*. Consequently, it is possible that striatal GABAergic efferents to the entopeduncular nucleus are stimulated by the dopamine agonist administration and, thus, produce a strong inhibition of the entopeduncular efferents to the LHb, which are also largely GABAergic. The reduced activity in the terminals of these GABA projections to LHb would result in both the reduction of 2DG uptake and the disinhibition of habenular cells.

This hypothesis, reviewed by Wirtshafter et al. in 1994 *(58)*, is supported by their finding that dopamine agonists induce Fos-like immunoreactivity in cells in the most lateral LHb. In fact, the pattern of induction produced by amphetamine in their study was almost identical to the pattern of cells staining for degeneration (Fig. 7). Wirshafter et al. *(58)* further found that this Fos-like induction could be abolished by 6-hydroxydopamine (6-OHDA) lesions of the nigrostriatal bundle. In studies in our laboratory conducted by J. Petrie, and also in collaboration studies with Miles Herkenham, we have obtained virtually identical findings. Acute injections of cocaine led to a substantial increase in Fos-like labeling in cells of the most lateral portions of lateral habenula; this effect was somewhat larger in animals sacrificed after 5 d of continuous cocaine, although variability was very great in these animals. These findings suggest that the neurotoxicity induced by continuous amphetamine or cocaine in the LHb and FR may be owing to the prolonged hyperactivity in the LHb cells produced by the removal of GABAergic inhibitory influences, which then lead to a prolonged overstimulation of these cells and eventual excitotoxic neurotoxicity.

This is an unusual kind of neurotoxicity, however. Almost all of the silver-stained degeneration in LHb and FR is of axons, whereas with prolonged amphetamine, the degeneration is predominantly dopamine terminals in caudate, and with prolonged phencyclidine and other NMDA antagonists, the predominant silver staining is of entire cells, with all of their processes stained.

9. AUTORADIOGRAPHIC STUDIES OF RECEPTORS FOLLOWING CONTINUOUS COCAINE

In a recent study by Keys and Ellison *(59)*, this selective neurotoxicity in habenula was validated using an entirely different methodology. Rats were given two bouts of cocaine spaced 10 d apart, followed by a 14-d recovery period, and then sacrificed for autoradiographic studies using a variety of ligands for GABA, muscarinic, AMPA, serotonergic, and other receptors, as well as for the dopamine transporter. The largest change in receptor density for any ligand and any brain region was in the GABA (flunitrazepam) receptors in habenula (Fig. 8). This is a striking confirmation of the neurotoxicity in habenula using an entirely independent measure, and one that is much more amenable to quantification than silver staining.

10. IMPLICATIONS OF THESE STUDIES FOR FUTURE RESEARCHERS: BINGES

In each of the examples of neurotoxic effects described in this chapter, there is a very consistent and rather surprising finding. Giving a psychotomimetic drug continuously over a prolonged period induces profound neurotoxic effects, whereas giving the same total dose of the drug, but in once daily injections produces little or no toxic effects. However, when the same dose of the drug is given in a single injection, the transient levels of drug in plasma (and brain) are at least an order of magnitude higher than with the 24-h regimen.

It is doubtful that this is a general rule of toxicology, or even of neurotoxicology, for there are many toxic agents that have immediate effects that are dose-dependant. However, it may be a more general phenomenon in cases modeling psychiatriac disorders. Prolonged hyperactivity seems to produce generally long-term effects very different from those produced by intermittent activity and, I hypothesize, reveals the "weak links" in brain that underlie psychoses, i.e., the neural pathways that are especially susceptible to failure. During the acute psychotic break, there is often a prolonged period of progressively developing symptomology. An underlying metabolic abnormality would be expressed continuously over days, rather an intermittently once daily.

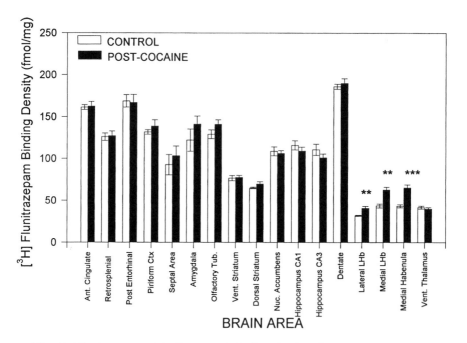

Fig. 8. Flunitrazepam binding at 14 d after final pellet removal. The largest changes in all of brain, for six different ligands, are of increased flunitrazepam binding in all three subdivisions of the habenula. Long-term flunitrazepam binding alterations. ☐ Control, ▆ postcocaine.

11. DEGENERATION PATTERNS AFTER OTHER PSYCHOTOMIMETIC DRUGS OF ABUSE

These findings suggest that roles of LHb and FR, and the dorsal diencephalic system in general, need to be reconsidered in the generation of stimulant-induced and other psychotic states, such as schizophrenia. It can be argued that alterations in these pathways are ideal candidates for producing the behaviors that occur during psychosis, and that future considerations of the circuitry underlying psychoses need to include this highly important, but relatively neglected system. Because these structures are not large in humans, it is presently very difficult to resolve them in scanning studies, but the clear prediction is of alterations in these structures in cocaine addicts and perhaps in schizophrenics.

These findings based on stimulant psychosis models should be clearly distinguished from those relating to the second drug model of psychosis: that produced by phencyclidine (PCP) and the other NMDA antagonists, such as ketamine and perhaps dizocilpine. The model psychoses that PCP and ketamine induce mimic a variety of schizophrenic symptoms, including flattened affect, a dissociative thought disorder, depersonalization, and catatonic states. These

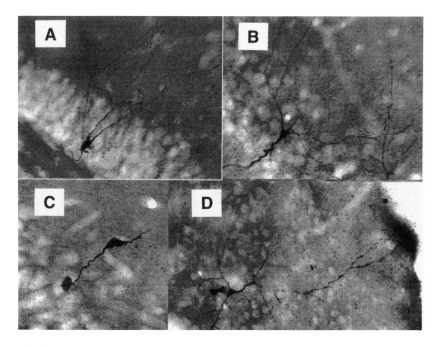

Fig. 9. Photomicrographs showing four degenerating cells in hippocampus following 5 d of continuous phencyclidine. **(A)** Granule cell in CA3. **(B)** Pyramidal cell in CA3. **(C)** Granule cell in CA1. **(D)** Cell in entorhinal cortex, with second cell in background. In each case, the entire cell, with all of its processes, are stained. Note the markedly hypertrophied process in (C), and also the extent to which the entorhinal cells in (D) are in a markedly advanced state of degeneration, with highly segmented processes.

symptoms can persist for prolonged periods, and there is evidence in chronic PCP and ketamine addicts of persisting memory deficits.

PCP, ketamine, and dizocilpine are quite similar in many of their effects, and they all have a neurotoxic effect on neurons in the most posterior cingulate cortex *(60)*. When we administered PCP or dizocilpine to rats in a 5-d binge regimen, we found minimal degeneration in LHb and FR. However both of these drugs induced neuronal degeneration in a variety other limbic structures. These included not only posterior cingulate ("retrosplenial") cortex, but also brain regions of rats related to olfaction, such as the olfactory tubercle, anterior olfactory nucleus, piriform cortex, and tenia tecta. Additional limbic structures affected were what some consider to be the most important structures in all of the brain for the organization of memory and orientation: the most posterior regions of entorhinal cortex and its projections, through the perforant pathway, to dentate gyrus and other cells in ventral hippocampus. This finding suggests anatomical substrates for a second drug model of psychosis, for most of these same structures are among the clearest areas where anatomical alterations occur in dementias, such as schizophrenia and Alzheimer's *(56)*.

Again, to obtain this more extensive degeneration, the NMDA antagonists must be given over a prolonged period, but what is stained in this case is different from the terminals in caudate following continuous amphetamine, or the fasciculus retroflexus axons with continuous amphetamine or cocaine. Rather, the predominately degenerating elements are entire cells, with all of their processes stained (Fig. 9), and this is probably an important clue regarding the mechanisms involved in this neurotoxicity. Glucose metabolism is markedly increased by NMDA antagonists in all of the brain regions where degeneration is observed. Excitotoxic lesions typically produce this same type of degeneration: complete cells, with all of their processes stained.

These studies have led to a number of surprises. Some are pharmacological, such as the finding that continuous, low-level amphetamines are much more neurotoxic to dopamine terminals than the same amount of drug given in large, single injections. Others are anatomical, such as the finding that fasciculus retroflexus is the dopamine-related pathway that is the "weak link" in stimulant binges. Another is the finding that NMDA antagonists, touted as the way to prevent neurotoxicity, are actually extraordinarily neurotoxic. The continuing evolution of these kinds of studies will hopefully clarify the mechanisms involved in "psychotomimetic neurotoxicology."

REFERENCES

1. Ellison, G. D. and Eison, M. S. (1983) Continuous amphetamine intoxication: An animal model of the acute psychotic episode, *Psychological Med.* **13,** 751–761.
2. Ellison, G. (1991) Animal models of hallucinations: continuous stimulants, in *Neuromethods*, vol. 18, *Animal Models in Psychiatry* (Boulton, A., Baker, G., and Martin-Iverson, M., eds.), Humana, Clifton, NJ, pp. 151–196.
3. Ellison, G. (1994) Stimulant-induced psychosis, the dopamine theory, and the habenula. *Brain Res. Rev.* **19,** 223–239.
4. Connell, P. (1958) *Amphetamine Psychosis.* Maudsley Monographs No. 5. Oxford University Press, London.
5. Bell, D. (1965) Comparison of amphetamine psychosis and schizophrenia. *Am. J. Psychiatry* **111,** 701–707.
6. Ellinwood, E. H., Jr. (1967) Amphetamine psychosis: I. Description of the individuals and the process. *J. Nervous and Ment. Dis.* **144,** 273–283.
7. Siegal, R. K. (1977) Cocaine: Recreational use and intoxication, in *NIDA Research Monograph* 13 (Petersen R. C. and Stillman, R. C., eds.), U.S. Government Printing Office, Washington, DC, pp. 55–67.
8. Lesko, L. M., Fischman, M., Javaid, J., and Davis, J. (1982) Iatrogenous cocaine psychosis. *N. Engl. J. Med.* **307,** 1153–1156.
9. Gawin, F. H. (1986) Neuroleptic reduction of cocaine-induced paranoia but not euphoria? *Psychopharmacology* **90,** 142,143.

10. Manschreck, T. C., Laughery, J. A., Weisstein, C. C., Allen, D., Humblestone, B., Neville, M., Podlewski, H., and Mitra, N. (1988) Characteristics of freebase cocaine psychosis. *Yale J. Biol. Med.* **61**, 115–122.

11. Brady, K., Lydiard, R., Malcolm, R., and Ballenger, J. (1991) Cocaine-induced psychosis. *J. Clin. Psychiatry* **52**, 509–512.

12. Elpern, D. (1988) Cocaine abuse and delusions of parasitosis. *Cutis* **42**, 273,274.

13. Mitchell, J. and Vierkant, A. (1991) Delusions and hallucinations of cocaine abusers and paranoid schizophrenics: a comparative study. *J. Psychiatry* **125**, 301–310.

14. Kramer, J. C., Gischman, V., and Littlefield, D. (1967) Amphetamine abuse: Pattern and effects of high doses taken intravenously. *J. Am. Med. Assoc.* **201**, 89–93.

15. Griffith, J., Cavanaugh, J., Held, N., and Oates, J. (1972) D-amphetamine: Evaluation of psychotomimetic properties in man. *Arch. Gen. Psychiatry* **26**, 97–100.

16. Angrist, B., Lee, H. K., and Gershon, S. (1974) The antagonism of amphetamine-induced symptomatology by a neuroleptic. *Am. J. Psychiatry* **131**, 817–819.

17. Bell, D. (1973) The experimental reproduction of amphetamine psychosis. *Arch. Gen. Psychiatry* **29**, 35–40.

18. Satel, S., Southwick, S., and Gawin, F. (1992) Clinical features of cocaine-induced paranoia. *Am. J. Psychiatry* **148**, 495–498.

19. Ellison, G. and Morris, W. (1981) Opposed stages of continuous amphetamine administration: Parallel alterations in motor sterotypies and in vivo spiroperidol accumulation. *Eur. J. Pharmacol.* **74**, 207–214.

20. Ellison, G. D., Eison, M. S., and Huberman, H. (1978) Stages of constant amphetamine intoxication: delayed appearance of abnormal social behaviors in rat colonies. *Psychopharmacology* **56**, 293–299.

21. Ellison, G. D., Nielsen, E. B., and Lyon, M. (1981) Animal models of psychosis: Hallucinatory behaviors in monkeys during the late stage of continuous amphetamine intoxication. *J. Psychiatry Res.* **16**, 13–22.

22. Nielsen, E., Lee, T., and Ellison, G. (1980) Following several days of continuous administration d-amphetamine acquires hallucinogen-like properties. *Psychopharmacology* **68**, 197–200.

23. de Leon, J., Antelo, R., and Simpson, G. (1992) Delusion of parasitosis or chronic hallucinosis; hypothesis about their brain physiopathology. *Compr. Psychiatry* **3**, 325–33.

24. Nielsen, E. B., Neilsen, M., Ellison, G., and Braestrup, E. (1980) Decreased spiroperidol and LSD binding in rat brain after continuous amphetamine. *Eur. J. Pharmacol.* **66**, 149–154.

25. Ellison, G. D., Eison, M., Huberman, H., and Daniel, F. (1978) Long term changes in dopaminergic innervation of caudate nucleus after continuous amphetamineetamine administration. *Science* **201**, 276–278.

26. Nwanze, E. and Jonsson, G. (1981) Amphetamineetamine neurotoxicity on dopamine nerve terminals in the caudate nucleus of mice. *Neurosci. Lett.* **26**, 163–168.

27. Ryan, L., Martone, M., Linder, J., Groves, P. (1990) Histological and ultrastructural evidence that d-amphetamine causes degeneration in neostriatum and frontal cortex of rats. *Brain Res.* **518**, 67–77.

28. Ryan, L. J., Martone, M., Linder, J., and Groves, P. M. (1988) Cocaine, in contrast to *d*-amphetamine, does not cause axonal terminal degeneration in neostriatum and agranular frontal cortex of long-evans rats. *Life Sci.* **43**, 1403–1409.

29. Hotchkiss, A. and Gibb, J. (1980) Long-term effects of multiple doses of methamphetamine on tryptophan hydroxylase and tyrosine hydroxylase activity in rat brain. *J. Pharmacol. Exp. Ther.* **214,** 257–262.

30. Ricaurte, G. A., Schuster, C. R., Seiden, L. S. (1980) Long-term effects of repeated methylamphetamine administration on dopamine and serotonin neurons in the rat brain: A regional study. *Brain Res.* **193,** 153–163.

31. Steranka, L. and Sanders-Bush, E. (1980) Long-term effects of continuous exposure to amphetamine on brain dopamine concentration and synaptosomal uptake in mice. *Eur. J. Pharmacol.* **65,** 439–443.

32. Fuller, R. and Hemrick-Luecke, S. (1980) Long-lasting depletion of striatal dopamine by a single injection of amphetamine in iprindole-treated rats. *Science* **209,** 305–306.

33. Wagner, G., Lucot, J., Chuster, C., and Seiden, L. (1983) Alpha-methyltyrosine attenuates and reserpine increases methamphetamine-induced neuronal changes. *Brain Res.* **270,** 285–288.

34. Fuller, R. and Hemrick-Luecke, S. (1982) Further studies on the long-term depletion of striatal dopamine in iprindole-treated rats by amphetamine. *Neuropharmacology* **21,** 433–438.

35. Hanson, G. R., Matsuda, L., and Gibb, J. W. (1987) Effects of cocaine on methamphetamine-induced neurochemical changes: characterization of cocaine as a monoamine uptake blocker. *J. Pharmacol. Exp. Ther.* **242,** 507–513.

36. Sonsalla, P., Nicklas, W., and Heikkila, R. (1989) Role for excitatory amino acids in methamphetamine-induced nigrostriatal dopaminergic toxicity. *Science* **243,** 398–400.

37. Fuller, R., Hemrick-Luecke, S., and Ornstein, P. (1992) Protection against amphetamine-induced neurotoxicity toward striatal dopamine neurons in rodents by LY274614, an excitatory amino acid antagonist. *Neuropharmacology* **31,** 1027–1032.

38. Seiden, L. and Ricaurte, G. (1987) Neurotoxicity of methamphetamine and related drugs, in *Psychopharmacology: The Third Generation of Progress* (Meltzer, H., ed), Raven, New York, pp. 359–366.

39. Ellison, G. and Switzer, R. III (1994) Dissimilar patterns of degeneration in brain following four different addictive stimulants. *NeuroReport* **5,** 17–20.

40. Lipton, J., Zeigler, S., Wilkins, J., and Ellison, G. (1991) Silicone pellet for continuous cocaine administration: heightened late-stage behaviors compared to continuous amphetamine. *Pharmacol. Biochem. Behavior* **38,** 927–930.

41. Zeigler, S., Lipton, J., Toga, A., and Ellison, G. (1991) Continuous cocaine produces persistent changes in brain neurochemistry and behavior different from amphetamineetamine. *Brain Res.* **552,** 27–35.

42. Switzer, R. C. (1991) Strategies for assessing neurotoxicity. *Neurosci. Biobehav. Rev.* **15,** 89–93.

43. de Olmos, J., Ebbesson, S., and Heimer, L. (1981) Silver methods for the impregnation of degenerating axoplasm, in *Neuroanatomical Tract-Tracing Methods* (Heimer, L. and Robards, N., eds.), Plenum, New York, pp. 117–168.

44. Ellison, G. (1992) Continuous amphetamine and cocaine have similar neurotoxic effects in lateral habenular nucleus and fasciculus retroflexus. *Brain Res.* **598,** 353–356.

45. Herkenham, M. and Nauta, W. J. H. (1977) Afferent connections of the habenular nuclei in the rat. *J. Comp. Neurol.* **173,** 123–146.

46. Herkenham, M. and Nauta, W. J. H. (1979) Efferent connections of the habenular nuclei in the rat. *J. Comp. Neurol.* **187,** 19–48.

47. Sutherland, R. J. (1982) The dorsal diencephalic conduction system: a review of the anatomy and functions of the habenular complex. *Neurosci. Biobehav. Rev.* **6,** 1–13.

48. London, E., Wilkerson, G., Goldberg, S., and Risner, M. (1986) Effects of L-cocaine on local cerebral glucose utilization in the rat. *Neurosci. Lett.* **68,** 73–78.

49. Sasaki, K., Suda, H., Watanabe, H., and Yagi, H. (1990) Involvement of the entopeduncular nucleus and the habenula in methamphetamine-induced inhibition of dopamine neurons in the substantia nigra of rats. *Brain Res. Bull.* **25,** 121–127.

50. Lisoprawski, A., Herve, D., Blanc, G., Glowinski, J., and Tassin, J. (1980) Selective activation of the mesocortico-frontal dopaminergic neurons induced by lesions of the habenula in the rat. *Brain Res.* **183,** 229–234.

51. Nishikawa, T., Fage, D., and Scatton, B. (1986) Evidence for, and nature, of the tonic inhibitory influence of habenulointerpeduncular pathways upon cerebral dopaminergic transmission in the rat. *Brain Res.* **373,** 324–336.

52. Christoph, C., Leonzio, R., and Wilcox, K. (1986) Stimulation of the lateral habenula inhibits dopamine-containing neurons in the substantia nigra and ventral tegmental area of the rat. *J. Neurosci.* **6,** 613–619.

53. Keys, A. and Ellison, G. (1994) Continuous cocaine induces persisting alterations in dopamine overflow in caudate following perfusion with a D1 agonist. *J. Neural. Trans.—General Section* **97,** 225–233.

54. Corodimas, K., Rosenblatt, J., and Morrell, J. (1992) The habenular complex mediates hormonal stimulation of maternal behavior in rats. *Behav. Neurosci.* **106,** 853–865.

55. Araki, M., McGeer, P., and Kimura, H. (1988) The efferent projections of the rat lateral habenular nucleus revealed by the PHA-L anterograde tracing method. *Brain Res.* **441,** 319–330.

56. Ellison, G. (1995) The NMDA antagonists phencyclidine, ketamine, and dizocilpine as both behavioral and anatomical models of the dementias. *Brain Res. Rev.* **20,** 250–267.

57. Sharp, F., Sagar, S., and Swanson, R. (1993) Metabolic mapping with cellular resolution: c-*fos* vs. 2-deoxyglucose. *Crit. Rev. Neurobiol.* **679,** 205–228.

58. Wirtshafter, D., Asin, K., and Pitzer, M. (1994) Dopamine agonists and stress produce different patterns of Fos-like immunoreactivity in the lateral habenula. *Brain Res.* **633,** 21–26.

59. Keys, A. and Ellison, G. (1996) Autoradiographic analysis of enduring receptor and dopamine transporter alterations following continuous cocaine. *Brain Res.*, in press.

60. Olney, J., Labruyere, J., and Price, M. (1989) Pathological changes induced in cerebrocortical neurons by phenclidine and related drugs. *Science* **244,** 1360–1362.

9

Evidence for and Mechanism of Action of Neurotoxicity of Amphetamine Related Compounds

R. Lew, J. E. Malberg, George A. Ricuarte, and Lewis S. Seiden

1. NEUROTOXICITY OF METHAMPHETAMINE AND RELATED COMPOUNDS

The purpose of this chapter is to review the evidence that substituted amphetamines (AMPHs), such as methamphetamine (METH), have neurotoxic potential, discuss critical determinants of AMPH neurotoxicity, and consider possible mechanisms underlying the neurotoxic action of AMPH and some of its derivatives. This evidence indicates that METH, AMPH, methylenedioxyamphetamine (MDA), 3,4-methylenedioxymethamphetamine (MDMA), *m*-trifluoromethyl-*N*-ethylamphetamine fenfluramine (FEN), and the combination of phentermine (PHEN) and FEN can cause long-lasting deficits in brain dopamine (DA) and/or serotonin (5HT) systems. For many of these drugs, the pharmacological properties are quite similar across species, including humans. Therefore, the possibility that the neurotoxic effects of AMPH derivatives may generalize from nonhuman mammals to humans must be considered.

2. DEFINITION OF SPECIFIC NEUROTOXICITY

Drug-induced neurotoxicity can be defined as a deleterious structural or functional change caused by chemicals and drugs in the central nervous system *(1)*. In this chapter, the term "neurotoxicity" specifically refers to damage of central monoamine-containing axons and terminals caused by administration of AMPH derivatives. According to this definition, after a putative neurotoxic compound has been administered, discontinued, and eliminated from the body, the neurotoxic effect on a given neuronal system will result in long-lasting reductions of:

1. Neurotransmitter content and its associated metabolites;
2. The levels of rate-limiting enzymes of that neurotransmitter;

From: Highly Selective Neurotoxins: Basic and Clinical Applications *Edited by: R. M. Kostrzewa.*
Humana Press Inc., Totowa, NJ

3. The density of transporter sites; and

4. Specific morphological changes consistent with destruction of axons and axon terminals *(2)*.

Although individually such effects would not be enough to document neurotoxicity, collectively they strongly suggest neurotoxicity.

Among the first specific chemical compounds to qualify for this definition was 6-hydroxydopamine *(3,4)*, which was found to be neurotoxic to the DA and norepinephrine (NE) systems. 5,6-Dihydroxytryptamine and 5,7-dihydroxytryptamine were also shown to be neurotoxic to the 5HT system *(5)*. In addition, parachloroamphetamine (PCA) was discovered to be a highly specific neurotoxin to the 5HT terminals *(6)*.

3. PHARMACOLOGY OF THE AMPHETAMINES

The amphetamines (AMPH, METH, MDMA, FEN, PHEN/FEN) exert their pharmacological and perhaps toxicological effects by interacting with monoamine transporters. AMPH and related compounds induce release of cytoplasmically stored amines through a transporter-mediated exchange diffusion mechanism and also block neurotransmitter uptake *(7)*. There are two primary types of neurotransmitter release: first, vesicular release, which is impulse-dependent *(7)*, calcium-dependent *(8)*, and stimulated by K+ and action potentials *(9)*, and second, transporter-mediated release, which is sodium-dependent, calcium-independent, and blocked by uptake inhibitors *(10)*. AMPH-induced DA release has been shown to be blocked by DA uptake inhibitors *(11)*, providing evidence that amphetamines release amines through a transporter-mediated release rather than an impulse-dependent release *(7)*. The process by which AMPH releases DA is theorized to be a reversal of the normal functioning of the uptake transporter *(12)*. The normal function of the uptake transporter is to bind extracellular DA or HT on the outside of the cell for uptake into the neuron. In this reverse uptake process, the opposite function of the transporter occurs. The transporter binds extracellular AMPH rather than DA and transports AMPH into the cell. Once inside the cell, the transporter is thought to change configuration and release AMPH to the inside of the cell. The transporter then binds intracellular DA and transports it to the cell surface, where it releases the DA into the extracellular space. On release of DA, the transporter is then available to bind AMPH once again. The cell must be polarized, and maintenance of the charge is energy-dependent. Therefore, this is an active transport process *(7)*.

AMPH-induced release is from cytoplasmic DA or 5HT, rather than vesicular DA or 5HT, that is, the normal vesicular stores do not directly exchange from the vesicle into the extracellular space. However, there is evidence that vesicular stored DA is involved in AMPH-induced DA release *(13)*. AMPH can enter the vesicle by diffusion, and this action may cause alkalization and eventual disruption of the vesicle. On disruption of the cell vesicle, leakage of amines from the vesicle occurs, thereby making more DA and 5HT available for release via the cell membrane reverse transporter mechanism *(14)*. According to Rudnick and Wall *(15)*, MDMA is also taken up into the cell by the 5HT transporter and releases 5HT into the synapse via the reverse transporter mechanism. Rudnick and Wall *(15)* also provided evidence that MDMA stimulates efflux from the vesicles into the cytoplasm.

The DA and 5HT releasing actions of MDMA, MDA, FEN, and METH have also been investigated. MDMA, MDA, FEN, and METH have all been shown to be potent 5HT-releasing agents *(18–25)*. Many pharmacological actions of these drugs are blocked by 5HT uptake inhibitors, indicating that the uptake transporters are a critical site of action of METH and related drugs *(16)*.

4. NEUROTOXICOLOGY OF AMPHETAMINE DERIVATIVES

4.1. Methamphetamine

4.1.1. Use and Abuse

METH's high abuse liability is evidenced by the fact that METH is self-administered by experimental animals and abused by humans *(26,27)*. However, an effort to assess their long-term effects in the central nervous system (CNS) was only made in the last 20 years. This effort was prompted by epidemics of METH abuse between 1950 and 1970 in Japan, Sweden, Great Britain, and the United States *(28–30)*. The social and medical problems caused by abuse of these drugs may be caused or compounded by their neurotoxic effects; these effects may also engender behaviors that may lead to further drug abuse, aggression, depression, mood swings, deleterious changes in motor function, memory, cognitive ability, sexual function, and sleep.

4.1.2. METH Neurotoxicity

A large body of evidence indicates that METH is selectively toxic to DA and 5HT nerve terminals in the CNS. METH-induced DA and 5HT neurotoxicity is evidenced by:

1. Long-lasting depletion of DA and 5HT in the CNS *(31,32)*;
2. Reduction of V_{max} for the rate-limiting enzymes tyrosine hydroxylase and/or tryptophan hydroxylase;
3. Reduction in the number of DA and 5HT transporters *(33)*; and
4. Morphological evidence that DA and 5HT axon terminals are damaged by METH *(34,35)* and undergo degeneration, as revealed by silver degeneration methods *(33,36–38)*.

An early observation suggesting neurotoxicity was made on rhesus monkeys who had received escalating doses of METH eight times a day with the cumulative daily dose being between 12.5 and 25 mg/kg *(39)*. Three to 6 mo after the cessation of METH administration, the caudate nucleus in these animal had only 30% of the normal level of DA. At approximately the same time, Gibb and his coworkers *(41,44)* reported deficits in brain tyrosine hydroxylase and trypto-phan hydroxylase in the rat. Their dosing regimen involved administering METH five times to rats in a 24-h period to mimic the type of bingeing that is common among humans that abuse METH. Ricaurte et al. *(40)* and Hotchkiss and Gibb *(41,46)* have also reported that METH caused long-term changes in brain 5HT systems in the rat. Similar deficits in neurochemical markers for the DA and 5HT system have also been obtained in other species, including mouse, guinea pig, cat, and several types of nonhuman primates *(37–39,42–47)*.

On the basis of histofluoresence studies using dosing regimes that cause depletion of DA, Lorez *(48)* concluded that METH administration at doses sufficient to produce reductions in DA were neurotoxic. Using silver-staining techniques, Ricaurte et al. *(49)* found evidence of terminal degeneration in the striatum within days of the last injection of a 4-d dosing regimen of METH. The silver staining observed in the striatum was strongly suggestive of DA nerve terminal degeneration and was consistent with the data that indicated a major loss in striatal DA.

However, absolute identification of the degenerating fibers as DAergic was not possible. A critical study was carried out by Ryan et al. *(36)*. These investigators labeled nerve endings with an antibody for tyrosine hydroxylase and silver-staining techniques. Under light microscopy, after continuous AMPH administration, they observed degeneration in neostriatum with silver staining. Furthermore, when they examined tissue slices under the electron microscope, they observed nerve terminals that were both silver-positive and contained tyrosine hydroxylase, thus providing strong evidence that the DA nerve cells were degenerating. Collectively, these data indicate that relatively high doses of AMPH (but not outside that of the abused dose range) can engender specific DA and 5HT neurotoxicity.

The degree of recovery of DA and 5HT systems from high-dose METH neurotoxicity is not clear. In the rat, these effects persist for at least 8 wk *(46)*. In the rhesus monkey, reductions in DA levels by METH have been observed to persist at least 3 yr *(50)*.

METH can induce long-term changes in behavior after dosing resulting in neurotoxicity. Walsh and Wagner *(51)* and Richards et al. *(52)* have found that large doses of METH cause depletion of DA and 5HT, and cause an impediment in the acquisition of a motor task and operant performance. Changes also occur in the sensitivity to various behaviorally active drugs after METH exposure *(53–55,57,59)*. However, using schedule-controlled behavior (FR and DRL performance) and motor coordination tasks (force-lever and eye-tracking studies), we have not found behavioral changes in monkeys whose monoamines had been depleted by METH *(53–56,58)*. Although neurotoxicity may be behaviorally "silent" under some conditions, it may become apparent when specific demands are placed on a neuronal system that has been damaged. Pharmacological challenge has been used to demonstrate functional changes in the CNS that result from exposure to neurotoxins *(60,61)*. A change in sensitivity to the effect of a drug that acts via a particular mechanism may be evidence of a functional change in that system in the CNS.

4.2. 3,4-Methylenedioxymethamphetamine (MDMA)

4.2.1. Use

MDMA, also known as "Ecstasy" or "X," is an AMPH analog that is recreationally abused *(62,63)* and is believed by some to have what are termed "empathogenic" or "entactogenic" effects *(64)*. These subjective effects include feelings of closeness toward others and enhanced communication abilities *(64,65)*, as well as visual hallucinations at high doses *(66)*. Its therapeutic use was originally advocated as an adjunct to psychotherapy *(67)*. However, because of its abuse potential, lack of rigorously demonstrated therapeutic value, and potential for neurotoxicity *(62)*, it was classified as a Schedule I drug by the DEA in 1985 *(63)*. However, its use and abuse have been increasing in the United States and Europe *(62,63,68,69)*, and MDMA is currently a popular illegal recreational drug on college campuses *(70)* and at dance parties known as "raves" *(63,71)*.

4.2.2. MDMA Neurotoxicity

In 1985, Ricaurte et al. *(205)* described the neurotoxic properties of MDA, the *N*-desmethyl derivative of MDMA, on the 5HT systems. Findings with MDA were extended to MDMA by Schmidt et al. *(25)*, who reported that MDMA had a greater effect on 5HT release than on DA release. Schmidt et al. *(25)* also reported a biphasic effect of MDMA's neurotoxic actions. An early phase of 5HT depletion was seen within 3–6 h after drug administration with a return to control levels of 5HT by 24 h, and a second phase where transmitter levels were again depleted 1 wk after administration. Other investigators have found a long-term decrease in 5HT levels as soon as 3 d after drug administration *(72)*.

MDMA produces a long-term decrease in 5HT and 5HIAA levels in a number of species, including rats *(25,73)*, mice *(74)*, guinea pigs *(75)*, and monkeys *(76–78)*, and also produces a decrease in tryptophan hydroxylase *(79,80)*. In binding studies, MDMA is observed to cause a decrease in the number of 5HT transporters *(75,81–83a)*. Furthermore, there are morphological signs of terminal degeneration as seen using Fink-Heimer silver staining *(81)*, loss of 5HT axons as visualized by immunocytochemistry *(35,82a,83)* and an increase in glial fibrillary acidic protein (GFAP) staining in the rat *(84)*. MDMA-induced long-term decreases in 5HT levels and 5HT uptake sites (at 12 mo) and morphological changes in 5HT axons (12–18 mo) have also been reported *(82,82a,83,85,86)*. As mentioned above, these effects individually would not be enough to indicate 5HT terminal destruction by MDMA, but when taken together, the argument can be made that MDMA is neurotoxic to 5HT nerve terminals *(31)*.

The neurotoxic effects of MDMA can be induced by either single dose or multiple-dose regimens, with the usual route of administration being subcutaneous, especially in monkeys *(87)*. In a comparison of subcutaneous vs oral doses in the primate, it was shown that orally administered MDMA was approximately as effective as subcutaneous administration in reducing 5HT concentrations *(88)*. This study also compared single vs multiple dosing, so it could be determined if the primate data could be extrapolated to the single oral doses taken by humans *(88)*. It was shown that multiple doses caused more neurotoxicity than single doses, but in squirrel monkeys, a single low dose of MDMA (5 mg/kg) still produced a 5HT depletion *(87,88)*.

Long-term studies with MDMA in rat have shown recovery of 5HT axonal markers in rat *(82,83,83a,86,89)*. In particular, Lew et al. *(82)* have demonstrated that recovery of 5HT markers (i.e., 5HT and 5HT transporters) in terminal field regions is dependent on the region's proximity to 5HT cell bodies in the dorsal and median raphe. For example, recovery of 5HT markers occurs in the substantia nigra, ventral tegmental area and hypothalamus within a few weeks after cessation of MDMA treatment *(82,86)*. At 8–16 wk after MDMA treatment, recovery of 5HT markers occurs in the nucleus accumbens, olfactory tubercle, caudate putamen, and frontal cortex. Recovery of 5HT axons in more distant regions, such as frontal-parietal cortex, occipital-temporal cortex, and occipital cortex, does not occur until approx 32 wk or later *(82,86)*.

In human studies, MDMA has been shown to reduce 5HIAA levels in cerebrospinal fluid *(90)*. When combined with the data from other species, these results indicate that MDMA may also be neurotoxic to the 5HT system in humans.

4.3. Fenfluramine

4.3.1. Use

FEN is a phenethylamine derivative that has similar anorectic, but not the stimulant properties of AMPH *(91,92)*. The *d*-isomer of FEN is more potent that the *l*-isomer *(93)*. Racemic *(dl)*-FEN has been available in the United States as an anorectic agent since 1971. However, in 1996 the *d*-isomer, dexfenfluramine (redux™), gained approval by the FDA for weight reduction. Considerable debate has been concerned with the safety and benefits of both *d*-and *dl*-FEN. Supporters for FEN claim that both *d*- and *dl*-FEN can be taken for extended periods of time without serious side effects, and argue that the benefits of weight reduction resulting from FEN are greater than any possible side effects *(94)*. Evidence, however, is presented below that demonstrates FEN at doses that cause weight loss can also cause neurotoxicity to 5HT systems.

4.3.2. Neurotoxicity

Considerable evidence demonstrating FEN to be neurotoxic to 5HT neurons has been reported since the 1970s. Using either single- or multiple-dose paradigms, investigators have reported that FEN causes long-term reductions in 5HT, but not DA or NE levels in several rat brain regions, including frontal cortex, hippocampus, hypothalamus, and striatum *(93,95–102)*. These reductions persist for several weeks to several months after FEN administration. In addition, FEN-induced neurotoxicity of 5HT neurons has also been observed in several animal species, including mice *(103)*, guinea pigs *(100)*, and nonhuman primates *(100,104,105)*. More recent studies employing immunohistochemical methods have demonstrated a decrease in the number of 5HT nerve terminals following FEN administration in the rat *(106,107)*. Radioligand binding and autoradiography studies have demonstrated decreases in the number of 5HT transporters in terminal brain regions, including striatum, cerebral cortex, hippocampus, thalamus, medial and lateral hypothalamus, septum, and amygdala *(102,108,109)*. It should be noted that to date, FEN-induced loss of 5HT cell bodies in the dorsal or median raphe as measured by neurochemical, pharmacological, and morphological means have not been observed. This is in keeping with previous studies on other AMPH derivatives, such as METH and MDMA, which do not have any effect on the dorsal and median raphe *(31,82)*.

Based on the pattern of recovery for 5HT neurons following MDMA treatment, it would be anticipated that a similar pattern of recovery of 5HT neurons would occur following FEN treatment. In support, several studies have reported recovery of 5HT levels in rat frontal cortex occurring within several weeks following FEN administration *(18,110,111)*.

In spite of the overwhelming evidence implicating FEN to be neurotoxic to 5HT axons, controversy regarding whether FEN is a neurotoxin remains *(94)*. It has been argued that the observed neurochemical and pharmacological deficits may be attributed to other mechanisms. For example, the decrease in 5HT levels may be simply owing to the 5HT releasing effects of FEN from intact nerve terminals. Detection of 5HT axons by immunofluorescence is dependent on the presence of the antigen 5HT. Thus, absence of immunofluorescent axons after FEN administration may simply reflect release of 5HT from intact neurons *(110)*. However, it is difficult to accept release of 5HT by FEN as a plausible explanation, since the reduction in 5HT levels following FEN administration lasts for several weeks *(18,98,100)* to several months *(102)*, which is considerably longer than the biological half-life of FEN ($t_{1/2} \sim 2$ h) in the rat *(112)*.

Arguments have also been raised against the loss of 5HT transporters following FEN administration as an indicator of neurotoxicity. Since FEN binds to the 5HT transporter, it has been argued that repeated FEN administration could cause downregulation of the 5HT transporter in intact nerve axons *(113–115)* as opposed to loss of 5HT transporters owing to neurodegeneration of 5HT axons. Battaglia et al. *(116)* has reported that recovery of downregulated $5HT_2$ receptors occurs at a rate of approx 9.2%/d. Since 5HT transporters and $5HT_2$ receptors are both membrane-bound proteins, it would be reasonable to assume that the recovery rate of 5HT transporters would be similar and would recover within 7–10 d. However, studies by our laboratory have shown that after repeated FEN administration in rats (12.5 mg/kg $\times$ 4 over 3 h), the density of 5HT transporters in striatum at 7 and 28 d after cessation of treatment is 20 and 40% of saline control, respectively *(109)*. Obviously, the recovery of 5HT transporters does not occur in the time frame as predicted for receptor down regulation. In fact, axonal recovery occurs at a rate that has been previously observed for regenerating axons following MDMA treatment *(82)*. In agreement, Zacek et al. *(102)* has also observed the density of 5HT transporters in rat brain to be significantly less than in saline controls at 8 mo after cessation of FEN treatment.

Irreversible 5HT uptake blockers have also been employed to assess if FEN causes a loss of 5HT transporters by degeneration of nerve terminals or loss of 5HT transporters by downregulation *(113–115)*. In such studies, rats were treated with either *dl*-FEN or *N*-ethoxycarbonyl-2-ethoxy-1,2-dihydroquinonoline (EEDQ), an alkylating agent that produces irreversible 5HT uptake block. In rats treated with EEDQ, 5HT uptake sites were noncompetitively blocked by EEDQ and, as such, are effectively lost from otherwise intact nerve terminals. In whole-brain synaptosomal preparations from EEDQ-treated rats, there is a reduction in the number of 5HT uptake sites and a reduction in the V_{max} of ^{14}C-5HT uptake, but no change in ^{14}C-5HT loading compared to those from saline-treated animals. In whole-brain synaptosomal preparations from *dl*-FEN-treated rats, similar reductions in the density of 5HT uptake sites and V_{max} of ^{14}C-5HT uptake were also observed. However, reduced ^{14}C-5HT loading was also observed in synaptosomes from *dl*-FEN-treated animals, thus indicating that FEN causes loss of 5HT transporters via neuronal degeneration *(113–115)*.

In addition to using 5HT content, 5HT transporters, and morphological changes as markers of AMPH neurotoxicity, other parameters have been employed to assess FEN neurotoxicity. Reactive gliosis, as measured by expression of GFAP, has been reported to be a sensitive indicator for chemical-induced neuronal damage *(117)*. In mice, O'Callaghan and Miller *(118,118a)* have reported that repeated administration of either *d*-METH, *d*-MDA, or MDMA significantly increases striatal and cortical GFAP concentrations with parallel decreases in DA levels and tyrosine hydroxylase levels. In contrast, repeated administration of *d*-FEN did not affect GFAP expression, but did reduce cortical 5HT levels for up to 21 d *(118,118a)*. Similarly, in the rat animal model, repeated administration of either *d*- or *dl*-FEN produces loss of cortical 5HT, 5HT uptake sites, and loss of 5HT immunoreactivity. However, these deficits were not accompanied by parallel increases in GFAP production *(119)*. It would appear from these studies that expression of GFAP may not be as sensitive an indicator for neurotoxicity induced by FEN as observed with other AMPH derivatives.

One major criticism of the FEN studies on neurotoxicity is that the doses used to induce FEN neurotoxicity in animals are excessively larger than that required to cause weight reduction *(94)*. However, studies by our laboratory show that the ED$_{50}$ of *dl*-FEN to inhibit milk intake is approx 3 mg/kg *(98)* and is considered to be a reliable estimate for the anorectic dose of *dl*-FEN in rats. The doses of FEN used to induce neurotoxicity in rats are comparable to the dose required to cause weight loss in rats. Repeated administration of

3.0 mg/kg *dl*-FEN causes a 20–25% reduction in 5HT levels and uptake sites in rat brain *(98,109)*. The question of the appropriate scaling for the comparison between animals and humans is discussed below.

4.3.3. FEN Toxicity and Human Use

Although evidence has been presented that demonstrates FEN to be neurotoxic to central 5HT neurons in various animal models, evidence for 5HT neurotoxicity in humans is yet to be reported. However, low 5HT levels have been associated with clinical symptoms, such as memory loss, sexual dysfunction, sleep disorders, and mood changes *(120)*. Since evidence has been presented in which FEN causes long-lasting reductions in 5HT levels in animals at doses comparable to that which also causes weight loss, caution should be exercised in the prescription of FEN for weight loss in humans

4.4. Phentermine

4.4.1. Use

PHEN is a phenethylamine derivative of AMPH that is currently being used as an anorectic agent *(109)*. PHEN exerts its anorectic effects via DA release *(121)*. Studies by our laboratory show that the ED_{50} of PHEN that inhibits milk intake in rats is 5.0 mg/kg *(122)* and is considered to be a suitable index of PHEN's anorectic properties in rats.

4.4.2. Neurotoxicity

In studies examining the neurotoxic potential of PHEN, rats were injected with either saline or 100 mg/kg PHEN twice daily for 4 d and sacrificed 2 wk after the final injection *(122)*. Analysis of DA and 5HT levels in several brain regions showed that PHEN reduced DA levels in striatum by ~27% and 5HT levels in hippocampus by ~50%. In another study, Lew et al. *(109)* administered 20 mg/kg PHEN to rats once an hour, for a total of four injections. Rats were allowed 7 d of recovery before sacrifice. Although a transient weight loss was observed in PHEN-treated rats compared to saline-treated rats, DA, 5HT, and NE levels were unaffected by PHEN. It would appear that at least in rat, PHEN induces neurotoxicity on DA and/or 5HT at doses severalfold higher than that which produces weight loss.

4.5. PHEN and FEN

4.5.1. Use

Combined administration of PHEN and FEN (PHEN/FEN) has been reported to be more efficacious than either drug alone in the treatment of obesity *(123–129)*. Weintraub et al. *(128)* observed that combined administration of half-doses of PHEN and FEN caused greater weight reduction and less side effects than administration of full doses of either PHEN or FEN alone. The effectiveness of combined PHEN/FEN administration is thought to be owing to the combined effects of PHEN causing DA release and FEN causing 5HT release in hypothalamus *(130)*. However, there is little information available concerning the possible long-term adverse effects from consuming PHEN/FEN.

4.5.2. Neurotoxicity

Recently, our laboratory compared the neurotoxic potential of combined administration of PHEN/FEN with administration of either drug alone in the rat animal model *(109)*. Rats were injected once each hour for a total of four injections with either saline (0.9% NaCl), 20 mg/kg PHEN, 3.125 mg/kg FEN, 12.5 mg/kg FEN, or combined PHEN and FEN (20 and 3.125 mg/kg). Seven days after cessation of treatment, rats were sacrificed, and brain regions dissected and assayed for either neurotransmitter content or for uptake sites. The weights of saline- and drug-treated rats were recorded each day of recovery until sacrifice. As observed in clinical studies, PHEN and FEN administered alone caused a transient weight reduction. However, combined administration of PHEN/FEN caused a greater weight reduction than either drug alone.

Neurochemical analysis showed that 5HT levels was unaffected by repeated administration of PHEN. However, FEN at 3.125 and 12.5 mg/kg reduced 5HT levels in striatum, nucleus accumbens/olfactory tubercle, hippocampus, hypothalamus, amygdala, and frontal-parietal cortex in a dose-dependent manner. More importantly, combined administration of PHEN/FEN (20/3.125 mg/kg) reduced 5HT levels in the above regions by a significantly greater amount than either drug alone. The reduction in 5HT levels by PHEN/FEN (20/3.125 mg/kg) was comparable to that observed with the high dose of FEN (12.5 mg/kg) administered alone. In contrast, DA and NE levels in the above regions were not affected by either drug alone or in combination. It would appear from these studies that combined administration of PHEN and FEN specifically enhances the neurotoxicity of either drug on central 5HT systems.

The effect of combined administration of PHEN/FEN (20/3.125 mg/kg) on the density of 5HT uptake sites in various brain regions was also examined. Radioligand binding studies showed combined PHEN/FEN administration reduced the density of 5HT uptake sites in striatum by a significantly greater amount than either drug alone and was found to persist up to 28 d following treatment. Consequently, the reduction in 5HT levels and density of 5HT transporters by combined administration of PHEN/FEN demonstrates that combined administration enhances the neurotoxic effects of these anorectic agents.

4.5.3. PHEN/FEN Toxicity and Human Use

To date, clinical evidence for PHEN/FEN neurotoxicity has not been observed. However, in a 4-yr study by Weintraub and colleagues, approx 16% of the participants treated with PHEN/FEN (15/60 mg) withdrew from the study because of side effects, including sleep disturbances, fatigue, vivid dreams, and depression *(124–127,129)*. Since alterations in 5HT activity have been associated with sleep disturbance, depression, and memory loss *(120)*, it is possible that the side effects experienced by the participants in the above study were owing to 5HT neurotoxicity induced by combined PHEN/FEN administration. Caution should therefore be exercised in the prescription and use of this drug combination. The benefits of long-term pharmacotherapy with PHEN/FEN on weight loss must be carefully considered against the drug-engendered health risks.

5. RELATIONSHIP BETWEEN HUMAN DOSES AND ANIMAL DOSES

Without exception, the doses of substituted AMPHs (expressed in mg/kg) required to produce neurotoxicity in small animals are higher than those required in larger animal species. This common (and expected) phenomenon can be explained by the principles of interspecies scaling *(131,132)*, which dictate that in order to calculate a comparable dose of drug in different species, one needs to correct for differences in pharmacokinetics and/or body surface area:body mass ratio.

Expression of dosage in mg/kg does not take into account the differences in pharmacokinetics. If, however, the dosages are expressed in terms of mg/surface area, i.e., mg/m^2, a more accurate and reliable comparison can be made. Body surface area is a more reliable unit, because the surface area is proportional to size of the liver, which is the main tissue for metabolism *(133)*. The surface areas of a 400-g rat and a 90-kg human are approx 0.054 and 2.0 m^2, respectively *(134)*. Expressing the clinical doses of *dl*-FEN (60–110 mg) used for weight reduction and that known to cause weight loss and 5HT neurotoxicity in rats (i.e., 5 mg; assuming 3.125 mg/kg) in terms of surface area reveals that the dose of FEN administered to rats is only 1.5–3 times that in humans. In light of this, it appears that the dose of FEN used to induce neurotoxicity in animal models is of more clinical relevance than previously determined.

When corrections for differences in body surface area/body mass ratio are done for *d*-FEN, the ED_{50} dose that produces a prolonged (i.e., at least 2-wk) depletion of brain 5HT axonal markers in the mouse is 6.8 mg/kg (given subcutaneously twice daily for 4 d), compared to 3.72 mg/kg in the rat, using the same schedule and route of drug administration *(103)*. In the squirrel monkey, only 1.25 mg/kg of *d*-FEN twice daily for 4 d is sufficient to produce a 50–60% reduction in brain cortical 5HT axon markers 2 wk later *(104)*. Although the lowest neurotoxic dose of *d*-FEN in primates has not been determined, it is possible, using the interspecies scaling method *(131,132)*, to take a known neurotoxic dose of *d*-FEN in a squirrel monkey (1.25 mg/kg BID) *(104)* and estimate an equivalent dose in humans. Estimating that the squirrel monkey weighs approx 1 kg and the human weighs approx 80 kg, using an accepted adjustment for the relative body surface areas in the two species *(131,132)* yields an equivalent dose in the human of 0.32 mg/kg or approx 26 mg. Since the typical starting dose of *d*-FEN is 15 mg twice daily *(135)*, data from nonhuman primates suggest that the dose of *d*-FEN used by humans may well be neurotoxic.

6. PHARMACOLOGICAL PROTECTION AGAINST METH AND RELATED DRUG-INDUCED NEUROTOXICITY

Several different classes of drugs have been reported to protect against METH-induced neurotoxicity. The primary mechanisms by which these drugs exert their neuroprotective effects seem to be:

1. Attenuation of hydroxy radical and neurotoxin formation;
2. Interference with an intact DA system, since it has been inferred that DA release is necessary for 5HT neurotoxicity *(136)*;
3. Interference with the excitatory loop hypothesized by Carlsson *(137)*; and
4. Prevention by blocking calcium channels.

6.1. α-Methyl-Para-Tyrosine (AMPT)

AMPT, an inhibitor of tyrosine hydroxylase, blocks METH-induced depletion of DA and 5HT *(34,136,138)*. AMPT also blocks MDMA-induced depletion of 5HT *(80)* and attenuates PCA depletion of 5HT *(139)*. An interpretation of these findings is that an intact DA system is necessary for METH or MDMA-induced neurotoxicity of DA and 5HT neurons *(136)*. The data obtained with AMPT are also consistent with the idea that DA is important in driving a potentially toxic, feed-forward striatal–thalamic–cortical loop *(137)*. Finally, the AMPT results are consistent with the proposal that the release of DA engenders the formation of neurotoxic metabolites of DA *(81,140–144)*. Therefore, AMPT pretreatment has been shown to decrease AMPH-induced DA release *(145)*. AMPT, therefore, decreases the availability of DA, which may engender hydroxy radical formation or reactions that lead to formation of a toxic metabolite *(see* Section 8.).

6.2. DA Antagonists

DA receptor antagonists block METH and MDMA-induced neurotoxicity *(146)*. An explanation for protection from neurotoxicity is that the antagonist alters output of the striatal–thalamic–cortical circuit. It is known that there are DA receptors in the substantia nigra as well as the striatum *(147)*. By blocking DA receptors at either location, one could theoretically interrupt the DAergic influence on the striatal–thalamic–cortical loop.

The protection afforded by DA antagonists is difficult to integrate with other theories of neurotoxicity. Haloperidol does not block AMPH-induced DA release *(136)*, and in fact, it increases DA synthesis *(148)*. Haloperidol, therefore, does not fit well with the hydroxy radical theory, because haloperidol still promotes DA release. However, haloperidol causes a reduction in core temperature, which in turn could affect the rate at which destructive radical reactions occur *(see* Section 7. for details on temperature and neurotoxicity). The neuroprotective effect of haloperidol, however, is consistent with the idea that DA may project to raphe cell bodies *(149)*, and may be an important link between the DA system and 5HT neurotoxicity induced by METH or MDMA.

6.3. 5HT$_2$ Antagonists

The 5HT$_2$ receptor antagonists ketanserin and ritanserin have been shown to protect against METH- and MDMA-induced neurotoxicity. Nash *(150)* first demonstrated the protective effect of ketanserin against MDMA-induced damage to 5HT nerve terminals. These neuroprotective effects of ketanserin have been reproduced with a variety of 5HT$_2$ antagonists *(151–153)*. It has been shown that ketanserin inhibits MDMA-induced DA release *(150)* and that MDMA-induced decreases in DA cell firing can be antagonized by 5HT$_2$ antagonists *(151,152)*. Nash *(150)* has suggested that MDMA-induced neurotoxicity involves the activation of DA neurons via 5HT$_2$ receptors on DA cell bodies. These data are supportive of the view that (1) DA mediates the MDMA-induced damage to 5HT terminals, and (2) the 5HT$_2$ blocking agents prevent this neurotoxicity by interacting with DAergic activity.

The neuroprotective effects of 5HT$_2$ antagonists are also consistent with a CA^{2+} theory of METH neurotoxicity. 5HT$_2$ receptors are linked to the second messenger inositol-1-4-5-trisphosphate (IP$_3$) *(154)*. IP$_3$ in turn, stimulates the release of intracellular Ca^{2+} from intracellular sequestration compartments *(155,156)*. Blockade of the 5HT$_2$ receptor could therefore diminish the amount of intracellular free Ca^{2+}, and decrease the likelihood of Ca^{2+}-induced cell death.

6.4. MK-801

Sonsalla et al. *(157)* first reported that MK-801 (a noncompetitive antagonist at the NMDA receptor site) could antagonize the METH-induced neurotoxicity to DA neurons. It was also shown that MK-801 protects against MDMA-induced damage to the 5HT system *(158,159)*. Further research determined that other noncompetitive as well as competitive NMDA antagonists protected against METH-induced neurotoxicity. The protective effects of MK-801 are consistent with the Ca^{2+} theory of METH neurotoxicity. MK-801 blocks Ca^{2+} entry into the cell. The blockade of Ca^{2+} entry may be important for two reasons. Keeping extracellular Ca^{2+} from entering the neuron would diminish the probability of Ca^{2+}-induced cell death. In addition, by blocking Ca^{2+} entry into the cell, subsequent Ca^{2+}-induced Ca^{2+} release from intracellular stores could also be blocked *(160,161)*.

The protective effects of MK-801 are also consistent with the hydroxy radical theory of METH neurotoxicity. MK-801 has been shown to decrease METH-induced DA release in vivo *(162)*, diminishing the availability of DA for conversion into a neurotoxic metabolite of DA. However, Kashihara et al. *(163)* failed to replicate this finding in vivo, and Bowyer et al. *(164)* did not block METH-induced DA release in vitro. This issue, therefore, remains controversial.

6.5. Antioxidants

Ascorbic acid and other antioxidants can block METH-induced neurotoxicity *(165)*. These data are consistent with the idea that DA and/or 5HT released by METH or MDMA reacts with hydroxy radicals to form 6-hydroxydopamine or 5,6-dihydroxytryptamine, respectively *(143,166)*. If, on the other hand, formation of a toxic metabolite of DA or 5HT occurs intracellularly, then enzymatically formed neurotoxic metabolites could develop. Whether auto-oxidation occurs enzymatically or nonenzymatically, the antioxidants could function in a similar manner by forming a nonreactive complex with the hydroxy radical or protecting DA from quinone formation.

6.6. Quinolinic Acid

Quinolinic acid unilaterally injected into the rat striatum blocks METH-induced neurotoxicity ipsilateral to the injection *(167)*. Quinolinic acid is an excitatory amino acid neurotoxin, which, when injected into the striatum, selectively destroys cell bodies, leaving DAergic fibers intact. This finding is supportive of the idea that METH and MDMA-induced neurotoxicity requires an intact striatal–thalamic–cortical circuit. Quinolinic acid lesions of the striatum disrupt this circuitry.

6.7. 5HT and DA Uptake Blockers

DA and 5HT transporter inhibitors block METH neurotoxicity. The ability of DA uptake inhibitors to block DA neurotoxicity is consistent with the hydroxy radical theory. DA uptake inhibitors block AMPH-induced DA release *(145)*, making less DA available for conversion into a neurotoxic metabolite of DA. The inhibition of METH-induced neurotoxicity with uptake inhibitors is also consistent with the idea that METH-induced neurotoxicity is dependent on a striatal–thalamic–cortical loop. Decreasing METH-induced DA release diminishes DA's influence on this circuit, resulting in protection against METH- or MDMA-induced neurotoxicity.

METH-induced toxicity to dopaminergic nerves is attenuated by amfonelic acid and other blockers of the DA transporter *(168)*. However, 5HT neuronal damage is not prevented *(169)*. Amfonelic acid seems to afford protection even when it is injected several hours after the last dose of METH. The failure of DA uptake blockers to protect against METH-induced 5HT neurotoxicity is not consistent with the theory that DA mediates METH- or MDMA-induced 5HT neurotoxicity. In parallel with DA uptake inhibitors, 5HT uptake inhibitors block METH-induced 5HT neurotoxicity, but not DA neurotoxicity *(170)*.

6.8. GABA Transaminase Inhibitors

GABA transaminase inhibitors and GABA agonists protect against METH-induced neurotoxicity *(41,171)*. Amino-oxyacetic acid inhibits GABA transaminase, an enzyme responsible for GABA inactivation. Chlormethiazole, an agonist at the GABA A receptor, also protects against METH-induced DA and 5HT damage *(171)*. GABA is an important inhibitory transmitter in the striatal–thalamic–cortical circuit. It can be postulated that as the levels of GABA increase, the toxic overexcitation of this circuit is diminished, allowing for protection against METH treatment.

6.9. Transgenic Mice

Transgenic mice that have a random insertion primer of human Mg/Cu superoxide dismutase (SOD) show no neurotoxicity *(172,173)* from MDMA, indicating that neurotoxicity may result from the MDMA-induced oxidative stress or free radical generation. Protection may then be owing to prevention of the generation of free radicals.

6.10. Spin-Trapping Agent

Several studies have reported that metabolism of MDMA can lead to the formation of the corresponding catecholamine 3,4-dihydroxymethamphetamine, which in turn is oxidized by superoxide to form quinones and subsequent free radicals *(174)*. Since free radicals cause neurodegeneration *(175)* and have been implicated in MDMA's neurodegenerative action on 5HT neurons, investigators have examined whether agents that protect against free radical-induced degeneration are also protective against neurotoxicity induced by METH and related compounds. In particular, investigators have reported α-phenyl-*N-tert*-butylnitrone (PBN), a spin-trapping agent, to protect against MDMA and PCA *(176,177)*. Using ^{3}H-paroxetine to assess the density of 5HT uptake sites in cortex and hippocampus, Colado and Green *(178)* observed that pretreatment with PBN could block the loss of 5HT uptake sites induced by either MDMA or PCA. Similarly, Che et al. *(179)* reported that pretreatment with PBN also blocked the neurotoxic effect of MDMA on tryptophan hydroxylase activity in hippocampus, striatum, and frontal cortex. Since PBN is known to protect against oxidative damage *(180)*, it could be inferred from the above observations that oxidative stress is involved in the neurotoxic mechanism of METH and related compounds. However, PBN also causes hypothermia, and at sufficient dosage (i.e., 200 mg/kg), can reverse MDMA-induced hyperthermia and attenuate the MDMA-induced decrease in tryptophan hydroxylase activity. Thus, it would appear that the hypothermic effect of PBN may contribute toward its neuroprotective actions.

7. TEMPERATURE AND NEUROTOXICITY

It has been shown that METH or MDMA administration produces an increase in rat core temperature (CORE TEMP). This relationship between core temp and drug-induced neurotoxicity has been further investigated *(72,158,181–183)*. A decrease in core temperature, whether produced by drugs or by adjusting the ambient temperature (AMB TEMP), protects against neurotoxicity. Conversely,

increases in CORE TEMP induce or exacerbate MDMA-induced neurotoxicity
(72,158,181,182,184).

7.1. Pharmacological Changes in Body Temperature Affect Neurotoxicity

Despite the variety of the sites of action of drugs that protect against neurotoxicity, a common effect is that many of these drugs cause a decrease in
CORE TEMP when coadministered with METH or MDMA *(72,118a,182)*.
Farfel et al. *(158)* looked at the effect of dizocilpine, a drug shown to protect
against METH neurotoxicity *(157)*, on CORE TEMP of the rat. When administered with METH or MDMA, dizocilpine caused hypothermia in the rat and
also protected against MDMA-induced neurotoxicity. When rats were warmed
to prevent hypothermia, protection against the neurotoxicity was no longer
observed, strongly suggesting that drug-induced hypothermia plays a role in
the protection. In agreement, Malberg et al. *(183)* have shown that ketamine
(KET) and AMPT protect against MDMA neurotoxicity by producing
hypothermia. Similarly, prevention of the hypothermia reverses the protection
against the MDMA neurotoxicity *(183)*. What is striking about these findings
is that dizocilpine, KET, and AMPT administered alone do not produce
hypothermia; only the combination of the protective agents plus the MDMA
produces hypothermia *(183)*. Although it is not known what the signal for the
production of hypothermia would be, it has been suggested that an increase in
oxidative stress may produce the signal for the generation of the hypothermia
(183). Drugs, such as ethanol, pentobarbital, and diethydithiocarbamate, also
protect against neurotoxicity by inducing hypothermia in the rat *(118a)*. Both
Bowyer et al. *(84)* and Farfel and Seiden *(182)* reported that the amount of
hypothermia induced by protective agents, such as haloperidol, diazepam, and
dizocilpine, is positively correlated to the amount of protection against METH
or MDMA neurotoxicity. Che et al. *(179)* reported that the spin-trapping agent
n-tert-butyl-α–phenylnitrone protects against the MDMA-induced decrease in
trytophan hydroxylase activity by inducing hypothermia.

There are exceptions to the findings that drugs that protect against
neurotoxicity do so by lowering body temperature. Colado and Green *(176)* and
Hewitt and Green *(185)* reported that chlormethiazole protected against MDMA
neurotoxicity and attenuated the MDMA-induced hyperthermia, but in their
study, the rats did not become hypothermic. Interestingly, fluoxetine protects
against MDMA-induced neurotoxicity, but has no effect on the MDMA-induced
hyperthermia; fluoxetine-pretreated rats show the same hyperthermia as rats that
were given a saline pretreatment *(183)*. This indicates that although production of a CORE TEMP hypothermia may be the mechanism by which
many neuroprotective drugs work, there are also other mechanisms by which
protection occurs.

7.2. AMB TEMP and Neurotoxicity

The environmental temperature in which the potentially neurotoxic drug is administered can affect the CORE TEMP of the rats as well as protect or exacerbate drug-induced neurotoxicity. Administration of MDMA at a low environmental temperature of 10°C decreased the animals' CORE TEMP by 2°C *(186)* and administration of MDMA at 15°C decreased the amount of neurotoxicity seen *(118a)*. Administration of METH at 4°C also reduced the effect of METH on striatal DA levels and striatal tyrosine hydroxylase activity *(187)*. Conversely, administration of MDMA at a high environmental temperature (30°C) increased the animals' body temperature by 3.2°C *(186)*. This treatment in a hot environment also increases the animals' CORE TEMP to the point of lethality *(84)*. More recently, Malberg and Seiden *(184)* constructed an apparatus that controls AMB TEMP and noninvasively records CORE TEMP, so that the interaction of CORE TEMP, AMB TEMP, and neurotoxicity could be systematically investigated. At an ambient dosing temperature of 20–24°C, a CORE TEMP hypothermia and protection against MDMA-induced 5HT neurotoxicity were observed. At AMB TEMPs of 28–30°C, a significant CORE TEMP hyperthermia and an exacerbation of 5HT neurotoxicity occurred. This study indicated that very small (2°C) increments in ambient temperature can have a large effect on the body temperature (CORE TEMP) of the animal as well as subsequent MDMA neurotoxicity.

AMB TEMP also affects FEN-induced neurotoxicity and CORE TEMP *(118a,182,184)*. Although FEN causes a hypothermia with neurotoxicity, it can be seen that the AMB TEMP affects the CORE TEMP and FEN-induced neurotoxicity as well. Animals treated with FEN (3–12 mg/kg) at a normal lab temperature (24°C) showed hypothermia and neurotoxicity. However, at 30°C AMB TEMP, rats became hyperthermic, and in certain brain regions, there was neurotoxicity above that seen in the hypothermic rats *(188)*. Again, these studies indicate that the AMB TEMP may be a large determinant of both CORE TEMP and neurotoxicity.

7.3. MDMA Administration and Thermoregulation

The increase and decrease in CORE TEMP as determined by a high or low AMB TEMP indicates that after MDMA or METH administration, the rat's thermoregulatory mechanisms may be compromised *(186,189)*. A number of studies have determined that the colonic temperature of rats treated with *d*-AMPH or FEN varies markedly with the AMB TEMP *(190–192)*. AMPHs have also been shown to affect normal thermoregulatory mechanisms. Gordon et al. *(186)* showed that rats treated with a high dose of MDMA (30 mg/kg) in high AMB TEMP (30°C) had increased CORE TEMP, but did not have a corresponding increase in blood flow to the tail, which would have been expected in response to an increase in CORE TEMP. Evaporative water loss was also found to be increased by MDMA administration, and this effect was potentiated by increasing the AMB TEMP *(186)*. Metabolic rate of MDMA-treated rats was proportional to the temperature at 10, 20, and 30°C, again indicating a dependence on AMB TEMP of the normal autonomic function of these animals. It has been reported that an open wire cage (with more air flow) is more effective than an acrylic (closed) cage in helping the rats maintain a normal CORE TEMP. Rats in the acrylic cage have CORE TEMPs that are higher and have greater evaporative water loss than animals in the wire cage *(193)*. These studies indicate that MDMA may impair the animals' ability to thermoregulate, making other variables, such as AMB TEMP, much more important than in normal animals on determination of CORE TEMP.

7.3.1. Possible Reasons Why Temperature Should Affect Neurotoxicity

A hyperthermic CORE TEMP does not in itself produce neurotoxicity. Rats given SAL injections and kept in a controlled high temperature of 40°C do not show any change in transmitter levels of 5HT, 5HIAA, NE, DA, 3,4-dihydroxyphenylacetic acid (DOPAC), or homovanillic acid (HVA) (Malberg and Seiden, in preparation), and it has been shown in rabbits that pyrogen-induced hyperthermia does not alter extracellular 5HT levels *(192)*. These studies lead us to hypothesize that it is not the change in CORE TEMP that causes the neurotoxicity, but rather that a change in CORE TEMP can affect the reactions leading to neurotoxicity.

It is known that AMB TEMP affects the CORE TEMP of the animal. This CORE TEMP change may then affect the reactions leading to MDMA-induced neurotoxicity. At high AMB TEMP, which induces high CORE TEMP, reactions leading to neurotoxicity may increase, whereas at low CORE TEMP, reactions leading to neurotoxicity may be slowed or even stopped completely *(194)*. For example, Bowyer et al. *(195)* have shown that in a cold environment of 4°C, rats become hypothermic, and the METH-induced release of extracellular DA and DOPAC is reduced in the caudate putamen as measured by in vivo microdialysis. Therefore, it is not the inability to thermoregulate that affects the neurotoxicity, but the CORE TEMP-mediated increase or decrease in the reactions that lead to neurotoxicity. It may be that MDMA produces a change in thermoregulatory ability that affects neurotoxicity, but only as a secondary effect of the change in CORE TEMP affecting the reactions leading to neurotoxicity. In order for this hypothesis to be fully tested, all reactions currently theorized to lead to neurotoxicity would have to be examined for their dependence on CORE TEMP.

8. MECHANISMS OF AMPH-INDUCED NEUROTOXICITY

The mechanisms underlying the neurotoxic actions of substituted AMPHs have yet to be identified. One proposed mechanism of neurotoxicity is that the AMPHs themselves are metabolized to toxic metabolites. However, our lab has injected several metabolites of METH and found that these metabolites did not produce toxicity (Seiden, unpublished data), indicating that a metabolite does not engender neurotoxicity.

A second theory is that the administration of AMPH may cause oxidative stress, leading to the formation of reactive oxygen species in such quantities as to be neurotoxic *(144)*. Alternatively, a Fenton-type reaction that could occur in the synaptic cleft could generate hydroxy radicals that oxidize released DA and 5HT to neurotoxic quinones and/or 6-OHDA or 5,6 DHT, respectively. These toxic metabolites can be recognized by DA or 5HT transporters and thereby selectively transported into the presynaptic neuron. Given the nature of the presumed reaction, it also explains why ascorbate and other antioxidants protect against the neurotoxicity *(165,196)*.

Cohen and Heikkila *(197)* showed that DA could be converted to one of three trihydroxyphenethylamines via the Fenton-Huber-Weiss reactions in an in vitro system where there was Fe^{2+}, hydrogen peroxide, EDTA, and DA. Fenton-Haber-Weiss reactions:

$$Fe^{2+}\text{-EDTA} + H_2O_2 \rightarrow Fe^{3+}\text{-EDTA} + OH^- + OH^{\cdot} \qquad (1)$$

$$2Fe^{3+}\text{-EDTA} + (H_2)\text{-ascorbate} \rightarrow 2Fe^{2+}\text{-EDTA-dehydroascorbate} + 2H^+ \qquad (2)$$

$$2Fe^{2+}\text{-EDTA} + 2H^+ + O_2 \rightarrow 2(Fe^{3+}\text{-EDTA}) + H_2O_2 \qquad (3)$$

$$2H^+ + Fe^{2+}\text{-EDTA} \rightarrow H_2O_2 + Fe^{3+}\text{-EDTA} \qquad (4)$$

$$O_2 + Fe^{3+}\text{-EDTA} \rightarrow O_2 + Fe^{2+}\text{-EDTA} \qquad (5)$$

$$O_2 + H_2O_2 \rightarrow O_2 + OH^- + OH^{\cdot} \qquad (6)$$

Based on the work of Senoh and Wiktop *(198)* and Cohen and Heikkila *(197)*, we reasoned that injections of large doses of METH could result in the formation of a toxic metabolite of DA. Riederer et al. *(199)* and Halliwell and Gutteridge *(200)* showed that there is Fe^{2+} stored in many regions of the brain. Hydrogen peroxide is a byproduct of monoamine oxidase metabolism, and its concentration is normally kept small by catalase. If there is excess hydrogen peroxide, however, it could undergo Fe^{2+} catalysis and result in hydroxy radical formation. These hydroxy radicals are highly reactive *(197,200)*. The hydroxy radical, once formed, could react with DA to form 6-OHDA. It is possible that with large amounts of DA in the synaptic cleft after high-dose METH treatment, a small proportion of DA could be metabolized to 6-OHDA and be transported back into the DA neuron through the DA transporter. Once back inside the neuron, it could be converted to a reactive semiquinone. This semiquinone looks for an electron donor, such as the sulfhydryl groups on cysteine or methionine (components of long-chain proteins). When the semiquinone and long-chain proteins are crosslinked through sulfur–oxygen bonds, the proteins are denatured and no longer functional *(201,202)*. This loss of protein function could then lead to cell damage (*see* nonenzymatic reactions).

We *(143)* have detected 6-OHDA after a single large dose of METH in the striatum of rats, and 5,6 dihydroxytryptamine in the hippocampus *(166)*. We have made the assumption that both conversions were proceeding according to a Fenton-type reaction. Our attempts to replicate this work have proven difficult, and we have found that there were instances where we could not detect 6-OHDA or 5,6-DHT in any of the rats treated with METH. Rollema et al. *(203)* failed to detect extracellular 6-OHDA in rats treated with METH using the in vivo dialysis technique. In addition, certain colleagues have tried to measure tissue concentrations of 6-OHDA after METH treatment, but either found the results inconsistent from rat to rat or could not detect any of the hydroxylated derivatives of DA (Cohen and Gibb, personal communication). Recently, Wagner et al. *(138)* reported the formation of 6-OHDA in the microgram range after rats were treated with METH. In this experiment, a monoamine oxidase inhibitor and a catechol-*O*-methyltransferase inhibitor were administered before treatment with METH. Marek et al. *(204)* obtained similar results with the use of a monoamine oxidase inhibitor.

Zigmond and colleagues *(141,144)* investigated the role of endogenous DA in the DA neurotoxicity induced by METH. They performed a series of experiments examining the oxidation of DA and the formation of cysteinyl-DA adducts using both in vitro and in vivo systems. These experiments were based on the finding that peroxidase enzymes are capable of catalyzing the conversion of DA to reactive DA quinones. Although peroxidase enzymes are not present in brain, they found a similar enzyme, prostaglandin (PG) synthase, which is present in brain. When purified PG synthase was combined with DA and bovine serum albumin, they identified a DA quinone and a cysteinyl-DA adduct. It was inferred from this reaction that hydroxy radicals could be formed (*see* enzymatic reactions). They concluded that DA oxidation could be catalyzed by PG synthase and, importantly, that the oxidized quinone was a potential mechanism for cytotoxicity.

Another hypothesis that has been proposed is that there is an excitatory feed-forward loop that coincides with the extrapyramidal tract. This loop is excited by monoamines, and the continued driving of 5HT and DA neurons in this loop demands excess energy. The cell cannot maintain homeostasis, and therefore, cellular death occurs *(137)*.

9. SUMMARY AND CONCLUSIONS

Evidence has been cited that substituted AMPHs (METH, MDMA, FEN, and PHEN/FEN) that are structurally related to METH cause specific neurotoxicity to DA and/or 5HT nerve endings in the central nervous system as indicated by a loss of neurotransmitter levels, loss of enzymes involved in neurotransmitter synthesis, and loss of the number of monoamine transporters. In addition, evidence has been presented that there are morpholigical changes indicative of neuronal degeneration during or shortly after drug administration. Toxic responses have been documented after METH, MDMA, FEN, and the combination of PHEN and FEN (PHEN/FEN). The pharmacological and/or physiological agents that protect against or enhance neurotoxicity were discussed, and both neurotoxic responses and protections were related to possible neurochemical mechanisms that may mediate the drug-engendered neurotoxicity. There exists sufficient evidence that the compounds are neurotoxic, but the mechanism by which they exert this neurotoxicity is not yet known and remains to be elucidated.

REFERENCES

1. Cavanagh, J. B. (1994) Principles of neurotoxicology, in *Principles of Neurotoxicology* (Chang, L. W., ed.), Marcel Decker, New York, pp. xv–xviii.
2. Seiden, L. S., Sabol, K. E., and Dyer, R. S. (1995) Neurotoxicity of methamphetamine-related drugs and cocaine, in *Handbook of Neurotoxicology* (Chang, L. W., ed.), Marcel Dekker, New York, pp. 825–843.
3. Uretsky, N. J. and Snodgrass, S. R. (1977) Studies on the mechanism of stimulation of dopamine synthesis by amphetamine in striatal slices. *J. Pharmacol. Exp. Ther.* **202,** 565–580.
4. Hedreen, J. C. and Chalmers, J. P. (1972) Neuronal degeneration in rat brain induced by 6-hydroxydopamine; a histological and biochemical study. *Brain Res.* **47,** 1–36.
5. Baumgarten, H. B. and Zimmerman, B. (1992) Neurotoxic phenylalkalyamines and indolealkylamines, in *Handbook of Experimental Pharmacology: Selective Neurotoxicity* (Herken, H. and Hucho, F., eds.), Springer-Verlag, New York, 225–276.
6. Sanders-Busch, E., Bushing, J. A., and Sulser, F. (1972) Long-term effects of p-chloroamphetamine on tryptophan hydroxylase activity and on the levels of 5-hydroxytryptamine and 5-hydroxyindole acetic acid in brain. *Eur. J. Pharm.* **20,** 385–388.
7. Seiden, L. S., Sabol, K. E., and Ricaurte, G. A. (1993) Amphetamine: effects on catecholamine systems and behavior [Review]. *Annu. Rev. Pharmacol. Toxicol.* **33,** 639–677.
8. Raiteri, M., Cerrito, F., Cervoni, A. M., and Levi, G. (1979) Dopamine can be released by two mechanisms differentially affected by the dopamine transport inhibitor nomifensine. *J. Pharmacol. Exp. Ther.* **208,** 195–202.
9. Kandel, E. R. and Schwartz, J. H. (1985) *Principles of Neural Science*, Elsevier, New York.
10. Wichems, C. H., Hollingsworth, C. K., and Bennet, B. A. (1995) Release of serotonin induced by 3,4-methylenedioxymethamphetamine (MDMA) and other substituted amphetamines in cultured fetal raphe neurons: further evidence for calcium-independent mechanisms of release. *Brain Res.* **695,** 10–18.
11. Heikkila, R. E., Orlansky, H., and Cohen, G. (1975) Studies on the distinction between uptake inhibition and release of (3H)dopamine in rat brain tissue slices. *Biochem. Pharmacol.* **24,** 847–852.
12. Fischer, J. F. and Cho, A. K. (1979) Chemical release of dopamine from striatal homogenates: evidence for an exchange diffusion model. *J. Pharmacol. Exp. Ther.* **208,** 203–209.
13. Parker, E. M. and Cubeddu, L. X. (1986) Effects of *d*-amphetamine and dopamine synthesis inhibitors on dopamine and acetylcholine neurotransmission in the striatum.

I. Release in the absence of vesicular transmitter stores. *J. Pharmacol. Exp. Ther.* **237**, 179–192.

14. Liang, N. Y. and Rutledge, C. O. (1982) Comparison of the release of [3H]dopamine from isolated corpus striatum by amphetamine, fenfluramine and unlabelled dopamine. *Biochem. Pharmacol.* **31**, 983–992.

15. Rudnick, G. and Wall, S. C. (1992) The molecular mechanism of "ecstasy" [3,4-methylenedioxymethamphetamine (MDMA)]: serotonin transporters are targets for MDMA-induced serotonin release. *Proc. Acad. Sci. USA* **89**, 1817–1821.

16. Berger, U. V., Gu, X. F., and Azmitia, E. C. (1992) The substituted amphetamines 3,4-methylenedioxymethamphetamine, methamphetamine, *p*-chloroamphetamine and fenfluramine induce 5-hydroxytryptamine release via a common mechanism blocked by fluoxetine and cocaine. *Eur. J. Pharmacol.* **215**, 153–160.

17. Johnson, M. P., Conarty, P. F., and Nichols D. (1991) [3H] Monoamine releasing and uptake inhibition properties of 3,4-methylenedioxymethamphetamine and *p*-chloramphetamine analogues. *Eur. J. Pharm.* **200**, 9–16.

18. Laferrere, B. and Wurtman, R. J. (1989) Effect of D-fenfluramine on serotonin release in brains of anaesthetized rats. *Brain Res.* **504**, 258–263.

19. McKenna, D. J., Guan, X. M., and Shulgin, A. T. (1991) 3,4-Methylenedioxyamphetamine (MDA)) analogues exhibit differential effects on synaptosomal release of 3H-dopamine and 3H-5-hydroxytryptamine. *Pharmacol. Biochem. Behav.* **38**, 505–512.

20. Nichols, D. E., Lloyd, D. H., Hoffman, A. J., Nichols, M. B., and Yim, G. K. W. (1982) Effects of certain hallucinogenic amphetamine analogues on the release of [3H]serotonin from rat brain synaptosomes. *J. Med. Chem.* **25**, 530–535.

21. Sabol, K. E., Richards, J. B., and Seiden, L. S. (1992) Fenfluramine-induced increases in extracellular hippocampal serotonin and progressively attentuated in vivo during a four-day fenfluramine regimen in rats. *Brain Res.* **571**, 64–72.

22. Mennini, T., Garattini, S., and Caccia, S. (1985) Anorectic effect of fenfluramine isomers and metabolites: relationship between brain levels and in vitro potencies on serotonergic mechanisms. *Psychopharmacology* **85**, 111–114.

23. Sarkissian, C. F., Wurtman, R. J., Morse, A. N., and Gleason, R. (1990) Effects of fluoxetine or D-fenfluramine on serotonin release from, and levels in, rat frontal cortex. *Brain Res.* **529**, 294–301.

24. Schmidt, C. J. (1987) Acute administration of methylenedioxymethamphetamine: comparison with the neurochemical effects of its N-desmethyl and N-ethyl analogs. *Eur. J. Pharmacol.* **136**, 81–88.

25. Schmidt, C. J., Wu, L., and Lovenberg, W. (1986) Methylenedioxymethamphetamine: a potentially neurotoxic amphetamine analogue. *Eur. J. Pharmacol.* **124**, 175–178.

26. Ellinwood, E. H. (1969) Amphetamine psychosis: a multidimensional process. *Simin. Psychiatry* **1**, 208–226.

27. Schuster, C. R. and Thompson, T. (1969) Self administration of and behavioral dependence on drugs. *Annu. Rev. Pharmacol.* **9**, 483–502.

28. Angrist, B. M., Schweitzer, J. W., Gershon, S., and Friedhoff, A. J. (1970) Mephentermine psychosis: misuse of the Wyamine inhaler. *Am. J. Psychiatry* **126**, 1315–1317.

29. Griffith, J. (1966) A study of illicit amphetamine drug traffic in Oklahoma City. *Am. J. Psychiatry* **123** 560–569.

30. Kramer, J. C., Fischman, V. S., and Littlefield, D. C. (1967) Amphetamine abuse. Pattern and effects of high doses taken intravenously. *JAMA* **201,** 305–309.

31. Seiden, L. S. and Ricaurte, G. (1987) Neurotoxicity of methamphetamine and related drugs, in *Psychopharmacology: The Third Generation of Progress* (Meltzer, H. Y., ed.), Raven, New York, pp. 359–365.

32. Seiden, L. S., Fischman, M. W., and Schuster, C. R. (1977) Changes in brain catecholamines induced by long-term methamphetamine administration in rhesus monkeys, in *Cocaine and Other Stimulants* (Ellinwood, E. H., ed.), Plenum, New York, pp. 179–185.

33. Wagner, G. C., Ricaurte, G. A., Seiden, L. S., Schuster, C. R., Miller, R. J., and Westley, J. (1980) Long-lasting depletions of striatal dopamine and loss of dopamine uptake sites following repeated administration of methamphetamine. *Brain Res.* **181,** 151–160.

34. Axt, K. J., Commins, D. L., Vosmer, G., and Seiden, L. S. (1990) Alpha-methyl-*p*-tyrosine pretreatment partially prevents methamphetamine-induced endogenous neurotoxin formation. *Brain Res.* **515,** 269–276.

35. O'Hearn, E., Battaglia, G., Desouza, E. B., Kuhar, M. J., and Molliver, M. E. (1988) Methylenedioxyamphetamine (MDA) and methylenedioxymethamphetamine (MDMA) cause selective ablation of serotonergic axon terminals in forebrain: immunocytochemical evidence for neurotoxicity. *J. Neurosci.* **8,** 2788–2803.

36. Ryan, L. J., Linder, J. C., Martone, M. E., and Groves, P. M. (1990) Histological and ultrastructural evidence that D-amphetamine causes degeneration in neostriatum and frontal cortex of rats. *Brain Res.* **518,** 67–77.

37. Steranka, L. R. and Sanders, B. E. (1980) Long-term effects of continuous exposure to amphetamine on brain dopamine concentration and synaptosomal uptake in mice. *Eur. J. Pharmacol.* **65,** 439–443.

38. Wagner, G. C., Ricaurte, G. A., Johanson, C. E., Schuster, C. R., and Seiden, L. S. (1980) Amphetamine induces depletion of dopamine and loss of dopamine uptake sites in caudate. *Neurology* **30,** 547–550.

39. Seiden, L. S., Fischman, M. W., and Schuster, C. R. (1976) Long-term methamphetamine induced changes in brain catecholamines in tolerant rhesus monkeys. *Drug Alcohol Depend.* **1,** 215–219.

40. Ricaurte, G. A., Schuster, C. R., and Seiden, L. S. (1980) Long-term effects of repeated methylamphetamine administration on dopamine and serotonin neurons in the rat brain: a regional study. *Brain Res.* **193,** 153–163.

41. Hotchkiss, A. and Gibb, J. W. (1980) Blockade of methamphetamine-induced depression of tyrosine hydroxylase by GABA transaminase inhibitors. *Eur. J. Pharmacol.* **66,** 201–205.

42. Wagner, G. C., Schuster, C. R., and Seiden, L. S. (1979) Methamphetamine induced changes in brain catecholamines in rats and guinea pigs. *Drug Alcohol Depend.* **4,** 435–439.

43. Levine, M., Hull, C. D., Garcia-Rill, E., Erinoff, L., Buchwald, N. A., and Heller, A. (1980) Long-term decreases in spontaneous firing of caudate neurons induced by amphetamine in cats. *Brain Res.* **194,** 263–268.

44. Hotchkiss, A. J. and Gibb, J. W. (1980) Long-term effects of multiple doses of methamphetamine on tryptophan hydroxylase and tyrosine hydroxylase activity in rat brain. *J. Pharmacol. Exp. Ther.* **214,** 257–262.

45. Nwanze, E. and Jonsson, G. (1980) Amphetamine toxicity on dopamine nerve terminals in the caudate nucleus of mice. *Neurosci. Lett.* **26,** 163–168.
46. Wagner, G. C., Ricaurte, G. A., Seiden, L. S., Schuster, C. R., Miller, R. J., and Westley, J. (1980) Long-lasting depletions of striatal dopamine and loss of dopamine uptake sites following repeated administration of methamphetamine. *Brain Res.* **181,** 151–160.
47. Wagner, G. C., Seiden, L. S., and Schuster, C. R. (1979) Methamphetmine-induced changes in brain catecholamines in rats and guinea pigs. *Drug Alcohol Depend.* **4,** 435–438.
48. Lorez, H. (1981) Fluorescence histochemistry indicates damage of striatal dopamine nerve terminals in rats after multiple doses of methamphetamine. *Life Sci.* **28,** 911–916.
49. Ricaurte, G. A., Guillery, R. W., Seiden, L. S., Schuster, C. R., and Moore, R. Y. (1982) Dopamine nerve terminal degeneration produced by high doses of methylamphetamine in the rat brain. *Brain Res.* **235,** 93–103.
50. Woolverton, W. L., Ricaurte, G. A., Forno, L. S., and Seiden, L. S. (1989) Long-term effects of chronic methamphetamine administration in rhesus monkeys. *Brain Res.* **486,** 73–78.
51. Walsh, S. L. and Wagner, G. C. (1990) The effects of methamphetamine-induced neurotoxicity on motor performance in the rat (56.8) [Abstract]. *Soc. Neurosci. Abstracts* **16**.
52. Richards, J. B., Baggott, M. J., Sabol. K. E., and Seiden, L. S. (1993) A high-dose methamphetamine regimen results in long lasting deficits on the performance of a reaction time task. *Brain Res.* **627,** 254–260.
53. Ando, K., Johanson, C. E., and Schuster, C. R. (1986) Effects of dopaminergic agents on eye tracking before and after repeated methamphetamine. *Pharmacol. Biochem. Behav.* **24,** 693–699.
54. Ando, K., Johanson, C. E., Seiden, L. S., and Schuster, C. R. (1985) Sensitivity changes to dopaminergic agents in fine motor control of rhesus monkeys after repeated methamphetamine administration. *Pharmacol. Biochem. Behav.* **22,** 737–743.
55. Finnegen, K. T., Ricaurte, G., Seiden, L. S., and Schuster, C. R. (1982) Altered sensitivity to d-methylamphetamine, apomorphine, and haloperidol in rhesus monkeys depleted of caudate dopamine by repeated administration of *d*-methylamphetamine. *Psychopharmacology (Berlin)* **77,** 43–52.
56. Fischman, M. W. and Schuster, C. R. (1977) Long-term behavioral changes in the rhesus monkey after multiple daily injections of d-methylamphetamine. *J. Pharmacol. Exp. Ther.* **201,** 593–605.
57. Nencini, P., Woolverton, W. L., and Seiden, L. S. (1988) Enhancement of morphine-induced analgesia after repeated injections of methylenedioxymethamphetamine. *Brain Res.* **457,** 136–142.
58. Johanson, C. E., Aigner, T. G., Seiden, L. S., and Schuster, C. R. (1979) The effects of methamphetamine on fine motor control in rhesus monkeys. *Pharmacol. Biochem. Behav.* **11,** 273–278.
59. Preston, K. L. and Schuster, C. R. (1982) A comparison of the central and peripheral effects of atropine on force lever performance. *Pharmacol. Biochem. Behav.* **16,** 423–427.
60. Carter, C. J. and Pycock, C. J. (1978) Differential effects of central serotonin manipulation on hyperactive and stereotyped behaviour. *Life Sci.* **23,** 953–960.

61. Zenick, H. and Goldsmith, M. (1981) Drug discrimination learning in lead-exposed rats. *Science* **212**, 569–571.

62. Green, A. R., Cross, A. J., and Goodwin, G. M. (1995) Review of the pharmacology and clinical pharmacology of 3,4-methylenedioxymethamphetamine (MDMA or "ecstacy"). *Psychopharmacology* **119**, 247–260.

63. Steele, T. D., McCann, U. D., and Ricaurte, G. A. (1994) 3,4-Methylenedioxymethamphetamine (MDMA, "Ecstasy"): pharmacology and toxicology in animals and humans [Review]. *Addiction* **89**, 539–551.

64. Nichols, D. (1986) Differences between the mechanism of action of MDMA, MBDB and the classic hallucinogens. Identification of a new therapeutic class: entactogens. *J. Psychoactive Drugs* **18**, 305–313.

65. Eisner, B. (1989) *Ecstacy: The MDMA Story*. Ronin Publications, Berkeley, CA.

66. Peroutka, S. J., Newman, H., and Harris, H. (1988) Subjective effects of 3,4-methylenedioxymethamphetamine in recreational users. *Neuropsychopharmacology* **1**, 273–277.

67. Grinspoon, L. and Bakalar, J. B. (1986) Can drugs be used to enhance the psychotherapeutic process? *Am. J. Psychother.* **40**, 393–404.

68. Henry, J. A. (1992) Toxicity and deaths from 3,4-methylenedioxymethamphetamine ("ecstacy"). *Lancet* **340**, 384–387.

69. Peroutka, S. J. (1987) Incidence of recreational use of 3,4-methylenediomethoxymethamphetamine (MDMA, "ecstacy") on an undergraduate campus [Letter]. *N. Engl. J. Med.* **317**, 1542–1543.

70. Cuomo, M. J., Dyment, P. G., and Gammino, V. M. (1994) Increasing use of "Ecstasy" (MDMA) and other hallucinogens on a college campus. *J. Am. Coll. Health* **42**, 271–274.

71. Randall, T. (1992) Ecstacy-fueled "Rave" parties become dances of death for english youths. *JAMA* **268**, 1505–1506.

72. Farfel, G. M. and Seiden, L. S. (1995) Role of hypothermia in the mechanism of protection against serotonergic toxicity. I. Experiments using 3,4-methylenedioxymethamphetamine, dizocilpine, CGS 19755 and NBQX. *J. Pharmacol. Exp. Ther.* **272**, 860–867.

73. Battaglia, G., Yeh, S. Y., and Desouza, E. B. (1988) MDMA-induced neurotoxicity: parameters of degeneration and recovery of brain serotonin neurons. *Pharmacol. Biochem. Behav.* **29**, 269–274.

74. Stone, D. M., Hanson, G. R., and Gibb, J. W. (1987) Differences in the central serotonergic effects of methylenedioxymethamphetamine (MDMA) in mice and rats. *Neuropharmacology* **26**, 1657–1661.

75. Battaglia, G., Brooks, B. P., Kulsakdinun, C., and De S. E. (1988) Pharmacologic profile of MDMA (3,4-methylenedioxymethamphetamine) at various brain recognition sites. *Eur. J. Pharmacol.* **149**, 159–163.

76. Insel, T. R., Battaglia, G., Johanssen, J., Marra, S., and Desouza, E. B. (1989) 3,4-Methylenedioxymethamphetamine ("Ecstasy") selectively destroys brain serotonin nerve terminals in rhesus monkeys. *J. Pharmacol. Exp. Ther.* **249**, 713–720.

77. Ricaurte, G. A., Martello, A. L., Katz, J. L., and Martello, M. B. (1992) Lasting effects of (+-)-3,4-methylenedioxymethamphetamine (MDMA) on central serotonergic neurons in nonhuman primates: neurochemical observations. *J. Pharmacol. Exp. Ther.* **261**, 616–622.

78. Ricaurte, G. A. and McCann, U. D. (1992) Neurotoxic amphetamine analogues: effects in monkeys and implications for humans [Review]. *Ann. NY Acad. Sci.* **648,** 371–382.

79. Schmidt, C. J. and Taylor, V. L. (1987) Depression of rat brain tryptophan hydroxylase activity following the acute administration of methylenedioxymethamphetamine. *Biochem. Pharmacol.* **36,** 4095–4102.

80. Stone, D. M., Johnson, M., Hanson, G. R., and Gibb, J. W. (1988) Role of endogenous dopamine in the central serotonergic deficits induced by 3,4-methylenedioxymethamphetamine. *J. Pharmacol. Exp. Ther.* **247,** 79–87.

81. Commins, D. L., Vosmer, G., Virus, R. M., Woolveerton, W. L., Schuster, C. R., and Seiden, L. S. (1987) Biochemical and histological evidence that methylenedioxymethamphetamine (MDMA) is toxic to neurons in the rat brain. *J. Pharmacol. Exp. Ther.* **241,** 338–345.

82. Lew, R., Sabol, K. E., Chou, C., Vosmer, G. L., Richards, J., and Seiden, L. S. (1996) Methylenedioxymethamphetamine (MDMA)-induced serotonin deficits are followed by partial recovery over a 52 week period. Part II: Radioligand binding and autoradiographic studies. *J. Pharmacol. Exp. Ther.* **276,** 855–865.

82a. Molliver, M. E., Mamounas, L. A., and Wilson, M. A. (1989) Effects of neurotoxic amphetamines on serotonergic neurons: immunocytochemical studies, in *Pharmacology and Toxicology of Amphetamine and Related Designer Drugs* (Asghar, K. and De Souza, E., eds.), NIDA Research Monograph, US Department of Health and Human Service, Washington, DC, pp. 270–305.

83. Scanzello, C. R., Hatzidimitriou, G., Martello, A. L., Katz, J. L., and Ricaurte, G. A. (1993) Serotonergic recovery after (+/−)3,4-(methylenedioxy) methamphetamine injury: observations in rats. *J. Pharmacol. Exp. Ther.* **264,** 1484–1491.

83a. Scheffel, U. and Ricaurte, G. A. (1990) Paroxetine as an in vivo indicator of 3,4-methylenedioxymethamphetamine neurotoxicity: a presynaptic serotonergic positron emission tomography ligand? *Brain Res.* **527,** 89–95.

84. Bowyer, J. F., Davies, D. L., Schmued, L., Broening, H. W., Newport, G. D., Slikker, W. J., and Holson, R. R. (1994) Further studies of the role of hyperthermia in methamphetamine neurotoxicity. *J. Pharmacol. Exp. Ther.* **268,** 1571–1580.

85. Fischer, C., Hatzidimitriou, G., Wlos, J., Katz, J., and Ricaurte, G. (1995) Reorganization of ascending 5-HT axon projections in animals previously exposed to the recreational drug (+/−)3,4-methylenedioxymethamphetamine (MDMA, "ecstasy"). *J. Neuroscience* **15,** 5476–5485.

86. Sabol, K. E., Lew, R., Richards, J. B., Vosmer, G. L., and Seiden, L. S. (1996) Methylenedioxymethamphetamine (MDMA)-induced serotonin deficits are followed by partial recovery over a 52 week period. Part I: Synaptosomal uptake and tissue concentrations. *J. Pharmacol. Exp. Ther.* **276,** 846–854.

87. Ricaurte, G. A., Forno, L. S., Wilson, M. A., Delanney, L. E., Irwin, I., Molliver, M. E., and Langston, J. W. (1988) (+/−)3,4-Methylenedioxymethamphetamine selectively damages central serotonergic neurons in nonhuman primates. *JAMA* **260,** 51–55.

88. Ricaurte, G. A., Delanney, L. E., Irwin, I., and Langston, J. W. (1988) Toxic effects of MDMA on central serotonergic neurons in the primate: importance of route and frequency of drug administration. *Brain Res.* **446,** 165–168.

89. Battaglia, G., Yeh, S. Y., O'Hearn, E., Molliver, M. E., Kuhar, M. J., and De Souza, E. B. (1987) 3,4-Methylenedioxymethamphetamine and 3,4-methylenedioxyampheta-

mine destroy serotonin terminals in rat brain: quantification of neurodegeneration by measurement of [3H]paroxetine-labeled serotonin uptake sites. *J. Pharmacol. Exp. Ther.* **242,** 911–916.

90. Ricaurte, G. A., Finnegan, K. T., Irwin, I., and Langston, J. W. (1990) Aminergic metabolites in cerebrospinal fluid of humans previously exposed to MDMA: Preliminary observations. *Ann. NY Acad. Sci.* **600,** 699–710.

91. Atkinson, R. L. and Hubbard, V. S. (1994) Report on the NIH workshop on pharmacologic treatment of obesity. *Am. J. Clin.. Nutr.* **60,** 153–156.

92. Le Douarec, P., Neveu, C., and Garattini, S. (1970) Pharmacology and biochemistry of fenfluramine, in *Amphetamine and Related Compounds* (Costa, E., ed.), Raven, New York, pp. 75–105.

93. Kleven, M. S. and Seiden, L. S. (1989) D-, L- and DL-fenfluramine cause long-lasting depletions of serotonin in rat brain. *Brain Res.* **505,** 351–353.

94. Nicolaidis, S. (1997) *Obesity Management & Redux.* Academic, San Diego.

95. Clineschmidt, B. V., Zacchei, A. G., Totaro, J. A., Pfluger, A. B., McGuffin, J. C., and Wishousky, T. I. (1978) Fenfluramine and brain serotonin. *Ann. NY Acad. Sci.* **305,** 222–241.

96. Harvey, J. A., McMaster, S. E., and Fuller, R. W. (1977) Comparison between the neurotoxic and serotonin depleting effects of various halogenated derivatives of amphetamine in the rat. *J. Pharmacol. Exp. Ther.* **202,** 581–589.

97. Harvey, J. A. and McMaster, S. E. (1975) Fenfluramine: evidence for a neurotoxic action on a long-term depletion of serotonin. *Psychopharmacol. Commun.* **1,** 217–228.

98. Kleven, M. S., Schuster, C. R., and Seiden, L. S. (1988) Effect of depletion of brain serotonin by repeated fenfluramine on neurochemical and anorectic effects of acute fenfluramine. *J. Pharmacol. Exp. Ther.* **246,** 822–828.

99. Sanders-Bush, E., Bushing, J. A., and Sulser, F. (1975) Long-term effects of p-chloroamphetamine and related drugs on central serotonergic mechanisms. *J. Pharmacol. Exper. Ther.* **192,** 33–41.

100. Schuster, C. R., Lewis, M., and Seiden, L. S. (1986) Fenfluramine: neurotoxicity. *Psychopharmacol. Bull.* **22,** 148–151.

101. Steranka, L. R. and Sanders-Bush, E. (1979) Long-term effects of fenfluramine on central serotonergic mechanisms. *Neuropharmacology* **18,** 895–903.

102. Zaczek, R., Battaglia, G., Culp, S., Appel, N. M., Contrera, J. F., and DeSouza, E. B. (1990) Effects of repeated fenfluramine administration on indices of monoamine function in rat brain: pharmacokinetic, dose response, regional specificity and time course data. *Pharmacol. Exp. Ther.* **253,** 104–112.

103. McCann, U., Hatzidimitriou, G., Ridenour, A., Fischer, C., Yuan, J., Katz, J., and Ricaurte, G. (1994) Dexfenfluramine and serotonin neurotoxicity: further preclinical evidence that clinical caution is indicated. *J. Pharmacol. Exp. Ther.* **269,** 792–798.

104. Ricaurte, G. A., Molliver, M. E., Martello, M. B., Katz, J. L., Wilson, M. A., and Martello, A. L. (1991) Dexfenfluramine neurotoxicity in brains of non-human primates [*see* comments]. *Lancet* **338,** 1487–1488.

105. Scheffel, U., Szabo, Z., Mathews, W. B., Finley, P. A., Yuan, J., Callahan, B., Hatzidimitriou, G., Dannals, R. F., Ravert, H. T., and Ricaurte, G. A. (1996) Fenfluramine-induced loss of serotonin transporters in baboon brain visualized with PET. *Synapse.*

106. Appel, N. M., Contrera, J. F., and De Souza, E. B. (1989) Fenfluramine selectively and differentially decreases the density of serotonergic nerve terminals in rat brain: evidence from immunocytochemical studies. *J. Pharmacol. Exp. Ther.* **249,** 928–943.

107. Molliver, D. C. and Molliver, M. E. (1990) Anatomic evidence for a neurotoxic effect of (+/−)-fenfluramine upon serotonergic projections in the rat. *Brain Res.* **511,** 165–168.

108. Appel, N. M., Mitchell, W. M., Contrera, J. F., and De Souza, E. B. (1990) Effects of high-dose fenfluramine treatment on monoamine uptake sites in rat brain: assessment using quantitative autoradiography. *Synapse* **6,** 33–44.

109. Lew, R., Weisenberg, B., Vosmer, G., and Seiden, L. S. (1997) Combined phentermine/fenfluramine administration enhances depletion of serotonin from central terminal fields. *Synapse* **26,** 36–45.

110. Kalia, M. (1991) Reversible, short-lasting, and dose-dependent effect of (+)-fenfluramine on neocortical serotonergic axons. *Brain Res.* **548,** 111–125.

111. Sotelo, C. (1991) Immunohistochemical study of short- and long-term effects of DL-fenfluramine on the serotonergic innervation of the rat hippocampal formation. *Brain Res.* **541,** 309–326.

112. Caccia, S., Ballabio, M., Guiso, G., Rocchetti, M., and Garattini, S. (1982) Species diffrences in the kinetics and metabolism of fenfluramine isomers. *Arch. Int. Pharmacodyn.* **258,** 15–28.

113. Westphalen, R. I. and Dodd, P. R. (1995) The nature of d,l-fenfluramine-induced 5-HT reuptake transporter loss in rats. *Mol. Neurobiol.* **11,** 165–175.

114. Westphalen, R. I. and Dodd, P. R. (1993) New evidence for a loss of serotonergic nerve terminals in rats treated with *d,l*-fenfluramine. *Pharmacol. Toxicol.* **72,** 249–255.

115. Westphalen, R. I. and Dodd, P. R. (1993) The regeneration of *d,l*-fenfluramine-destroyed serotonergic nerve terminals. *Eur. J. Pharmacol.* **238,** 399–402.

116. Battaglia, G., Norman, A. B., Newton, P. L., and Creese, I. (1986) In vitro and in vivo irreversible blockade of cortical S2 serotonin receptors by *N*-ethoxycarbonyl-2-ethoxy-1,2-dihydroquinoline: a technique for investigating S2 serotonin receptor recovery. *J. Neurochem.* **46,** 589–593.

117. Norton, W. T., Aquino, D. A., Hozumi, I., Chiu, F. C., and Brosnan, C. F. (1992) Quantitative aspects of reactive gliosis: a review. *Neurochem. Res.* **17,** 877–885.

118. O'Callaghan, J. P. and Miller, D. B. (1994) Neurotoxicity profiles of substituted amphetamines in the C57BL/6J mouse. *J. Pharmacol. Exp. Ther.* **270,** 741–751.

118a. Miller, D. B. and O'Callaghan, J. P. (1994) Environment-, drug- and stress-induced alterations in body temperature affect the neurotoxicity of substituted amphetamines in the C57BL/6J mouse. *J. Pharmacol. Exp. Ther.* **270,** 752–760.

119. Rowland, N. E., Kalehua, A. N., Li, B. H., Semple-Rowland, S. L., and Streit, W. J. (1993) Loss of serotonin uptake sites and immunoreactivity in rat cortex after dexfenfluramine occur without parallel glial cell reactions. *Brain Res.* **624,** 35–43.

120. Roth, B. L. (1994) Multiple serotonin receptors: clinical and experimental aspects [Review]. *Ann. Clin. Psychiatry* **6,** 67–78.

121. Hoebel, B. G., Hernandez, L., Schwartz, D. H., Mark, G. P., and Hunter, G. A. (1989) Microdialysis studies of brain norepinephrine, serotonin, and dopamine release during ingestive behavior. Theoretical and clinical implications [Review]. *Ann. of Acad. Sci.* **575,** 171–191.

122. Kleven, M. S., Woolverton, W. L., and Seiden, L. S. (1991) Evaluation of potential neurotoxic effects of amphetamine-related anorectic agents on brain serotonin and dopamine in the rat [Abstract]. *Proc. Soc. Neurosci.* **17**.

123. Weintraub, M., Sundaresan, P. R., Madan, M., Schuster, B., Balder, A., Lasagna, L., and Cox, C. (1992) Long-term weight control study. I (weeks 0 to 34). The enhancement of behavior modification, caloric restriction, and exercise by fenfluramine plus phentermine versus placebo. *Clin. Pharmacol. Ther.* **51**, 586–594.

124. Weintraub, M., Sundaresan, P. R., Schuster, B., Ginsberg, G., Madan, M., Balder, A., Stein, E. C., and Byrne, L. (1992) Long-term weight control study. II (weeks 34 to 104). An open-label study of continuous fenfluramine plus phentermine versus targeted intermittent medication as adjuncts to behavior modification, caloric restriction, and exercise. *Clin. Pharmacol. Ther.* **51**, 595–601.

125. Weintraub, M., Sundaresan, P. R., Schuster, B., Moscucci, M., and Stein, E. C. (1992) Long-term weight control study. III (weeks 104 to 156). An open-label study of dose adjustment of fenfluramine and phentermine. *Clin. Pharmacol. Ther.* **51**, 602–607.

126. Weintraub, M., Sundaresan, P. R., Schuster, B., Averbuch, M., Stein, E. C., Cox, C., and Byrne, L. (1992) Long-term weight control study. IV (weeks 156 to 190). The second double-blind phase. *Clin. Pharmacol. Ther.* **51**, 608–614.

127. Weintraub, M., Sundaresan, P. R., Schuster, B., Averbuch, M., Stein, E. C., and Byrne, L. (1992) Long-term weight control study. V (weeks 190 to 210). Follow-up of participants after cessation of medication. *Clin. Pharmacol. Ther.* **51**, 615–618.

128. Weintraub, M., Hasday, J. D., Mushlin, A. I., and Lockwood, D. H. (1984) A double-blind clinical trial in weight control. Use of fenfluramine and phentermine alone and in combination. *Arch. Int. Med.* **144**, 1143–1148.

129. Weintraub, M. (1992) Long-term weight control study: Conclusions. *Clin. Pharmacol. Ther.* **51**, 642–646.

130. Silverstone, T. (1992) Appetite suppressants. A review [Review]. *Drugs* **43**, 820–36.

131. Chappell, W. R. (1989) Interspecific scaling of toxicity data: A question of interpretation. *Risk Anal.* **9**, 13–14.

132. Mordenti, J., Chen, S. A., Moore, J. A., Ferraiolo, B. L., and Green, J. D. (1991) Interspecies scaling of clearance and volume of distribution data for five therapeutic proteins. *Pharm. Res.* **8**, 1351–1359.

133. Alvares, A. P., Pratt, W. B., and Taylor, P. (1990) Pathways of drug metabolism, in *Principles of Drug Action: The Basis of Pharmacology* (Pratt, W. B., ed.), pp. 365–422, Churchill Livingstone, New York.

134. Freireich, E. J., Gehan, E. A., Rall, D. P., Schmidt, L. H., and Skipper, H. E. (1966) Quantitative comparison of toxicity of anticancer agents in mouse, rat, hamster, dog, monkey and man. *Cancer Cherother. Rep.* **50**, 219–226.

135. Guy-Grand, B. (1992) Clinical studies with d-fenfluramine. *Am. J. Clin. Nutr.* **55**, 173S-176S.

136. Schmidt, C. J., Ritter, J. K., Sonsalla, P. K., Hanson, G. R., and Gibb, J. W. (1985) Role of dopamine in the neurotoxic effects of methamphetamine. *J. Pharmacol. Exp. Ther.* **233**, 539–544.

137. Carlsson, A. (1993) Search for the neuronal circuitries and neurotransmitters involved in "Positive" and "Negative" schizophrenic symptomatology. *Fidia Research Foundation Lecture Series* **7**.

138. Wagner, G. C., Lowndes, H. E., and Kita, T. (1993) Methamphetamine-induced 6-hydroxydopamine formation following MAO and COMT inhibition [Abstract]. *Soc. Neurosci. Abstracts* **19**.

139. Axt, K. J. and Seiden, L. S. (1990) alpha-Methyl-*p*-tyrosine partially attenuates *p*-chloroamphetamine-induced 5-hydroxytryptamine depletions in the rat brain. *Pharmacol. Biochem. Behav.* **35**, 995–997.

140. Giovanni, A., Hastings, T. G., Liang, L. P., and Zigmond, M. J. (1992) Metamphetamine increases hydroxyl radicals in rat striatum: Role of dopamine [Abstract] *Soc. Neurosci. Abstracts* **18**.

141. Hastings, T. G., and Zigmond, M. J. (1992) Prostaglandin synthase-catalyzed oxidation of dopamine [Abstract]. *Soc. Neurosci. Abstracts* **18**.

142. Liang, L. P., Hastings, T. G., Zigmond, M. J., and Giovanni, A. (1992) Use of salicylate to trap hydroxyl radicals in rat brain: a methodological study [Abstract]. *Soc. Neurosci. Abstracts* **18**.

143. Seiden, L. S. and Vosmer, G. (1984) Formation of 6-hydroxydopamine in caudate nucleus of the rat brain after a single large dose of methylamphetamine. *Pharmacol. Biochem. Behav.* **21**, 29–31.

144. Zigmond, M. J. and Hastings, T. G. (1992) A method for measuring dopamine-protein conjugates as an index of dopamine oxidation [Abstract]. *Soc. Neurosci. Abstracts* **18**.

145. Butcher, S. P., Fairbrother, I. S., Kelly, J. S., and Arbuthnott, G. W. (1988) Amphetamine-induced dopamine release in the rat striatum: an in vivo microdialysis study. *J. Neurochem.* **50**, 346–355.

146. Sonsalla, P. K., Gibb, J. W., and Hanson, G. R. (1986) Roles of D1 and D2 dopamine receptor subtypes in mediating the methamphetamine-induced changes in monoamine systems. *J. Pharmacol. Exp. Ther.* **238**, 932–937.

147. Creese, I., Sibley, D. R., Hamblin, M. W., and Leff, S. E. (1983) Dopamine receptors in the central nervous system [Review] *Adv. Biochem. Psychopharmacol.* **36**, 125–134.

148. Carlsson, A. and Lindqvist, M. (1963) Effect of chlorpromazine or haloperidol on formation of 3-methoxytyramine and normetanephrine in mouse brain. *Acta Pharmacol. Toxicol.* **20**, 140–144.

149. Brownstein, M. J. and Palkovits, M. (1984) Catecholamines, serotonin, acetylcholine, and γ-aminobutyric acid in the rat brain: biochemical studies, in *Handbook of Chemical Neuroanatomy: Classical Transmitters in the CNS* (Bjorklund, A. and Hokfelt, T., eds.), Elsevier, Amsterdam, pp. 23–54.

150. Nash, J. F. (1990) Ketanserin pretreatment attenuates MDMA-induced dopamine release in the striatum as measured by in vivo microdialysis. *Life Sci.* **47**, 2401–2408.

151. Schmidt, C. J., Black, C. K., Taylor, V. L., Fadayel, G. M., Hymphreys, T. M., Nieduzak, T. R., and Sorensen, S. M. (1992) The 5-HT2 receptor antagonist, MDL 28, 133A, disrupts the serotonergic-dopaminergic interaction mediating the neurochemical effects of 3,4-methylenedioxymethamphetamine. *Eur. J. Pharmacol.* **220**, 151–159.

152. Schmidt, C. J., Fadayel, G. M., Sullivan, C. K., and Taylor, V. L. (1992) 5-HT2 receptors exert a state-dependent regulation of dopaminergic function: studies with MDL 100,907 and the amphetamine analogue, 3,4-methylenedioxymethamphetamine. *Eur. J. Pharmacol.* **223**, 65–74.

153. Schmidt, C. J., Taylor, V. L., Abbate, G. M., and Nieduzak, T. R. (1991) 5-HT2 antagonists stereoselectively prevent the neurotoxicity of 3,4-methylenedioxymethampheta-

mine by blocking the acute stimulation of dopamine synthesis: reversal by L-dopa. *J. Pharmacol. Exp. Ther.* **256,** 230–235.

154. Minchin, M. C. W. (1985) Inositol phospholipid breakdown as an index of serotonin receptor function, in *Neuropharmacology of Serotonin* (Green, A. R., ed.), Oxford, New York, pp. 117–130.

155. Berridge, M. J. and Galione, A. (1988) Cytosolic calcium oscillators. *FASEB J* **2,** 3074–3082.

156. Gandhi, C. R. and Ross, D. H. (1988) Characterization of a high-affinity Mg^{2+}-independent Ca^{2+}-ATPase from rat brain synaptosomal membranes. *J. Neurochem.* **50,** 248–256.

157. Sonsalla, P. K., Nicklas, W. J., and Heikkila, R. E. (1989) Role for excitatory amino acids in methamphetamine-induced nigrostriatal dopaminergic toxicity. *Science* **243,** 398–400.

158. Farfel, G. M., Vosmer, G. L., and Seiden, L. S. (1992) The *N*-methyl-D-aspartate antagonist MK-801 protects against serotonin depletions induced by methamphetamine, 3,4-methylenedioxymethamphetamine and *p*-chloroamphetamine. *Brain Res.* **595,** 121–127.

159. Johnson, M., Hanson, G. R., and Gibb, J. W. (1989) Effect of MK-801 on the decrease in tryptophan hydroxylase induced by methamphetamine and its methylenedioxy analog. *Eur. J. Pharm.* **165,** 315–318.

160. Frandsen, A. and Schousboe, A. (1991) Dantrolene prevents glutamate cytotoxicity and Ca^{2+} release from intracellular stores in cultured cerebral cortical neurons. *J. Neurochem.* **56,** 1075–1078.

161. Lei, S. Z., Zhang, D., Abele, A. E., and Lipton, S. A. (1992) Blockade of NMDA receptor-mobilization of intracellular Ca2+ prevents neurotoxicity. *Brain Res.* **598,** 196–202.

162. Weihmuller, F. B., O'Dell, S. J., Cole, B. N., and Marshall, J. F. (1991) MK-801 attenuates the dopamine-releasing but not the behavioral effects of methamphetamine: an in vivo microdialysis study. *Brain Res.* **549,** 230–235.

163. Kashihara, K., Okumura, K., Onishi, M., and Otsuki, S. (1991) MK-801 fails to modify the effect of methamphetamine on dopamine release in the rat striatum. *Neuroreport* **2,** 236–238.

164. Bowyer, J. F., Scallet, A. C., Holson, R. R., Lipe, G. W., Slikker, W., and Ali, S. F. (1991) Interactions of MK-801 with glutamate-, glutamine- and methamphetamine-evoked release of [3H]dopamine from striatal slices. *J. Pharmacol. Exp. Ther.* **257,** 262–270.

165. Wagner, G. C., Carelli, R. M., and Jarvis, M. F. (1986) Ascorbic acid reduces the dopamine depletion induced by methamphetamine and the 1-methyl-4-phenyl pyridinium ion. *Neuropharmacology* **25,** 559–561.

166. Commins, D. L., Axt, K. J., Vosmer, G., and Seiden, L. S. (1987) 5,6-Dihydroxytryptamine, a serotonergic neurotoxin, is formed endogenously in the rat brain. *Brain Res.* **403,** 7–14.

167. O'Dell, S. J., Weihmuller, F. B., McPherson, R. J., and Marshall, J. F. (1992) Excitotoxic lesions in rat striatum protect against subsequent methamphetamine-induced dopamine terminal damage [Abstract]. *Soc. Neurosci. Abstracts* **18.**

168. Fuller, R. W. and Hemrick-Luecke, S. (1980) Long-lasting depletion of striatal dopamine by a single injection of amphetamine in iprindole-treated rats. *Science* **209,** 305–307.

169. Marek, G. J., Vosmer, G., and Seiden, L. S. (1990) Dopamine uptake inhibitors block long-term neurotoxic effects of methamphetamine upon dopaminergic neurons. *Brain Res.* **513,** 274–279.

170. Ricaurte, G. A., Fuller, R. W., Perry, K. W., Seiden, L. S., and Schuster, C. R. (1983) Fluoxetine increases long-lasting neostriatal dopamine depletion after administration of d-methamphetamine and d-amphetamine. *Neuropharmacology* **22,** 1165–1169.

171. Green, A. R., DeSouza, R. J., Williams, J. L., Murray, T. K., and Cross, A. J. (1992) The neurotoxic effects of methamphetamine on 5-hydroxytryptamine and dopamine in brain: Evidence for the protective effect of chlormethiazole. *Neuropharmacology* **31,** 315–321.

172. Cadet, J. L., Ladenheim, B., Baum, I., Carlson, E., and Epstein C. (1994) CuZn-superoxide dismutase (CuZnSOD) transgenic mice show resistance to the lethal effects of methylenedioxyamphetamine (MDA) and of methylenedioxymethamphetamine (MDMA). *Brain Res.* **655,** 259–262.

173. Cadet, J. L., Sheng, P., Ali, S., Rothman, R., Carlson, E., and Epstein, C. (1994) Attenuation of methamphetamine-induced neurotoxicity in copper/zinc superoxide dismutase transgenic mice. *J. Neurochem.* **62,** 380–383.

174. Hiramatsu, M., Kumagai, Y., Unger, S. E., and Cho, A. K. (1990) Metabolism of methylenedioxymethamphetamine: formation of dihydroxymethamphetamine and a quinone identified as its glutathione adduct. *J. Pharmacol. Exp. Ther.* **254,** 521–527.

175. Chiueh, C. C., Miyake, H., and Peng, M. T. (1993) Role of dopamine autoxidation, hydroxyl radical generation, and calcium overload in underlying mechanisms involved in MPTP-induced parkinsonism. *Adv. Neurol.* **60,** 251–258.

176. Colado, M. I. and Green, A. R. (1994) A study of the mechanism of MDMA ("ecstasy")-induced neurotoxicity of 5-HT neurones using chlormethiazole, dizocilpine and other protective compounds. *Br. J. Pharmacol.* **111,** 131–36.

177. Murray, T. K., Williams, J. L., Misra, A., Colado, M. I., and Green, A. R. (1996) The spin trap reagent PBN attenuates degeneration of 5-HT neurons in rat brain induced by *p*-chloroamphetamine but not fenfluramine. *Neuropharmacology* **35,** 1615–1620.

178. Colado, M. I. and Green, A. R. (1995) The spin trap reagent alpha-phenyl-*N-tert*-butyl nitrone prevents "ecstasy"-induced neurodegeneration of 5-hydroxytryptamine neurones. *Eur. J. Pharmacol.* **280,** 343–346.

179. Che, S., Johnson, M., Hanson, G. R., and Gibb, J. W. (1995) Body temperature effect on methylenedioxymethampheatmine-induced acute decrease in tryptophan hydroxylast acitvity. *Eur. J. Pharmacol.* **293,** 447–453.

180. Carney, J. M. and Floyd, R. A. (1991) Protection against oxidative damage to CNS by alpha-phenyl-tert-butyl nitrone (PBN) and other spin-trapping agents: a novel series of nonlipid free radical scavengers. *J. Mol. Neurosci.* **3,** 47–57.

181. Albers, D. S. and Sonsalla, P. K. (1995) Methamphetamine-induced hyperthermia and dopaminergic neurotoxicity in mice: Pharmacological profile of protective and non-protective agents. *J. Pharmacol. Exp. Ther.* **275,** 1104–1114.

182. Farfel, G. M. and Seiden, L. S. (1995) Role of hypothermia in the mechanism of protection against serotonergic toxicity. II. Experiments with methamphetamine, p-chloroamphetamine, fenfluramine, dizocilpine and dextromethorphan. *J. Pharmacol. Exp. Ther.* **272,** 868–875.

183. Malberg, J. E., Sabol, K. E., and Seiden, L. S. (1996) Co-administration of MDMA with drugs that protect against MDMA neurotoxicity produces different effects on body temperature. *J. Pharmacol. Exp. Ther.* **278**, 258–267.

184. Malberg, J. E. and Seiden, L. S. (1996) 3,4-Methylenedioxymethamphetamine (MDMA) 5HT neurotoxicity is a function of ambient temperature and core body temperature in rats [Abstract]. *Soc. Neurosci. Abstracts* **22**.

185. Hewitt, K. E. and Green, A. R. (1994) Chlormethiazole, dizocilpine and haloperidol prevent the degeneration of serotonergic nerve terminals induced by administration of MDMA (Ecstacy) to rats. *Neuropharmacology* **33**, 1589–1595.

186. Gordon, C. J., Watkinson, W. P., O'Callaghan, J. P., and Miller, D. B. (1991) Effects of 3,4-methylenedioxymethamphetamine on autonomic thermoregulatory responses of the rat. *Pharmacol. Biochem. Behav.* **38**, 339–344.

187. Ali, S. F., Newport, G. D., Holson, R. R., Slikker, W. J., and Bowyer, J. F. (1994) Low environmental temperatures or pharmacologic agents that produce hypothermia decrease methamphetamine neurotoxicity in mice. *Brain Res.* **658**, 33–38.

188. Malberg, J. E. and Seiden, L. S. (1997) Administration of Fenfluramine at different ambient temperatures produces different core temperature and 5HT neurotoxicity profiles. *Brain Res.*, in press.

189. Dafters, R. I. (1994) Effect of ambient temperature on hyperthermia and hyperkinesis induced by 3,4-methylenedioxymethamphetamine (MDMA or "ecstasy") in rats. *Psychopharmacology* **114**, 505–508.

190. Frey, H. H. (1975) Hyperthermia induced by amphetamine, *p*-chloroamphetamine and fenfluramine in the rat. *Pharmacology* **13**, 163–176.

191. Preston, E., Ma, S., and Hass, N. (1990) Ambient temperature modulation of fenfluramine-induced thermogenesis in the rat. *Neuropharmacology* **29**, 277–283.

192. Wilkinson, L. O., Auerbach, S. B., and Jacobs, B. L. (1991) Extracellular serotonin levels change with behavioral state but not with pyrogen-induced hyperthermia. *J. Neurosci.* **11**, 2732–2741.

193. Gordon, C. J. and Fogelson, L. (1994) Metabolic and thermoregulatory responses of the rat maintained in acrylic or wire-screen cages: implications for pharmacological studies. *Physiol. Behav.* **56**, 73–79.

194. Lehninger, A. L. (1975) *Biochemistry*. Worth Publishers, New York.

195. Bowyer, J. F., Gough, B., Slikker, W. J., Lipe, G. W., Newport, G. D., and Holson, R. R. (1993) Effects of a cold environment or age on methamphetamine-induced dopamine release in the caudate putamen of female rats. *Pharmacol. Biochem. Behav.* **44**, 87–98.

196. Pileblad, E., Slivka, A., Bratvold, B., and Cohen, G. (1988) Studies on the autoxidation of dopamine: Interaction with ascorbate. *Arch. Biochem. Biophys.* **263**, 447–452.

197. Cohen, G. and Heikkila, R. E. (1974) The generation of hydrogen peroxide, superoxide radical, and hydroxyl radical by 6-hydroxydopamine, dialuric acid, and related cytotoxic agents. *J. Biol. Chem.* **249**, 2447–2452.

198. Senoh, S. and Wiktop, B. (1959) Non-enzymatic conversions of dopamine to norephinephrine and trihydroxy phenethylamines. *J. Am. Chem. Soc.* **81**, 6222–6235.

199. Riederer, P., Dirr, A., Goetz, M., Sofic, E., Jellinger, K., and Youdim, M. B. (1992) Distribution of iron in different brain regions and subcellular compartments in Parkinson's disease [Review]. *Ann. Neuro.* **32(Suppl)**, S101–104.

200. Halliwell, B. and Gutteridge, J. M. (1984) Role of iron in oxygen radical reactions. *Methods Enzymol.* **105,** 47–56.

201. Fornstedt, B. and Carlsson, A. (1989) A marked rise in 5-S-cysteinyl-dopamine levels in guinea-pig striatum following reserpine treatment. *J. Neural Transm.* **76,** 155–161.

202. Fornstedt, B., Rosengren, E., and Carlsson, A. (1986) Occurrence and distribution of 5-S-cysteinyl derivatives of dopamine, dopa and dopac in the brains of eight mammalian species. *Neuropharmacology* **25,** 451–454.

203. Rollema, H., De, V. J. B., Westerink, B. H., Van, P. F. M., and Horn, A. S. (1986) Failure to detect 6-hydroxydopamine in rat striatum after the dopamine releasing drugs dexamphetamine, methylamphetamine and MPTP. *Eur. J. Pharmacol.* **132,** 65–69.

204. Marek, G. J., Vosmer, G., and Seiden, L. S. (1990) Pargyline increases 6-hydroxy-dopamine levels in the neostriatum of methamphetamine-treated rats. *Pharm. Biochem. Behav.* **36,** 187–190.

205. Ricaurte, G., Bryan, G., Strauss, L., Seiden, L., and Schuster, C. (1985) Hallucinogenic amphetamine selectively destroys brain serotonin nerve terminals. *Science* **229,** 986–988.

10

Molecular Mechanisms of Action of 5,6- and 5,7-Dihydroxytryptamine

Tahereh Tabatabaie and Glenn Dryhurst

1. INTRODUCTION

Disruption of neuronal function has frequently been employed to obtain morphological, physiological, biochemical, and behavioral information about different neuronal types and pathways. Surgical or electrolytic denervation of particular neurotransmitter systems represents methods to accomplish such disruptions. However, in an organ as complex as the brain, these techniques suffer from inherent limitations particularly with respect to selectivity. In the mid-1960s, several groups reported that a hydroxylated analog of the neurotransmitter dopamine (DA), namely 6-hydroxydopamine (6-OHDA), induced degeneration of peripheral and central nerve terminals that was remarkably selective for catecholaminergic (particularly noradrenergic) neurons (1–3). This discovery introduced the concept of chemical lesioning to the fields of neurochemical and neurobiological research.

2. HISTORY AND DISCOVERY OF 5,6- AND 5,7-DIHYDROXYTRYPTAMINE

Following the discovery of the catecholaminergic neurotoxin 6-OHDA, Baumgarten et al. (4) initiated studies of dihydroxytryptamine derivatives. These studies were based on the idea that such compounds might selectively damage central serotonergic neurons in a fashion similar to that by which 6-OHDA affected catecholaminergic neurons. Among the dihydroxy derivatives of tryptamine, 5,6-dihydroxytryptamine (5,6-DHT) seemed to hold the most promise as a selective serotonergic neurotoxin, because 6-hydroxytryptamine had earlier been shown to possess high affinity for the uptake mechanism of the neurotransmitter 5-hydroxytryptamine (5HT; serotonin) (5,6). Indeed, it was subsequently demonstrated that when injected into the cerebrospinal fluid (CSF) of rats in a dose range of 25–75 µg, 5,6-DHT evoked a significant, selective,

From: Highly Selective Neurotoxins: Basic and Clinical Applications Edited by: R. M. Kostrzewa.
Humana Press Inc., Totowa, NJ

and long-lasting reduction of 5HT levels in different regions of the brain and spinal cord *(4)*. This depletion of 5HT strongly suggested that 5,6-DHT mediated the degeneration of serotonergic axons and axon terminals, a conclusion that was further supported by the observation that 5,6-DHT also evokes a significant reduction in the uptake of [3H]5HT by rat brain *(7,8)*. At dose levels ranging from 25–75 µg, 5,6-DHT evoked no significant alterations of DA or norepinephrine (NE) in any region of the rat brain *(9)*. These preliminary results stimulated considerable interest in the use of 5,6-DHT as a chemical agent for the selective destruction of serotonergic neurons. However, further studies have revealed a number of limitations associated with the use of this compound for such purposes. To illustrate, intracerebroventricular (icv) injection of 5,6-DHT (25–75 µg) into the rat results in a profound, long-term depletion of 5HT in the spinal cord (>80%), but more modest decreases in the brain (30–50%) *(4,8)*. Higher doses (>75 µg) do not increase the neurodegenerative effects of 5,6-DHT on serotonergic pathways in the brain and are very poorly tolerated by animals (i.e., rats). Furthermore, such high doses of 5,6-DHT cause damage not only to serotonergic neurons, but also other neurons, including myelinated axons and glial cells, and have been reported to result in a general shrinkage of the striatum on the side of the injection *(10)*. The nonselectivity of 5,6-DHT at doses ≥75 µg (icv) is apparent from the significant reductions of DA and NE levels in various regions of the brain *(10,11)*.

These limitations associated with 5,6-DHT prompted an investigation into the properties of 5,7-dihydroxytryptamine (5,7-DHT) in the hope of discovering a more selective serotonergic neurotoxin. Preliminary studies *(12)* revealed that 5,7-DHT induced morphological changes to serotonergic axon terminals in the brains of rats that were very similar to those caused by 5,6-DHT. Furthermore, doses of up to 300 µg of 5,7-DHT are generally well tolerated by rats provided that the injection is made by the icv route, i.e., avoiding direct injection of the drug into brain tissue *(12)*. However, at icv doses of 5,7-DHT > 100 µg, rats often develop convulsions. These can be prevented by prior treatment with barbiturates *(12)*. 5,7-DHT is not a specific serotonergic neurotoxin, and also evokes decrements of NE, but not DA following icv administration to the rat *(12)*. Numerous subsequent investigations have provided unequivocal evidence for the neurodegenerative properties of 5,7-DHT. Baumgarten et al. *(13)* have shown that icv doses of 5,7-DHT as low as 10 µg are capable of evoking significant decrements (ca. 25%) of both 5HT and NE levels in rat brain. With increasing doses up to 50 µg, brain NE and 5HT levels were both reduced to approx 50% of control values. However, with even higher doses (75–200 µg), 5HT levels were further decreased, reaching 25% of control levels after a 200-µg icv dose of 5,7-DHT, but no further reductions in NE levels were observed. Even large doses of 5,7-DHT have no effect on brain levels of DA. The neurodegenerative action of 5,7-DHT is not uniform through the central nervous system (CNS). To illustrate, almost complete depletion of 5HT (ca. 85%) has been reported in the spinal cord following injection of 10 µg of

5,7-DHT *(13)*. Ten days after icv injection of 200 μg of 5,7-DHT to the rat, 5-HT reductions ranged from 55% in the medulla and pons, to about 85% in forebrain regions and septum *(13)*. The 5HT depletions were most profound in brain and spinal cord regions rich in serotonergic axons and terminals, particularly in structures located close to the lateral ventricles, such as the septum, striatum, and hypothalamus. By contrast, 5HT depletions evoked by 5,7-DHT were much lower in regions containing serotonergic cell bodies, such as the mesencephalon, and medulla/pons *(13)*. During the first 3 h following icv injection of a 200-μg dose of 5,7-DHT, an initial, rapid depletion of 5HT occurs ranging from 35–65% in different brain regions and spinal cord *(13)*. A somewhat variable, partial recovery of 5HT levels then occurs in areas rich in serotonergic cell bodies after 45 d. However, other regions exhibit either no significant recovery (spinal cord, striatum) or only a late and slow recovery (mesencephalon, medulla/pons, hypothalamus, septum) between 16 and 45 d. Studies using titrated 5HT and NE and slices of cortex, hypothalamus, and spinal cord from the brains of rats treated with 5,7-DHT (200 μg, icv) have demonstrated massive losses of both 5HT (85–90%) and NE (40–57%) uptake sites *(13)*. These and fluorescence microscopy studies clearly indicate extensive damage evoked by 5,7-DHT to serotonergic axons and nerve terminals *(13)*.

Similar doses (50 or 75 μg) of 5,6-DHT and 5,7-DHT evoke almost the same depletions of 5HT levels in the brain or spinal cord *(13,14)*. However, based on their IC_{50} values for inhibition of 5HT uptake into serotonergic neurons, it has been concluded that the affinity of 5,7-DHT for the uptake site ($IC_{50} = 4$ μM) is approximately seven times lower than that of 5,6-DHT ($IC_{50} = 0.6$ μM). Thus, despite its lower affinity for 5HT uptake sites, 5,7-DHT is as toxic toward serotonergic neurons as 5,6-DHT, indicating the higher intraneuronal neurotoxic potency of 5,7-DHT. Furthermore, more complete denervation of serotonergic neurons can be accomplished by administration of higher doses of 5,7-DHT. By contrast, high doses of 5,6-DHT cannot be tolerated by laboratory animals. However, a disadvantage of 5,7-DHT is that it evokes the degeneration not only of serotonergic, but also of noradrenergic terminals. Pargyline and other monoamine oxidase (MAO) inhibitors, such as iproniazid and pheniprazine, all block the neurodegenerative effects of 5,7-DHT on noradrenergic neurons without affecting its ability to reduce 5-HT levels in the rat *(15)*. Interestingly, pargyline also potentiates the neurotoxicity of 6-OHDA toward dopaminergic neurons and provides some protection for noradrenergic terminals *(16)*. Desipramine, which inhibits uptake of both 5HT and NE, also blocks the neurotoxic effects of 5,7-DHT on noradrenergic neurons, but not on serotonergic neurons *(17–19)*.

3. HYPOTHESES CONCERNING THE SELECTIVE NEUROTOXICITY OF 5,6-DHT AND 5,7-DHT

Selectivity of 5,6-DHT and 5,7-DHT is believed to be related to their uptake into serotonergic (and perhaps noradrenergic) terminals by the active transport systems designed for reuptake of the structurally similar endogenous neuro-transmitter 5HT *(20)*. On reaching some critical intraneuronal concentration, these compounds then evoke neurotoxic effects. Two major hypotheses have been advanced concerning the molecular mechanisms that might underlie the toxicity of 5,6-DHT and 5,7-DHT. Both hypotheses stress an inherent chemical property of these compounds, i.e., ease of oxidation by molecular oxygen without catalysis by an enzyme (commonly referred to by the misnomer autoxidation), as a key factor contributing to their neurotoxicity *(21)*. One hypothesis suggests that autoxidation of 5,6-DHT or 5,7-DHT within the cytoplasm of serotonergic axon terminals might result in formation of highly reactive (electrophilic) quinonoid products that covalently bind to proteins and other macromolecules essential for neuronal function. The second hypothesis invokes the intraneuronal generation of cytotoxic reactive oxygen species (ROS) formed as byproducts of autoxidation of 5,6-DHT and 5,7-DHT. ROS include superoxide radical anion ($O_2\cdot-$), hydrogen peroxide (H_2O_2), and the hydroxyl radical ($HO\cdot$). These hypotheses derive from studies initially carried out on 6-OHDA that undergoes an oxygen-dependent covalent interaction with several model proteins in vitro *(21)*. Reducing agents (antioxidants) or the absence of molecular oxygen inhibit this binding. Formation of ROS in the process of autoxidation of 6-OHDA is well documented *(22,23)*. Both 5,6-DHT and 5,7-DHT are also autoxidized at physiological pH *(24)*. Accordingly, it is logical to suggest that the neurotoxicity of these dihydrotryptamines is related to their intraneuronal oxidation. The following sections of this chapter, therefore, will summarize the oxidation chemistry and biochemistry of these serotonergic neurotoxins. Evidence in support of the hypotheses advanced to explain the neurotoxicity of 5,6-DHT and 5,7-DHT will also be discussed in more detail.

4. OXIDATION CHEMISTRY AND BIOCHEMISTRY OF 5,7-DHT

Autooxidation of 5,7-DHT in buffered aqueous solution at pH 7.4 results in formation of a pink solution *(24–27)*. This reaction is first-order with respect to both 5,7-DHT and molecular oxygen, and is therefore overall second-order *(28)*. Studies with methylated analogs of 5,7-DHT demonstrated that the 4-methyl- and 4,6-dimethyl-derivatives were oxidized by molecular oxygen, but did not give colored reaction products *(29)*. By contrast, 6-methyl-5,7-DHT formed colored product solutions on autoxidation that exhibited spectral features similar to those for the autoxidation product(s) of unsubstituted 5,7-DHT. These observations led to the conclusion that the primary site of oxygen incorporation into 5,7-DHT is at the C-4 position *(29)*. During the autoxidation of 5,7-DHT, small quantities of H_2O_2 are formed *(25,30)*, although this has not always been observed *(28)*. The electron paramagnetic resonance (EPR) spectrum of an autoxidizing solution of 5,7-DHT containing the spin-trapping agent dimethyl-1-pyrroline-1-oxide (DMPO) exhibits features of spin adducts of both

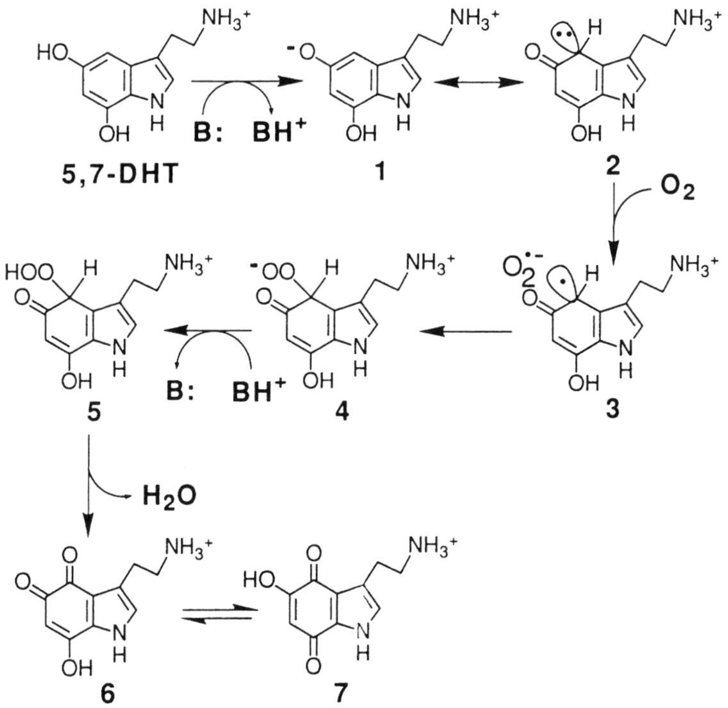

Scheme 1

a carbon-centered radical (derived from 5,7-DHT) and the hydroxyl radical (HO·) *(30)*. The latter adduct would originate either from the well-documented facile decomposition of the DMPO-O_2^- spin adduct into the DMPO-HO· spin adduct at physiological pH *(31)* or from direct trapping of HO· formed by the decomposition of H_2O_2 catalyzed by trace levels of transition metal ions (i.e., Fenton reaction). Formation of O_2^- may also be inferred indirectly based on inhibition of the autoxidation of 5,7-DHT by superoxide dismutase (SOD) *(30)*.

Autoxidation of 5,7-DHT at pH 7.4 yields two major products, 5-hydroxytryptamine-4,7-dione (**7**, Scheme 1) and its dimer 6,6′-bi(5-hydroxytryptamine-4,7-dione) (**19**, Scheme 3) in equimolar yields *(30)*. In the absence of any catalytic effects deriving from O_2^- and trace transition metal ions, the autoxidation of 5,7-DHT is very slow *(30)*. Under these conditions, the reaction pathway originally proposed by Sinhababu and Borchardt *(29)* probably accounts for formation of **7**. This pathway proposes that the initial step in the reaction involves deprotonation of 5,7-DHT to yield anion **1** and, thence, carbanion **2** (Scheme 1). Carbanion **2** then acts as the primary electron donor to molecular oxygen to form the free radical superoxide complex **3**. Recombination of the superoxide residue of **3** with the incipient C(4)-radical then yields the hydroperoxy anion **4**, which, on protonation, generates hydroperoxide **5**. Loss of the elements of water from **5** then gives *o*-quinone **6**, which in turn rearranges to the more stable *p*-quinone **7**. It is important to note that the reaction between molecular oxygen and carbanion **2** to give hydroperoxy anion **4** must almost certainly occur via two consecutive steps, because spin preservation considerations do not permit a concerted two-electron transfer from singlet carbon to triplet oxygen *(29,30)*. Scheme 1 provides an elegant rationale for the formation of **7** by the very slow, uncatalyzed autoxidation of 5,7-DHT. However, it has been found that trace levels of transition metal ions (Fe^{2+}, Cu^{2+}, Fe^{3+}, Mn^{2+}) catalyze the autoxidation of 5,7-DHT to **7** and dimer **19** as the major reaction products *(30)*. Both catalase and SOD inhibit the rate of this reaction, and although added O_2^- potentiates the oxidation of 5,7-DHT, H_2O_2 has no effect on the reaction. The inhibitory effect of catalase, but the lack of stimulatory effect by H_2O_2 suggests that an organic hydroperoxide that would be decomposed by catalase *(32,33)* plays a key role in the transition metal-catalyzed autoxidation of 5,7-DHT. There is also abundant evidence that transition metal ions also react with organic hydroperoxy compounds analogous to **5** that are formed in many autoxidation reactions, the actual pathway catalyzed being dependent on the oxidation state of the transition metal. Thus, it has been proposed *(30)*, for example, that Fe^{3+} reacts with putative hydroperoxy intermediate **5** to yield peroxy radical **8** with concomitant formation of Fe^{2+} (Scheme 2). Oxidation of 5,7-DHT by **8** then yields radical **9** and regenerates hydroperoxide **5**. Direct attack by molecular oxygen on the C(4)-centered radical **9** then generates peroxy radical **8**, and hence, a catalytic cycle is established. The catalytic effect of Fe^{2+} probably proceeds by reduction of hydroperoxide **5** to oxy radical **10** with concomitant formation of Fe^{3+} (Scheme 2). By analogy with similar autoxidation reactions *(34–36)*, oxy radical **10** oxidizes 5,7-DHT to, predominantly, radical **9** with formation 4,5,7-trihydroxytryptamine (**12**) via **11**. Again, attack of molecular oxygen on **9** generates key peroxy radical **8**, so that a second catalytic cycle is established (Scheme 2). The trihydroxyindoleamine **12** is very easily autoxidized to form biradical **13** and, thence, **7** with concomitant formation of O_2^-. Experimentally, Fe^{2+} evokes a much stronger initial catalytic effect on the autoxidation of 5,7-DHT, indicating that the reaction between hydroperoxide **5** and Fe^{2+} is more facile than that with Fe^{3+}. Catalase causes a profound inhibitory effect on the rate of autoxidation of 5,7-DHT, because it catalyzes the conversion of **5** into 4,5,7-trihydroxytryptamine (**12**), which is rapidly autoxidized to **7**. Decomposition of **5** by this route terminates the catalytic cycles driven by trace levels of transition metal ions and, hence, decreases that rate of autoxidation of 5,7-DHT.

The rate of autoxidation of 5,7-DHT is inhibited by SOD, whereas addition of O_2^- greatly potentiates the rate of oxidation of the neurotoxin *(30)*. These observations imply that O_2^- is not only formed as a byproduct of the autoxidation of 5,7-DHT, but that it plays a key role in potentiating the rate of the reaction. Since neither HO· scavengers nor H_2O_2 has influence on the rate of autoxidation of 5,7-DHT, it is clear that O_2^- is directly involved in the reaction.

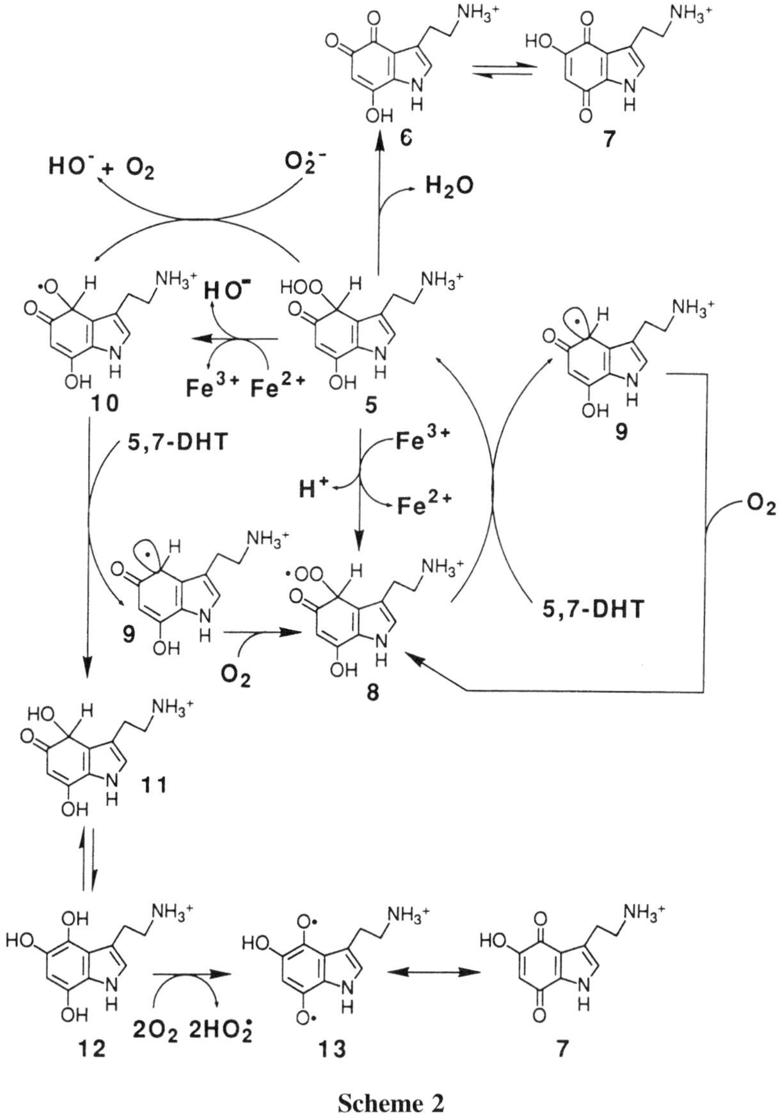

Scheme 2

However, it is extremely unlikely that O_2^- directly oxidizes 5,7-DHT, because it is a remarkably weak oxidizing agent *(39)*. Two different mechanisms may account for the role of O_2^- in potentiating the autoxidation of 5,7-DHT. Thus, based on reports *(37,38)* that O_2^- reacts with organic hydroperoxides to yield the corresponding alkoxyl radical, it is possible that O_2^- similarly reacts with **5** to form the phenoxyl radical **10** (Scheme 2) and contributes to the free radical chain reaction leading to **7**. It has also been proposed *(30)* that O_2^- potentiates deprotonation of 5,7-DHT to form carbanion **2**, the initial step in the autoxidation reaction (Scheme 1). However, although O_2^- is indeed an extraordinarily powerful Bronsted base in aprotic media *(40,41)*, it is only a rather weak base in aqueous solution *(42)*.

There are, in principle, at least three routes that could lead to dimer **19** as a result of autoxidation of 5,7-DHT. These include direct dimerization of monomer **7**, intermediary formation of the 6,6-linked dimer of 5,7-DHT that is subsequently oxidized, or attack by **7** on 5,7-DHT. At pH 7.4 under aerobic conditions, the conversion of **7** to **19** is too slow to account for the very rapid appearance of the latter dimer in the autoxidation of 5,7-DHT *(30)*. The 6,6-dimer of 5,7-DHT is indeed formed as an intermediate in some oxidations of the neurotoxin, but, in fact, is not a precursor of dimer **19**. As a result, the most plausible route leading to **19** involves an initial nucleophilic attack (Michael addition) by the C(6)-centered carbanion **14** on the C(5) = C(6) double bond of **7** to form dimer **15**, which on protonation, yields **16**, a compound containing residues of 4,5,7-trihydroxytryptamine and 5,7-DHT (Scheme 3). Autoxidation of **16** then yields dimer **17**, with concomitant formation of O_2^- as a byproduct. Oxidation of the 5,7-DHT residue of **17** to **19** has been proposed to follow the same pathways proposed for the free base in Schemes 1 and 2, a reaction that also generates O_2^- as a byproduct.

Once sequestered within a serotonergic neuron, 5,7-DHT must be exposed to a number of intraneuronal enzymes and mitochondria that could play functional roles in catalyzing the oxidation of this neurotoxin. Furthermore, there is evidence to suggest that mitochondria may be sites for the accumulation of and subsequent damage by 5,7-DHT. To illustrate, following 5,7-DHT administration, electron-dense material in the intermembranous compartment and ultrastructural damage to brain mitochondria have been noted *(43)*. Furthermore, radioactivity derived from [^{14}C]5,7-DHT is concentrated in the mitochondria-rich fraction of brain homogenates of rats treated with the compound *(44)*. Isolated rat brain mitochondria also stimulate the rate of oxygen consumption by 5,7-DHT *(45)*. Cyanide, a cytochrome oxidase inhibitor, inhibits such stimulation, whereas the MAO inhibitor pargyline has no effect. Thus, it has been suggested that 5,7-DHT might donate electrons to the cytochrome-*c* segment of the respiratory chain *(45)*. Similar effects have been observed using rat liver mitochondria *(44)*. The nature of the metabolites formed as a result of such putative intraneuronal enzyme-mediated oxidations is, therefore, clearly of interest. In the presence of the one-electron oxidants peroxidase/H_2O_2, ceruloplasmin/O_2, ferri cytochrome-*c*, or Cu^{2+}, the rate of aerobic oxidation of 5,7-DHT is greatly increased *(30)*. However, although **7** and dimer **19** are the major products of all these oxidations, three additional significant products are formed. These are the asymmetric dimers 3-(2-aminoethyl)-6-[3-(2-aminoethyl)-1,7-dihydro-5-hydroxy-7-oxo-*6H*-indol-6-ylidene]-*1H*-indole-5,7(*4H,6H*)-dione (**22**, Scheme 4) and two interconvertible tautomers of 3-(2-aminoethyl)-6-[3-(2-aminoethyl)-1,7-dihydro-5-hydroxy-7-oxo-*4H*-indol-4-ylidene]-*1H*-indole-4,7-(*4H,6H*)-dione (**25a/25b**). It has been concluded that the key intermediate in this oxidation by

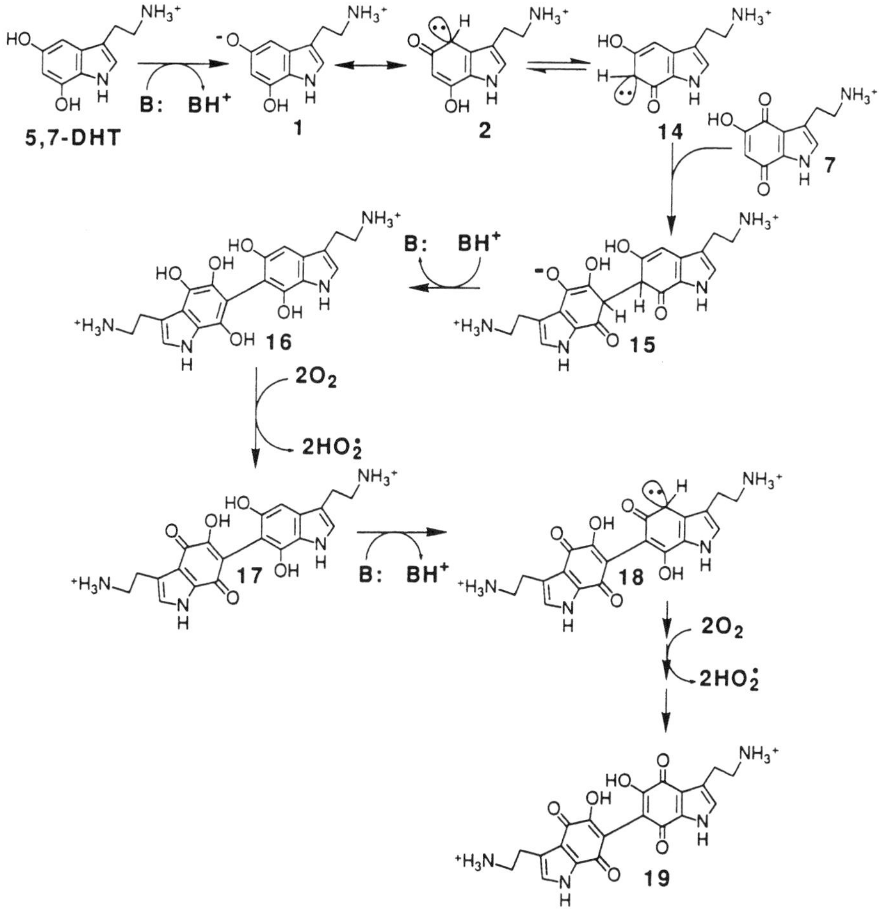

Scheme 3

one-electron oxidants in the presence of molecular oxygen is radical **9**, several forms of which react to give the various products by the pathways conceptualized in Scheme 4 *(30)*. Peroxidase also catalyzes the oxidation of 5,7-DHT in the absence of H_2O_2 to give **7** and **19** as major products along with a minor amount of **22** *(30)*. Under anaerobic conditions, the peroxidase/H_2O_2-mediated oxidation of 5,7-DHT gives **22**, **25a**, and **25b**, and two additional, rather unstable products that appear to be dimers **21** and **24** (Scheme 4) *(30)*.

5. INTERACTIONS OF 5,7-DHT WITH INTRANEURONAL NUCLEOPHILES

Creveling and Rotman *(46)* reported a slow and incomplete binding of 5,7-DHT when incubated in vitro with bovine serum albumin. However, the

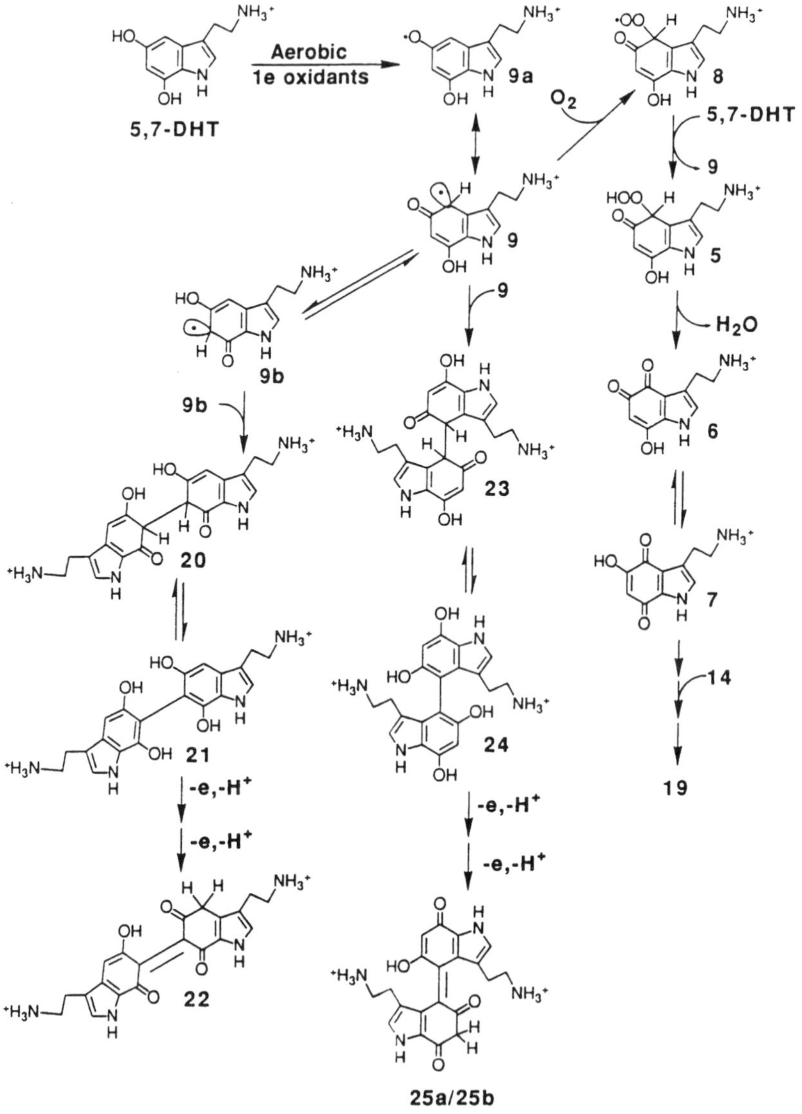

Scheme 4

extent of protein binding is appreciably greater in vivo *(47)*. Thus, in vivo oxidative metabolism or autoxidation of 5,7-DHT with resultant binding of its oxidative intermediates and/or products with intraneuronal proteins has been suggested to contribute to the neurotoxicity of this compound *(47)*. Some potential insights into such reactions have been obtained by studies of the autoxidation of 5,7-DHT in the presence of glutathione (GSH), which yields, in addition to **7** and **19**, 6-*S*-

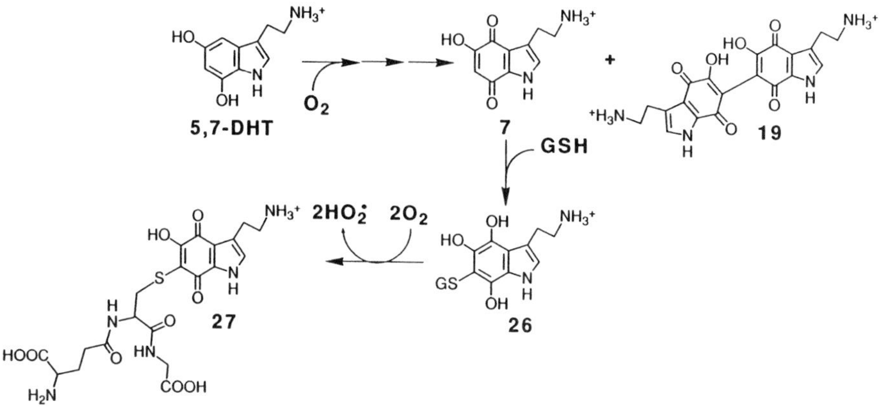

Scheme 5

glutathionyl-5-hydroxytryptamine-4,7-dione (**27**, Scheme 5) *(48)*. With increasing concentrations of GSH, the rate of the autoxidation of 5,7-DHT is decreased, but ultimately, the yields of **27** increase relative to those of **7** and **19** *(48)*. Glutathionyl conjugate **27** is also formed when **7** is incubated with GSH. Thus, nucleophilic addition of GSH to **7** yields, initially, the 6-*S*-glutathionyl conjugate of 4,5,7-trihydroxytryptamine, i.e., **26**, which is very easily autoxidized to **27** (Scheme 5) *(48)*. Dimer **19** does not react with GSH *(48)*.

6. REDOX CYCLING PROPERTIES OF THE OXIDATION PRODUCTS OF 5,7-DHT

The $E^{\circ\prime}$ values of **7**, **19**, and **27**, the major products of autoxidation of 5,7-DHT, at pH 7.4 are –540 (–300), –510 (–270), and –540 (–300) mV vs the saturated calomel electrode (standard hydrogen electrode), respectively *(48)*, which fall within the optimum range for cellular redox cycling *(49)*. Indeed, when incubated with representative intracellular reducing agents, such as GSH, L-cysteine (CySH), ascorbic acid (AA), or with rat brain homogenates **7**, **19** *(49)* and **27** *(30)* redox cycle with concomitant formation of O_2^{-} and, thence, other ROS.

7. POSSIBLE MECHANISMS UNDERLYING THE NEUROTOXICITY OF 5,7-DHT

Presumably owing to its structural similarity to the neurotransmitter 5HT, 5,7-DHT is efficiently taken up into and concentrated within serotonergic nerve terminals *(20)*. The intraneuronal fate of 5,7-DHT remains to be determined, although biochemical studies in vivo *(44)* and in vitro *(47)* suggest that it is oxidized. Autoxidation and many in vitro enzyme-mediated oxidations of 5,7-DHT give **7** and **19** as the major reaction products *(26,30,48)*. Both **7** and **19** are toxic (lethal) when administered into the ventricular system of mice having LD_{50} values of 30 and 25 µg, respectively *(48)*. Thus, **7** and **19** are appreciably more toxic than 5,7-DHT (LD_{50} = 52 µg in mouse) *(50)*. However, both **7** and **19** evoke a quite different neurobehavioral response than 5,7-DHT *(48)*. Furthermore, the neurotransmitter/metabolite changes that occur in mouse brain following intracerebral administration of **7** or **19** are transient and quite different from those evoked by 5,7-DHT *(48)*. In particular, neither **7** nor **19** evokes the profound and long-lasting depletion of 5HT that occurs following administration of 5,7-DHT. Taken together, therefore, it appears clear that if 5,7-DHT is oxidized in the brain, it cannot occur to any extent in the extraneuronal space, i.e., oxidation of this neurotoxin must occur intraneuronally. Intraneuronal autoxidation of 5,7-DHT, perhaps catalyzed by mitochondria, would be expected to yield **7** and **19** as the major initial metabolites *(26,28,30,44,45,48)*. This oxidation would provide an initial route for intraneuronal formation of O_2^- and H_2O_2, which in the presence of trace levels of transition metal ions, particularly low-mol-wt Fe^{2+} species, can generate highly cytotoxic HO· by well-known Fenton and Haber-Weiss chemistry *(51)*. However, intraneuronal redox cycling reactions of **7** and **19** would be expected to provide yet another source of ROS (Scheme 6). Indirect support for the preceding hypothesis is provided from the observation that HO· scavengers provide some protection for sympathetic neurons against destruction by 5,7-DHT *(25)*. Increased consumption of molecular oxygen as a result of both the intraneuronal autoxidation of 5,7-DHT and redox cycling of its putative metabolites **7** and **19** would also be expected to lead to hypoxia, a condition that evokes gross changes in intermediary metabolism and cellular damage *(52)*. A further, potentially cytotoxic mechanism, derives from the ability of **7** to react with GSH, a significant constituent of neuronal axons and terminals *(53)*, to form **27**, which can redox cycle forming ROS (Scheme 6). Intraneuronal formation of **27** would also deplete GSH, which, being a cellular antioxidant, would tend to exacerbate damage caused by ROS. The reaction of **7** with GSH might also indicate that similar reactions occur with protein nucleophiles and, hence, contribute to the neurodegenerative effects of 5,7-DHT.

8. OXIDATION CHEMISTRY AND BIOCHEMISTRY OF 5,6-DHT

The autoxidation of 5,6-DHT at physiological pH ultimately results in formation of a black, very insoluble, melanin-like polymer *(28,54)*. The reactions leading to this polymer are undoubtedly extremely complex, and only recently have even the initial steps been elucidated. The reaction is initiated by direct oxidation of 5,6-DHT by molecular oxygen to form a quinoid intermediate represented by **30a** and **30b** in Scheme 7 *(54,55)*. The byproduct of the reaction is

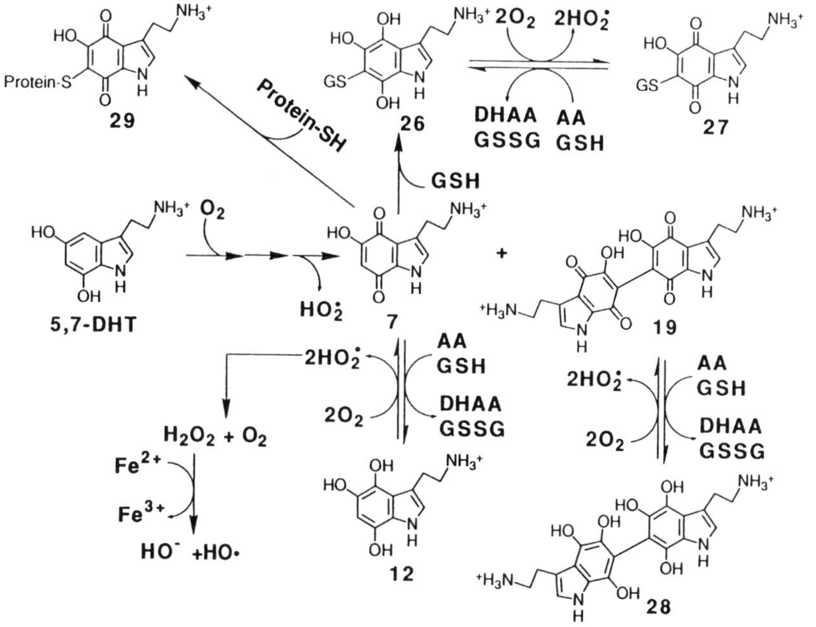

Scheme 6

H_2O_2 and is formed in high yield *(28,55)*. A rapid nucleophilic attack by 5,6-DHT on **30a/30b** then leads to 2,7'-bi(5,6-dihydroxytryptamine) (**32**). The latter dimer can be isolated *(54)*, although it is more rapidly autoxidized than 5,6-DHT *(55)* to give the diquinone **33** (Scheme 7) along with 2 mol of H_2O_2. The accumulation of dimer **32** in the reaction solution during the autoxidation of 5,6-DHT despite its more facile autoxidation has been interpreted to indicate that diquinone **33** chemically oxidizes 5,6-DHT (2 mol) to **30a/30b** such that an autocatalytic cycle is established *(55)*. Catalase, SOD, and iron-complexing agents all inhibit the rate of autoxidation of 5,6-DHT, indicating that H_2O_2, O_2^-, and trace levels of transition metal ions, respectively, probably play active roles in the overall autoxidation. However, when transition metal ions are securely complexed and molecular oxygen is excluded, 5,6-DHT and dimer **32** are not oxidized by H_2O *(55)* as proposed by Klemm et al. *(28)*. Additions of very low concentrations of Fe^{3+}, Cu^{2+}, and O_2^- accelerate the oxidation of 5,6-DHT. Accordingly, it appears that H_2O_2, formed as a byproduct of these autoxidation reactions, undergoes transition metal ion-catalyzed Fenton chemistry to form HO·, which directly oxidizes 5,6-DHT to putative radical **34a/34b** that is then attacked by molecular oxygen to give **30a/30b** and O_2^- which, in turn, further potentiates HO· formation by reducing Fe^{3+} to Fe^{2+}. The latter conclusion is supported by the inhibitory influence of SOD on the autoxidation of 5,6-DHT. A minor side reaction of 5,6-DHT with HO· leads to formation of at

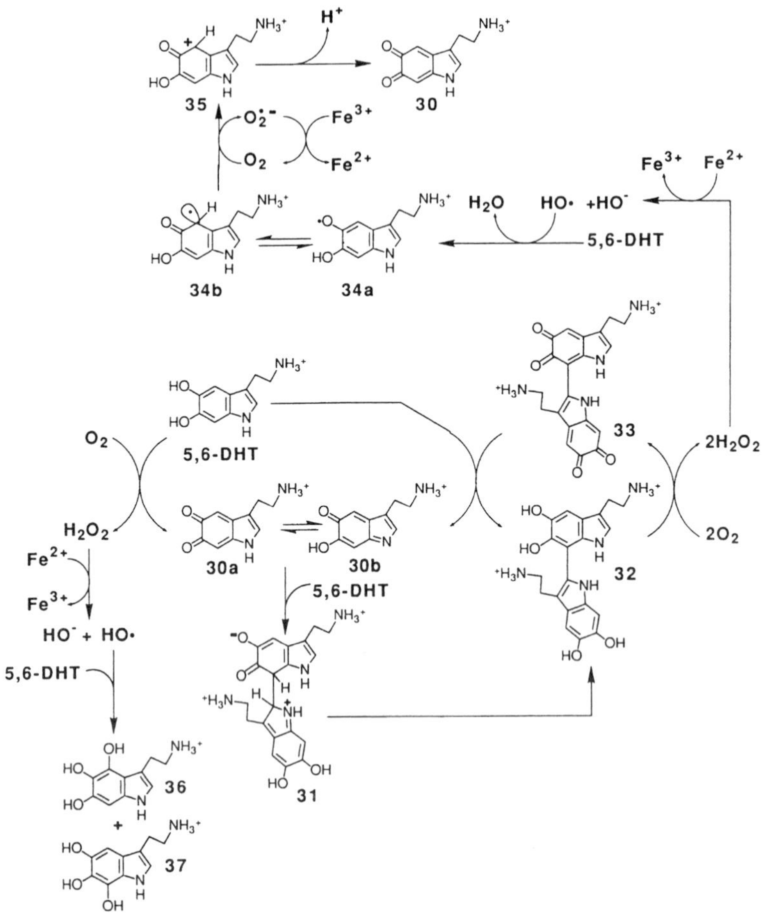

Scheme 7

least two trihydroxytryptamines, perhaps **36** and **37**, that are also very easily autoxidized *(55)*. Diquinone **33** ultimately reacts with **32**, 5,6-DHT and perhaps trihydroxytryptamines leading, via a sequence of coupling and oxidation reactions, to black indolic melanin polymer *(55)*. Enzyme systems, such as tyrosinase/O_2, ceruloplasmin/O_2, and peroxidase/H_2O_2 *(55)* and rat brain mitochondrial preparations *(25,28,55)* catalyze the oxidation of 5,6-DHT to dimer **32** and, ultimately, to indolic melanin.

The antioxidants GSH, CySH, and AA protect 5,6-DHT against autoxidation *(28,56)*. This protective action derives from the ability of these antioxidants to maintain 5,6-DHT in its reduced form, i.e., to reduce *o*-quinone **30**. The result of this process is that these antioxidants are oxidized. Indeed, it has been experimentally demonstrated that 5,6-DHT acts as a catalyst to convert GSH to its oxidized form, glutathione disulfide (GSSG), as conceptualized in Scheme 8 *(56)*. At the stage when GSH is depleted, 5,6-DHT is autoxidized to dimer **32** and thence to melanin polymer by the pathway shown in Scheme 7. However, in the presence of an oxidizing enzyme system, such as ceruloplasmin/O_2 or tyrosinase/O_2 and, presumably, mitochondria *(55)*, the oxidation of 5,6-DHT is greatly accelerated. Under such conditions, significantly higher concentrations of *o*-quinone **30** are generated, and GSH functions not only as an antioxidant (i.e., reducing agent), but also as a nucleophile. Thus, **30** is attacked by GSH to give, initially, 4-*S*-glutathionyl-5,6-dihydroxytryptamine (**38**) which is further oxidized to the corresponding *o*-quinone **39** (Scheme 8) *(56)*. Nucleophilic addition of GSH to **39** gives 4,7-di-*S*-glutathionyl-5,6-dihydroxytryptamine (**40**). Reaction between **40** and *o*-quinone **39** yields 4,7,4-tri-*S*-glutathionyl-2,7-bi(5,6-dihydroxytryptamine) (**42**), whereas attack by **38** on **39** yields 4,4-di-*S*-glutathionyl-2,7-bi(5,6-dihydroxytryptamine) (**41**) *(56)*. 5,6-DHT also catalyzes the autoxidation of CySH to cystine *(57)*. In the presence of oxidative enzymes, however, CySH initially attacks *o*-quinone **30** to give 4-*S*-cysteinyl-5,6-dihydroxytryptamine (**43**), which subsequently oxidatively polymerizes *(57)*.

9. POSSIBLE MECHANISMS UNDERLYING THE NEUROTOXICITY OF 5,6-DHT

Similar to 5,7-DHT, the selectivity of 5,6-DHT derives from the facile uptake of this neurotoxin into serotonergic nerve terminals *(20)*. The intraneuronal fate of 5,6-DHT is largely unknown, and hence, hypotheses concerning the mechanisms that underlie its neurotoxicity are based to a considerable extent on the results of in vitro studies. Experiments with radiolabeled 5,6-DHT provide some support for the conclusion that its autoxidation products undergo covalent binding with protein nucleophiles both in vitro *(46)* and in vivo *(44)*. However, autoxidation of 5,6-DHT eventually leads to formation of an extremely insoluble, high-mol-wt, melanin-like polymer. Thus, it is very likely that a considerable fraction of the radioactivity assumed to be protein-bound could, in fact, have been derived to some appreciable extent from melanin polymer. In an attempt to probe the role of covalent binding of the autoxidation product(s) of 5,6-DHT in expressing the neurodegenerative effects of this neurotoxin, Sinhababu et al. *(58)* investigated the activity of a number of analogs that were methylated at the C(4)- and/or C(7)-positions. This strategy was designed to block either simultaneously or independently the sites believed to be the loci of attack by protein nucleophiles on putative *o*-quinone intermediate **30** (Scheme 7). Substitution of methyl groups had no deleterious effects on the cytotoxicity of 5,6-DHT, although some reduction in uptake affinity was noted compared to that of the unsubstituted compound. Furthermore, 4,7-dimethyl-5,6-DHT was found to be at least 50 times more cytotoxic than 5,6-DHT. Since both of the electrophilic sites in the *o*-quinone proximate autoxidation product of

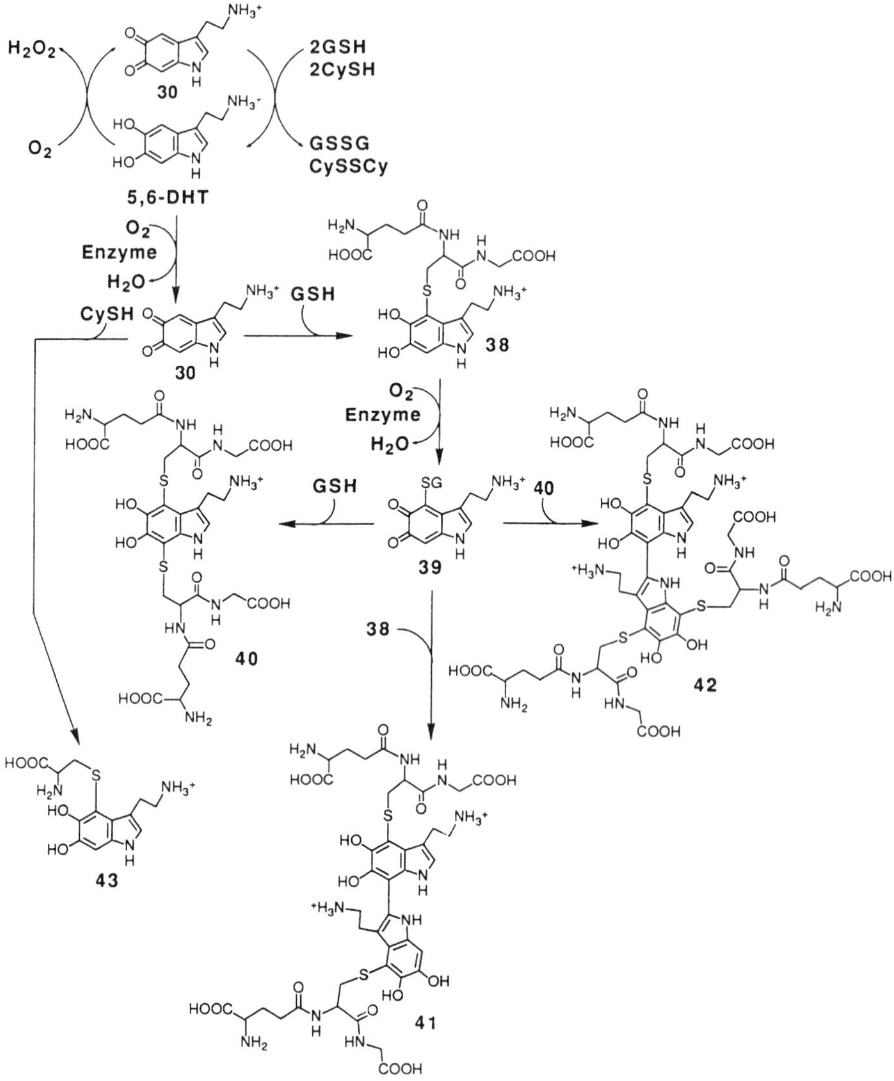

Scheme 8

4,7-dimethyl-5,6-DHT are blocked, the latter observation tends to indicate that covalent binding of protein is not essential in expressing the cytotoxic properties of either this compound or 5,6-DHT itself. However, unlike 5,6-DHT and its 4-methyl- and 7-methyl-derivatives, 4,7-dimethyl-5,6-DHT did not appear to polymerize on autoxidation *(58)*. Thus, in the presence of intraneuronal antioxidants and molecular oxygen, redox cycling of the latter compound and its *o*-quinone with concomitant formation of cytotoxic ROS could in principle occur indefinitely and, perhaps, account for its enhanced neurotoxic properties.

The catalytic effect of 5,6-DHT autoxidation on the oxidation of GSH might also play an important role in the neurodegenerative action of this compound. Thus, once sequestered within a serotonergic neuron, it could deplete endogenous GSH and other intraneuronal antioxidants with concomitant formation of ROS. Together, such 5,6-DHT-mediated depletion of intraneuronal antioxidants and generation of ROS would be expected to result in oxidative damage to cellular lipids, proteins, and DNA.

Based on the results of in vitro experiments, its seems unlikely that intraneuronal autoxidation of 5,6-DHT would proceed at a rate necessary to generate levels of *o*-quinone **30** that would alkylate protein nucleophiles to any significant extent *(56,57)*. However, it has been established that rat brain mitochondria *(25,28,55)* and a number of enzyme systems *(55)* catalyze the oxidation of 5,6-DHT. Under these more rapid oxidation conditions, sufficient concentrations of **30** are generated such that nucleophilic addition reactions occur, for example, with GSH *(56)* and CySH *(57)* (Scheme 8). Formation of compounds, such as **40–42** (Scheme 8), when 5,6-DHT is incubated with oxidative enzymes and GSH provides important models for the ability of the oxidation products of this neurotoxin not only to alkylate but also to crosslink proteins. In vivo, such reactions with essential intraneuronal proteins could well have lethal consequences.

10. ENDOGENOUS FORMATION OF DIHYDROXYTRYPTAMINE NEUROTOXINS

The fact that 5,6-DHT and 5,7-DHT evoke the selective degeneration of serotonergic neurons (and with certain dose regimens, catecholaminergic neurons) and their close structural similarity to 5HT have prompted much speculation that aberrant oxidation of the latter neurotransmitter might generate these neurotoxins, which in turn might play roles in a number of neurodegenerative brain disorders. Similarly, there has been much speculation that endogenous formation of 6-OHDA from the neurotransmitter DA might also be of relevance in certain neurodegenerative processes. There have been a number of recent studies concerning the fundamental oxidation of chemistry of 5HT at physiological pH. These have included the electrochemically driven *(59)*, enzyme-mediated *(60)*, and HO·-mediated *(61)* oxidations in addition to the autoxidation of this neurotransmitter *(62)*. Formation of 5,6-DHT has only been observed in the HO·-mediated oxidation of 5HT at pH 7.4 *(61)*. 5,7-DHT has not been observed as a product of any of these in vitro oxidation reactions. Interestingly, the catecholaminergic neurotoxin 6-OHDA is formed as a product of the in vitro HO·-mediated oxidation of DA at neutral pH *(63)*. It is therefore of potential relevance that there is strong evidence for oxygen radical species being directly or indirectly involved in a number of neurodegenerative brain disorders, such as

Alzheimer's disease (AD) and ischemia-reperfusion (stroke) *(64,65)*. Similarly, the neurodegeneration evoked by methamphetamine (MA) and related amphetamine drugs also appears to be related to the generation of abnormal levels of oxygen radical species *(66–68)*. Furthermore, certain serotonergic pathways undergo severe degeneration in AD *(69–72)*, and following an ischemic insult *(73)* or a neurotoxic dose of MA *(74–76)*. Although it is widely believed that ROS evoke neuronal damage to membrane lipids, proteins, and DNA *(77,78)*, from a chemical perspective, it is extremely unlikely that 5HT or the catecholaminergic neurotransmitters DA and NE would escape oxidation by oxygen radical species. In particular, the unsheltered (i.e., cytoplasmic) pools of these neurotransmitters would be expected to be very susceptible to oxygen radical-mediated oxidation. It is of interest, therefore, that following neurotoxic doses of MA *(79)* or *p*-chloroamphetamine *(80)* to the rat, 5,6-DHT has been detected in the cortex and hippocampus. Similarly, 6-OHDA has been detected in the rat striatum shortly following a single, large dose of MA *(81)*. These observations suggest that 5HT and DA are oxidized by HO· *(62,63)*, which is formed as a result of MA administration. Furthermore, it has been proposed *(79–81)* that 5,6-DHT and 6-OHDA are responsible for the MA-induced degeneration of serotonergic and dopaminergic nerve terminals, respectively. However, subsequent efforts to replicate the detection of 5,6-DHT and 6-OHDA in rat brain following MA administration have either been unsuccessful *(82–84)* or difficult *(76)*. Thus, at this time, endogenous formation of 5,6-DHT (and 6-OHDA) in 5HT- (and DA-) rich regions of the brain as a result of MA administration or in other neurodegenerative brain disorders that involve the degeneration of serotonergic (and dopaminergic) neurons is somewhat questionable.

11. SUMMARY

Since their discovery more than 20 years ago, 5,6-DHT and 5,7-DHT have become very important tools for the selective lesioning of serotonergic neuronal pathways, and as such, are widely employed in neurochemical, neurobiological, and neurobehavioral research. However, neither 5,6-DHT nor 5,7-DHT is a specific serotonergic neurotoxin, and under appropriate conditions, can also mediate damage to catecholaminergic and, perhaps, other pathways. Thus, caution must always be exercized in the interpretation of results based on the assumption that 5,6-DHT or 5,7-DHT has damaged only serotonergic pathways.

The fact that oxygen radical species (oxidative stress) appear to play important roles in a number of neurodegenerative brain disorders and that serotonergic neurons are often vulnerable clearly raises the possibility that dihydroxytryptamine neurotoxins could be formed by oxidation of 5HT. The reports that 5,6-DHT (and 6-OHDA) are formed in rat brain following administration of amphetamines provided very plausible explanations for the degeneration of serotonergic (and dopaminergic) pathways evoked by these drugs. The initial excitement generated by these reports, however, has been tempered by the difficulties in duplicating these results.

REFERENCES

1. Tranzer, J. P. and Thoenen, H. (1968) An electron-microscopic study of selective acute degeneration of sympathetic nerve terminals after administration of 6-hydroxydopamine. *Experientia* **24**, 155–156.
2. Ungerstedt, U. (1968) 6-Hydroxydopamine-induced degeneration of central monoamine neurons. *Eur. J. Pharmacol.* **5**, 107–110.
3. Uretsky, N. J. and Iversen, L. (1969) Effects of 6-hydroxydopamine on noradrenaline-containing neurons in the rat brain. *Nature* **221**, 557–559.
4. Baumgarten, H. G., Björklund, A., Lachenmayer, L., Nobin, A., and Stenevi, U. (1971) Long-lasting selective depletion of brain serotonin by 5,6-dihydroxytryptamine. *Acta Physiol. Scand. Suppl.* **373**, 1–16.
5. Jonsson, G., Fuxe, K., Hamberger, B., and Hokfelt, T. (1969) 6-Hydroxytryptamine. Tool in monoamine fluorescence histochemistry. *Brain Res.* **13**, 190–194.
6. Baumgarten, H. G. and Schlossberger, H. G. (1973) Effects of 5,6-dihydroxytryptamine on brain monoamine neurons in the rat, in *Serotonin and Behavior* (Barchas, J. and Usdin, E., eds.), Academic, New York, pp. 209–224.
7. Björklund, A., Nobin, A., and Stenevi, U. (1973) Effects of 5,6-dihydroxytryptamine on nerve terminal serotonin and serotonin uptake in the rat brain. *Brain Res.* **53**, 117–127.
8. Daly, J. W., Fuxe, K., and Jonsson, G. (1973) Effects of intracerebral injections of 5,6-dihydroxytryptamine on central monoamine neurons. Evidence for selective degeneration of central 5-hydroxytryptamine neurons. *Brain Res.* **49**, 476–482.
9. Baumgarten, H. G., Evetts, K. D., Holman, R. B., Iversen, L. L., Vogt, M., and Wilson, G. (1972) Effects of 5,6-dihydroxytryptamine on monoaminergic neurons in the central nervous system of the rat. *J. Neurochem.* **19**, 1587–1597.
10. Baumgarten, H. G., Björklund, A., Holstein, A. F., and Nobin, A. (1972) Chemical degeneration of indoleamine axons in rat brain by 5,6-dihydroxytryptamine. Ultrastructural study. *Z. Zellforsch.* **129**, 259–271.
11. Björklund, A., Nobin, A., and Stenevi, U. (1973) Use of neurotoxic dihydroxytryptamines as tools for morphological studies and localized lesioning of central indol amine neurons. *Z. Zellforsch.* **145**, 479–501.
12. Baumgarten, H. G. and Lachenmeyer, L. (1972) 5,7-Dihydroxytryptamine. Improvement in chemical lesioning of indoleamine neurons in the mammalian brain. *Z. Zellforsch.* **135**, 399–414.
13. Baumgarten, H. G., Björklund, A., Lachenmayer, L., and Nobin, A. (1973) Evaluation of the effects of 5,7-dihydroxytryptamine on serotonin and catecholamine neurons in the CNS. *Acta Physiol. Scand. Suppl.* **391**, 1–19.
14. Horn, A., Baumgarten, H. G., and Schlossberger, H. G. (1973) Inhibition of the uptake of 5-hydroxytryptamine, noradrenaline, and dopamine into rat brain homogenates by various hydroxylated tryptamines. *J. Neurochem.* **21**, 233–236.
15. Breese, G. R. and Cooper, B.R. (1975) Biochemical and behavioral interactions of 5,7-dihydroxytryptamine with various drugs when administered intracisternally to adult and developing rats. *Brain Res.* **98**, 517–527.
16. Sachs, C. and Jonsson, G. (1975) Mechanisms of action of 6-hydroxydopamine. *Biochem. Pharmacol.* **24**, 1–8.

17. Wuttke, W., Björklund, A., Baumgarten, H. G., Lachenmayer, L., Fenske, M., and Klemm, H. P. (1977) De- and regeneration of brain serotonin neurons following 5,7-dihydroxytryptamine treatment: Effects on serum LH, FSH and prolactin levels in male rats. *Brain Res.* **134,** 317–331.

18. Björklund, A. and Baumgarten, H. G. (1975) 5,7-Dihydroxytryptamine: improvement of its selectivity for serotonin neurons in the CNS by pretreatment with desipramine. *J. Neurochem.* **24,** 833–835.

19. Gerson, S., Baldessarini, R. J., and Wheeler, S. C. (1974) Biochemical effects of dihydroxylated tryptamines on central indoleamine neurons. *Neuropharmacology.* **13,** 987–1004.

20. Baumgarten, H. G., Klemm, H. P., Lachenmayer, L., and Schlossberger, H. G. (1978) Effect of drugs on the distribution of [14C]-5,6-dihydroxytryptamine and [14C]-5,7-dihychoxytryptamine in rat brain. *Ann. NY Acad. Sci.* **305,** 107–118.

21. Rotman, A., Daly, J. W., and Creveling, C. R. (1976) Oxygen-dependent reaction of 6- hydroxydopamine, 5,6-dihydroxytryptamine, and related compounds with proteins in vitro: a model for cytotoxicity. *Mol. Pharmacol.* **12,** 877–899.

22. Heikkila, R. E. and Cohen, G. (1973) 6-Hydroxydopamine. Evidence for superoxide radical as an oxidative intermediate. *Science* **181,** 456–457.

23. Cohen, G. and Heikkila, R. E. (1974) The generation of hydrogen peroxide, superoxide radical, and hydroxyl radical by 6-hydroxydopamine, dialuric acid, and related cytotoxic agents. *J. Biol. Chem.* **249,** 2447–2452.

24. Schlossberger, H. G. (1978) Synthesis and chemical properties of some indole derivatives. *Ann. NY Acad. Sci.* **305,** 25–35.

25. Cohen, G. and Heikkila, R. E. (1978) Mechanisms of action of hydroxylated phenylethylamine and indoleamine neurotoxins. *Ann. NY Acad. Sci.* **305,** 74–84.

26. Tabatabaie, T., Wrona, M. Z., and Dryhurst, G. (1990) Autoxidation of the serotonergic neurotoxin 5,7-dihydroxytryptamine. *J. Med. Chem.* **33,** 667–672.

27. Wrona, M. Z., Lemordant, D., Blank, C. L., and Dryhurst, G. (1986) Oxidation of 5-hydroxytryptamine and 5,7-dihydroxytryptamine. A new oxidation pathway and formation of a novel neurotoxin. *J. Med. Chem.* **29,** 499–505.

28. Klemm, H. P., Baumgarten, H. G., and Schlossberger, H. G. (1980) Polarographic measurements of spontaneous and mitochondria-promoted oxidation 5,6- and 5,7-dihydroxytryptamine. *J. Neurochem.* **35,** 1400–1408.

29. Sinhababu, A. K. and Borchardt, R. T. (1985) Mechanism and products of autoxidation of 5,7-dihydroxytryptamine. *J. Am. Chem. Soc.* **107,** 7618–7627.

30. Tabatabaie, T. and Dryhurst, G. (1992) Chemical and enzyme-mediated oxidation of the serotonergic neurotoxin 5,7-dihydroxytryptamine: Mechanistic insights. *J. Med. Chem.* **35,** 2261–2273.

31. Finkelstein, E., Rosen, G. M., and Raukman, E. J. (1980) Spin trapping. Kinetics of the reaction of superoxide and hydroxyl radicals with nitrones. *J. Am. Chem. Soc.* **102,** 4994–4999.

32. Sayo, H. and Hosokawa, M. (1983) Kinetics of the organic hydroperoxide-supported oxidation of aminopyrine catalyzed by catalase. *Chem. Pharm. Bull.* **30,** 2161–2168.

33. Kadlubar, F. F., Morton, K. C., and Ziegler, D. M. (1973) Microsomal-catalyzed hydroperoxide-dependent C-oxidation of amines. *Biochem. Biophys. Res. Commun.* **54,** 1255–1261.

34. Howard, J. A. (1973) Homogeneous liquid-phase autoxidations, in *Free Radicals,* vol. 2 (Kochi, J. K., ed.), John Wiley, New York, pp. 3–62.
35. Sheldon, R. A. and Kochi, J. K. (eds.) (1981) Introduction to metal-catalyzed oxidations, in *Metal Catalyzed Oxidation of Organic Compounds,* Academic, New York, pp. 1–68.
36. Nonhebel, D. C., Tedder, J. M., and Walton, J. C. (1979) Autoxidation, in *Radicals,* (Nonhebel, D. C., Tedder, J. M., and Walton, J. C. eds.), Cambridge University Press, New York, pp. 150–159.
37. Lynch, R. E. and Fridovich, I. (1978) Effects of superoxide on the erythrocyte membrane. *J. Biol. Chem.* **253,** 1838–1845.
38. Peters, J. W. and Foote, C. S. (1976) Chemistry of superoxide ion. II. Reaction with hydroperoxides. *J. Am. Chem. Soc.* **98,** 873–875.
39. Sawyer, D. T., Gibian, M. J., Morrison, M. M., and Seo, E. T. (1978) On the chemical reactivity of superoxide ion. *J. Am. Chem. Soc.* **100,** 627,628.
40. Chin, D. H., Chiericato, E. J., Nanni, E. J., and Sawyer, D. T. (1982) Proton-induced disproportionation of superoxide ion in aprotic media. *J. Am. Chem. Soc.* **104,** 1296–1299.
41. Nanni, E. J., Stallins, M. D., and Sawyer, D. T. (1980) Does superoxide ion oxidize catechol, α-tocopherol, and ascorbic acid by direct electron transfer? *J. Am. Chem. Soc.* **102,** 4481–4485.
42. Behar, D., Czapski, G., Rabini, J., Dorfman, L. M., and Schwartz, H. A. (1970) Acid dissociation constant and decay kinetics of the perhydroxyl radical. *J. Phys. Chem.* **74,** 3209–3213.
43. Møllgard, K., Lundberg, J. J., Lachenmayer, L., Wiklund, L., and Baumgarten, H. G. (1978) Morphologic consequences of serotonin neurotoxin administration: Neuron-target cell interaction in the rat subcommissural organ. *Ann. NY Acad. Sci.* **305,** 262–288.
44. Baumgarten, H. G., Klemm, H. P., Lachenmayer, L., Björklund, A., Lovenberg, W., and Schlossberger, H. G. (1978) Mode and mechanism of action of neurotoxic indoleamines: A review and progress report. *Ann. NY Acad. Sci.* **305,** 3–24.
45. Tamir, H. and Rapport, M. M. (1978) Effects of neurotoxins *in vitro* on the binding of serotonin to serotonin-binding protein. *Ann. NY Acad. Sci.* **305,** 85–95.
46. Creveling, C. R. and Rotman, A. (1978) Mechanism of Action of Dihydroxytryptamines. *Ann. NY Acad. Sci.* **305,** 57–73.
47. Klemm, H. G. and Baumgarten, H. G. (1978) Interaction of 5,6-and 5,7-dihydroxytryptamine with tissue monoamine oxidase. *Ann. NY Acad. Sci.* **305,** 36–56.
48. Tabatabaie, T., Goyal, R. N., Blank, C. L., and Dryhurst, G. (1993) Further insights into the molecular mechanisms of action of the serotonergic neurotoxin 5,7-dihydroxytryptamine. *J. Med. Chem.* **36,** 229–236.
49. Frank, D. M., Arora, P. K., Blumer, J. L. and Sayre, L. M. (1987) Model study on the bioreduction of paraquat, MPP+, and analogs. Evidence against a "redox cycling" mechanism in MPTP neurotoxicity. *Biochem. Biophys. Res. Commun.* **147,** 1095–1104.
50. Massotti, M., Scotti de Carolis, A., and Longo, V. G. (1974) Effects of three dihydroxylated derivatives of tryptamine on the behavior and on brain amine content in mice. *Pharmacol. Biochem. Behav.* **2,** 769–775.
51. Halliwell, B. (1992) Reactive oxygen species and the central nervous system. *J. Neurochem.* **59,** 1609–1623.

52. Kappus, H. (1986) Overview of enzyme systems involved in bioreduction of drugs and redox cycling. *Biochem. Pharmacol.* **35**, 1–6.

53. Slivka, A., Mytilineou, C., and Cohen, G. (1987) Histochemical evaluation of glutathione in brain. *Brain Res.* **409**, 275–284.

54. Singh, S., Jen, J-F., and Dryhurst, G. (1990) Autoxidation of the indolic neurotoxin 5,6-dihydroxytryptamine. *J. Org. Chem.* **55**, 1484–1489.

55. Singh, S. and Dryhurst, G. (1990) Further insights into the oxidation chemistry and biochemistry of the serotonergic neurotoxin 5,6-dihydroxytryptamine. *J. Med. Chem.* **33**, 3035–3044.

56. Singh, S. and Dryhurst, G. (1991) Reactions of the serotonergic neurotoxin 5,6-dihydroxytryptamine with glutathione. *J. Org. Chem.* **56**, 1767–1773.

57. Singh, S. and Dryhurst, G. (1991) Interactions between 5,6-dihydroxytryptamine and cysteine. *Bioorg. Chem.* **19**, 274–282.

58. Sinhababu, A. K., Ghosh, A. K. and Borchardt, R. T. (1985) Molecular mechanism of action of 5,6-dihydroxytryptamine. Synthesis and biological evaluation of 4-methyl-7-methyl- and 4,7-dimethyl-5,6-dihydroxytryptamines. *J. Med. Chem.* **28**, 1273–1279.

59. Wrona, M. Z. and Dryhurst, G. (1990) Electrochemical oxidation of 5- hydroxytryptamine in aqueous solution at physiological pH. *Bioorg. Chem.* **18**, 291–317.

60. Wrona, M. Z. and Dryhurst, G. (1991) Interactions of 5-hydroxytryptamine with oxidative enzymes. *Biochem. Pharmacol.* **41**, 1145–1162.

61. Wrona, M. Z., Yang, Z., McAdams, M., O'Connor-Coates, S., and Dryhurst, G. (1995) Hydroxyl radical-mediated oxidation of serotonin: Potential insights into the neurotoxicity of methamphetamine. *J. Neurochem.* **64**, 1390–1400.

62. Wrona, M. Z., Yang, Z., Waskiewicz, J., and Dryhurst, G. (1996) Oxygen radical-mediated oxidation of serotonin: Potential relationship to neurodegenerative diseases, in *Neurodegenerative Disease* (Fiskum, G., ed.), Plenum, New York, pp. 285–297.

63. Slivka, A. and Cohen, G. (1985) Hydroxyl radical attack on dopamine. *J. Biol. Chem.* **260**, 15,466–15,472.

64. Floyd, R. A. (1990) Role of oxygen free radicals in carcinogenesis and brain ischemia. *FASEB J.* **4**, 2587–2597.

65. Simonian, N. A. and Coyle, J. T. (1996) Oxidative stress in neurodegenerative diseases. *Annu. Rev. Pharmacol. Toxicol.* **36**, 83–106.

66. DeVito, M. J. and Wagner, G. C. (1989) Methamphetamine-induced neuronal damage: A possible role for free radicals. *Neuropharmacology* **28**, 1145–1150.

67. Cadet, J. L., Sheng, P., Ali, S., Rothman, R., Carlson, E., and Epstein, C. (1994) Attenuation of methamphetamine-induced neurotoxicity in copper/zinc superoxide dismutase transgenic mice. *J. Neurochem.* **62**, 380–383.

68. Kondo, T., Sugita, Y., Kanazawa, A., Ito, T., and Mizuno, Y. (1992) Free dopamine and iron contribute to the hydroxyl radical generation in methamphetamine-induced experimental parkinsonism. *Movement Disord.* **17 (Suppl. 1)**, 73–79.

69. Bowen, D. M., Allen, S. J., Benton, J. S., Goodhart, M. J., Haan, E. A., Palmer, A. M., Sims, N. R., Smith, C. C. T., Spillane, J. A., Esiri, M. M., Neary, D., Snowden, J. S., Wilcock, G. K., and Davison, A. W. (1983) Biochemical assessment of serotonergic and cholinergic dysfunction in cerebral atrophy in Alzheimer's disease. *J. Neurochem.* **41**, 266–272.

70. Mann, D. M. A. (1988) Neuropathological and neurochemical aspects of Alzheimer's disease, in *Psychopharmacology of the Aging Nervous System* (Iverson, L. L., Iversen, S. D., and Snyder, S. H., eds.), Plenum, New York, pp. 1–67.
71. Tejani-Butt, S. M., Yang, J., and Pawlyk, A. C. (1995) Altered serotonin transporter sits in Alzheimer's disease raphe and hippocampus. *NeuroReports* **6**, 1207–1210.
72. Palmer, A. M., Francis, P. T., Benton, J. S., Sims, N. R., Mann, D. M. A., Neary, D., Snowden, J. S., and Bowen, D. M. (1987) Presynaptic serotonergic dysfunction in patients with Alzheimer's disease. *Brain Res.* **48**, 8–15.
73. Weinberger, J., Cohen, G., and Nieves-Rosa, J. (1983) Nerve terminal damage in cerebral ischemia: Greater susceptibility of catecholamine nerve terminals relative to serotonin nerve terminals. *Stroke* **14**, 986–989.
74. Gibb, J. W., Hanson, G. R., and Johnson, M. (1994) Neurochemical mechanisms of toxicity, in *Amphetamine and Its Analogs: Psychopharmacology, Toxicology, and Abuse* (Cho, A. K. and Segal, D. S., eds.), Academic, New York, pp. 269–295.
75. Bowyer, J. F. and Holson, R. R. (1995) Methamphetamine and Amphetamine neurotoxicity, in *Handbook of Neurotoxicology* (Chang, L. W. and Dyer, R. S., eds.), Marcel Dekker, New York, pp. 845–870.
76. Seiden, L. S. and Sabol, K. E. (1995) Neurotoxicity of methamphetamine-related drugs and cocaine, in *Handbook of Neurotoxicology* (Chang, L. W. and Dyer, R. S., eds.), Marcel Dekker, New York, pp. 825–843.
77. Gutteridge, J. M. C. (1996) Hydroxyl radicals, iron, oxidative stress, and neurodegeneration. *Ann. NY Acad. Sci.* **738**, 201–213.
78. Halliwell, B. (1992) Oxygen radicals as key mediators in neurological disease: Fact or fiction? *Ann. Neurol.* **32**, S10–S15.
79. Commins, D. L., Axt, K. J., Vosmer, G., and Seiden, L. S. (1987) 5,6-Dihydroxytryptamine, a serotonergic neurotoxin, is formed endogenously in rat brain. *Brain Res.* **403**, 7–14.
80. Commins, D. L., Axt, K. J., Vosmer, G., and Seiden, L. S. (1987) Endogenously produced 5,6-dihydroxytryptamine may mediate the neurotoxic effects of para-chloroamphetamine. *Brain Res.* **419**, 253–261.
81. Seiden, L. S. and Vosmer, G. (1984) Formation of 6-hydroxydopamine in caudate nucleus of the rat brain after a single large dose of methylamphetamine. *Pharmacol. Biochem. Behav.* **21**, 29–31.
82. Rollema, H., De Vries, J. B., Westerink, B. H. C., Van Putten, F. M., and Horn, A. S. (1986) Failure to detect 6-hydroxydopamine in rat striatum after the dopamine-releasing drugs dexamphetamine, methylamphetamine and MPTP. *Eur. J. Pharmacol.* **132**, 65–69.
83. Karoum, F., Chrapusta, S. J., Egan, M. F., and Wyatt, R. J. (1993) Absence of 6-hydroxydopamine in the rat brain after treatment with stimulants and other dopaminergic agents: A mass fragmentographic study. *J. Neurochem.* **61**, 1369–1375.
84. Yang, Z., Wrona, M. Z., and Dryhurst, G. (1997) 5-Hydroxy-3-ethylamino-2-oxindole is not formed in rat brain following a neurotoxic dose of methamphetamine: evidence that methamphetamine does not induce hydroxyl radical-mediated oxidation of serotonin. *J. Neurochem.* **68**, 1929–1941.

11

Use of 5,6- and 5,7-Dihydroxytryptamine to Lesion Serotonin Neurons

Michael R. Pranzatelli

1. INTRODUCTION

The most potent and useful dihydroxylated tryptamines, 5,6- and 5,7-dihydroxytryptamine (DHT), have been employed as serotonin (5HT) neurotoxins for more than 20 years *(1,2)*. This chapter is a summary of research highlights since 1978, when the topic was reviewed comprehensively in the *Annals of the New York Academy of Sciences*.

2. CHEMISTRY

2.1. Synthesis

DHTs are indole derivatives rendered neurotoxic by the addition of one more hydroxy group on the benzene nucleus of the indole ring than contained by 5HT. The synthesis of DHTs utilizes the corresponding dibenzyloxyindole with protection of the free hydroxyl groups *(3)*.

2.2. Structure–Activity

The numbering system for tryptamine derivatives is shown in Fig. 1. The full action of dihydroxylated analogs requires active uptake and interaction with monoamine-containing neurons *(4)*. Various other ring substitutions do not improve on the ratio of selectivity to potency *(2)*.

3. PHARMACOKINETICS

3.1. Dose

The dose of 5,6- or 5,7-DHT depends on route of injection, age of the animal, and desired degree of 5HT depletion. For adult rats, intracranial doses of 50–200 µg free base are usually dissolved in 0.9% saline with 0.1% ascorbic acid as an antioxidant for an injection volume of 100 µL. For neonatal intracranial injec-

From: Highly Selective Neurotoxins: Basic and Clinical Applications *Edited by: R. M. Kostrzewa.*
Humana Press Inc., Totowa, NJ

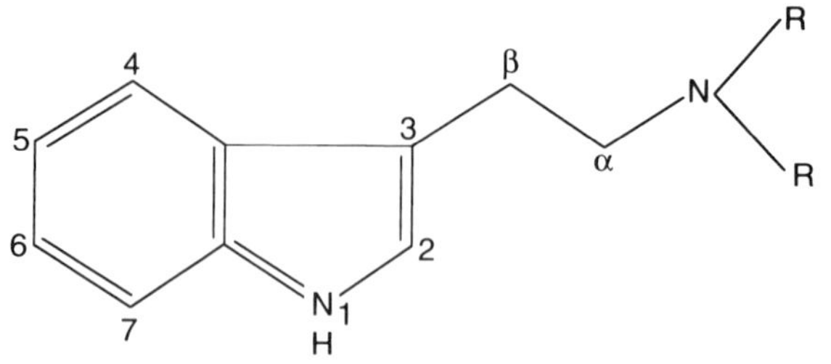

Fig. 1. The numbering system for tryptamine derivatives.

tions, 5–10 µL are used. For systemic injections in neonatal rats, 100 mg of 5,7-DHT are used. The recommended chemical form of 5,7-DHT for animal injections is creatinine sulfate, which is readily water-soluble.

3.2. Route of Administration

Both 5,6- and 5,7-DHT have been administered intracerebroventricularly (icv), intracerebrally, intracisternally (ic), and intrathecally. The regional distributional pattern is similar after administration of either drug, but penetration of [^{14}C]5,7-DHT into brain from icv or ic injection occurs preferentially *(2)*.

3.3. Age-Related Factors

Peripheral injections can be used to induce brain lesions only during the first week of life when 5,7-DHT is able to cross the blood–brain barrier *(5)*. 5,7-DHT is often given intraperitoneally (ip) to rat pups.

4. MECHANISMS OF ACTION

4.1. Uptake

Dihydroxylated tryptamines are taken up by 5HT and NA fibers and produce a fluorescing autoxidation in vivo. They bind rapidly and irreversibly to tissue proteins only in the presence of oxygen. The mechanisms of neurotoxicity of 5,6-DHT and 5,7-DHT differ largely because of differences in the reactivity and nature of the products formed by interaction with molecular oxygen *(2)*. Nucleophilic 4-addition of the *p*-quinoneimine of 5,7-DHT to functionally critical groups in proteins may explain the neurotoxicity of 5,7-DHT *(2)*. The requirement that 5,7-DHT be metabolized by MAO for neurotoxicity to central monoaminergic axons is disputed. Uptake site affinity is not the only determinant of the potency and specificity of neurotoxic indoleamines *(4)*.

4.2. Axotomy/Denervation

A major advantage of dihydroxylated tryptamines is that denervation is chemically induced without the need for electrolytic or knife lesions, or stereotactic injections, all of which would induce glial scar formation. Absence of a glial scar is essential for the regrowth of axons through denervated areas, which has been clearly demonstrated in brainstem after 5,7-DHT *(6,7)*.

4.3. Cell Death/Denervation

Dihydroxylated tryptamines induce dose-dependent cell death at doses of 150 µg or higher through a dying-back phenomenon *(8)*.

4.4. Age-Related Factors

The difference in the neurochemical response of neonatal and adult rats to 5,7-DHT is striking *(6,7)*. If the neonate is injected ic, the degree of 5HT cell death is profound *(9)*, especially in the dorsal or median raphe nuclei, and serotonergic axons do not regenerate, whereas cell death is minimal in rats treated as adults with 5,7-DHT despite significant destruction of terminals *(10)*. Compared to 5,7-DHT lesions in adult rats, neonatal lesions induced significantly greater 5HT depletions in brainstem *(11)*.

5. NEUROCHEMICAL EFFECTS

5.1. Regional Distribution

The regional profile of 5,7-DHT lesions in adult rats is maximal depletion of 5HT in spinal cord and selected forebrain structures, and little effect in diencephalon and midbrain. Multiple 5,7-DHT injections produce more consistent and more complete 5HT depletions than single injections *(12)*. After an icv injection of 75 µg 5,6-DHT, 5HT depletion is maximal in 3–24 h in periventricular areas, whereas in remote areas, such as cortex, maximal depletion occurs in 10 d presumably because of anterograde degeneration of axons lesioned near cell bodies *(7)*.

5.2. Specificity

The greater potency of 5,7-DHT as a neurotoxin for 5HT fibers than NA fibers may be explained by its higher affinity for 5HT transport sites. The affinity of 5,7-DHT for 5HT and NA uptake differed by a factor of about 16 in [^{14}C]5,7-DHT incorporation studies, but was nearly equal in uptake studies according to IC_{50} values *(2,4)*. Use of reuptake blockers, such as desipramine, as a pretreatment protects catecholamine neurons from injury *(13)*. In neonatal rats, MAO inhibitors, such as pargyline, have been used by some investigators instead *(9,10)*. 5,6-DHT better discriminates between 5HT and NA neurons, but is more generally toxic. 5,7-DHT can be given at higher doses and, thereby, achieves greater 5HT depletions *(2)*. 5,6-DHT was less effective and was soon replaced by 5,7-DHT. The small, significant increase in hippocampal NA in rats with neonatal 5,7-DHT lesions made by the ip route is similar to the 20% increase *(14)* found in forebrain and brainstem after icv 5,7-DHT administration in adult rats.

5.3. Cotransmitters

After an icv injection of 200 µg 5,7-DHT in adult rats, thyroid-releasing hormone (TRH) and adrenocorticotrophic hormone (ACTH) are codepleted in brain regions where they are colocalized with 5HT *(15)*. TRH was reduced 80% in lumbar spinal cord, 55% in nucleus accumbens, and 38% in septal nuclei *(15)*. 5,7-DHT lesions evoke no myoclonic supersensitivity to TRH *(16,17)*. After neonatal ic 5,7-DHT injection, TRH content of adult rat spinal cord was reduced 50–65% when 5HT was reduced 90% *(18,19)*. Nearly all substance P- and TRH-immunoreactive cells in ventral pons-medulla, but not other brain areas were destroyed by 50 µg ic 5,7-DHT *(19)*. Substance P content was not altered in any brain region *(19)*. Substance P injected ic in neonatal rats counteracted the 5HT denervation and hyperinnervation *(20)*.

5.4. Species Differences

Guinea pigs tolerated ic 5,7-DHT less well than rats, with a higher mortality, although immediate postinjection convulsions were less severe and did not require phenobarbital prophylaxis *(21)*. Staged lower doses of 5,7-DHT (100–200 µg) were better tolerated than a single high dose of neurotoxin (400 µg). Depletions in the guinea pig were less selective for 5HT using desipramine pretreatment than in the rat. The neurochemical effects of 5,6- and 5,7-DHT in mice are similar to those reported in rats *(22)*.

In pigeons, 600 µg of 5,7-DHT delivered by intra-acqueductal terminal fields rather than brainstem areas showed 5HT depletions *(23)*.

6. BEHAVIORAL EFFECTS

6.1. Locomotor Activity (LMA)

Increased spontaneous LMA in adult rats has been reported 1–3 wk after ic *(13)* and icv *(24)* 5,7-DHT, but locomotor hyperactivity during the first week after 5,7- DHT injection was much greater in magnitude *(25)*. Stereotactic injections of 5,7-DHT in nucleus accumbens septi (bilateral), substantia nigra, or striatum *(24,26)*, but not cortex *(27)* also stimulate LMA and suggest a basal forebrain locus *(28)*. Increased locomotor responses after median raphe nucleus (MRN) lesions, which cause chiefly nocturnal hyperactivity, have been correlated with hippocampal 5HT depletion, suggesting a limbic mechanism *(29)*. Mediation of inhibitory 5HT effects on LMA by ascending 5HT projections and of excitation by descending projections has been proposed *(1)*. Forebrain sites have been implicated in the locomotor response to serotonergic manipulations *(26,29)*.

The acute decline of locomotor hyperactivity after 5,7-DHT lesioning and the disappearance of several other transient motor abnormalities represent a recovery of function. The "recovery" paradigm concerns any CNS insult followed by a period of aberrant behavior and a subsequent return to a normal level of functioning. Of the several proposed mechanisms of recovery *(30)*, functional reorganization and supersensitivity *(31,32)* seem most applicable to the 5,7-DHT lesion in the adult rat *(13,33)*. LMA recovers, whereas supersensitivity to 5-hydroxy-L-tryptophan (L-5-HTP) may persist for at least 12 mo after multiple 5,7-DHT injections. The 5HT syndrome inhibits LMA at the time of its greatest severity (15–30 min), based on the inverse correlation of LMA and 5HT syndrome behaviors. These two behaviors may be mutually exclusive. 5,7-DHT lesions also induce a failure of habituation to the effects of L-5-HTP *(34)*. Increased tilt cage activity also has been reported after 5,7-DHT in adult rats *(35)*.

6.2. Seizure Activity

5,7-DHT induced transient postinjection convulsions in rats injected ic, but not ip *(36)*. Adult rats injected with 5,7-DHT (200 µg) icv exhibited increased sensitivity to seizures induced by the convulsant pentylenetetrazol or maximal electroshock 1 mo after treatment *(37)*.

6.3. 5HT Syndrome

DHTs rapidly induce behavioral supersensitivity to 5HT precursors and agonists, which is long-lasting if not permanent *(38,39)*. In 5,7-DHT-lesioned rats, L-5-HTP evokes complex motor behaviors (stereotyped lateral head movements, hindlimb abduction, rigid arching of back), autonomic responses (salivation, ejaculation, piloerection), and myoclonus of several types (forepaw, truncal, convulsive) *(1,37,40)*. The behaviors are dose-related and last for about 60–90 min following ip L-5-HTP injection. The brain regions responsible for "serotonergic behaviors" have not been identified, but indirect evidence suggests brainstem and spinal cord *(41)*.

6.4. Species Differences

In the guinea pig, 5,7-DHT lesions significantly increased the severity of myoclonic response to L-5-HTP (150 mg/kg) compared to rats, resulting in lethal convulsions *(21)*. Guinea pigs treated with 5,7-DHT did not develop spontaneous myoclonus or, when treated with L-5-HTP, other serotonergic behaviors, such as lateral head weaving, hindlimb abduction, and forepaw tapping. These data suggest that the functional integrity of serotonergic neurons is not requisite for the expression of myoclonus induced by L-5-HTP in the guinea pig.

6.5. Other Actions

5,6- and 5,7-DHT lesions also affect sleeping behavior, feeding and drinking, sexual behavior, aggressive behavior, responses to novel or noxious stimuli, and temperature regulation *(42)*.

7. EFFECTS ON 5HT RECEPTORS

7.1. Receptor Binding Studies

Dihydroxylated tryptamines are the only 5HT neurotoxins known to induce behavioral supersensitivity to 5HT agonists *(13,39)*. It has been suggested that proliferation of postsynaptic 5HT receptors (denervation supersensitivity) *(43)* after 5,7-DHT lesions is the mechanism of behavioral supersensitivity to 5HT agonists *(13,33)*. An extension of this hypothesis is that the receptor changes also underlie recovery from injury *(31)*. Many studies have been done in attempting to test that hypothesis without entirely consistent results *(44–63)* (Table 1). The large number of 5HT receptor subtypes, most of which had not been discovered at the time much of the receptor binding studies were done, probably plays a role in apparent discrepancies.

7.1.1. Adult 5,7-DHT Lesions

Route of 5,7-DHT injection in adult rats appears to be one crucial determinant in studies of 5HT receptor changes after 5,7-DHT. Intracerebral injection with selective destruction of 5HT fibers of the fornix fimbria and cingulum bundle *(64)* was associated with increased $5HT_1$ binding sites *(44)*, as was intrahemispheric injection *(47)*. Most studies of effects of 5,7-DHT on $5HT_2$ receptors have been negative. In adult rats after 5,7-DHT lesions made as adults, upregulation of $5HT_{2C}$ sites was observed using $[^3H]5HT$. However, $5HT_{1A}$ and $5HT_{1B}$ sites were likewise increased *(48)*. A 5,7-DHT-induced increase in B_{max} in hippocampus was found using $[^3H]$mianserin, but not $[^3H]$ketanserin *(57)* prior to the identification of the $5HT_{2C}$ binding site. Although mianserin binds to $5HT_{2C}$ receptors, $5HT_{2C}$ sites would not be measured under those assay conditions *(65)*.

7.1.2. Neonatal 5,7-DHT Lesions

Prior to the recognition of multiple 5HT receptors and agonist-labeled $5HT_2$ receptors, unsubtyped $5HT_1$ and $5HT_2$ binding studies showed no changes *(66,67)*. Moreover, no changes were found when 5,7-DHT was injected ic instead of ip *(68)*. Spinal $[^3H]$mesulergine-labeled $5HT_{2C}$ sites measured 4 or 14 wk after 5,7-DHT injection exhibited a significant increase (+35% for ip and 27% for ic) in B_{max} without changes in K_d or n_H *(69)*. Spinal 5HT content was significantly reduced (−80 to 89%) by either route of 5,7-DHT injection. There were no changes in $5HT_{2C}$ sites in the brain *(65)*. Spinal $5HT_{2C}$ receptor upregulation may contribute to the behavioral supersensitivity to L-5-HTP in rats with 5,7-DHT lesions, but it does not explain the behavioral recovery only after ip 5,7-DHT injection.

7.2. Quantitative Autoradiography

7.2.1. Adult 5,7-DHT Lesions

There is consensus in autoradiographic studies of 5HT receptors in adult rats that dorsal raphe $5HT_{1A}$ sites decreased *(49–51)*. Regarding $5HT_{1B}$ sites,

Table 1
Summary of 5HT Receptor Changes in Adult Rats Given 5,7-DHT

Receptor	5,7-DHT route	Radioligand	Change	Brain region	Reference
5-HT$_1$	icer	[^{3}H]5HT	Increase (38%)	Midbrain	*(44)*
	ic	[^{3}H]5HT	None	Multiple	*(45)*
	icv	[^{3}H]5HT(ar)	Decrease	Dentate gyrus, CA3/4, dosal raphe	*(46)*
	icer	[^{3}H]5HT	Increase (39%)	Hippocampus	*(47)*
5HT$_{1A}$	icv	[^{3}H]5-HT	Increase	Cerebral cortex	*(48)*
	icv	[^{3}H]8-OH-DPAT(ar)	Decrease	Dorsal and median Raphe and CA$_2$/CA$_3$	*(49)*
	icv	[^{3}H]8-OH-DPAT(ar)	Decrease	Raphe nuclei	*(50)*
	icer	[^{3}H]8-OH-DPAT(ar)	Decrease	Dorsal raphe (−60%)	*(51)*
5-HT$_{1B}$	icv	[^{3}H]5HT	Increase (56%)	Cerebral cortex	*(48)*
	icer	[^{3}H]5HT(ar)	Decrease (37%)	Substantia nigra	*(52)*
	icv	[^{3}H]5HT(ar)	Increase (40%)	Substantia nigra	*(53)*
	icv	[^{125}I]ICYP	Increase (40%)	Striatum	*(54)*
	ic	[^{3}H]5HT	None	Spinal cord	*(55)*
	it	[^{3}H]5HT	Increase (42%)[a]	Spinal cord	*(56)*
5HT$_{2A}$	icv	[^{3}H]ketanserin	None	Hippocampus	*(57)*
	icv	[^{3}H]mianserin	Increase (30%)	Hippocampus	*(57)*
	icv	[^{3}H]ketanserin	Increase	Brain	*(58)*
	icv	[^{3}H]ketanserin	None	Cerebral cortex	*(59)*
	icv	[^{3}H]ketanserin	None	Cerebral cortex	*(60)*
	icv	[^{3}H]ketanserin	None	Multiple	*(61)*
	icer	[^{3}H]spiroperidol	None	Midbrain	*(44)*
	icv	[^{3}H]ketanserin	None	Multiple	*(46)*
	ic	[^{3}H]ketanserin	None	Multiple	*(62)*
5HT$_{1E}$	icv	[^{3}H]5HT	None	Cortex and basal ganglia	*(63)*
5HT$_{2C}$	icv	[^{3}H]5HT	Increase	Cerebral cortex	*(48)*
5HT$_{UT}$	icv	[^{3}H]CN-IMI	Decrease	All regions	*(49)*

[a]Not a saturation study.
icer, intracerebral; ar, autoradiography; 5-HT$_{UT}$, 5HT uptake site.

increases or decreases in substantia nigra have been reported, but using different routes of 5,7-DHT injection *(52,53)*. Reduction of 5HT uptake sites has been demonstrated *(49)*. No changes in 5HT$_2$ sites were found *(46)*.

7.2.2. Neonatal 5,7-DHT Lesions

5HT$_{1A}$ binding sites labeled autoradiographically with [^{3}H]8-OH-DPAT 4 mo after 5,7-DHT in the first week postnatal significantly decreased (–87%) in the dorsal raphe after ic-made 5,7-DHT lesions *(70)*. No reductions were found after 5,7-DHT lesions made by ip injection compared to controls, but rather a 246% increase in area of 5HT$_{1A}$ binding extending from the dorsal raphe was observed. These changes in 5HT$_{1A}$ binding sites in the dorsal raphe in the chronic phase of 5,7-DHT lesions may contribute to the different behavioral consequences of the route of neonatal 5,7-DHT injection.

For 5HT$_{1B}$ sites labeled autoradiographically with [^{3}H]5-HT in the same rats, changes were confined to the subiculum and substantia nigra, regions with the most 5HT$_{1B}$ specific binding and projection areas of structures with high mRNA expression *(71)*. Both routes of 5,7-DHT injection were associated with increases in specific binding in subiculum (24% for ip and 47% for ic route). In contrast, there was a 32% increase in specific binding in substantia nigra in rats with lesions made ic, but not ip. No significant differences were found in 22 other brain areas. These data indicate differential effects of route of neonatal 5,7-DHT injections on plasticity of 5HT$_{1B}$ receptor recognition sites and suggest the presence of a subpopulation of postsynaptically located 5HT$_{1B}$ sites, which increases in response to denervation. The data also suggest that sprouting of 5HT neurons after neonatal 5,7-DHT lesions does not involve 5HT$_{1B}$ sites.

7.3. 5HT Receptor Signal Transduction

5HT-stimulated phosphoinositide hydrolysis is mediated by the 5HT$_{2A}$ receptor in cortex, but not in brainstem *(61)*. 5,7-DHT lesions in adult rats did not affect basal levels of [^{3}H]inositol phosphate accumulation, but significantly increased 5HT-stimulated [^{3}H]inositol phosphate accumulation in the brainstem (+27%) and (+23%) *(62)*. Because brainstem rather than cortex is involved in L-5-HTP-evoked myoclonus, increased 5-HT-stimulated phosphoinositide hydrolysis in brainstem following 5,7-DHT lesions in the rat may be relevant to serotonergic behavioral supersensitivity. The increases in phosphoinositide hydrolysis were seen with submaximal doses of 5HT. In another study, 5HT-stimulated phosphoinositide hydrolysis was increased in cortex of 5,7-DHT-lesioned animals at submaximal doses, which became nonsignificant when maximal doses were included, possibly owing to a ceiling effect *(61)*.

In adult rats injected icv with 5,7-DHT, the 5HT$_{2C}$-mediated phosphoinositide hydrolysis response in choroid plexus became supersensitive *(72)*. In mice, 5,7-DHT lesions had no effect on 5HT-stimulated phosphoinositide turnover *(73)*. After ip neonatal administration of 5,7-DHT, 5HT-sensitive adenylate cyclase activity increased slightly in hippocampus and brainstem *(66)*.

7.4. Functional Receptor Correlates

Locomotor activity provides one functional measure of the effects of 5,7-DHT on 5HT receptors *(33,37,38,74–81)* (Table 2). In naive rats, evidence sug-

Table 2
Summary of Behavioral Evidence of Supersensitivity in Adult Rats Given 5,7-DHT

Receptor	Parameter	Route of 5,7-DHT	Test drug	Response	Reference
$5HT_2$	Head twitch	icv	Quipazine (0.5–2 mg/kg)	No change	*(74)*
			L-5-HTP (25–50 mg/kg)	Increased	
$5HT_{1B}$	LMA	icv	RU 24969 (3.5 mg/kg)	No change	*(75)*
		icv	RU 24966 (5 mg/kg)	Increased	*(76)*
$5HT_{1A}$	5HT syndrome	icv	5-MeO-DMT (dose–response)	Increased	*(77)*
		icv	5-MeO-DMT (dose–response)	Increased	*(33)*
		icv	5-MeO-DMT 2 (2.5 mg/kg))	No change	*(75)*
		icv	L-5-HTP (dose–response)	Increased	*(33)*
		ic	L-5-HTP (dose–response)	Increased	*(37)*
		ic	5-HIAA (icv)	Increased	*(38)*
			5-HT (icv)	Increased	
			5-MeO-DMT	Increased	
—	Electrophysiology	—	5-HT	Increased	*(78–80)*
—	Antinociception	icv	5-MeO-DMT (2 mg/kg)	Increased	*(81)*
			8-OH-DPAT (0.5 mg/kg)		
			L-5-HTP (200 mg/kg)		

gests that stimulation of $5HT_1$ and $5HT_2$ receptors facilitates or inhibits, respectively, DA-dependent LMA. The effect of RU24969 of increasing LMA in naive rats has been attributed to activity at the $5HT_{1B}$ receptor. The $5HT_{2C}$ agonist *m*-chlorophenylpiperazine (m-CPP) decreases locomotor activity in the rat. In rats with 5,7-DHT lesions, RU24969 and 8-OH-DPAT increased LMA, whereas the suppressive effects of the $5-HT_{2A/2C}$ agonist DOI were reversed by ritanserin at a dose that did not itself reduce LMA and may therefore represent a $5HT_2$ agonist effect *(82)*.

Respiratory depression exerted by 5HT agonists has also been used as a measure of the functional effects of 5,7-DHT. After neonatal 5,7-DHT injection, 5-MeO-DMT and L-5-HTP produced greater respiratory depression *(83)*. Early ic 5,7-DHT injection did not affect spatial memory *(84)*.

Electrophysiologic studies in adult rats demonstrate that 5,7-DHT lesions increase responsiveness to iontophoretically applied 5HT in the amygdala and lateral geniculate *(78,79)* and cortical neurons *(80)*.

Regarding the 5HT syndrome as a functional receptor correlate, the main consequence of ip compared to ic 5,7-DHT injection in neonatal rats was increased presynaptic $5HT_{1A}$ responses and decreased $5HT_{2A/2C}$ responses *(85)*. 8-OH-DPAT- evoked behaviors, such as forepaw myoclonus and head weaving, were enhanced more by the ip route. The effects of DOI, such as shaking behavior and forepaw myoclonus, were enhanced by 5,7-DHT lesions made ic, not ip. These data support the interpretation that 5,7-DHT-induced dysregulation of 5HT receptors, including both presynaptic and postsynaptic changes and altered interactions between receptor subtypes, better explains the data than postsynaptic changes alone.

In rats treated with 5,7-DHT as adults, $5HT_2$ antagonists suppressed the 5HT syndrome induced by L-5-HTP more than other neuroleptics, but the nonselective 5HT antagonist methysergide was also effective *(86)*.

7.5. Changes in Cotransmitter and Non-5HT Receptors

Specific corticosterone binding was reduced by 50–70% in 5,7-DHT-treated adrenalectomized rats *(87)*. An increase in cortical β-adrenoceptors that was restricted to the low-affinity conformation was found in rats injected icv with 5,7-DHT and pretreated with desipramine *(88)*.

In rats injected with 5,7-DHT as neonates, opiate receptor binding in the spinal cord was reduced by 22% *(89)*. After ic 5,7-DHT, a 30–40% increase in [^{3}H]MeTRH receptor binding was found in spinal cord in one study *(18)*, but not in a nonsaturation study *(90)*.

5,7-DHT-treated adult rats were behaviorally supersensitive to 10 mg/kg MK-771, a TRH analog, as indicated by shortened latency of wet-dog shakes and greater frequency of abnormal forepaw movements *(90)*.

8. EFFECTS ON NEUROPLASTICITY

8.1. Neonatal Lesions

If the neonatal rat is injected ip with 5,7-DHT, which crosses the blood–brain barrier only during the first week postnatal, severe "pruning" of distal 5HT terminals and "sprouting" of 5HT terminals proximal to cell bodies occur rapidly, denervating terminal-rich regions and hyperinnervating the brainstem *(5,36,67,70,91,92)* (Table 3). If the neonate is injected ic, however, the degree of 5HT cell death is profound in the dorsal and median raphe nuclei, and serotonergic axons do not regenerate. This route-dependent difference in response to

Table 3
Evidence Supporting Brainstem Sprouting After ip Neonatal 5,7-DHT

Observation	Reference
Increased brainstem 5HT concentration	*(5,36)*
Increased B_{max} of 5HT uptake sites	*(91)*
Increased V_{max} of 5HT uptake	*(67,92)*
Increase area of $5HT_{1A}$ binding in dorsal raphe	*(70)*

5,7-DHT lesions raises many interesting neurobiological questions about the response of developing brain to injury. Are both types of early 5,7-DHT lesions functionally significant?

Rat pups injected ic or ip with 5,7-DHT displayed supersensitive behavioral responses to L-5-HTP *(36,69,70,71,93)* (Table 4). However, rats lesioned by ip injections exhibited significantly greater shaking behavior (+1445%) in response to L-5-HTP than their ic counterparts, who instead showed more forepaw myoclonus (+250%) and head weaving (+270%), the core features of the 5HT syndrome. Differences in 5HT syndrome behaviors were already present 2 wk after lesioning, whereas the difference in shaking behavior was not. After 14 wk, 5HT was selectively depleted (–43% to –92%) in hippocampus, spinal cord, and frontal cortex, and differences between ic and ip 5,7-DHT routes were insignificant, except in frontal cortex. Brainstem 5HT concentrations were significantly increased (+35%) after ip 5,7-DHT injections in contrast to reduction (–89%) after ic 5,7-DHT. The ratio of 5-hydroxyindoleacetic acid (5HIAA) to 5HT was decreased (–20%) with either route. These data suggest that brainstem 5HT hyperinnervation following ip 5,7-DHT injection modifies the functional consequences of injury in abating the 5HT syndrome, but does not result in complete recovery, since shaking behavior is enhanced. Loss of presynaptically mediated autoregulation or receptor dysregulation may play a major role in behavioral supersensitivity induced by L-5-HTP in rats with 5,7-DHT lesions. To the extent that the 5HT syndrome is mediated by $5HT_{1A}$ receptors and shaking behavior by $5HT_2$ sites, differential responses to injury of $5HT_{1A}$ and $5HT_2$ receptors may contribute to these behavioral differences.

Increased brainstem 5HT is a manifestation of brainstem hyperinnervation in rats with neonatal 5,7-DHT lesions. The loss or reduction of behavioral supersensitivity manifested as the 5HT syndrome in response to L-5-HTP corresponds to the time-course of brainstem hyperinnervation *(5)*. The brainstem is also the apparent locus of 5HT syndrome behaviors, but less unequivocally of shaking behavior, in the rat. In brainstem, turnover, as estimated by 5HIAA/5HT ratios, was decreased *(36)* and in turnover studies of neonatal 5,7-DHT lesion made by the ic route *(94)*, although brainstem 5HT hyperinnervation did not affect 5HIAA/5HT ratios.

Table 4
Summary of Effects of Neonatal 5,7-DHT Lesions

Parameter	Route of 5,7-DHT injection	
	ip	ic
Behavior		
2 wk 5-HTP	Right-shift dose–response	High doses lethal
14 wks		
5-HTP	Shaking behavior	Forepaw myoclonus
8-OH-DPAT	↑ Forepaw myoclonus	↑ Forepaw myoclonus
DOI	—	↑ Shaking behavior + forepaw myoclonus
Neurochemistry		
14 wk	Increased BS 5HT	Decreased BS 5HT
	Decreased 5HIAA/5HT	Decreased 5HIAA/5HT
	Decreased 5HT + 5HIAA elsewhere	Decreased 5HT + 5HIAA elsewhere
	Increased 5HIAA/5HT in hippocampus	
Receptor binding		
2 wk + 14 wk	↑ Spinal $5HT_{2C}$ sites	↑ Spinal $5HT_{2C}$ sites
	↑ Reuptake sites in BS + Dien	↓ Reuptake sites
	↓ Reuptake sites elsewhere	
Autoradiography		
14 wk	$5HT_{1A}$ sites	$5HT_{1A}$ sites
	Dorsal raphe	Dorsal raphe
	↑ spread	↓ binding
	$5HT_{1B}$ sites	$5HT_{1B}$ sites
	↑ binding in subiculum	↑ binding subiculum
		↓ binding substantia nigra

Abbreviations: 5HIAA, 5-hydroxyindoleacetic acid; BS, brainstem; Dien, diencephalon.
References: *(36,69,70,71,93).*

However, how might brainstem 5HT hyperinnervation attenuate 5HT syndrome behaviors, which are purported to be $5HT_{1A}$ receptor-mediated, while increasing shaking behavior, mediated by $5-HT_2$ receptors? $5HT_{1A}$ receptors increased in brainstem 3 mo after 5,7-DHT lesions *(92)* The increase was interpreted as presynaptic, in the absence of postsynaptic receptor changes using [^{3}H]5HT. Increased presynaptic $5HT_{1A}$ sites would be expected to decrease the 5HT syndrome. L-5-HTP-evoked shaking behaviors may manifest then simply by removal of competing behaviors. L-5-HTP may not evoked shaking behavior at 2 wk either for ontogenic reasons or because $5HT_{1A}$ sites were not yet sufficiently increased. An alternative explanation to behaviorally significant changes in $5HT_{1A}$ receptors is that $5HT_2$ receptors were upregulated to account for increased shaking behavior and that increased $5HT_2$ responses inhibit the 5HT syndrome. However, no increases in $5HT_2$ receptor binding have been found.

8.2. Adult Lesions

Central 5HT neurons regenerate after 5,6- or 5,7-DHT-induced axotomy even in adult rats, most extensively at the medullary-cervical junction where sprouting from lesioned axon stumps occurs *(6)*. High DHT doses (150–200 µg) reduce or prevent the regeneration owing to retrograde cell loss. Regenerated 5HT neurons may re-establish functional innervation in once denervated proximal, but not distal terminal areas, and abnormal hyperinnervations, which continue for at least 1 yr occur in brainstem nuclei, such as the inferior olive. During regrowth, areas that normally lack 5HT fibers may become innervated *(7)*.

9. SUMMARY

Dihydroxylated tryptamines have been useful tools for study of the morphology and function, plasticity, and growth of 5HT neurons in experimental animals. Pretreatment with a reuptake inhibitor to protect catecholamine neurons allows for selective lesions of 5HT neurons. 5,7-DHT, the preferred neurotoxin, produces a chemical axotomy with dose-dependent cell loss via uptake into 5HT terminals and autoxidation. The effects of 5,7-DHT on behavior and 5HT receptors are complex, owing probably to the extreme number of variables in lesion studies, the many physiological functions of 5HT, and a multiplicity of 5HT receptor subtypes. The dihydroxylated tryptamines are unique among 5HT neurotoxins in inducing behavioral supersensitivity to 5HT precursors or agonists. Although molecular biological techniques have supplanted 5,7-DHT as a method for studying 5HT receptors and genetic abnormalities of the 5HT system, 5,7-DHT is relevant to acquired human neurological disorders in which denervation, regrowth, and recovery of function continue to be important issues.

ACKNOWLEDGMENTS

This work was funded by grants from the National Institutes of Health (1 K08 NSO1158), the Myoclonus Research Foundation, and the Food and Drug Administration Orphan Products Development (FD-U-000746). The author thanks George Mandel, Chairman, Department of Pharmacology, The George Washington University for his continuing support. Gail Cain typed the manuscript.

REFERENCES

1. Gerson, S. C. and Baldessarini, R. J. (1974) Motor effects of serotonin in the central nervous system. *Life Sci.* **27,** 1435–1451.

2. Baumgarten, H. G., Klemm, H. P., and Lachenmayer, L. (1978) Mode and mechanism of action of neurotoxic indoleamines: a review and a progress report. *Ann. NY Acad. Sci.* **305,** 3–24.

3. Schlossberger, H. G. (1978) Synthesis and chemical properties of some indole derivatives. *Ann. NY Acad. Sci.* **305,** 25–35.

4. Horn, A. S. (1978) Structure-activity relationships of serotonin neurotoxins: effects on serotonin uptake. *Ann. NY Acad. Sci.* **305,** 128–133.

5. Sachs, C. and Jonsson, G. (1975) 5,7-Dihydroxytryptamine induced changes in the postnatal development of central 5-hydroxytryptamine neurons. *Med. Biol.* **53,** 156–164.

6. Björklund, A. and Wiklund, L. (1980) Mechanisms of regrowth of the bulbospinal serotonin system following 5,6-dihydroxytryptamine induced axotomy. I. Biochemical correlates. *Brain Res.* **191,** 109–127.

7. Wiklund, L., Descarries, L., and Mollgard, K. (1981) Serotonergic axon terminals in the rat dorsal accessory olive: normal ultrastructure and light microscopic demonstration of regeneration after 5,6-dihydroxytryptamine lesioning. *J. Neurocytol.* **10,** 1009–1027.

8. Fuxe, K., Ogren, S-O., and Agnati L. F. et al., (1978) 5,7-Dihydroxytryptamine as a tool to study the functional role of central 5-hydroxytryptamine neurons. *Ann. NY Acad. Sci.* **305,** 346–369.

9. Breese, G. R. and Mueller, R. A. (1978) Alterations in the neurocytoxicity of 5,7-dihydroxytryptamine by pharmacologic agents in adult and developing rats. *Ann. NY Acad. Sci.* **305,** 160–174.

10. Towle, A. C., Breese, G. R., and Mueller, R. A., et al. (1984) Early postnatal administration of 5,7-DHT: effects on serotonergic neurons and terminals. *Brain Res.* **310,** 67–75.

11. Pranzatelli, M. R., Dollison, A. M., and Hibczuk, V. (1989) Intracisternal 5,7- dihydroxytryptamine lesions in neonatal and adult rats: Comparison of response to 5-hydroxytryptophan. *Dev. Neurosci.* **11,** 205–211.

12. Pranzatelli, M. R. and Snodgrass, S. R. (1986) Enhanced selective 5-HT depletions in the DHT rat model: Denervation supersensitivity and recovery of function. *Psychopharmacology* **89,** 449–455.

13. Stewart, R. M., Campbell, A., and Sperk, G., et al. (1979) Receptor mechanisms in increased sensitivity to serotonin agonists after dihydroxytryptamine shown by electronic monitoring of muscle twitches in the rat. *Psychopharmacology* **60,** 690–692.

14. Nobin, A. and Björklund, A. (1978) Degenerative effects of various neurotoxic indoleamines on central monoamine neurons. *Ann. NY Acad. Sci.* **305,** 305–327.

15. Lighton, C., Marsden, C. A., and Bennett, G. W. (1984) The effects of 5,7-dihydroxytryptamine and p-chlorophenylalanine on thyrotropin-releasing hormone in regions of rat brain and spinal cord. *Neuropharmacology* **23,** 53–60.

16. Drust, E. G. and Connor, J. D. (1983) Pharmacological analysis of shaking behavior induced by enkephalins, thyrotropin-releasing hormone or serotonin in rats: evidence for different mechanisms. *J. Pharmacol. Exp. Ther.* **224,** 148–154.

17. Pranzatelli, M. R., Dailey, A., and Markush, S. (1988) The regulation of TRH and serotonin receptors: Chronic TRH and analog administration in the rat. *J. Rec. Res.* **8,** 667–681.

18. Sharif, N. A., Burt, D. R., Towle, A. C., Mueller, R. A., and Breese, G. R. (1983) Codepletion of serotonin and TRH induces apparent supersensitivity of spinal TRH receptors. *Eur. J. Pharmacol.* **95,** 301–304.

19. Towle, A. C., Breese, G. R., Mueller, R. A., Hunt, R., and Lauder, J. M. (1986) Early postnatal administration of 5,7-dihydroxytryptamine: effects on substance P and thyrotropin-releasing hormone neurons and terminals in rat brain. *Brain Res.* **363,** 38–46.

20. Jonsson, G. and Hallman, H. (1983) Effect of substance P on the 5,7- dihydroxytryptamine induced alteration of postnatal development of central serotonin neurons. *Med. Biol.* **61,** 105–112.

21. Pranzatelli, M. R. and Snodgrass, S. R. (1987) The guinea pig myoclonic model: Behavioral supersensitivity to 5-hydroxytryptophan induced by intracisternal 5,7-dihydroxytryptamine. *Eur. J. Pharmacol.* **143,** 237–242.

22. Herman, Z. S. and Bonczek, A. (1978) Model of central chemical 'serotoninectomy' in mice. *Pharmacology* **17,** 8–14.

23. Alesci, R. and Bagnoli, P. (1988) Endogenous levels of serotonin and 5-hydroxyindoleacetic acid in specific areas of the pigeon CNS: effects of serotonin neurotoxins. *Brain Res.* **450,** 259–271.

24. Lyness, W. H. and Moore, K. E. (1981) Destruction of 5-hydroxytryptaminergic neurons and the dynamics of dopamine in nucleus accumbens septi and other forebrain regions of the rat. *Neuropharmacology* **20,** 327–334.

25. Pranzatelli, M. R. and Snodgrass, S. R. (1986) Serotonin-lesion myoclonic syndromes. II. Analysis of individual syndrome elements, locomotor activity, and behavioral correlations. *Brain Res.* **364,** 67–76.

26. Carter, C. J. and Pycock, C. J. (1981) The role of 5-hydroxytryptamine in dopamine-dependent stereotyped behaviour. *Neuropharmacology* **20,** 261–265.

27. Black, R. S. and Robinson, R. G. (1985) Intracortical 5,7-dihydroxytryptamine depletes brain serotonin concentration without affecting spontaneous activity. *Pharmacol. Biochem. Behav.* **22,** 327–331.

28. Dickinson, A. L., Andrews, C. D., and Curzon G. (1984) The effect of lesions produced by 5,7-dihydroxytryptamine on 5-hydroxytryptamine-mediated behavior induced by amphetamine in large doses in the rat. *Neuropharmacology* **23,** 423–429.

29. Geyer, M. A., Puerto, A., Menkes, D. B., Segal, D. S., and Mandell, A. J. (1976) Behavioral studies following lesions of the mesolimbic and mesostriatal serotonergic pathways. *Brain Res.* **106,** 257–270.

30. St. James-Roberts, I. (1979) Neurological plasticity, recovery from brain insult, and child development. *Adv. Child Dev. Behav.* **14,** 253–319.

31. Glick, S. D. (1974) Changes in drug sensitivity of mechanism of functional recovery following brain damage, in *Plasticity and Recovery of Function in the Central Nervous System* (Stein, D. G., Rosen, and J. J., Butters, N., eds), Academic, New York, pp. 339–372.

32. Finger, S. and Almli, R. C. (1986) Brain damage and neuroplasticity: mechanisms of recovery or development? *Brain Res. Rev.* **10,** 177–186.

33. Trulson, M. E. and Jacobs, B. L. (1978) Behavioral evidence for denervation supersensitivity after destruction of central serotonergic nerve terminals. *Ann. NY Acad. Sci.* **305,** 497–499.

34. Pranzatelli, M. R. and Snodgrass, S. R. (1986) Motor habituation in the DHT model: Bin analysis of daytime and nocturnal locomotor activity. *Pharmacol. Biochem. Behav.* **24,** 1679–1686.

35. Mackenzie, R., Hobel, B., Norelli, C., and Trulson, M. (1978) Increased tilt cage activity after serotonin depletion by 5,7-dihydroxytryptamine. *Neuropharmacology* **17,** 957–963.

36. Pranzatelli, M. R., Huang, Y., and Dollison, A. M. et al., (1989) Brainstem serotonergic hyperinnervation modifies behavioral supersensitivity to 5-hydroxytryptophan in the rat. *Dev. Brain Res.* **56,** 89–100.

37. Browning, R. A., Hoffmann, W. E., and Simonton, R. L. (1978) Changes in seizure susceptibility after intracerebral treatment with 5,7-dihydroxytryptamine: role of serotonergic neurons. *Ann. NY Acad. Sci.* **305,** 437–456.

38. Stewart, R. M., Growdon, J. H., Cancian, D., and Baldessarini, R. J. (1976) Myoclonus after 5-hydroxytryptophan in rats with lesions of indoleamine neurons in the central nervous system. *Neurology* **26,** 690–692.

39. Warbritton, J. D., Stewart, R. M., and Baldessarini, R. J. (1980) Increased sensitivity to intracerebroventricular infusion of serotonin and deaminated indoles after lesion rat with dihydroxytryptamine. *Brain Res.* **183,** 355–366.

40. Sloviter, R. S., Drust, E. G., and Connor, J. D. (1978) Specificity of a rat behavioral model for serotonin receptor activation. *J. Pharmacol Exp. Ther.* **206,** 339–347.

41. Jacobs, B. L. and Klemfuss, H. (1976) Brainstem and spinal cord mediation of a serotonergic behavioral syndrome. *Brain Res.* **100,** 450–457.

42. Messing, R. B. (1978) Behavioral effects of serotonin neurotoxins: an overview. *Ann. NY Acad. Sci.* **305,** 480–496.

43. Cannon, W. B. and Rosenblueth, A. (1949) *The Supersensitivity Denervated Structures: A Law of Denervation.* MacMillan, NY.

44. Quik, M. and Azmitia, E. (1983) Selective destruction of the serotonergic fibers of the fornix fimbria and cingulum bundle increases in 5-HT$_1$ but not 5-HT$_2$ receptors in rat midbrain. *Eur. J. Pharmacol.* **90,** 377–384.

45. Pranzatelli, M. R., Rubin, G., and Snodgrass, S. R. (1986) Serotonin-lesion myoclonic syndromes. I. Neurochemical profile and S1 receptor binding. *Brain Res.* **364,** 57–66.

46. Fischette, C. T., Nock, B., and Renner, K. (1987) Effects of 5,7-dihydroxytryptamine on serotonin1 and serotonin2 receptors throughout the rat central nervous system using quantitative autoradiography. *Brain Res.* **421,** 263–279.

47. Nelson, D. L., Herbet, A., Bourgoin, S., Glowinsk, I. J., and Hamon, M. (1978) Characteristics of central 5-HT receptors and their adaptive changes following interacerebral 5,7-dihydroxytryptamine administration in the rat. *Mol. Pharmacol.* **14,** 983–995.

48. Blurton, P. A. and Wood, M. D. (1986) Identification of multiple binding sites for ^{3}H-5-hydroxytryptamine in the rat CNS. *J. Neurochem.* **46,** 1392–1398.

49. Hensler, J. G., Kovachich, G. B., and Frazer, A. (1991) A quantitative autoradiographic study of serotonin1A receptor regulation. Effect of 5,7-dihydroxytryptamine and antidepressant treatments. *Neuropsychopharmacology* **4,** 131–144.

50. Weissman-Nanopoulos, D., Mach, E., and Magare, J. et al. (1985) Evidence for the localization of 5-HT$_{1A}$ binding sites on serotonin-containing neurons in the raphe dorsalis and raphe centralis nuclei of the rat brain. *Neurochem. Int.* **7,** 1061–1072.

51. Verge, D., Daval, G., Patey, A., Gozlan, H., El Mestikawy, S., and Hamon, M. (1985) Presynaptic 5-HT autoreceptors on serotonergic cell bodies and/or dendrites but not terminals are of the 5-HT$_{1A}$ subtype. *Eur. J. Pharmacol.* **113,** 463–464.

52. Verge, D., Daval, G., Marcinkiewicz, M., Patey, A., El Mestikawy, S., Gozlan, H., and Hamon, M. (1986) Quantitative autoradiography of multiple 5-HT$_1$ receptor subtype in the brain of control or 5,7-dihydroxytryptamine-treated rats. *J. Neurosci.* **6,** 3474–3482.

53. Weissman, D., Mach, E., Oberlander, C., Demassey, Y., and Pujol, J. F. (1986) Evidence for hyperdensity of 5-HT$_{1B}$ binding sites in the substantia nigra of the rat after 5,7-dihydroxytryptamine intraventricular injection. *Neurochem. Int.* **9**, 191–200.

54. Offord, S. J., Ordway, G. A., and Frazer, A. (1988) Application of [^{125}I]liodocyanopindolol to measure 5-hydroxytryptamine^{1B} receptors in the brain of the rat. *J. Pharmacol. Exp. Ther.* **244**, 144–153.

55. Matsumoto, I., Combs, M. R., and Jones, D. J. (1992) Characterization of 5-hydroxytryptamine$_{1B}$ receptors in rat spinal cord via [^{125}I]iodocyanopindolol binding and inhibition of [^{3}H]-5-hydroxytryptamine release. *J. Pharmacol. Exp. Ther.* **260**, 614–626.

56. Brown, L. M., Smith, D. L., Williams, G. M., and Smith, D. J. (1989) Alterations in serotonin binding sites after 5,7-dihydroxytryptamine treatment in the rat spinal cord. *Neurosci. Lett.* **102**, 103–107.

57. Barbaccia, M. L., Gandolfi, O., Chuang, D-M., and Costa, E. (1983) Differences in the regulatory adaptation of the 5-HT$_2$ recognition sites labelled by [^{3}H]-mianserin and ^{3}H-ketanserin. *Neuropharmacology* **22**, 123–126.

58. Nabeshima, T., Ishikawa, K., Yamaguchi, K., Furukawa, H., and Kameyama, T. (1986) An ability of phencyclidine as an agonist for serotonin$_2$ (5-HT$_2$) receptors in rat brain. *Soc. Neurosci. Abstracts* **12**, 362.

59. Leysen, J. E., Van Gompel, P., Verwimp, M., and Niemegeers, C. J. E. (1983) Role and localization of serotonin$_2$ (S$_2$) receptor binding sites: effects of neuronal lesions, in *CNS Receptor—from Molecular Pharmacology to Behavior* (Mandel, P. and DeFeudis, F. V., eds.), Raven, New York, pp. 373–83.

60. Stockmeier, C. A. and Kellar, K. (1986) *In vivo* regulation of the serotonin-2 receptors in rat brain. *Life Sci.* **38**, 117–127.

61. Conn, P. J. and Sanders-Bush, E. (1986) Regulation of serotonin-stimulated phosphoinositide hydrolysis: relation to the serotonin 5-HT$_2$ binding site. *J. Neurosci.* **6**, 3669–3675.

62. Butler, P. D., Pranzatelli, M. R., and Barkai, A. L. (1990) Regional central serotonin-2 receptor binding and phosphoinositide turnover in rats with 5,7-dihydroxytryptamine lesions. *Brain Res. Bull.* **24**, 125–129.

63. Barone, P., Millet, S., Moret, C., Prudhomme, N., and Fillion, G. (1993) Quantitative autoradiography of 5-HT$_{1E}$ binding sites in rodent brains: effect of lesion of serotonergic neurons. *Eur. J. Pharmacol.* **249**, 221–230.

64. Azmitia, E. C., Buchan, A. M., and Williams, J. H. (1978) Structural and functional restoration by collateral sprouting of hippocampal 5-HT axons. *Nature* **274**, 374–376.

65. Pranzatelli, M. R. and Gregory, C. M. (1993) High and low affinity 5-HT$_2$ and 5-HT$_{1C}$ binding sites: responses to neonatal 5,7-DHT lesion in rat brain. *Cytobios* **75**, 197–209.

66. Hamon, M., Nelson, D. L., and Mallat, M., et al. (1981) Are 5-HT receptors involved in the sprouting of serotonergic terminals following neonatal 5,7-dihydroxytryptamine treatment in the rat? *Neurochem. Int.* **3**, 69–79.

67. Jonsson, G., Pollare, T., Hallman, H., and Sachs, Ch. (1978) Developmental plasticity of central serotonin neurons after 5,7-dihydroxytryptamine treatment. *Ann. NY Acad. Sci.* **305**, 328–345.

68. Bennett, J. P. and Snyder, S. H. (1976) Serotonin and lysergic acid diethylamide binding in rat brain membranes: relationship to postsynaptic serotonin receptors. *Mol. Pharmacol.* **12**, 373–379.

69. Pranzatelli, M. R. (1990) Neonatal 5,7-DHT lesions upregulate [^{3}H]mesulergine-labelled spinal 5-HT$_{1C}$ binding sites in the rat. *Brain Res. Bull.* **25,** 151–153.

70. Pranzatelli, M. R., Durkin, M. M., and Barkai, A. L. (1994) Quantitative autoradiography of 5-hydroxytryptamine$_{1A}$ binding sites in rats with neonatal 5,7-dihydroxytryptamine lesions. *Dev. Brain Res.* **80,** 1–6.

71. Pranzatelli, M. R., Durkin, M. M., and Farmer, M. (1994) Quantitative autoradiography of 5-hydroxytryptamine$_{1B}$ binding sites in rats with neonatal 5,7-dihydroxytryptamine lesions *Int. J. Devel. Neurosci.* **14,** 621–629.

72. Conn, P. J., Janowsky, A., and Sanders-Bush, E. (1987) Denervation supersensitivity of 5-HT$_{1C}$-receptors in rat choroid plexus. *Brain Res.* **400,** 396–398.

73. Godfrey, P. P., McClure, S. J., Young, M. M., and Heal, D. L. (1988) 5- hydroxytryptamine-stimulated inositol phospholipid hydrolysis in the mouse cortex has pharmacological characteristics compatible with mediation via 5-HT$_2$ receptors but this response does not reflect altered 5-HT$_2$ function after 5,7- dihydroxytryptamine-lesioning or repeated antidepressant treatments. *J. Neurochem.* **50,** 730–738.

74. Dall'Olio, R., Vaccheri, A., and Montamaro, N. (1985) Reduced head-twitch response to quipazine of rats previously treated with methiothepin: possible involvement of dopaminergic system. *Pharmacol. Biochem. Behav.* **23,** 43–48.

75. Nisbet, A. P. and Marsden, C. A. (1984) Increased behavioral response to 5-methoxy-*N*, *N*-dimethyltryptamine but not to RU 24969 after intraventricular 5,7-dihydroxytryptamine administration. *Eur. J. Pharmacol.* **104,** 177–180.

76. Oberlander, C., Demassey, Y., Verdu, A., Van de Velde, D., and Bardelay, C. (1987) Tolerance to the serotonin 5-HT$_1$ agonist RU 24969 and effects on dopaminergic behavior. *Eur. J. Pharmacol.* **139,** 205–214.

77. Ortmann, R., Martin, S., and Waldmeier, P. C. (1981) Supersensitivity to L-5-hydroxytryptophan after 5,7-dihydroxytryptamine injections in desmethylimipramine—and nomifensine-pretreated rats: behavioral evidence for postsynaptic supersensitivity. *Psychopharmacology* **74,** 109–114.

78. McCall, R. and Aghajanian, G. (1979) Denervation supersensivity to serotonin in the facial nucleus. *Neuroscience* **4,** 501.

79. Wang, R., de Montigny, C., and Cold, B., et al (1979) Denervation supersensitivity to serotonin in rat forebrain: single cell studies. *Brain Res.* **178,** 479–497.

80. Ferron, A., Decarries, L., and Reader, T. A. (1982) Altered neuronal responsiveness to biogenic amines in rat cerebral cortex after serotonin denervation or depletion. *Brain Res.* **231,** 93–108.

81. Eide, P. K., Hole, K., Berge, O-G., and Broch, O. J. (1988) 5-HT depletion with 5,7-DHT, PCA and PCPA in mice: differential effects on the sensitivity to 5-MeODMT, 8-OH-DPAT, and 5-HTP as measured by two nociceptive tests. *Brain Res.* **440,** 42–52.

82. Pranzatelli, M. R., Jappay, E., and Snodgrass, S. R. (1987) Effects of 5-HT receptor subtype-selective drugs on locomotor activity and motor habituation in the DHT adult rat model. *Pharmacol. Biochem. Behav.* **27,** 497–504.

83. Mueller, R. A., Lundberg, D., and Breese, G. R. (1980) Evidence that respiratory depression by serotonin agonists may be exerted in the central nervous system. *Pharmacol. Biochem. Behav.* **13,** 247–255.

84. Volpe, B. T., Hendrix, C. S., Park, D. H., Towle, A. C., and Davis, H. P. (1992) Early post-natal administration of 5,7-dihydroxytryptamine destroys 5-HT neurons but does not affect spatial memory. *Brain Res.* **589,** 262–267.

85. Pranzatelli, M. R., Dollison, A. M., and Huang, Y-Y. (1990) The functional significance of neonatal 5,7-dihydroxytryptamine lesions in the rats: Response to selective 5-HT$_{1A}$ and 5-HT$_{2,1C}$ agonists. *Brain Res. Bull.* **24,** 747–753.

86. Pranzatelli, M. R. and Snodgrass, S. R. (1986) Antimyoclonic properties of S-2 serotonin receptor antagonists in the rat. *Neuropharmacology* **25,** 5–12.

87. Siegel, R. A., Weidenfeld, J., Chen, M., Feldman, S., Melamed, E., and Chowers, I. (1983) Hippocampal cell nuclear binding of corticosterone following 5,7-dihydroxy-tryptamine. *Mol. Cell Endocrinol.* **31,** 253–259.

88. Manier, D. H., Gillespie, D. D., and Sulser, F. (1987) 5,7-dihydroxytryptamine-induced lesions of serotonergic neurons and desipramine-induced down-regulation of cortical beta adrenoceptors: a re-evaluation. *Biochem. Pharmacol.* **56,** 3308–3310.

89. Kirby, M. L. and Mattio, T. G. (1982) Developmental changes in serotonin and 5-hydroxyindoleacetic acid concentrations and opiate receptor binding in rat spinal cord following neonatal 5,7-dihydroxytryptamine treatment. *Dev. Neurosci.* **5,** 394–402.

90. Pranzatelli, M. R. (1988) The comparative pharmacology of the behavioral syndromes induced by TRH and by 5-HT in the rat. *Gen. Pharmacol.* **19,** 205–211.

91. Pranzatelli, M. R. and Martens, J. N. (1992) Plasticity and ontogeny of the central 5-HT transporter: effect of neonatal 5,7-dihydroxytryptamine lesions in the rat. *Dev. Brain Res.* **70,** 191–195.

92. Gozlan, H., El Mestikaway, S., and Pichat, L., et al. (1983) Identification of presynaptic serotonin autoreceptors using a new ligand: ^{3}H-PAT. *Nature* **305,** 140–142.

93. Pranzatelli, M. R. (1994) Dissociation of the plasticity of 5-HT$_{1A}$ sites and 5-HT transporter sites. *Neurochem. Res.* **19,** 311–315.

94. Mueller, R. A., Towle, A., and Breese, G. R. (1980) Serotonin turnover and supersensitivity after neonatal 5,7-dihydroxytryptamine. *Pharmacol. Biochem. Behav.* **22,** 221–225.

12

Selective Cholinergic Neurotoxins
AF64A and 192-IgG-Saporin

Thomas J. Walsh and Pamela E. Potter

1. HISTORY OF THE PROBLEM

The study of cholinergic function is richly intertwined with the history of modern neuroscience. In fact, acetylcholine (ACh) was the first compound shown to be a neurotransmitter. Otto Loewi demonstrated that the cardiac branch of the vagus nerve releases a chemical transmitter, then termed "vaggustoff," and now ACh, that decreases heart rate. This simple, yet elegant, experiment established the validity of chemical neurotransmission. The biology and function of cholinergic systems then became a major focus for neuroscientists from all disciplines. The precursors, uptake systems, and enzymatic reactions involved in ACh synthesis were established, as were the receptors and second messenger systems involved in the response to ACh. With the advent of selective pharmacological antagonists of cholinergic receptors, there came an effort to define the functional properties of cholinergic systems. However, muscarinic antagonists like scopolamine and atropine have limited utility since they:

1. Lack central nervous system specificity;
2. Affect multiple muscarinic receptor subtypes;
3. Can produce systemic toxicity; and
4. Have a short duration of action.

Cholinergic systems rapidly recover from transient perturbations, making it difficult to study the consequences of cholinergic loss and to model diseases related to cholinergic degeneration.

Neurotoxins selective for cholinergic neurons would be useful tools to explore the functional biology and plasticity of cholinergic systems and to model diseases *(1)*. A major impetus for the development of a cholinergic neurotoxin was the discovery that the cholinergic system was affected early in Alzheimer's disease (AD) and that the degree of damage is related to the prevailing symptoms *(2,3)*. Therefore, development of a reliable and selective cholinergic neu-

From: Highly Selective Neurotoxins: Basic and Clinical Applications *Edited by: R. M. Kostrzewa.*
Humana Press Inc., Totowa, NJ

rotoxin would allow for the study of the pathobiology, therapy, and long-term consequences of this disease in animal models.

1.1. Alzheimer's Disease

AD is the most prevalent of the age-related cognitive disorders. Incidence appears to double with each decade after the age of 65, and current estimates indicate that up to 45% of the population over the age of 85 is affected *(4,5)*. With no treatments to prevent the onset, attenuate the course of deterioration, or mitigate the cognitive deficits, the increase in incidence of AD as the population ages will represent a major and growing challenge for health care professionals and the biomedical research community. The emotional toll of the disease for its victims, family, and caregivers is immeasurable.

A number of neural processes at the systems, anatomical, neurotransmitter, and molecular levels are affected in AD. The variety of changes that occur makes it difficult to isolate biological substrates that might be targets for therapeutic intervention. However, the central cholinergic system appears to be affected very early in the course of disease and to be related to the symptoms that characterize the disorder. AD is associated with a pattern of cognitive deficits in which some types of memory are impaired and others are not *(6)*. Episodic memory—the ability to recall specific information and episodes (consciously) is affected early in the disease and appears to relate to the loss of cholinergic function. Procedural memory—the ability to acquire new motor or cognitive skills through retention—is not dependent on cholinergic mechanisms and is not affected in AD. This functional dissociation has helped to demonstrate that cholinergic systems are a critical substrate for episodic memory processes, and their compromise in AD results in the observed profile of cognitive deficits.

The "cholinergic hypothesis" of dementia is consistent with a vast animal literature that supports a critical involvement of brain cholinergic systems in cognition *(7)*. For example, there are alterations in the activity of basal forebrain cholinergic neurons associated with learning, and both pharmacological antagonism of the cholinergic system and damage to cholinergic neurons produce deficits in episodic memory, which resemble those observed in AD *(6,7)*.

Cholinergic innervation of the cortex and the hippocampus, derived from the basal forebrain, is affected early in the course of AD. All biochemical parameters reflecting ACh synthesis and release are decreased in hippocampus and neocortex, and there is a marked loss of cholinergic cell bodies in the basal forebrain. The functional significance of these alterations is highlighted by the well-established correlation between the degree of cholinergic hypofunction, the number of plaques and tangles, and the extent of the dementia *(8,9)*. Although other brain areas that can be affected include noradrenergic neurons in the locus cerulus and the serotonergic neurons in the raphe, these changes are evident in subpopulations of patients or late in the course of the disease, and they have not been shown to relate to the prevailing memory disorder *(10)*. Furthermore, the only drug to date that has significantly alleviated the cognitive deficits in AD is the cholinesterase inhibitor tetrahydroaminoacridine (THA), which has produced a modest improvement, particularly in patients in early stages of the disease *(11)*.

Both functional and binding studies indicate that cholinergic receptors are also altered in AD. A selective loss of M2-receptors, the autoreceptor that modulates ACh release, has been reported in AD *(12)*. There is also a decrease in the number of nicotinic receptors *(12)*. Although the number and affinity of M1 muscarinic receptors, which are primarily postsynaptic, are not changed, coupling of this receptor to signal transduction mechanisms appears to be impaired. This has been demonstrated with binding studies in which an antagonist is displaced with an agonist. Under these conditions, high- and low-affinity states of the receptor can be observed. A shift to the low-affinity state is induced by addition of a nonhydrolyzable GTP analog, such as GppNHp, which uncouples the muscarinic receptors from G-proteins. In tissues from Alzheimer brain, the magnitude of the affinity shift induced by GppNHp was decreased, suggesting that the receptors were already uncoupled *(13,14)*.

Since the effects of muscarinic receptors are mediated through coupling to G-proteins, the response to muscarinic agonists would be reduced. Muscarinic agonist stimulation of phosphoinositide hydrolysis and binding of GTPγS to G-proteins are significantly decreased *(15)*. This is important from a therapeutic standpoint, given the development of M1 agonists as potential treatments for AD. Drugs that are effective in a normal system may not work in an AD patient, because the fundamental pharmacological responsiveness of the brain has changed. If this alteration in receptor coupling could be reproduced in an animal model, it would permit a greater understanding of the mechanisms leading to changes in receptor function and allow us to test pharmacological approaches to circumvent this deficit.

Animal models must promote an integrated study of the brain–behavior relationships that are affected in neurodegenerative diseases like AD. A logical strategy for developing an animal model of AD is to focus on the symptoms of episodic memory loss and changes in pre- and postsynaptic cholinergic function *(1)*. For this effort, animal models offer several important advantages:

1. They simplify the system being studied. Animal models have a known cause and allow the study of a specific aspect of the disease state.
2. They reduce the number of uncontrollable variables. In humans, it is hard to control for other medical conditions, agonal state, and the amount of time between death and autopsy, whereas in animal models, this is not a problem.
3. They allow unlimited access to tissue. Animal models provide access to unlimited fresh tissue and also allow for experimental manipulations that would be impossible in humans.

The development of ethylcholinc aziridinium ion (AF64A) and more recently 192 IgG-saporin has provided the neuroscience community with cholinergic neurotoxins that can be used to model the cholinergic degeneration and subsequent behavioral impairments observed in AD.

2. AF64A: BIOLOGICAL AND BEHAVIORAL EFFECTS

ACh is synthesized from dietary choline and acetyl CoA derived from glucose metabolism in the mitochondria. The enzyme catalyzing this reaction is choline acetyltransferase (ChAT). The rate-limiting step in the synthesis of ACh is the high-affinity choline transport system (HAChT) localized on cholinergic terminals. The activity of this system is coupled to the function of the neuron, and increasing or decreasing the activity of cholinergic neurons produces parallel changes in the kinetics of the HAChT system *(16,17)*. The unique presence of the HAChT system on cholinergic terminals encouraged investigators to develop compounds that selectively target this site. A chronic disruption of HAChT should lead to a persistent compromise of ACh synthesis. A variety of analogs of choline have been examined as potential cholinergic-specific neurotoxins that interact with the HAChT system (reviewed in *18*).

Fisher and Hanin explored a series of choline analogs and found that one of these, ethylcholine aziridinium ion (AF64A), has cholinotoxic properties that make it useful for both in vivo and in vitro studies *(19)*. AF64A combines a choline structure (i.e., ethylcholine) that is recognized by the HAChT system, with a cytotoxic aziridinium moiety. Because of its structural similarity to choline, AF64A is accumulated by the HAChT system, and once inside the terminal, the highly reactive aziridinium induces a pathological cascade that results in cholinergic hypofunction and ultimate death of the cell *(20)*. Hortnagl and colleagues showed that the aziridinium is essential for neurotoxicity, since opening the aziridinium ring with thiosulfate eliminates the cholinotoxic effects of AF64A (21). AF64A represents a cytotoxic "Trojan horse" that masquerades as choline.

3. SPECIFICITY AND POTENCY OF AF64A

3.1. Presynaptic Changes

Following intracerebroventricular (icv) injection, AF64A produces a 30–60% reduction in all indices of presynaptic cholinergic function including:

1. Regional concentrations of ACh;
2. The activity of ChAT;
3. HAChT; and
4. Both K^+- and ouabain-stimulated release of ACh from hippocampal slices *(18,20–22)*.

AF64A has no effect on serotonin uptake into synaptosomes, or low-affinity choline transport (LAChT) in erythrocytes and kidney slices, even at doses up to 20-fold greater than those needed to inhibit HAChT in synaptosomes from rats and mice *(23)*. Rylett and Colhoun demonstrated that the related compound choline mustard aziridinium was an approx 30-fold more potent inhibitor of HAChT than LAChT *(24)*. The cholinergic hypofunction induced by AF64A develops over time with maximal decreases evident approximately 7 d following treatment and persisting for at least 1 yr *(25)*. These changes are regionally selective, affecting cholinergic input to the hippocampus, and sparing projections to the cortex and interneurons in the striatum *(26)*.

The AF64A-induced decreases in cholinergic function are not accompanied by persistent changes in the regional concentrations of catecholamines, serotonin, or glutamate *(22)*. However, AF64A causes a transient alteration in the turnover of serotonin and norepinephrine *(22)*. To determine whether these changes were caused by direct effects of AF64A or were secondary to cholinergic toxicity, hemicholinium-3 or its analog A-4 was administered prior to AF64A. These compounds are both potent inhibitors of HAChT with inhibitory rate constants 30–100 times greater than that of AF64A, and they prevent AF64A from gaining access to the HAChT system *(27)*. Preventing uptake of AF64A into cholinergic terminals attenuated the loss of cholinergic and serotonergic markers in the hippocampus *(28)*, indicating that the effects of AF64A on neurotransmitter systems other than ACh are secondary to the cholinergic hypofunction.

3.2. Postsynaptic Changes

The effects of AF64A lesions on cholinergic receptors and function have also been studied. There was no change in either the number or affinity of M1-receptors, nor was responsiveness to the M1 agonist McN-A-343 altered *(29)*. However, there were changes in M2- and nicotinic receptor properties and function. Feedback inhibition of ACh release via M2-autoreceptors was decreased, as was M2-receptor-mediated inhibition of evoked norepinephrine release *(30)*. Binding studies indicated a decrease in the affinity of the M2-receptor, which could account for the loss of function. There was a decrease in the number of nicotinic receptors 6 wk after AF64A treatment, as well as a reduction in the enhancement of evoked norepinephrine release by nicotine. These results indicate that a profile of alterations in cholinergic receptors, similar to that which has been described in AD, can be produced in an animal model in which hippocampal cholinergic pathways are lesioned.

3.3. Behavioral Changes

The effects of AF64A on learning and memory have been evaluated in a variety of paradigms. AF64A has been consistently shown to produce episodic memory impairments in a variety of spatial memory tasks *(6)*. In addition, the episodic memory impairments induced by AF64A are dose-dependent and related to the cognitive demands of the task *(31)*. Rats were bilaterally injected icv with 0, 0.75, or 1.5 nmol of AF64A, and tested with either 0-, 1-, or 2-h delays in an episodic/working memory task. Although none of the groups were impaired at the 0 delay condition, the 1.5-nmol group was impaired at both the 1- and the 2-h delay. However, the 0.75-nmol group was only impaired at the 2-h delay. Therefore, the memory impairments were related to the dose of AF64A and to the demand placed on the working memory system by the task.

This is particularly interesting owing to the delay-dependent memory impairments observed in patients with AD *(32)*. An additional question is whether AF64A produces a dissociation of memory processes that resembles the one observed in AD. To address this issue, we developed a T-maze task that independently assessed both episodic (working) and procedural (reference) memory within the same trial. In this task, each trial was initiated by placing the rat into either the left or the right start/goal box of the T-maze for 15 s. The food-deprived rat was then rewarded with a single chocolate-flavored pellet for turning into and traversing the stem of the maze and for subsequently returning to the alternate start/goal location. This task involves (1) a trial-independent component (procedural memory), which involves an invariant response of running down the stem of the maze and (2) a trial-dependent response (episodic memory) in which the animal must maintain a representation of the start location in order to perform accurately *(33)*. Rats injected icv with AF64A exhibited a persistent deficit in episodic memory that was evident throughout the 75 postoperative trials. Episodic memory was impaired in this group even when their procedural memory performance was comparable to controls and close to 100% accurate. These data demonstrate that AF64A can produce a dissociation of episodic and procedural memory processes similar to that observed in AD.

4. MOLECULAR MECHANISMS OF AF64A TOXICITY

AF64A induces a cascade of biochemical changes that disrupt the synthesis of ACh and then lead to death of cholinergic neurons. The first stage of toxicity involves a concentration-dependent interaction with the HAChT system located on cholinergic nerve terminals *(21,22)*. At high concentrations (> 22.5 μM), AF64A rapidly alkylates nucleophilic sites on the proteins comprising the uptake system. This rapidly inhibits HAChT and prevents AF64A from gaining access to the inside of the terminal. However, at low concentrations (< 5 μM), AF64A is taken into the terminal by HAChT along with choline in a competitive manner. The initial cholinotoxic effects of AF64A are dependent on its interaction with the HAChT system, since they can be prevented by competing compounds, such as choline or hemicholinium-3 *(28,34)*. Choline can inhibit the cholinotoxic effects of AF64A in synaptosomes and in neuron-enriched cell cultures *(22)*. In addition, Chrobak and colleagues reported that pretreatment of rats with hemicholinium-3 prevented AF64A-induced cholinotoxicity as well as toxin-induced behavioral impairments *(35)*.

The second phase of toxicity involves AF64A's interaction with a variety of enzymes that utilize choline as a substrate. It binds to these enzymes owing to its structural similarity to choline, but its aziridinium ring alkylates their catalytic sites; AF64A inhibits ChAT, choline kinase, choline dehydrogenase, and acetylcholinesterase *(36,37)*. This results in a persistent presynaptic cholinergic hypofunction in which all measures of ACh synthesis and release are affected. Sandberg and colleagues reported that concentrations of AF64A that almost completely inhibit ChAT, choline kinase, and acetylcholinesterase activity in a cholinergic cell line had no effect on enzymes that do not interact with choline, including alcohol dehydrogenase, lactate dehydrogenase, carboxypeptidase A, and chymotrypsinogen *(37)*.

Although the effects of AF64A on cholinergic parameters have been extensively studied, there has been little effort to understand how AF64A kills cells. Following icv injection, AF64A does produce a dose-related decrease in the number of cholinergic, but not GABAergic, neurons in the medial septum, with different cell groups being uniquely vulnerable *(26,38)*. Recent studies have focused on the contribution of AF64A-induced genotoxicity and oxidative stress to the selective neurodegeneration induced by this compound.

The third and final phase of AF64A toxicity (i.e., cell death) might relate to the effects of this compound on the integrity of nuclear DNA and the transcription of genes needed to produce enzymes involved in ACh synthesis (i.e., ChAT). AF64A is structurally similar to several antineoplastic agents, such as nitrogen mustard, that produce cytotoxic effects by alkylating specific sites on nuclear DNA, thereby inhibiting gene expression and cell replication. Futscher and colleagues *(39)* recently demonstrated that AF64A does produce dose-dependent DNA strand breaks and premature termination of RNA transcription in cultured mouse leukemia cells. In addition, AF64A produced dose-dependent termination of RNA transcription and alkylation of the N-7 guanine site in DNA fragments exposed to AF64A in vitro (cited in *20*). This mechanism is also common to a variety of alkylating antineoplastic agents that covalently bind to DNA and impede replication. Therefore, the final and terminal phase of toxicity might be the result of a disruption of fundamental cellular processes involving the translation and transcription of genes that are necessary for acetylcholine synthesis and/or neuronal survival. Finally, there is evidence that oxidative stress might be a critical component of AF64A-induced cholinotoxicity.

4.1. AF64A and Oxidative Stress

There is accumulating evidence that oxidative stress contributes to the neurodegeneration induced by AF64A. A number of studies have now demonstrated that the prototype antioxidant vitamin E can attenuate both the cholinergic hypofunction and the memory impairments induced by AF64A *(40,41)*. Vitamin E is an endogenous antioxidant that minimizes the biological impact of free radicals, protects polyunsaturated lipids from free radicals, and stabilizes membranes. Acute injection of 50 mg/kg of vitamin E 24 h and 15 min prior to injection of AF64A prevented the spatial memory impairments in a Morris water maze task and the cholinergic hypofunction induced by the toxin.

A recent report provides more direct evidence that icv injection of AF64A induces oxidative stress. Gulyaeva and colleagues reported that bilateral icv injection of 3 nmol of AF64A increased a number of direct indices of oxidative stress measured in cerebral cortex, hippocampus, and the rest of the brain, 1, 3, or 5 d after surgery *(42)*. Thiobarbituric acid reactive species (TBARS) were assessed under basal conditions, and again 30 and 60 min following the addition of $FeSO_4$ (10 μM) and sodium ascorbate (0.5 mM) to the supernatant with incubation at 37°C. The stress of surgery increased the basal levels of TBARS in both the vehicle and AF64A-injected groups. However, the increase was more pronounced in the AF64A group, and it appeared earlier. The increase in TBARS production induced by iron/ascorbate was significantly greater in the AF64A group in all regions examined. Finally, superoxide scavenging activity was significantly increased in the hippocampus of AF64A-treated rats 5 d after surgery, and this increase was still evident 4 mo later. Superoxide scavenging is one of the primary free radical scavenging mechanisms in the brain, and it reflects both enzymatic (i.e., superoxide dismutase) and nonenzymatic processes. The increased superoxide scavenging activity observed in the hippocampus of AF64A-treated rats reflects a compensatory response to oxidative stress.

These data provide direct evidence that AF64A can induce oxidative stress, and together with the reports that vitamin E can prevent cholinergic hypofunction, suggest that AF64A-induced oxidative stress contributes to neurodegeneration. In addition, the oxidative stress might also contribute to the nucleic acid damage and dysfunction induced by AF64A. Oxidative stress might be a potential common cause of the cholinergic cell death observed in normal aging, AD, and AF64A-induced cholinergic hypofunction *(43)*. Cholinergic neurons in the basal forebrain contain the lowest levels of endogenous vitamin E in the brain, making them uniquely susceptible to free radicals and lipid peroxidative processes *(44)*. In addition, this inherent vulnerability is ultimately compounded by age-related decreases in the activity of endogenous antioxidant systems (ie., glutathione, vitamin E, and vitamin C) *(45,46)*. These changes occur in normal aging, but are markedly exaggerated in AD *(47)*.

5. BIOLOGICAL SIGNIFICANCE OF THE AF64A MODEL

Animal models can be used to examine the relationship between neurotransmitter function and behavior, and to model neurological diseases. All models represent an experimental compromise in which a simple system (animals) is used to address a more complex system (i.e., humans). It is important to appreciate the advantages and limitations of a specific model. No model captures a neurological disease in all of its dimensions: etiology, time-course, selective neurodegeneration, functional changes, and responsiveness to therapy. A good model should provide insights into some fundamental aspects of a neurological disorder.

AF64A appears to be a useful tool for examining the relationship between the activity of the cholinergic septohippocampal pathway and cognitive function, and for modeling diseases with a prominent cholinergic hypofunction *(1,6)*. The specificity of AF64A depends on a multitude of interdependent variables including:

1. The purity of the starting compound;
2. The dose and concentration;

3. The rate and total volume of injection; and
4. The site of administration.

Intraventricular injection of AF64A seems to be an appropriate way to produce a selective compromise of the cholinergic septohippocampal pathway. However, injection of AF64A into discrete sites can produce extensive nonspecific damage *(see 48)*. A very high concentration of AF64A in a restricted area will probably result in the compound being taken up into a variety of noncholinergic cells by the ubiquitous LAChT system. The result would be a general cytotoxicity evidenced by nonspecific morphological damage.

The use of AF64A, like any neurotoxin, is appropriate for addressing a restricted set of specific questions. There are conceptual and experimental limitations that dictate the appropriate uses of AF64A. In particular, based on the relevant literature and our own experience, we believe that AF64A is an extremely powerful tool to compromise selectively the cholinergic innervation of the hippocampus. This can be exploited to:

1. Explore the biological and behavioral properties of this brain system;
2. Examine strategies to limit, attenuate, or reverse the functional consequences of damage to this system;
3. Determine the plasticity of this and interacting brain systems following selective insult; and finally
4. Model the functional deficits that occur following damage to this particular system in neurological disorders, such as AD.

6. FUTURE DIRECTIONS: THE ANTINEURONAL IMMUNOTOXIN 192 IgG-SAPORIN (SAP)

Neurotoxins used to lesion the cholinergic basal forebrain lack neurochemical specificity (e.g., excitotoxins) or have proven to be of limited use in creating models of cholinergic degeneration. The cholinergic toxin AF64A does selectively lesion cholinergic input to the hippocampus. However, it produces an incomplete lesion (<50%), it cannot be injected directly into tissues owing to a small window of specificity, there is considerable variability in results between different laboratories, and the potency may vary in different preparations from the commercial supplier. The development of the antineuronal immunotoxin SAP provides a model of cholinergic degeneration that can be used to ask fundamental questions about cholinergic biology and to evaluate new and innovative therapies to prevent or treat the functional consequences of cholinergic damage.

Immunotoxins are conjugates of a monoclonal antibody (MAb) targeting a specific antigen combined with a ribosome-inactivating protein (RIP) *(49)*. These toxins irreversibly halt protein synthesis, which results in cell death. An immunotoxin's antibody component recognizes and attaches to a highly specific membrane-associated antigen. The antibody and its coupled RIP are internalized by receptor-mediated endocytosis and then transported to the cell body.

Since immunotoxins recognize and destroy only antibody-targeted cells, it is possible to create highly selective lesions, which can mimic neurodegenerative disorders and/or address fundamental neurobiological questions.

Cholinergic neurons in the cholinergic basal forebrain express p75 neurotrophin receptors, which mediate the effects of nerve growth factor (NGF) *(50)*. SAP combines the 192 IgG MAb to the p75 low-affinity neurotrophin receptor with saporin, a potent RIP derived from *Saponaria officinalis*. The immunotoxin targets the p75 receptor localized on cholinergic nerve terminals in neocortex and hippocampus and on cholinergic cell bodies in the basal forebrain, but not on those cholinergic cell groups found in the upper brainstem *(51)*. This regional selectivity is important, since the upper brainstem complex of cholinergic neurons is spared in AD *(see 51)*. Site-specific injection of SAP produces a selective loss of cholinergic neurons and neurochemical markers of ACh synthesis. Since all ChAT-positive cells within the medial septum express p75 receptors, site-specific injection of SAP selectively destroys this population of cells *(52)*.

There is a great deal of interest in SAP, but we must evaluate the advantages and limitations of this immunotoxin. This lack of programmatic evaluation has plagued the study of other potentially useful compounds, such as AF64A. It has taken close to 15 yr to appreciate fully the kind of experimental questions that AF64A can be used to address, or not to address. The essential issues here are:

1. How selective is SAP?
2. How can it be used?
3. How should it be used?

The studies described below are beginning to address these questions.

One of the first experimental questions concerns the effects of SAP-induced cholinergic lesions on memory. Working memory was assessed in a variable-delay nonmatch to sample radial-arm maze task following injection of SAP into the medial septum. Rats were tested in an RAM task with delays of 15 min, 1 h, 4 h, and 8 h interpolated between the first 4 and second 4-arm choices. Control performance decreased from 90% accuracy at the 15-min and 1-h delays to 85% at the 4-h and 70% at the 8-h delay. Although only the 375-ng group was impaired at the 15-min delay condition, the 100-n and 237.5-ng groups were impaired at the 1-, 4- and 8-h delays. The memory impairments were related to the dose of SAP and to the demand placed on the working memory system by the task *(52)*.

A second question concerns the effects of SAP on cholinergic receptor function in the hippocampus. Administration of SAP either icv or into the medial septum does not alter binding to M1-receptors, which are primarily postsynaptic *(53,54)*. In the cerebral cortex, there is also no alteration in muscarinic stimulation of phosphoinositide hydrolysis. However, after intraseptal administration, SAP reduced the M1-receptor-mediated enhancement of evoked release of norepinephrine in hippocampal slices. This decrease in function may be owing to uncoupling of M1-receptors from G-proteins. This is indicated by a shift to the right of the binding curve for displacement of antagonist by agonist to the extent seen in control tissues after addition of GppNHp. These findings suggest that the M1 muscarinic receptor in hippocampus becomes uncoupled from G-proteins, and is very similar to findings that have been reported in AD. Understanding of the effects of SAP on muscarinic receptor subtypes may suggest new therapeutic strategies to alleviate the biochemical and behavioral consequences of cholinergic hypofunction, in lesion models, and by analogy, in AD.

7. CLINICAL RELEVANCE OF SELECTIVE CHOLINERGIC NEUROTOXINS

Cholinergic neurotoxins are being used to address a variety of issues related to cholinergic biology and function. As tools, AF64A and SAP each have distinct advantages, as well as limitations. One of the advantages of AF64A is that it can be used to study the cellular processes that contribute to ongoing degeneration of septo-hippocampal cholinergic neurons, such as oxidative stress, DNA damage, apoptosis, etc. Understanding these processes may lead to the identification of novel targets for therapeutic intervention. The limitation is that AF64A only affects the septo-hippocampal pathway and only induces an incomplete lesion. One of the advantages of SAP is that it produces a pronounced cholinergic lesion with an acute phase of degeneration and a known mechanism of action (i.e., irreversible inhibition of protein synthesis). This toxin produces dose-dependent cholinergic hypofunction and can be injected into cholinergic nuclei (i.e., components of the cholinergic basal forebrain) and specific terminal fields (hippocampus, cortex). Since both the septo-hippocampal and basal forebrain-cortical cholinergic pathways can be lesioned with SAP, this toxin can produce a profile of cholinergic pathology that is more consistent with that observed in AD. The limitation of SAP is that the degeneration is acute, not ongoing. Once a cell is exposed to SAP it will die. Therefore, SAP does not provide a good model to evaluate the effectiveness of neuroprotective strategies. The appropriate use and comparison of AF64A and SAP should provide complementary information about the function of cholinergic systems and their involvement in neurodegenerative diseases. AF64A models dysfunction and the degenerating neuron; SAP models the chemical, anatomical, and behavioral consequences of a completed cholinergic lesion. Since they provide complementary information about degeneration of cholinergic neurons, they may also be used to develop different, and again hopefully complementary strategies to prevent cholinergic cell loss or promote recovery of function.

One of the goals of developing animal models for a disease is to provide a system that can be used to evaluate new and innovative therapies for that disease. In the case of AD, the cholinergic neurotoxins model the degeneration of cholinergic neurons that is a hallmark of the disease. The models reproduce to a large degree the behavioral deficits, i.e., loss of episodic memory, as well as the biochemical deficits, i.e., loss of cholinergic markers and alterations in function of cholinergic receptors. Understanding the mechanisms involved in alterations in receptor function may lead to therapies that prevent these changes and increase the time period in which drugs are effective in alleviating symptoms. For example, synaptic plasticity is aberrant in AD, as well as in lesioned rats *(1)*. Replacing cholinergic inputs pharmacologically might "stabilize" the system, so that changes in function and plasticity are prevented. The effects of drugs may be different in a system that is adapting to a long-term loss of cholinergic input than in the normal brain. The lesion models provide a way to evaluate the effects of cholinergic drugs in a disrupted system. Drugs that are effective in the lesion model are more likely to be effective in treating the symptoms of AD. Cholinergic therapy might not be the only drug strategy needed to promote complete functional recovery, but could be a necessary component of a combined drug approach.

ACKNOWLEDGMENTS

This chapter was written while T. J. W. was supported by a grant from NSF (IBN95145557) and a gift in memory of Colonel Norman C. Kalmar. P. E. P. was supported by NIH grant AG11384.

REFERENCES

1. Walsh, T. J. and Opello, K. D. (1994) The use of AF64A to model Alzheimer's disease, in *Toxin-Induced Models of Neurological Disorders* (Woodruff, M. and Nonneman, A., eds.), Plenum, New York, pp. 259–279.
2. Coyle, J. T., Price, D. L., and DeLong, M. R. (1983) Alzheimer's Disease: a disorder of cortical cholinergic innervation. *Science* **219,** 1184–1190.
3. Bierer, L., Haroutunian, V., Gabriel, S., Knott, P., Carlin, L., Purohit, D. P., Perl, D., Schmeiodler, J., Kanof, P., and Davis, K. (1995) Neurochemical correlates of dementia severity in Alzheimer's disease: Relative importance of the cholinergic deficits. *J. Neurochem.* **64,** 749–760.
4. Evans, D. A., Funkenstein, H. H., Albert, M. S., Scherr, P. A., Cook, N. R., Chown, M. J., Herbert, L. E., Hennekens, C. H., and Taylor, J. O. (1989) Prevalence of Alzheimer's disease in a community population of older persons. *J. Am. Med. Assoc.* **262,** 2551–2556.
5. Cooper, B. (1991) The epidemiology of primary degenerative dementia and related neurological disorders. *Eur. Arch. Psychiatr. Clin. Neurosci.* **240,** 223–233.
6. Walsh, T. J. and Chrobak, J. J. (1991) Animals models of Alzheimer's disease: role of hippocampal cholinergic systems in working memory, in *Current Topics in Animal Learning: Brain, Emotion, and Cognition* (Dachowski, L. and Flaherty, C., eds.), Lawrence Erlbaum, Hillsdale, NJ, pp. 347–379.
7. Bartus, R. T., Dean, R. L., Beer, B., and Lippa, A. S. (1982) The cholinergic hypothesis of geriatric memory dysfunction. *Science* **217,** 408–417.
8. Perry, E. K., Tomlinson, B. E., Blessed, G., Bergmann, K., Gibson, P. H., and Perry, R. H. (1978) Correlation of cholinergic abnormalities with senile plaques and mental test scores in senile dementia. *Br. Med. J.* **2,** 1457–1159.

9. Wilcock, G. K., Esiri, M. M., Bowen, D. M., and Smith, C. C. T. (1982) Alzheimer's disease: Correlation of cortical choline acetyltransferase activity with the severity of dementia and histological abnormalities. *J. Neurol. Sci.* **57**, 407–417.

10. Mann, J. J., Stanley, M., Neophitides, A., de Leon, M., Ferris, S. H., and Gershon, S. (1981) Central amine metabolism in Alzheimer's disease: in vivo relationship to cognitive deficit. *Neurobiol. Aging* **2**, 57–60.

11. Sahakian, B. J., Owen, A. M., Morant, N. J., Eagger, S. A., Boddington, S., Crayton, L., Crockford, H. A., Crooks, M., Hill, K., and Levy, R. (1993). Further analysis of the cognitive effects of THA in Alzheimer's disease: Assessment of attentional and mnemonic function using CANTAB. *Psychopharmacology* **110**, 395–401.

12. Kellar, K. J., Whitehouse, P. J., Martino-Barrows, A. M., Marcus, K., and Price, D. L. (1987) Muscarinic and cholinergic binding sites in Alzheimer's disease cerebral cortex. *Brain Res.* **436**, 62–68.

13. Ladner, C. L., Celesia, G. G., Magnison, D. J., and Lee, J. M. (1995) Regional alterations in M1 muscarinic receptor-G protein coupling in Alzheimer's disease. *J. Neuropathol. Exp. Neurol.* **54**, 783–789.

14. Wang, H.-Y. and Friedman, E. (1994) Receptor-mediated activation of G proteins is reduced in postmortem brains from Alzheimer's disease patients. *Neurosci. Lett.* **173**, 37–39.

15. Jope, R. S., Song, L., Li, X., and Powers, R. (1994) Impaired phosphoinositide hydrolysis in Alzheimer's disease brain. *Neurobiol. Aging* **15**, 221–226.

16. Atweh, S. Simon, J. R., and Kuhar, M. J. (1976) Utilization of sodium-dependent high affinity choline uptake in vitro as a measure of the activity of cholinergic neurons in vivo. *Life Sci.* **17**, 1535–1544.

17. Jope, R. S. (1979) High affinity choline transport and acetyl CoA production in brain and their roles in the regulation of acetylcholine synthesis. *Brain Res. Rev.* **180**, 313–344.

18. Hanin, I., Fisher, A., Hortnagl, H., Leventer, S. M., Potter, P. E., and Walsh, T. J. (1987) Ethylcholine mustard aziridinium (AF64A; ECMA) and other potential cholinergic neuron-specific neurotoxins, in *Psychopharmacology—The Third Generation of Progress* (Meltzer, H. Y. ed.), Raven, NY, pp. 341–349.

19. Fisher, A. and Hanin, I. (1980) Minireview: choline analogs as potential tools in developing selective animal models of central cholinergic hypofunction. *Life Sci.* **27**, 1615–1643.

20. Hanin, I. (1996) The AF64A model of cholinergic hypofunction: An update. *Life Sci.* **58**, 1955–1964.

21. Hortnagl, H., Potter, P. E., Happe, K., Goldstein, S., Leventer, S., Wulfert, E., and Hanin, I. (1988) Role of the aziridinium moiety in the *in vivo* cholinotoxicity of ethylcholine aziridinium ion (AF64A). *J. Neurosci. Methods* **23**, 107–113.

22. Hortnagl, H. and Hanin, I. (1992) Toxins affecting the cholinergic system, in *Handbook of Experimental Pharmacology: Selective Neurotoxicity,* vol. 102, (Herken, H. and Hucho, F., eds.), Springer-Verlag, Berlin, pp. 293–331.

23. Uney, J. B. and Marchbanks, R. M. (1987) Specificity of ethylcholine mustard aziridinium ion as an irreversible inhibitor of choline transport in cholinergic and non-cholinergic tissue. *J. Neurochem.* **48**, 1673–1676.

24. Rylett, R. J. and Colhoun, E. H. (1984) An evaluation of irreversible inhibition of synaptosomal high-affinity choline transport by choline mustard aziridinium ion. *J. Neurochem.* **43**, 787–794.

25. Leventer, S. M., Wulfert, E., and Hanin, I. (1987) Time course of ethylcholine aziridinium ion (AF64A)-induced cholinotoxicity in vivo. *Neuropharmacology* **26,** 361–365.

26. Chrobak, J. J., Hanin, I., Schmechel, D. E., and Walsh, T. J. (1988) AF64A-induced working memory impairment: Behavioral, neurochemical and histological correlates. *Brain Res.* **463,** 107–117.

27. Barker, L. A. and Mittag, T. W. (1975) Comparative studies of substrates and inhibitors of choline transport and choline acetyltransferase. *J. Pharmacol. Exp. Ther.* **192,** 86–94.

28. Potter, P. E., Tedford, C. E., Kindel, G. H., and Hanin, I. (1987) Inhibition of high affinity choline transport attenuates both cholinergic and noncholinergic effects of ethylcholine aziridinium (AF64A). *Brain Res.* **13,** 238–244.

29. Thorne, B. and Potter, P. E. (1995) Lesions with the neurotoxin AF64A alters hippocampal cholinergic receptor function. *Brain Res. Bull.* **38,** 121–127.

30. Potter, P. E., Thorne, B., and Gaugham, C. (1997) Modulation of hippocampal norepinephrine release by cholinergic agonists is altered by AF64A lesion. *Brain Res. Bull.* **42,** 153–160.

31. Chrobak, J. J. and Walsh, T. J. (1991) Dose and delay dependent working/episodic memory impairments following intraventricular administration of ethylcholine aziridinium ion (AF64A). *Behav. Neural. Biol.* **56,** 200–212.

32. Hart, R. P., Kwentus, J. A., Harkins, S. W., and Taylor, J. R. (1988) Rate of forgetting in mild Alzheimer's-type dementia. *Brain Cogn.* **7,** 31–38.

33. Chrobak, J. J., Hanin, I., and Walsh, T. J. (1987) AF64A (ethylcholine aziridinium ion), a cholinergic neurotoxin, selectively impairs working memory in a multiple component T-maze task. *Brain Res.* **414,** 15–21.

34. Curti, D. and Marchbanks, R.M. (1984) Kinetics of irreversible inhibition of choline transport in synaptosomes by ethylcholine mustard aziridinium. *J. Membrane Biol.* **82,** 259–268.

35. Chrobak J. J., Spates M. J., Stackman R. W., and Walsh, T. J. (1989) Hemicholinium-3 prevents the working memory impairments and the cholinergic hypofunction induced by ethylcholine aziridinium ion (AF64A). *Brain Res.* **504,** 269–275.

36. Barlow, P. and Marchbanks, R. M. (1984) Effect of ethylcholine mustard on choline dehydrogenase and other enzymes of choline metabolism. *J. Neurochem.* **43,** 1568–1573.

37. Sandberg, K., Schnaar, R. L., McKinney, M., Hanin, I., Fisher, A., and Coyle, J. T. (1985) AF64A: an active site directed irreversible inhibitor of choline acetyltransferase. *J. Neurochem.* **44,** 439–445.

38. Lorens, S. K., Kindel, G, Dong, X. W., Lee, J. M., and Hanin, I. (1991) Septal choline acetyltransferase immunoreactive neurons: Dose-dependent effects of AF64A. *Brain Res. Bull.* **26,** 965–971.

39. Futscher, B. W., Pieper, R. O., Barnes, D. M., Hanin, I., and Erickson, L. C. (1992) DNA-damaging and transcription terminating lesions induced by AF64A in vitro. *J. Neurochem.* **58,** 1504–1509.

40. Johnson, G. V., Simanato, M., and Jope, R. S. (1988) Dose- and time-dependent hippocampal cholinergic lesions induced by ethylcholine mustard aziridinium ion: Effects of nerve growth factor, GM1 ganglioside, and vitamin E. *Neurochem. Res.* **13,** 685–692.

41. Wortwein, G., Stackman, R. W., and Walsh, T. J. (1994) Vitamin E prevents the place learning deficit and the cholinergic hypofunction induced by AF64A. *Exp. Neurol.* **125,** 15–21.

42. Gulyaeva, N. V., Lazareva, N. A., Libe, M. L., Mitrokhina, M. V., Yu, M., and Walsh, T. J. (1996) Oxidative stress in the brain following intraventricular administration of ethylcholine aziridinium (AF64A). *Brain Res.* **726**, 174–180.

43. Bondy, S. C. (1992) Reactive oxygen species: Relation to aging and neurotoxic damage. *Neurotoxicology* **13**, 87–100.

44. Fariello, R. G., Ghilardi, O., Peschechera, A., Ramacci, M. T., and Angelucci, L. (1988) Regional distribution of ubiquinones and tocopherols in the substantia nigra. *Neuropharmacology* **27**, 1077–1080.

45. Ravindranath, V., Shivakumar, B. R., and Anandatheerthavarda, H. K. (1989) Low glutathione levels in brain regions of aged rats. *Neurosci. Lett.* **101**, 187–190.

46. Mizuno, Y. and Otha, K. (1986) Regional distribution of thiobarbituric acid-reactive products, activities of enzymes regulating the metabolism of oxygen free radicals, and some of the related enzymes in adult and aged rat brains. *J. Neurochem.* **46**, 1344–1352.

47. Jeandel, C., Nicolas, M. B., Dubois, F., Nabet-Belleville, F., Penin, F., and Cuny, G. (1989) Lipid peroxidation and free radical scavengers in Alzheimer's disease. *Gerontology* **35**, 275–282.

48. McGurk, S. R., Hartgraves, S. L., Kelly, P. H., Gordons, M. N., and Butcher, L. L. (1987) Is ethylcholine aziridinium ion a specific cholinergic neurotoxin? *Neuroscience* **22**, 215–224.

49. Wiley, R. G. (1992) Neural lesioning with ribosome-inactivating proteins: Suicide transport and immunolesioning. *Trends Neurosci.* **15**, 285–290.

50. Springer, J. E. (1988) Nerve growth factor receptors in the central nervous system. *Exp. Neurol.* **102**, 354–365.

51. Woolf, N. J., Jacobs, R. W., and Butcher, L. L. (1989) The pontomesencephalotegmental cholinergic system does not degenerate in Alzheimer's disease. *Neurosci. Lett.* **96**, 277–282.

52. Walsh, T. J., Herzog, C., Gandhi, C., Stackman, R. W., and Wiley, R. G. (1996) Injection of IgG 192-saporin into the medial septum produces cholinergic hypofunction and dose-dependent working memory deficits. *Brain Res.* **726**, 69–79.

53. Rossner, S., Schliebs, R., Perez-Polo, J. R., Wiley, R. G., and Bigl, V. (1995) Differential changes in cholinergic markers from selected brain regions after specific immuno-lesion of the rat cholinergic basal forebrain system. *J. Neurosci. Res.* **40**, 31–43.

54. Potter, P. E., Gaugham, C., and Assouline, Y. (1996) 192 IgG-Saporin lesions produce uncoupling of the M1 muscarinic receptor. *Soc. Neurosci. Abstracts* **22**, 1258.

13
Glutamatergic Receptor Agonists and Brain Pathology

Ewa M. Urbanska, Andrzej Dekundy, Zdzislaw Kleinrok, Waldemar A. Turski, and Stanislaw J. Czuczwar

1. INTRODUCTION

Excitatory amino acids (EAAs) represent a major group of neurotoxic agents, primarily affecting the nerve cell body. The beginning of the excitotoxicity era can be traced back to early 1950s when Hayashi described convulsant properties of L-glutamic acid (GLU) and L-aspartic acid (ASP) following their administration into the cerebral cortex of dogs and monkeys (1). Later, Lucas and Newhouse reported that systemic GLU destroyed inner layers of immature mouse retina (2). Excitatory properties of GLU within the central nervous system (CNS) were described by Curtis and Watkins in 1959 (3,4). This inspired further studies on GLU and its derivatives, which ultimately led to consensus that GLU is a principal excitatory neurotransmitter acting at a majority of EAA synapses, which are present on practically every neuron in the CNS.

In 1969 Olney described GLU-induced neuronal loss, which was restricted to brain areas lacking the protective blood–brain barrier (5,6). Soon, in pioneering sets of experiments, Olney with his colleagues established the correlation among excitatory properties of GLU, its analogs, and their neurodegenerative potential. The concept of excitotoxicity was thus introduced and in a short time became a classic in neurobiology (7).

Data accumulated over the last several decades indicate that EAAs have an essential role in inducing neuronal cell loss and overall neuropathology that occurs acutely, e.g., with stroke, subacutely, e.g., during viral infections, and in the course of slowly progressing neurodegenerative disorders, such as Huntington's (HD) or Alzheimer's disease (AD). The diverse nature of endogenous EAA neurotransmitters might be described by their special ability to transform under certain conditions into deleterious agents causing acute or chronic

From: Highly Selective Neurotoxins: Basic and Clinical Applications *Edited by: R. M. Kostrzewa.*
Humana Press Inc., Totowa, NJ

neurodegeneration. The functional variety of EAAs can be attributed to at least three groups of factors: first, the balance between inhibitory and excitatory influences; second, the multiplicity of receptor types and subtypes occurring in CNS; and third, the number of modulators affecting glutamatergic functions. Homeostasis between excitatory and inhibitory brain systems remains beyond the scope of this chapter and will not be discussed here.

2. GLUTAMATERGIC RECEPTORS

2.1. Basics

Diverse combinations of receptors are located either pre- or postsynaptically on neuronal membranes and on glial cells near the synaptic space. EAA or glutamatergic receptors are currently classified into two major groups: ionotropic ligand-gated ion channels and metabotropic receptors coupled to the second messenger system via GTP-binding protein (G protein) *(8,9,10)*.

Ionotropic receptors are subdivided into three types, so named because of their respective preferential agonists: *N*-methyl-D-aspartate (NMDA), α-amino-3-hydroxy-5-methyl-isoxazole-4-propionate (AMPA), and kainate *(11)*. The latter two are also given a common designation, non-NMDA receptors. Numerous data from autoradiographic, immunohistochemical, and electrophysiological experiments have revealed that EAA receptors are distributed with different densities throughout brain. Moreover, their location on cellular membranes of both glial cells and neurons is highly heterogeneous *(8–11)*.

NMDA receptors can occur both pre- and postsynaptically *(8,12)* and are prevalent in hippocampal stratum radiatum, CA1 and dentate gyrus, layers 1–3 of frontal cortex, pyriform cortex, basal ganglia, and amygdala *(13,14)*. AMPA receptor distribution generally parallels the occurrence of NMDA binding sites. AMPA receptors are also situated on presynaptic membranes *(15)*. AMPA receptors are found in the highest density on hippocampal pyramidal cells, especially on cell bodies and dendritic areas, and also on cells in the molecular layer of the cerebellum *(13)*. Kainate receptors are prevalent in those areas where NMDA receptors are least numerous, such as stratum lucidum of hippocampal CA3, inner part of dentate gyrus, layer 5 and 6 of cerebral cortex, hypothalamus, and certain thalamic nuclei. Presynaptic kainate receptors have been described in hippocampus *(16,17)*. Metabotropic glutamate (mGLU) receptors are reported to occur on neurons, pre- and postsynaptically, and on glial cells, e.g., in hippocampus and cerebral cortex *(10)*.

The majority of all fast synaptic transmission seems to be mediated via AMPA receptors, whereas NMDA stimulation is involved in synaptic formation, neuronal plasticity during ontogeny, and such processes as long-term potentiation related to some form of learning and memory *(8,9)*. Presynaptic NMDA and AMPA receptors can affect the exocytosis of noradrenaline, dopamine, acetylcholine, and 5-hydroxytryptamine *(15,18–21)*. Kainate receptors probably participate in fast synaptic transmission in concordance with AMPA receptors

(9). They were also implicated in the presynaptic control of GLU and dopamine release *(16,17,22)*. Metabotropic receptors are rarely directly involved in the process of synaptic transmission. They seem to fulfill a modulatory function toward both excitatory and inhibitory neurotransmission. Recently, it has been shown that metabotropic receptors can modulate release of GABA and glutamate *(23–25)*. All of the EAA receptors can be activated by endogenous agonists and exogenous ligands, many of which are considered to be potent neurotoxins (*see* Table 1).

2.2. Ionotropic Receptors

2.2.1. NMDA

The NMDA receptor is a multiple unit complex coupled to a high-conductance ion channel permeable to Na^+, K^+, and Ca^{2+}. The immense progress in molecular neurobiology during the last several years has resulted in the cloning of two main subunits of NMDA receptors *(26,27)*. The core NMDAR1 (or NR1) subunit displays virtually all properties of the NMDA ionophore, since it is equipped with all of the binding sites ascribed to NMDA receptors. It can occur in eight functional isoforms: $NMDAR1_A$–$NMDAR1_H$. However, whole-cell currents induced by binding of an agonist to the NMDAR1 unit are potently increased when NMDAR1 is coexpressed with one of the NMDAR2 subunits. Currently, there are four known NMDAR2 subunits, $NMDAR2_A$–$NMDAR2_D$. It appears that NMDAR2 subunits exert regulatory actions, because they do not form functional receptors on their own *(26,27)*. The third type of NMDA receptor subunit, called C-1 or NMDA-L, was recently cloned *(28,29)*. The expression of different combinations of NMDA units can result in varying properties and distinct effects within the CNS.

Binding of an agonist to an NMDA receptor evokes cell responses that depend on membrane potential and the extracellular presence of Mg^{2+} ions. At the resting potential Mg^{2+} blocks the ion channel associated with the NMDA receptor. Partial depolarization of the cell membrane leads to a reduction of Mg^{2+} block and an increase of Ca^{2+} and Na^+ influx, with K^+ efflux. Depolarization progresses and induces the excitation of a nerve cell. Broadly documented voltage-dependent channel block represents a unique property of NMDA receptors, determined by the necessity of simultaneous agonist binding and postsynaptic depolarization for the activation to occur *(8,9,26)*.

Numerous lines of evidence suggest that there are several binding sites within the NMDA receptor complex *(8,26,30)*. The neurotransmitter binding site is susceptible to a variety of endo- and exogenous ligands (Table 1). Antagonists, such as DL-2-amino-7-phosphonoheptanoic acid (DL-AP7), 3-((R)-2-carboxy-piperazin-4-yl)propyl-1-phosphonic acid (CPP), or cis-4-phoshponomethyl-2-piperidine carboxylic acid (CGS 19755), bind competitively to an agonist domain. Three separate sites for noncompetitive antagonists were identified within the ion channel. Inactivation of NMDA receptor function may result from

Table 1
Some of the Agonists of EAA Ionotropic and Metabotropic Receptors

Receptor type	Endogenous agonists	Exogenous agonists
Ionotropic		
NMDA[a]	Homocysteine sulfinate	NMDA
(8, 152–154)	Quinolinate	Ibotenate
	Aspartate	
	Glutamate	
	Cysteine sulfinate	
	L-Serine-*O*-phosphate	
AMPA	Glutamate	AMPA
(8,152,153,155,156)	Aspartate	Fluorowillardiine
		Kainate
		Quisqualate
		BOAA
Kainate	Glutamate	Kainate
(152,153,157)		SYM 2081
		Diomonate
Metabotropic		
Group I	Glutamate	Quisqualate
(10,23,43)	Cysteine sulfinate	Ibotenate
	Homocysteine sulfinate	DHPG
		(1S,3R)-ACPD
		BMAA
Group II	Glutamate	CCG-I
(10,49)		(1S,3R)-ACPD
		Ibotenate
		Quisqualate
Group III	Glutamate	L-AP4
(10,49,158)		L-serine-*O*-phosphate
		HomoAMPA
Group IV(?)[b]	Cysteine sulfinate	Quisqualate
(phospholipase		(1S,3R)-ACPD
D-coupled) *(51)*		L-AP3

the action of Mg^{2+} or Zn^{2+} or phencyclidine (PCP)/dizocilpine (MK-801) at their respective binding sites. PCP-like substances display their activity following channel opening as a consequence of an agonist binding, and therefore evoke use-dependent NMDA receptor channel block *(26)*.

The existence of a coagonist site, on which glycine binding is obligatory for NMDA activation, is very well documented *(31)*. This strychnine-insensitive glycine site can also bind L-serine and L-alanine, although in high concentrations, as well as D-serine *(32,33)*. It was suggested that D-serine might be released from glial cells in the vicinity of NMDA receptors in order to synergize with synaptic glutamate *(34)*.

Kynurenic acid, the only known endogenous broad-spectrum antagonist for all the ionotropic glutamate receptors, also displays high affinity for the glycine-binding site of the NMDA complex *(35)*. Kynurenic acid, which is produced and released preferentially from glial cells, displays potent inhibitory effects both in vivo and in vitro *(35)*.

Recent studies suggest the existence of several other modulatory sites within the NMDA receptor complex. Their physiological significance is still being elucidated. The first reports on polyamine-induced modulation of NMDA receptor action showed increased binding of MK-801 to rat brain membranes in the presence of spermine and spermidine *(36)*. It was further demonstrated that spermine, spermidine, and putrescine can enhance NMDA currents because of interaction with perhaps two or more extracellular regulatory sites *(37)*. NMDA receptor function can also be altered via a pH sensitive, H^+ binding site located outside the ion channel and through modulation of the whole receptor redox state *(38)*. The latter one is probably influenced by thiol derivatives, and/or ascorbic acid and/or free radical formation *(38)*.

Ca^{2+}/calmodulin-dependent protein kinase II has now been shown to regulate NMDA receptor function *(39–41)*. The action of calmodulin is exerted via binding to the NMDAR1 subunit, which reduces the probability of channel opening by a factor of four *(42)*. Ca^{2+}/calmodulin also antagonizes binding of α-actinin-2, cytoskeletal protein discovered to immobilize NMDAR1 and NMDAR2B subunits at postsynaptic sites *(43)*.

Table 1 *(see opposite page)*

[a]Abbreviations: **NMDA**: *N*-Methyl-D-aspartic acid, **AMPA**: α-amino-2,3-hydroxy-5-methyl-3-oxo-isoxazole-propanoic acid, **BOAA**: β-*N*-oxalylamino-L-alanine, **SYM 2081**: (2S,4R)-4-methylglutamic acid, **DHPG**: 3,5-dihydroxyphenylglycine, **CCG-I**: (2S,1′S,2′S)-2-(carboxycyclopropyl)glycine, **(1S,3R)-ACPD**: (1S,3R)-1-aminocyclopentane-1,3-dicarboxylic acid, **BMAA**: β-*N*-methyloamino-L-alanine, **L-AP4**: L-2-amino-4-phosphonopropionate, **homoAMPA**: 2-amino-4-(3-hydroxy-5-methylisoxazol-4-yl)butyric acid, **L-AP3**: L-2-amino-3-phosphonopropionate.

[b]Group IV(?)-suggested new class of metabotropic receptors.

2.2.2. Non-NMDA Receptors

AMPA-activated channels display fast kinetics and are associated with increased Na^+ influx and increased K^+ efflux. In contrast to that observed for NMDA receptors, agonist binding to the AMPA receptor usually results in very low Ca^{2+} permeability. AMPA-mediated transmission is voltage-insensitive and leads to rapid depolarization of neurons, electrophysiologically manifested by occurrence of an excitatory postsynaptic potential (EPSP). Classical AMPA-induced currents decay rapidly. Kainate may trigger a response that persists with its continued presence on AMPA binding sites. When kainate is used repeatedly, the high-affinity recognition site desensitizes rapidly *(9)*.

Molecular studies suggest that AMPA receptors are heteromeric complexes made up from a variety of units. Assemblies of GluR1–4 (A–D) subunits construct functional AMPA receptors. Cloned homomeric GluR1, R3, and R4 are all Ca^{2+}-permeable. However, when coexpressed with GluR2, new heteromeric AMPA channels lose their permissiveness toward Ca^{2+} entry *(44)*. Brain cells may thus control the amount of Ca^{2+} load mediated via AMPA receptors, by differential expression of GluR2 gene products.

Two groups of kainate receptor subunits comprise the recently cloned GluR5-7 and KA1-2. Homomeric assembles of GluR5 and R6 but not GluR7, KA1, or KA2 display properties characteristic for the kainate receptor ion channel. Heteromeric channels show more rapid desensitization in comparison to homomeric ones *(9)*.

AMPA and kainate receptors possess an agonist site competitively antagonized by quinoxalinediones, such as 6-nitro-7-sulfamoylbenzo(f)-quinoxaline-2,3-dione (NBQX) *(45)*. A negative allosteric site at non-NMDA receptors binds noncompetitive antagonists, such as 1-(4-aminophenyl)-4-methyl-7,8-methyl-enedioxy-5H-2,3-benzodiazepine (GYKI 52466) *(46)*. Some types of Ca^{2+}-permeable AMPA/kainate receptors might be potently modified by spermine. This polyamine-evoked inward current rectification occurs intracellularly, contrary to NMDA receptor modulation displayed by extracellular polyamines *(37)*. AMPA receptors were recently found to interact with GRIP (GLU receptor interacting protein) that links receptor to other membrane proteins and thus appears to cluster it at EAA synapses in the brain *(47)*.

2.3. Metabotropic Receptors

The presence of metabotropic GLU receptor (mGLU) modulating the synthesis of intracellular second messengers was first reported in 1986 by Sladeczek and colleagues *(48)*. The following studies soon revealed that glutamate activates a whole family of receptors coupled to G-proteins *(10)*.

mGLU receptors are divided into three main groups comprising eight recently cloned receptor subtypes. Group I receptors, $mGLU_1$ and $mGLU_5$, are known to stimulate phospholipase C and thus initiate phosphoinositide (PI) hydrolysis and release of Ca^{2+} from internal stores of both neurons and glial cells *(49)*. Astrocytic Ca^{2+} waves seem to be mediated via mGLU class I receptors *(50)*. Group I binding sites were also demonstrated to affect K^+ channel function *(10,49)*. Receptors described as group II, $mGLU_{2-3}$ and group III, $mGLU_{4, 6-8}$ are negatively coupled to adenylate cyclase. It was noted that formation of cAMP might be enhanced because of mGLU-evoked potentiation of stimulatory adenosine effects *(10)*. All metabotropic agonists inhibit voltage-dependent L-type Ca^{2+} channels *(49)*, whereas Group I and II agonists inhibit N-type Ca^{2+} channels *(10)*. A fourth type of mGLU receptor, coupled to phospholipase D, was recently suggested *(51)*. L-cysteine potently stimulates this mGLU receptor *(10)*.

Mechanisms associated with regulating the function of mGLU receptors are not well understood. Some insights have come from experiments that reveal the specific distribution of mGLU receptors within the synaptic cleft. Presynaptic mGLU receptors are located at peripheries of the synaptic junction. During initial exocytosis of GLU these receptors might be silent, but as GLU levels rise, presynaptic mGLU receptors appear to become activated. As a consequence, neurotransmitter efflux could be inhibited. Activation of presynaptic mGLU receptors would thus be use-dependent *(52)*.

Metabotropic glutamate signaling can be also controlled through processes responsible for spatial localization and clustering of receptoral proteins. Recent studies have identified the dendritic protein Homer, whose expression is regulated during development. Homer binds specifically to mGLU receptors and is thus able to control mGLU-mediated signaling *(53)*.

3. EXCITOTOXICITY

3.1. Selectivity

It is fortunate for neuronal cells with EAA receptors that susceptibility to toxic actions of GLU and other EAAs is not a simple consequence of glutamatergic system activity. Brain regions and neuronal subsets within defined brain areas basically differ in their sensitivity to EAA action and the outcome of receptor stimulation. However, differences in distribution of glutamatergic receptor types and subtypes do not fully account for this.

NMDA agonists display pronounced disparities when studied as neurotoxic and convulsant agents in vivo. The potent neurotoxin, ibotenic acid, is extremely weak in evoking seizures when applied intrahippocampally *(54)*. In contrast, the weaker excitotoxin NMDA is a potent convulsant in rodents *(55)*, whereas the endogenous neurotoxin, quinolinic acid, is less potent as a seizure-inducing agent *(56)*.

Disruption of afferent pathways can totally abolish neurotoxic effects of locally injected kainate, but not ibotenate *(57,58)*. This intriguing dependence of EAA toxicity on presynaptic input is still a poorly understood phenomenon.

3.2. Cellular Events

Neurodegeneration following the exposure to EAA agonists might be mediated by any of the three ionotropic receptors. Activation of mGLU receptors was also reported to induce neuronal loss *(59,60)*.

The earliest neuropathological evidence following exposure to excitotoxins was described as an acute swelling of postsynaptic dendrosomal elements, accompanied by mitochondrial edema, with eventual condensation and vacuolization of the cell. The presynaptic axonal part remains unchanged, and the lesions are therefore called "axon-sparing" *(61)*. A small fraction of cells may respond with the condensation of both cytoplasm and nucleus. It has also been observed that large doses of excitotoxins can induce glial swelling *(61)*.

When neurons are briefly exposed to the high levels of GLU, they may undergo transient edema, followed by a gradual deterioration over a period of several hours. It is believed that two components of neurodegeneration correspond to acute swelling and delayed neurodegeneration *(62)*. The depolarization of postsynaptic membrane increases Na^+ ion entry through voltage-dependent Na^+ channels; subsequently, passive Cl^- and water influx induces osmotic cellular edema.

Altered intracellular homeostasis of Ca^{2+} seems to be a final common pathway in excitotoxic phenomena. Studies initiated by Choi and colleagues in 1980s reveal that delayed neuronal loss depends on extracellular Ca^{2+} availability *(62)*. NMDA receptors and those AMPA receptors lacking GluR2 subunits are directly permeable to Ca^{2+} ions. However, EAA ionotropic receptors that do not directly stimulate Ca^{2+} channels can indirectly activate voltage-operated Ca^{2+} channels by promoting Na^+ influx via depolarization of the postsynaptic membrane.

Ca^{2+} influx can initiate a variety of events, including neurotoxicity if intracellular levels become high enough. The first step in the process is release of Ca^{2+} from intracellular stores in the endoplasmic reticulum *(63)*. High intracellular Ca^{2+} acts further, activating a number of enzymes, including proteases, such as calpain I. Calpain I is known to degrade structural proteins, such as tubulin, microtubule-associated proteins, neurofilamental polypeptides, and spectrin *(64)*. Ca^{2+}-evoked phospholipase A2 induction might cause membrane disintegration and subsequent release of arachidonic acid, which in turn could enhance GLU efflux, potentiate NMDA-evoked currents, and inhibit GLU uptake *(65)*. Free radicals produced in the arachidonic acid cascade, together with Ca^{2+}-induced nitric oxide formation, were postulated to cause further lipid peroxidation and cellular destruction *(66,67)*.

Elevated cytosolic Ca^{2+} can also activate protein kinase C, and thus indirectly affect the function of different cellular enzymes *(68)*. High levels of EAA agonists, especially GLU, may also activate presynaptic protein kinase C via phospholipase-coupled presynaptic mGLU receptors. This would facilitate GLU release, but only in the presence of arachidonic acid *(65)*.

New insights into the mechanisms of excitotoxicity were provided by reports on the involvement of cell suicide genes in the observed neuronal loss. A proportion of neuronal deaths following excitotoxic insult might occur because of Ca^{2+}-triggered transcriptional activation of cell death genes. This would lead to the induction of delayed programmed cell death, apoptosis *(69)*. Complex processes ultimately leading to neurodegeneration involve positive and negative gene regulation. In contrast to severe and sudden necrotic cell loss, apoptosis is characterized by remarkable cellular shrinking, pyknotic nuclear changes, and degradation of the nucleus into dense apoptotic bodies *(69)*. Mice deficient in p53 tumor supresor gene, identified as one promoting cell death, are virtually resistant to the neurotoxic action of kainate *(70)*. It has been suggested that degenerative neuronal phenotypes are dependent on the degree of brain maturation and the type of EAA receptor that is activated. Particularly after non-NMDA receptor stimulation, there is a morphological gradient of cell loss, ranging from apoptosis that predominates in newborn animals, to a decline in apoptotic-like changes with prevalent necrosis in adult animals *(71)*.

Once set in motion, the process of excitotoxicity becomes a vicious cycle. Injured neurons and swollen astrocytes release excessive quantities of GLU, which triggers further neurodegeneration. A variety of factors contributing to the function of EAA systems suggests that a counter approach aimed at protecting against EAA neurotoxicity should not be restricted to a single receptor type or to one subcellular event. Numerous mechanisms are implicated in EAA neurotoxicity, particularly processes related to cellular energy status. This topic was the focus of recent excellent reviews *(72,73)*.

4. CLINICAL RELEVANCE

4.1. General

Excitotoxins represent a valuable experimental tool in research concerning the pathogenesis of several human CNS disorders. It is well recognized that the peripheral or local application of acidic amino acids in animal brain might produce seizures and neuronal loss. An accumulating body of data elaborated in vitro and in vivo, and based on human tissue analysis, strongly implicates involvement of the glutamatergic system in several human brain disorders.

4.2. Exogenous Poisoning

The most direct evidence for EAA's involvement in brain pathologies comes from reports on the consequences of poisoning by foods containing high amounts of glutamatergic agonists. Mussels might contain the kainate agonist, domoic acid, which induces a syndrome in humans characterized by memory disturbance and seizures. In the most severe cases, hippocampal and amygdaloid lesions are observed *(74,75)*. The AMPA agonist β-*N*-amino-oxalylamino-L-alanine (BOAA), found in high amount in the chick-pea *Lathyrus sativus,* evokes lathyrism, a disorder characterized by motoneuron atrophy with spastic paraplegia *(76)*. Guam-type amyotrophic lateral sclerosis with parkinsonism and dementia is related to consumption of β-methyl-aminoalanine (BMAA), which is found in sago palm and described to interact with NMDA receptors and mGLU receptors class I *(10,77)*.

4.3. Ischemia/Anoxia

Hypoxic-ischemic brain damage may occur in all age groups as a consequence of focal or global impairment of blood flow, resulting from cerebral hemorrhage, trauma, cardiac arrest, respiratory failure, and so forth. Transient global ischemia induces selective loss preferentially of CA1 hippocampal pyramidal cells, striatal medium size neurons, and neuronal cells in cortical layers 3, 5, and 6. Regional sustained ischemia leads to infarction with pannecrosis involving neurons as well as glial and endothelial cells *(78)*.

The involvement of altered glutamatergic transmission in ischemic events is supported by numerous data from in vivo and in vitro studies. A massive, six- to eightfold rise in extracellular GLU and ASP concentrations was found in adult rat brain following a transient hypoxia *(79)*. Profound impairment of GLU and ASP uptake from extracellular compartment was observed in rodents *(80)*. The reperfusion period following transient ischemia is characterized by an augmented release of endogenous EAA agonists *(81)*. Hippocampal cell cultures undergo neurodegeneration when subjected to GLU in the absence of oxygen *(82)*. A lesion of the glutamatergic pathway into the hippocampus prevented ischemia/hypoxia-induced neuronal loss *(83)*. The decreased expression of NMDA receptors following administration of antisense oligonucleotides also abated the process of neurodegeneration after cerebral artery occlusion in rats *(84)*. Further indictment of the glutamatergic system in ischemic neuronal death originates from the studies based on the use of ionotropic EAA receptor antagonists. Many studies demonstrate successful neuroprotection by glutamatergic antagonists in experimental models of ischemia. NMDA blocking agents effectively protected against lesions resulting from experimental hypoxia both in vitro and in vivo in models of local but not global ischemic neuronal loss *(85,86)*. AMPA/kainate antagonists ameliorated the outcome of experimental ischemia and were effective in both focal and local ischemia models *(86)*. However, the therapeutic window is rather narrow and does not extend beyond a few hours postinsult. Antagonists at the NMDA glycine site, when administered before or within a short time after induction of focal ischemia, are also protective *(87)*. With these antagonists of the polyamine site there is a wider window for protection, up to 6 h after focal ischemia *(88)*. Redox modulation of NMDA receptors might constitute another approach to treat ischemia. An oxidizing agent affecting the NMDA redox site diminished the size of the ischemic infarct zone in rats *(38)*. Recently, mGLU receptors were suggested to be involved in the pathogenesis of ischemic neuronal loss. The altered expression of mGLU receptors was described in rat hippocampus following transient global ischemia *(89)*. As evaluated in vitro, class I mGLU agonists increased the recovery of hippocampal slices rendered anoxic, whereas class II agonist displayed deleterious effects *(90)*.

Future therapeutic strategies aimed toward treating ischemia might incorporate assorted ways to modulate EAA systems. Several agents influencing glutamatergic transmission are currently in clinical trials, evaluating their efficacy in different forms of ischemic neuropathology *(91)*.

4.4. Huntington's Disease

Despite the enormous scientific effort that has gone into discovery of the defective gene in HD *(92)*, the exact nature of pathomechanisms underlying this hereditary neurodegenerative disorder are still unknown. HD, which strikes usually in middle age, is characterized by motor symptoms, such as choreiform movements, and progressive mental disorder with psychiatric symptoms ranging from personality changes, through psychosis, to debilitating dementia. Neuropathological findings have consistently demonstrated that in the early stage of HD, neuronal death is confined to caudate and putamen, with preferential loss of medium spiny neurons *(93)*. Neuropeptide Y/NADPH diaphorase-containing large neurons are usually well preserved.

Pathological changes observed after intrastriatal administration of EAAs are remarkably similar to these observed in the course of HD. In fact, excitotoxic lesions of caudate-putamen were suggested as experimental models of the disease *(94)*. Described as a first, kainate-induced striatal neurodegeneration revealed certain morphological and biochemical similarities to HD; losses of GABA- and substance P-containing neurons, with relative sparing of large aspiny neurons *(95–97)*. When closely assessed, the kainate model was rendered as too inaccurate, mainly because of the occurrence of neurodegeneration in distant limbic structures, a finding inconsistent with the neuropathology of HD. Striatal lesions with quinolinic acid, an endogenous metabolite of tryptophan along the kynurenine pathway, were found to provide a better animal model of HD. There is no extrastriatal neuronal death following local application of quinolinate; moreover, the biochemical and histopathological profile more closely resembles HD-related alterations *(98–100)*. In the search for an endogenous compound(s) linked to the pathogenesis of HD, quinolinic acid was proposed a candidate. However, the initial report of enhanced activity of quinolinate synthesizing enzyme in brain specimens from HD victims *(101)* was not followed by observations of increased quinolinate levels either in putamen or cerebrospinal fluid *(102,103)*.

An involvement of altered glutamatergic transmission in the pathogenesis of HD has gained support from studies exploring its function in the most severely lesioned brain structures. Marked loss of glutamate binding sites, especially the NMDA subtype, was demonstrated in HD putamen *(104,105)*. Later, however, it was noted that binding of ligands for all EAA receptors is rather uniformly decreased in the course of HD *(106)*, most probably from the intense loss of EAA receptor-bearing neuronal populations. Increased vulnerability toward the excitotoxic action of GLU was shown in fibroblasts obtained from HD patients *(107)*. Prominent loss of GLU uptake sites was found in HD striatal tissue *(108)*. These findings led to the hypothesis that in HD there is not only an overabundance of endogenous EAA agonists, but also either an increased sensitivity to physiological levels of glutamate-related compounds, or an imbalance between endogenous excitatory and inhibitory substances. Indeed, we have observed that aminooxyacetic acid (AOAA) an inhibitor of the enzyme synthesizing the only known endogenous EAA antagonist, kynurenic acid *(109)*, induces HD-like sequelae in rat striatum *(110,111)*. Furthermore, AOAA-induced neuronal loss is attenuated by NMDA receptor antagonists *(110,111)*. The consequences of AOAA-induced striatal neurodegeneration are remarkably similar to quinolinate-evoked lesions (Table 2). Our report was confirmed and extended by others who revealed that the essential component of AOAA-related neurotoxicity seems to be failure of mitochondrial function and consequent impairment of cellular energy status *(112)*. In fact, markedly diminished glucose metabolism in the HD caudate-putamen had been previously described in several studies using positron emission tomography (PET) techniques *(113)*. The concept of cellular metabolism inhibition is now suggested in the pathomechanism of neurodegeneration *(72,73,114)*. Reduced activity of kynurenic acid-synthesizing enzyme in striatal tissue of HD victims *(115)* suggests that altered kynurenic acid availability may also contribute to HD pathogenesis.

In animal models of HD EAA antagonists effectively block the occurrence of neurodegeneration. Consequently, the first clinical trials with an NMDA channel blocker were undertaken in people with HD, despite the obvious difficulties in determining the real benefits *(91)*.

4.5. Alzheimer's Disease

Alzheimer's disease, considered to be the most common neurodegenerative disorder, is characterized clinically by progressive deterioration of cognitive functions and inevitable development of total dementia. Neuropathological alterations include extensive cell loss (mainly in the neocortex, amygdala, hippocampal CA1 area, nucleus basalis of Meynert, medial septal area, and diagonal band of Broca), accompanied by cerebral amyloidosis. The characteristic feature of AD consists of neurofibrillary tangles and senile plaques, the latter ones being insoluble aggregates of β-amyloid derived from β-amyloid precursor protein and deposited in gray matter *(116)*.

It has been speculated that EAA systems are involved in the neuropathology of AD. EAA lesions of rodent basal forebrain, despite some disadvantages, appear to induce learning and memory deficits that resemble the AD-related deficits observed in humans *(117)*. The highest density of neurofibrillary tangles was observed within the AD brain areas supplied by numerous glutamatergic afferents *(118)*. Prominent reductions in GLU binding sites, especially NMDA binding sites, correspond to the observed neuronal loss in AD brain tissue *(104)*. Also, AMPA receptor messenger RNA is markedly reduced in AD entorhinal

Table 2
Characteristic of Neurodegeneration Following Intrastriatal Application of Some EAA Agonists and Amino-Oxyacetic Acid

| | Occurrence of neuronal loss | | | |
Substance	Locally	In extrastriatal structures	In striatum of young rats	When coadministered with NMDA antagonist
Kainic acid (8,159–161)	+[a]	+	–	+
NMDA (8,161,162)	+	–	+	–
Ibotenic acid (163,164)	+	–	+	–
Quinolinic acid (98,161,165)	+	–	–	–
Amino-oxyacetic acid (110,111)	+	–	–	–

[a] +: Presence of neuronal loss, –: absence of neuronal loss, NMDA: *N*-methyl-D-aspartic acid.

cortex *(119)*. In an attempt to evaluate the interaction between β-amyloid and EAA receptor density, transgenic mice expressing human β-amyloid were generated. NMDA and AMPA binding sites remained unchanged in this model *(120)*. However, this does not exclude a role for EAA system abnormalities in the pathology of AD, since neurotoxicity might be the consequence of more than simple glutamatergic hyperfunction. In vitro, GLU was found to induce paired helical filaments that mimic neurofibrillary tangles *(121)* and to evoke simplification of the dendritic tree, another AD feature *(122)*. β-Amyloid potentiates the excitotoxicity of GLU in cultured cortical neurons *(123)*, enhances the release of GLU and ASP (especially from hippocampal slices of aged rats)*(124)*, and inhibits Na^+-dependent GLU uptake in rat hippocampal astrocytic cultures *(125)*. It has been suggested that β-amyloid destabilizes intracellular Ca^{2+} levels and thus increases neuronal susceptibility toward endogenous excitotoxins *(126)*. Processing of β-amyloid precursor protein into nonamyloidogenic peptides was recently demonstrated to depend on mGLU receptors function *(127)*. Neurodegeneration in the course of AD leads to GLU deficiency which, surprisingly, might in turn diminish protective actions mediated by mGLU receptors and further enhance the progress of disease.

4.6. Epilepsy

Epilepsy, affecting approx 0.5% of the population at any age, represents a group of disorders characterized by recurrent seizure activity. Seizures might be generalized, with widespread cortical hyperactivity; or partial, originating from an epileptogenic focus, usually within the temporal lobe. The most common neuropathological feature observed in about half the cases of focal epilepsy appears in the form of sclerosis of Ammon's horn, i.e., a hippocampal lesion accompanied by massive gliosis *(128)*.

A large body of evidence supports the hypothesis that altered GLU transmission is a contributing factor in the pathogenesis of epilepsy. A classical way to induce generalized tonic-clonic or limbic seizures in animals is to administer EAA agonists intracerebrally or locally *(55,56,129)*. Among the most potent convulsants, kainic acid plays a special role. The behavioral sequelae following kainate injection resemble the symptoms of temporal lobe epilepsy in humans *(129)*. After its local administration, kainate induces limbic seizures, easily progressing into status epilepticus, with both components also being manifested electroencephalographically. The subsequent neurodegeneration of limbic structures is remarkably similar to that observed in epileptic patients *(129,130)*. Other EAA agonists, such as NMDA or quinolinic acid, are also able to evoke seizures that occur in the form of generalized tonic-clonic convulsions or wild running episodes in experimental animals *(54,55)*.

There are inconsistent findings on levels of GLU and ASP in serum and cerebrospinal fluid of epileptic patients, with reports of either an increase or decrease in neurotransmitter levels *(131)*. In vivo microdialysis studies revealed that GLU levels rise during epileptic activity. However, it is not known if increased GLU levels are the consequence or the cause of seizure discharges *(131)*. In the hippocampus of a conscious person a sustained rise in GLU preceded the onset of seizures *(132)*. This strongly argues for the involvement of altered glutamatergic transmission in the pathogenesis of epilepsy. An epileptogenic role for GLU was also demonstrated by the development of seizures in rats with a knockout of neuronal GLU transporter *(133)*.

A correlation between seizure susceptibility and increased formation of quinolinic acid, another endogenous EAA agonist, was suggested by the presence of enhanced activity of the quinolinate-synthesizing enzyme in brain structures of epileptic mouse strains El and DBA/2 *(134)*. Apart from the overabundance of endogenous glutamatergic agonist, the insufficient neuroprotection afforded by reduced formation of kynurenic acid might also participate in seizure phenomena. Intracerebral or peripheral application of AOAA, known to inhibit kynurenic acid synthesis, evokes seizures susceptible to treatment with NMDA antagonists *(135)*.

Data from receptor binding studies suggest that there is an increased density of ionotropic EAA receptors in hippocampal dentate gyrus, parahippocampal gyrus, and entorhinal cortex; and reduced densities of receptors within severely lesioned hippocampal CA1 area *(131,136)*. An intriguing finding is the presence of autoantibodies against AMPA GluR3 subunit in focal epilepsy of Rasmussen's encephalitis *(137)*. It is thus possible that an autoimmune response to elements of the glutamatergic system contributes to epileptogenesis.

The role of mGLU receptors in epilepsy has gained increased attention. Agonists of group II and III mGLU receptors were reported to suppress excessive synaptic transmission in slices from kindled amygdala *(138)*. Preferential group II agonists were shown to increase the threshold for amygdala kindling *(139)*. Transient upregulation of mGLUR1 and persistent downregulation of the mGLUR5 receptor subtype, both belonging to group I, was noted in hippocampus of amygdala-kindled rats *(140)*.

Antagonists of ionotropic glutamatergic receptors are potent antiepileptic compounds in a variety of models of chemically induced convulsions *(136,141,142)*, genetically determined audiogenic seizures *(143)*, and to a lesser extent, in kindling models of epilepsy and electroconvulsions *(136,144)*. EAA antagonists may potentiate the anticonvulsant activity of conventional antiepileptic drugs, reducing the number of adverse effects, in some cases *(145,146)*. Initial studies on the simultaneous use of EAA antagonists and calcium-channel blockers are also very promising for the inhibition of seizure activity *(147)*. Some EAA antagonists have passed through clinical trials and are now being used to treat epilepsy. New antiepileptic drugs, felbamate and flupirtine, possess the ability to antagonize NMDA function; lamotrigine inhibits the release of GLU *(91)*.

4.7. Others

EAA transmitters are implicated as major causal factors in the brain pathologies, such as brain damage related to viral infections, hypoglycemic brain lesions, Parkinson's disease, and others. NMDA receptor activation is implicated in the pathogenesis of measles virus-induced neurodegeneration and AIDS-related dementia. Measles-infected mice display increased activity of quinolinic acid-synthesizing enzyme in hippocampus, which appears prior to selective neuronal loss and seizures *(148)*. HIV-1 virus coat protein, gp120, augments potent neurotoxic effects of NMDA. HIV-infected or gp120-activated macrophages secrete arachidonic acid, free radicals, and also GLU and quinolinic acid. Moreover, neuronal death seems to be the result of an excessive NMDA receptor activation, followed by Ca^{2+} influx *(69,149)*. Profound hypoglycemia-induced release of ASP and GLU is associated with severe brain damage in cerebral cortex, hippocampus, and striatum *(150)*; NMDA antagonists prevent this hypoglycemic neuronal loss *(151)*.

The primary deficit in Parkinson's disease is a loss of the majority of large, dopamine-containing neurons in substantia nigra and associated dopamine-denervation of basal ganglia. Clinical symptoms include hypokinesia, tremor, and rigidity. Of numerous neurotoxins able to induce selective lesions of nigrostriatal dopamine-containing neurons, 1-methyl-4-phenyl-1,2,3,6-tetrahydropyridine (MPTP) has gained widest attention. MPTP induces parkinsonian symptoms in experimental animals and humans. Notably, the action of

MPTP can be effectively blocked by NMDA antagonists. This topic is addressed more fully in Chapter 6.

5. CONCLUSIONS

Findings over the past few decades have vastly increased our knowledge and understanding of the involvement of EAAs in human neuropathology and their role in mechanisms underlying neurodegeneration. However, we are still in need of innovative ideas. The goal remains the same, to understand in order to find a cure. It could be suggested that modern therapeutic approaches should be targeted toward interfering with glutamatergic activity, including ionotropic and/or metabotropic receptor functions. In addition, the broad number of factors influencing the EAA-mediated neurotransmission appears to indicate that therapeutic actions could be more beneficial when not limited to one system or a single aspect of glutamatergic transmission.

REFERENCES

1. Hayashi, T. (1952) A physiological study of epileptic seizures following cortical stimulation in animals and its application to human clinics. *Jpn. J. Pharmacol.* **3,** 46–64.
2. Lucas, D. R. and Newhouse, J. P. (1957) The toxic effect of sodium L-glutamate on the inner layers of the retina. *Arch. Ophtalmol.* **58,** 193–201.
3. Curtis, D. R., Phillis, J. W., and Watkins, J. C. (1959) Chemical excitation of spinal neurons. *Nature* **183,** 611,612.
4. Curtis, D. R. and Watkins, J. C. (1960) The excitation and depression of spinal neurons by structurally-related amino acids. *J. Neurochem.* **6,** 117–141.
5. Olney, J. W. (1969) Brain lesions, obesity and other disturbances in mice treated with monosodium glutamate. *Science* **164,** 719–721.
6. Olney, J. W. and Sharpe, L. G. (1969) Brain lesions in an infant rhesus monkey treated with monosodium glutamate. *Science* **166,** 386–388.
7. Olney, J. W., Ho, O. L., and Rhee, V. (1971) Cytotoxic effects of acidic and sulphur containing amino acids on the infant mouse central nervous system. *Exp. Brain Res.* **14,** 61–76.
8. Monaghan, D. T., Bridges, R. J., and Cotman, C. W. (1989) The excitatory amino acid receptors: their classes, pharmacology and distinct properties in the function of central nervous system. *Annu. Rev. Pharmacol. Toxicol.* **29,** 365–402.
9. Seeburg, P. H. (1993) The TINS/TiPS lecture. The molecular biology of mammalian glutamate receptor channels. *Trends Neurosci.* **16,** 359–365.
10. Pin, J. P. and Duvoisin, R. (1995) The metabotropic glutamate receptors: structure and functions. *Neuropharmacology* **34,** 1–26.
11. Watkins, J. C. and Evans, R. H. (1981) Excitatory amino acid transmitters. *Annu. Rev. Pharmacol. Toxicol.* **21,** 165–204.
12. Stone, T. W. (1993) Subtypes of NMDA receptors. *Gen. Pharmacol.* **24,** 825–832.
13. Cotman, C. W., Monaghan, D. T., Ottersen, O. P., and Storm-Mathisen, J. (1987) Anatomical organization of excitatory amino acid receptors and their pathways. *Trends Neurosci.* **10,** 273–280.

14. Monaghan, D. T. and Cotman, C. W. (1985) Distribution of N-methyl-D-aspartate-sensitive L-[^{3}H]glutamate binding sites in rat brain. *J. Neurosci.* **5,** 2909–2919.

15. Desce, J. M., Godeheu, G., Galli, T., Artaud, F., Cheramy, A., and Glowinski, J. Presynaptic facilitation of dopamine release through D,L-α-amino-3-hydroxy-5-methyl-4-isoxazole propionate receptors on synaptosomes from the rat striatum. *J. Pharmacol. Exp. Ther.* **259,** 692–698.

16. Martin, D., Bustos, G. A., Bowe, M. A., Bray, S. D., and Nadler, J. V. (1991) Autoreceptor regulation of glutamate and aspartate release from slices of the hippocampal CA1 area. *J. Neurochem.* **56,** 1647–1655.

17. Chittajallu, R., Vignes, M., Dev, K. K., Barnes, J. M., Collingridge, G. L., and Henley, J. M. (1996) Regulation of glutamate release by presynaptic kainate receptors in the hippocampus. *Nature* **379,** 78–81.

18. Ohta, K., Araki, N., Shibata, M., Komatsumoto, S., Shimazu, K., Fukuuchi, Y. (1994) Presynaptic ionotropic glutamate receptors modulate in vivo release and metabolism of striatal dopamine, noradrenaline and 5-hydroxytryptamine: involvement of both NMDA and AMPA/kainate subtypes. *Neurosci. Res.* **21,** 83–89.

19. Fink, K., Bonisch, H., and Gothert, M. (1990) Presynaptic NMDA receptors stimulate noradrenaline release in the cerebral cortex. *Eur. J. Pharmacol.* **185,** 115–117.

20. Clow, D. W. and Jhamandas, K. (1989) Characterisation of L-glutamate action on the release of endogenous dopamine from the rat caudate-putamen. *J. Pharmacol. Exp. Ther.* **248,** 722–728.

21. Lehmann, J., Schaefer, P., Ferkany, J. W., and Coyle J. T. (1983) Quinolinic acid evokes [^{3}H]acetylcholine release in striatal slices: mediation by NMDA type excitatory amino acid receptors. *Eur. J. Pharmacol.* **96,** 111–115.

22. Fedele, E., Versace, P., and Raiteri, M. (1993) Evaluation of the mechanisms underlying the kainate-induced impairment of [^{3}H]dopamine release in the rat striatum. *Eur. J. Pharmacol.* **249,** 71–77.

23. Hayashi, Y., Momiyama, A., Takahashi, T., Ohishi, H., Ogawa-Meguro, R., Shigemoto, R., Mizuno, N., and Nakanishi, S. (1993) Role of a metabotropic glutamate receptor in synaptic modulation in the accessory olfactory bulb. *Nature* **366,** 687–690.

24. O'Connor, J. J., Rowan, M. J., and Anwyl, R. (1994) Long-lasting enhancement of NMDA receptor-mediated synaptic transmission by metabotropic glutamate receptor activation. *Nature* **367,** 557–559.

25. Di Iorio, P., Battaglia, G., Ciccarelli, R., Ballerini, P., Giuliani, P., Poli, A., Nicoletti, F., and Caciagli, F. (1996) Interaction between A1 adenosine and class II metabotropic glutamate receptors in the regulation of purine and glutamate release from rat hippocampal slices. *J. Neurochem.* **67,** 302–309.

26. McBain, C. J. and Mayer, M. L. (1994) N-methyl-D-aspartic acid receptor structure and function. *Physiol. Rev.* **74,** 723–760.

27. Hollmann, M. and Heinemann S. (1994) Cloned glutamate receptors. *Annu. Rev. Neurosci.* **17,** 31–108.

28. Ciabarra, A. M., Sullivan, J. M., Gahn, L. G., Pecht, G., Heinemann, S., and Sevarino, K. A. (1995) Cloning and characterization of chi-1: a developmentally regulated member of a novel class of the ionotropic glutamate receptor family. *J. Neurosci.* **15,** 6498–6508.

29. Sucher, N. J., Akbarian, S., Chi, C. L., Leclerc, C. L., Awobuluyi, M., Deitcher, D. L., Wu, M. K., Yuan, J. P., Jones, E. G., and Lipton, S. A. (1995) Developmental and regional expression pattern of a novel NMDA receptor-like subunit (NMDAR-L) in the rodent brain. *J. Neurosci.* **15,** 6509–6520.

30. Cunningham, M. D., Ferkany, J. W., and Enna, S. J. (1994) Excitatory amino acid receptors: a gallery of new targets for pharmacological intervention. *Life Sci.* **54,** 135–148.

31. Johnson, J. W. and Ascher, P. (1987) Glycine potentiates the NMDA response in cultured mouse brain neurons. *Nature* **325,** 529–533.

32. McBain, C. J., Kleckner, N. W., Wyrick, S., and Dingledine, R. (1989) Structural requirements for activation of the glycine coagonist site of N-methyl-D-aspartate receptors expressed in Xenopus oocytes. *Mol. Pharmacol.* **36,** 556–565.

33. Hashimoto, A., Nishikawa, T., Oka, T., and Takahashi, K. (1993) Endogenous D-serine in rat brain: N-methyl-D-aspartate receptor-related distribution and aging. *J. Neurochem.* **60,** 783–786.

34. Schell, M. J., Molliver, M. E., and Snyder, S. H. (1995) D-serine, an endogenous synaptic modulator: localization to astrocytes and glutamate-stimulated release. *Proc. Natl. Acad. Sci. USA* **92,** 3948–3952.

35. Stone, T. W. (1993) Neuropharmacology of quinolinic and kynurenic acids. *Pharmacol. Rev.* **45,** 309–379.

36. Ransom, R. W. and Stec, N. L. (1988) Cooperative modulation of [^{3}H]MK-801 binding to the N-methyl-D-aspartate receptor-ion channel complex by L-glutamate, glycine and polyamines. *J. Neurochem.* **51,** 830–836.

37. Williams, K. (1997) Modulation and block of ion channels: a new biology of polyamines. *Cell Signal.* **9,** 1–13.

38. Gozlan, H. and Ben-Ari, Y. (1995) NMDA receptor redox sites: are they targets for selective neuronal protection. *Trends Pharmacol. Sci.* **16,** 368–374.

39. Wang, L. Y., Orser, B. A., Brautigan, D. L., and MacDonald, J. F. (1994) Regulation of NMDA receptors in cultured hippocampal neurons by protein phosphatases 1 and 2A. Nature 369, 230–232.

40. Wang, Y. T. and Salter M. W. (1994) Regulation of NMDA receptors by tyrosine kinases and phosphatases. *Nature* **369,** 233–235.

41. Lieberman, D. N. and Mody, I. (1994) Regulation of NMDA channel function by endogenous Ca^{2+}-dependent phosphatase. *Nature* **369,** 235–239.

42. Ehlers, M. D., Zhang, S., Bernhadt, J. P., and Huganir, R. L. (1996) Inactivation of NMDA receptors by direct interaction of calmodulin with the NR1 subunit. *Cell* **84,** 745–755.

43. Wyszynski, M., Lin, J., Rao, A., Nigh, E., Beggs, A. H., Craig, A. M., and Sheng, M. (1997) Competitive binding of α-actinin and calmodulin to the NMDA receptor. *Nature* **385,** 439–442.

44. Jorgensen, M., Tygesen, C. K., and Andersen, P. H. (1995) Ionotropic glutamate receptors—focus on non-NMDA receptors. *Pharmacol. Toxicol.* **76,** 312–319.

45. Sheardown, M. J., Nielsen, E. O., Hansen, A. J., Jacobsen, P., Honore, T. (1990) 2,3-Dihydroxy-6-nitro-7-sulfamoyl-benzo(F)quinoxaline: a neuroprotectant for cerebral ischemia. *Science* **247,** 571–574.

46. Tarnawa, I. (1990) Reflex excitatory action of non-NMDA type excitatory amino acid antagonist, GYKI 52466. *Acta Physiol. Hung.* **75,** 277,278.

47. Dong, H., O'Brien, R. J., Fung, E. T., Lanahan, A. A., Worley, P. F., and Huganir, R. L. (1997) GRIP: a synaptic PDZ domain-containing protein that interacts with AMPA receptors. *Nature* **386,** 279–284.

48. Sladeczek, F., Pin, J. P., Recasens, M., Bockaert, J., and Weiss, S. (1985) Glutamate stimulates inositol phosphate formation in striatal neurones. *Nature* **317,** 717–719.

49. Nicoletti, F., Bruno, V., Copani, A., Casabona, G., and Knopfel, T. (1996) Metabotropic glutamate receptors: a new target for the therapy of neurodegenerative disorders? *Trends Neurosci.* **19,** 267–271.

50. Cornell-Bell, A. H., Finkbeiner, S. M., Cooper, M. S., and Smith, S. J. (1990) Glutamate induces calcium waves in cultured astrocytes: long-range glial signaling. *Science* **247,** 470–473.

51. Boss, V., Nutt, K. M., and Conn, P. J. (1994) L-cysteine sulfinic acid as an endogenous agonist of a novel metabotropic receptor coupled to stimulation of phospholipase D activity. *Mol. Pharmacol.* **45,** 1177–1182.

52. Scanziani, M., Salin, P. A., Vogt, K. E., Malenka, R. C., and Nicoll, R. A. (1997) Use-dependent increases in glutamate concentration activate presynaptic metabotropic glutamate receptors. *Nature* **385,** 630–634.

53. Brakeman, P. R., Lanahan, A. A., O'Brien, R., Roche, K., Barnes, C. A., Huganir, R. L., and Worley, P. F. (1997) Homer: a protein that selectively binds metabotropic glutamate receptors. *Nature* **386,** 284–288.

54. Zaczek, R. and Coyle, J. T. (1982) Excitatory amino acids analogues: neurotoxicity and seizures. *Neuropharmacology* **21,** 15–26.

55. Zaczek, R., Collins, J., and Coyle, J. T. (1981) N-methyl-D-aspartic acid: a convulsant with weak neurotoxic properties. *Neurosci. Lett.* **24,** 181–186.

56. Schwarcz, R., Brush, G. S., Foster, A. C., and French, E. D. (1984) Seizure activity and lesions after intrahippocampal quinolinic acid injection. *Exp. Neurol.* **84,** 1–17.

57. Contestabile, A., Migani, P., Poli, A., and Villani, L. (1984) Recent advances in the use of selective neuron-destroying agents for the neurobiological research.*Experientia* **40,** 524–534.

58. Biziere, K. and Coyle, J. T. (1979) Effects of cortical ablation on the neurotoxicity and receptor binding of kainic acid in striatum. *J. Neurosci. Res.* **4,** 383–398.

59. Thomas, R. J. (1995) Excitatory amino acids in health and disease. *J. Am. Geriatr. Soc.* **43,** 1279–1289.

60. Doble, A. (1995) Excitatory amino acid receptors and neurodegeneration. *Therapie* **50,** 319–337.

61. Olney, J. W. (1981) Kainic acid and other excitotoxins: a comparative analysis, in *Glutamate as Neurotransmitter* (DiChiara, G. and Gessa, G. J., eds.), Raven, New York, pp. 375–384.

62. Choi, D. W. (1988) Glutamate neurotoxicity and diseases of the nervous system. *Neuron* **1,** 623–634.

63. Mody, I. and MacDonald, J. F. (1995) NMDA receptor-dependent excitotoxicity: the role of intracellular Ca^{2+} release. *Trends Pharmacol. Sci.* **16,** 356–359.

64. Siman, R. and Noszek, J. C. (1988) Excitatory amino acids activate calpain I and induce structural protein breakdown in vivo. *Neuron* **1,** 279–287.

65. Sanchez-Prieto, J., Budd, D. C., Herrero, I., Vazquez, E., and Nicholls, D. G. (1996) Presynaptic receptors and the control of glutamate exocytosis. *Trends Neurosci.* **19,** 235–239.

66. Chan, P. H. and Fishman, R. A. (1980) Transient formation of superoxide radicals in polyunsaturated fatty acid-induced brain swelling. *J. Neurochem.* **35,** 1004–1007.

67. Bredt, D. S. and Snyder, S. H. (1989) Nitric oxide mediates glutamate-linked enhancement of cGMP levels in the cerebellum. *Proc. Natl. Acad. Sci. USA* **86,** 9030–9033.

68. Connor, J. A., Wadman, W. J., Hockberger, P. E., and Wong, R. K. (1988) Sustained dendritic gradients of Ca^{2+} induced by excitatory amino acids in CA1 hippocampal neurons. *Science* **240,** 649–653.

69. Charriaut-Marlangue, C., Aggoun-Zouaoui, D., Represa, A., and Ben-Ari, Y. (1996) Apoptotic features of selective neuronal death in ischemia, epilepsy and gp 120 toxicity. *Trends Neurosci.* **19,** 109–114.

70. Morrison, R. S., Wenzel, H. J., Kinoshita, Y., Robbins, C. A., Donehower, L. A., and Schwartzkroin, P. A. (1996) Loss of the p53 tumor supressor gene protects neurons from kainate-induced cell death. *J. Neurosci.* **16,** 1337–1345.

71. Portera-Cailliau, C., Price, D. L., and Martin, L. J. (1997) Non-NMDA and NMDA receptor-mediated excitotoxic neuronal deaths in adult brain are morphologically distinct: further evidence for an apoptosis-necrosis continuum. *J. Comp. Neurol.* **378,** 88–104.

72. Turski, L. and Turski, W. A. (1993) Towards an understanding of the role of glutamate in neurodegenerative disorders: energy metabolism and neuropathology. *Experientia* **49,** 1064–1072.

73. Greene, J. G. and Greenamyre, J. T. (1996) Bioenergetics and glutamate excitotoxicity. *Prog. Neurobiol.* **48,** 613–634.

74. Perl, T. M., Bedard, L., Kosatsky, T., Hockin, J. C., Todd, E. C., and Remis, R. S. (1990) An outbreak of toxic encephalopathy caused by eating mussels contaminated with domoic acid. *N. Engl. J. Med.* **322,** 1775–1780.

75. Teitelbaum, J. S., Zatorre, R. J., Carpenter, S., Gendron, D., Evans, A. C., Gjedde, A., and Cashman, N. R. (1990) Neurologic sequelae of domoic acid intoxication due to the ingestion of contaminated mussels. *N. Engl. J. Med.* **322,** 1781–1787.

76. Spencer, P. S., Roy, D. N., Ludolph, A., Hugon, J., Dwivedi, M. P., and Schaumburg, H. H. (1986) Lathyrism: evidence for role of the neuroexcitatory aminoacid BOAA. *Lancet* **2(8515),** 1066,1067.

77. Spencer, P. S., Nunn, P. B., Hugon, J., Ludolph, A. C., Ross, S. M., Roy, D. N, and Robertson, R. C. (1987) Guam amyotrophic lateral sclerosis-parkinsonism-dementia linked to a plant excitant neurotoxin. *Science* **237,** 517–522.

78. Rothman, S. M. and Olney, J. W. (1986) Glutamate and the pathophysiology of hypoxic-ischemic brain damage. *Ann. Neurol.* **19,** 105–111.

79. Benveniste, H., Drejer, J., Schousboe, A., and Diemer, N. H. (1984) Elevation of the extracellular concentrations of glutamate and aspartate in rat hippocampus during transient cerebral ischemia monitored by intracerebral microdialysis. *J. Neurochem.* **43,** 1369–1374.

80. Silverstein, F. S., Buchanan, K., and Johnston, M. V. (1986) Perinatal hypoxia-ischemia disrupts striatal high-affinity [^{3}H]glutamate uptake into synaptosomes. *J. Neurochem.* **47,** 1614–1619.

81. Simon, R. P., Griffiths, T., Evans, M. C., Swan, J. H., and Meldrum, B. S. (1984) Calcium overload in selectively vulnerable neurons of the hippocampus during and after ischemia: an electron microscopy study in the rat. *J. Cerebr. Blood Flow Metab.* **4,** 350–361.

82. Rothman, S. M. (1983) Synaptic activity mediates death of hypoxic neurons. *Science* **220**, 536,537.
83. Wieloch, T., Lindvall, O., Blomquist, P., and Gage, F. H. (1985) Evidence for amelioration of ischaemic neuronal damage in the hippocampal formation by lesions of the perforant path. *Neurol. Res.* **7**, 24–26.
84. Wahlestedt, C., Golanov, E., Yamamoto, S., Yee, F., Ericson, H., Yoo, H., Inturrisi, C. E., and Reis, D. J. (1993) Antisense oligodeoxynucleotides to NMDA-R1 receptor channel protect cortical neurons from excitotoxicity and reduce focal ischaemic infarctions. *Nature* **363**, 260–263.
85. Clark, G. D. and Rothman, S. M. (1987) Blockade of excitatory amino acid receptors protects anoxic hippocampal slices. *Neuroscience* **21**, 665–671.
86. Pulsinelli, W., Sarokin, A., and Buchan, A., (1993) Antagonism of the NMDA and non-NMDA receptors in global versus focal brain ischemia. *Prog. Brain Res.* **96**, 125–135.
87. Warner, D. S., Martin, H., Ludwig, P., McAllister, A., Keana, J. F., and Weber, E. (1995) In vivo models of cerebral ischemia: effects of parenterally administered NMDA receptor glycine site antagonists. *J. Cerebr. Blood Flow Metab.* **15**, 188–196.
88. Gotti, B., Duverger, D., Bertin, J., Carter, C., Dupont, R., Frost, J., Gaudilliere, B., MacKenzie, E. T., Rousseau, J., Scatton, B., and Wick, R. (1988) Ifenprodil and SL 82.0715 as cerebral anti-ischemic agents I. Evidence for efficacy in models of focal cerebral ischemia. *J. Pharmacol. Exp. Ther.* **247**, 1211–1221.
89. Iversen, L., Mulvihill, E., Haldeman, B., Diemer, N. H., Kaiser, F., Sheardown, M., and Kristensen, P. (1994) Changes in metabotropic glutamate receptor mRNA levels following global ischemia: increase of a putative presynaptic subtype (mGluR4) in highly vulnerable rat brain areas. *J. Neurochem.* **63**, 625–633.
90. Opitz, T., Richter, P., Carter, A. J., Kozikowski, A. P., Shinozaki, H., and Reymann, K. G. (1995) Metabotropic glutamate receptor subtypes differentially influence neuronal recovery from in vitro hypoxia/hypoglycemia in rat hippocampal slices. *Neuroscience* **68**, 989–1001.
91. Danysz, W., Parsons, C. G., Bresink, I., and Quack, G. (1995) A revived target for drug development. Glutamate in CNS disorders. *DN&P* **8**, 261–277.
92. The Huntington's Disease Collaborative Research Group (1993) A novel gene containing a trinucleotide repeat that is expanded and unstable on Huntington's disease chromosomes. *Cell* **72**, 971–983.
93. Ferrante, R. J., Kowall, N. W., Beal, M. F., Richardson, E. P., Jr., Bird, E. D., and Martin, J. B. (1985) Selective sparing of a class of striatal neurons in Huntington's disease. *Science* **230**, 561–563.
94. DiFiglia, M. (1990) Excitotoxic injury of the neostriatum: a model for Huntington's disease. *Trends Neurosci.* **13**, 286–289.
95. Coyle, J. T. and Schwarcz, R. (1976) Lesion of striatal neurones with kainic acid provides a model for Huntington's chorea. *Nature* **263**, 244–246.
96. McGeer, E. G. and McGeer, P. L. (1976) Duplication of biochemical changes of Huntington's chorea by intrastriatal injections of glutamic and kainic acids. *Nature* **263**, 517–519.
97. Araki, M., McGeer, P. L., and McGeer, E. G. (1985) Differential effect of kainic acid on somatostatin, GABAergic and cholinergic neurons in the rat striatum. *Neurosci. Lett.* **53**, 197–202.

98. Schwarcz, R., Whetsell, W. O., Jr., and Mangano, R. M. (1983) Quinolinic acid: an endogenous metabolite that produces axon-sparing lesions in rat brain. *Science* **219**, 316–318.
99. Beal, M. F., Kowall, N. W., Ellison, D. W., Mazurek, M. F., Swartz, K. J., and Martin, J. B. (1986) Replication of the neurochemical characteristics of Huntington's disease by quinolinic acid. *Nature* **321**, 168–171.
100. Koh, J. Y., Peters, S., and Choi, D. W. (1986) Neurons containing NADPH-diaphorase are selectively resistant to quinolinate toxicity. *Science* **234**, 73–76.
101. Schwarcz, R., Okuno, E., White, R. J., Bird, E. D., and Whetsell, W. O., Jr. (1988) 3-Hydroxyantranilic acid oxygenase activity is increased in the brains of Huntington disease victims. *Proc. Natl. Acad. Sci. USA* **85**, 4079–4081.
102. Reynolds, G. P., Pearson, S. J., Halket, J., and Sandler, M. (1988) Brain quinolinic acid in Huntington's disease. *J. Neurochem.* **50**, 1959,1960.
103. Heyes, M. P., Swartz, K. J., Markey, S. P., and Beal, M. F. (1991) Regional brain and cerebrospinal fluid quinolinic acid concentrations in Huntington's disease. *Neurosci. Lett.* **122**, 265–269.
104. Greenamyre, J. T., Penney, J. B., Young, A. B., D'Amato, C. J., Hicks, S. P., and Shoulson, I. (1985). Alterations in L-glutamate binding in Alzheimer's and Huntington's diseases. *Science* **227**, 1496–1499.
105. Young, A. B., Greenamyre, J. T., Hollingsworth, Z., Albin, R., D'Amato, C., Shoulson, I., and Penney, J. B. (1988) NMDA receptors losses in putamen from patients with Huntington's disease. *Science* **241**, 981–983.
106. Dure, L. S., IV, Young, A. B., and Penney, J. B. (1991) Excitatory amino acid binding sites in the caudate nucleus and frontal cortex of Huntington's disease. *Ann. Neurol.* **30**, 785–793.
107. Gray, P. N., May, P. C., Mundy, L., and Elkins, J. (1980) L-Glutamate toxicity in Huntington's disease fibroblasts. *Biochem. Biophys. Res. Commun.* **95**, 707–714.
108. Cross, A. J., Slater, P., and Reynold, G. P. (1986) Reduced high-affinity glutamate uptake sites in the brains of patients with Huntington's disease. *Neurosci. Lett.* **67**, 198–202.
109. Turski, W. A., Gramsbergen, J. B., Traitler, H., and Schwarcz, R. (1989) Rat brain slices produce and liberate kynurenic acid upon exposure to L-kynurenine. *J. Neurochem.* **52**, 1629–1636.
110. Urbanska, E., Ikonomidou, C., Sieklucka, M., and Turski, W. A. (1989) Aminooxyacetic acid produces excitotoxic lesions in the rat striatum. *Soc. Neurosci. Abstr.* **15**, 764.
111. Urbanska, E., Ikonomidou, C., Sieklucka, M., and Turski, W. A. (1991) Aminooxyacetic acid produces excitotoxic lesions in the rat striatum. *Synapse* **9**, 129–135.
112. Beal, M. F., Swartz, K. J., Hyman, B. T., Storey, E., Finn, S. F., and Koroshetz, W. (1991) Aminooxyacetic acid results in excitotoxin lesions by a novel indirect mechanism. *J. Neurochem.* **57**, 1068–1073.
113. Alavi, A., Dann, R., Chawluk, J., Alavi, J., Kushner, M., and Reivich, M. (1986) Positron emission tomography imaging of regional cerebral glucose metabolism. *Semin. Nucl. Med.* **16**, 2–34.
114. Beal, M. F., Hyman, B. T., and Koroshetz, W. (1993) Do defects in mitochondrial energy metabolism underlie the pathology of neurodegenerative diseases? *Trends Neurosci.* **16**, 125–131.

115. Jauch, D., Urbanska, E. M., Guidetti, P., Bird, E. D., Vonsattel J. P., Whetsell, W. O., Jr., and Schwarcz, R. (1995) Dysfunction of brain kynurenic acid metabolism in Huntington's disease: focus on kynurenine aminotransferases. *J. Neurol. Sci.* **130,** 39–47.

116. Gandy, S. and Greengard, P. (1992) Amyloidogenesis in Alzheimer's disease: some possible therapeutic oppotunities. *Trends Pharmacol. Sci.* **13,** 108–113.

117. Dunnett, S. B., Everitt, B. J., and Robbins, T. W. (1991) The basal forebrain-cortical cholinergic system: interpreting the functional consequences of excitotoxic lesions. *Trends Neurosci.* **14,** 494–501.

118. Pearson, R. C., Esiri, M. M., Hiorns, R. W., Wilcock, G. K., and Powell, T. P. (1985) Anatomical correlates of the distribution of the pathological changes in the neocortex in Alzheimer disease. *Proc. Natl. Acad. Sci. USA* **82,** 4531–4534.

119. Armstrong, D. M., Ikonomovic, M. D., Sheffield, R., and Wenthold, R. J. (1994) AMPA-selective glutamate receptor subtype immunoreactivity in the entorhinal cortex of non-demented elderly and patients with Alzheimer's disease. *Brain Res.* **639,** 207–216.

120. Sandhu, F. A., Porter, R. H., Eller, R. V., Zain, S. B., Salim, M., and Greenamyre, J. T. (1993) NMDA and AMPA receptors in transgenic mice expressing human β-amyloid protein. *J. Neurochem.* **61,** 2286–2289.

121. De Boni, U. and McLachlan, D. R. (1985) Controlled induction of paired helical filaments of the Alzheimer type in cultured human neurons, by glutamate and aspartate. *J. Neurol. Sci.* **68,** 105–118.

122. Mattson, M. P., Dou, P., and Kater, S. B. (1988) Outgrowth-regulating actions of glutamate in isolated hippocampal pyramidal neurons. *J. Neurosci.* **8,** 2087–2100.

123. Koh, J. Y., Yang, L. L., and Cotman, C. W. (1990) β-Amyloid protein increases the vulnerability of cultured cortical neurons to excitotoxic damage. *Brain Res.* **533,** 315–320.

124. Arias, C., Arrieta, J., and Tapia, R. (1995) β-Amyloid peptide fragment 25–35 potentiates the calcium-dependent release of excitatory amino acids from depolarized hippocampal slices. *J. Neurosci. Res.* **41,** 561–566.

125. Harris, M. E., Wang, Y., Pedigo, N. W., Jr., Hensley, K., Butterfield, D. A., and Carney, J. M. (1996) Amyloid β peptide (25–35) inhibits Na^+-dependent glutamate uptake in rat hippocampal astrocyte cultures. *J. Neurochem.* **67,** 277–286.

126. Mattson, M. P. (1994) Calcium and neuronal injury in Alzheimer's disease. Contributions of β-amyloid precursor protein mismetabolism, free radicals and metabolic compromise. *Ann. NY Acad. Sci.* **747,** 50–76.

127. Lee, R. K., Jimenez, J., Cox, A. J., and Wurtman, R. J. (1996) Metabotropic glutamate receptors regulate APP processing in hippocampal neurons and cortical astrocytes derived from fetal rats. *Ann. NY Acad. Sci.* **777,** 338–343.

128. Falconer, M. A. (1974) Mesial temporal (Ammon's horn) sclerosis as a common cause of epilepsy. Aetiology, treatment and prevention. *Lancet* **2(7883),** 767–770.

129. Ben-Ari, Y., Tremblay, E., and Ottersen, O. P. (1980) Injections of kainic acid into the amygdaloid complex of the rat: an electrographic, clinical and histological study in relation to the pathology of epilepsy. *Neuroscience* **5,** 515–528.

130. French, E. D., Aldinio, C., and Schwarcz, R. (1982) Intrahippocampal kainic acid, seizures and local neuronal degeneration: relationships assessed in unanesthesized rats. *Neuroscience* **7,** 2525–2536.

131. Bradford, H. F. (1995) Glutamate, GABA and epilepsy. *Prog. Neurobiol.* **47,** 477–511.

132. During, M. J. and Spencer, D. D. (1993) Extracellular hippocampal glutamate and spontaneous seizure in the conscious human brain. *Lancet* **341(8861),** 1607–1610.

133. Rothstein, J. D., Dykes-Hoberg, M., Pardo, C. A., Bristol, L. A., Jin, L., Kuncl, R. W., Kanai, Y., Hediger, M. A., Wang, Y., Schielke, J. P., and Welty, D. F. (1996) Knockout of glutamate transporters reveals a major role for astroglial transport in excitotoxicity and clearance of glutamate. *Neuron* **16,** 675–686.

134. Eastman, C. L., Urbanska, E. M., Chapman, A. G., and Schwarcz, R. (1994) Differential expression of the astrocytic enzymes 3-hydroxyantranilic acid oxygenase, kynurenine aminotransferase and glutamine synthetase in seizure-prone and non-epileptic mice. *Epilepsy Res.* **18,** 185–194.

135. Turski, W. A., Dziki, M., Urbanska, E., Calderazzo-Filho, L. S., and Cavalheiro, E. A. (1991) Seizures induced by aminooxyacetic acid in mice: pharmacological characteristics. *Synapse* **7,** 173–180.

136. Meldrum, B. S. (1994) The role of glutamate in epilepsy and other CNS disorders. *Neurology* **44,** S14–S23.

137. Rogers, S. W., Andrews, P. I., Gahring, L. C., Whisenand, T., Cauley, K., Crain, B., Hughes, T. E., Heinemann, S. F., and McNamara, J. O. (1994) Autoantibodies to glutamate receptor GluR3 in Rasmussen's encephalitis. *Science* **265,** 648–651.

138. Neugebauer, V., Keele, N. B., and Shinnick-Gallagher, P. (1997) Epileptogenesis in vivo enhances the sensitivity of inhibitory presynaptic metabotropic glutamate receptors in basolateral amygdala neurons in vitro. *J. Neurosci.* **17,** 983–995.

139. Attwell, P. J., Kaura, S., Sigala, G., Bradford, H. F., Croucher, M. J., Jane, D. E., and Watkins, J. C. (1995) Blockade of both epileptogenesis and glutamate release by (1S,3S)-ACPD, a presynaptic glutamate receptor agonist. *Brain Res.* **698,** 155–162.

140. Akbar, M. T., Rattray, M., Powell, J. F., and Meldrum, B. S. (1996) Altered expression of group I metabotropic glutamate receptors in the hippocampus of amygdala-kindled rats. *Brain Res. Mol. Brain Res.* **43,** 105–116.

141. Czuczwar, S. J. and Meldrum, B. S. (1982) Protection against chemically induced seizures by 2-amino-7-phosphonoheptanoic acid. *Eur. J. Pharmacol.* **83,** 335–338.

142. Turski, L., Klockgether, T., Sontag, K. H., Herrling, P. L., and Watkins, J. C. (1987) Muscle relaxant and anticonvulsant activity of 3-((±)-2-carboxypiperazin-4-yl)-propyl-1-phosphonic acid, a novel N-methyl-D-aspartate antagonist, in rodents. *Neurosci. Lett.* **73,** 143–148.

143. Meldrum, B. S., Croucher, M. J., Badman, G., and Collins, J. F. (1983) Antiepileptic action of excitatory amino acid antagonists in the photosensitive baboon, Papio papio. *Neurosci. Lett.* **39,** 101–104.

144. Czuczwar, S. J., Cavalheiro, E. A., Turski, L., Turski, W. A., and Kleinrok, Z. (1985) Phosphonic analogues of excitatory amino acids raise the threshold for maximal electroconvulsions in mice. *Neurosci. Res.* **3,** 86–90.

145. Czuczwar, S. J., Turski, W. A., and Kleinrok, Z. (1996) Interactions of excitatory amino acid antagonists with conventional antiepileptic drugs. *Metab. Brain Dis.* **11,** 143–152.

146. Czuczwar, S. J., Kleinrok, Z., and Turski, W. A. (1996) Interaction of calium channel blockers and excitatory amino acid antagonists with conventional antiepileptic drugs. *CNS Drug Rev.* **2,** 452–467.

147. Gasior, M., Borowicz, K., Starownik, R., Kleinrok, Z., and Czuczwar, S. J. (1996) Ca^{2+} channel blockade and the antielectroshock activity of NMDA receptor antagonists, CGP 40116 and CGP 43487, in mice. *Eur. J. Pharmacol.* **312**, 27–33.

148. Eastman, C. L., Urbanska, E., Love, A., Kristensson, K., and Schwarcz, R. (1994) Increased brain quinolinic acid production in mice infected with a hamster neurotropic measles virus. *Exp. Neurol.* **125**, 119–124.

149. Lipton, S. A. (1996) Similarity of neuronal cell injury and death in AIDS dementia and focal cerebral ischemia: potential treatment with NMDA open-channel blockers and nitric oxide-related species. *Brain Pathol.* **6**, 507–517.

150. Auer, R. N. and Siesjo, B. K. (1993) Hypoglycaemia: brain neurochemistry and neuropathology. *Baillieres Clin. Endocrinol. Metab.* **7**, 611–625.

151. Wieloch, T. (1985) Hypoglycemia-induced neuronal damage prevented by an N-methyl-D-aspartate antagonist. *Science* **230**, 681–683.

152. Watkins, J. C. (1984) Excitatory amino acids and central synaptic transmission. *Trends Pharmacol. Sci.* **5**, 373–376.

153. Fagg, G. F., Foster, A. C., and Ganong, A. H. (1996) Excitatory amino acid synaptic mechanisms and neurological function. *Trends Pharmacol. Sci.* **7**, 357–363.

154. Krogsgaard-Larssen, P. and Hansen, J. J. (1992) Naturally-occurring excitatory amino acids as neurotoxins and leads in drug design. *Toxicol. Lett.* **64/65**, 409–416.

155. Hawkins, L. M., Beaver, K. M., Jane, D. E., Taylor, P. M., Sunter, D. C., and Roberts, P. J. (1995) Characterization of the pharmacology and regional distribution of (S)-[3H]-5-fluorowillardiine binding in rat brain. *Br. J. Pharmacol.* **116**, 2033–2039.

156. Ross, S. M., Roy, D. N., and Spencer, P. S. (1989) Beta-*N*-oxalylamino-L-alanine action of glutamate receptors. *J. Neurochem.* **53**, 710–715.

157. Zhou, I. M., Gu, Z. Q., Costa, A. M., Yamada, R. A., Mansson, P. E., Giordano, T., Skolnick, P., and Jones, K. A. (1997) (2S,4R0-4-methylglutamic acid (SYM 2081): a selective, high-affinity ligand for kainate receptors. *J. Pharmacol. Exp. Ther.* **280**, 422–427.

158. Brauner-Osborne, H., Slok, F. A., Skjaerback, N., Ebert, B., Sekiyama, N., Nakanishi, S., and Krogsgaard-Larssen, P. (1996) A new highly selective metabotropic excitatory amino acid agonist; 2-amino-4-(3-hydroxy-5-methylisoxazol-4-yl)butyric acid. *J. Med. Chem.* **39**, 3188–3194.

159. Ben-Ari, Y., Tremblay, E., Ottersen, O. P., and Meldrum, B. S. (1980) The role of epileptic activity in hippocampal and "remote" cerebral lesions induced by kainic acid. *Brain Res.* **191**, 79–97.

160. Campochiaro, P. and Coyle, J. T. (1978) Ontogenetic development of kainate neurotoxicity: correlates with glutamatergic innervention. *Proc. Natl. Acad. Sci. USA* **75**, 2025–2029.

161. Schwarez, R., Foster, A. C., French, E., Whetsell, J. W., Jr., and Kohler, C. (1984) Excitoxic models for neurodegenerative disorders. *Life Sci.* **35**, 85–90.

162. McDonald, J. W., Silverstein, F. S., and Johnston, M. V. (1988) Neurotoxicity of N-methyl-D-aspartate is markedly enhanced in developing rat central nervous system. *Brain Res.* **459**, 200–203.

163. Guldin, W. O. and Markowitsch, H. J. (1981) No detectable remote lesions following massive intrastriatal injections of ibotenic acid. *Brain Res.* **225**, 446–451.

164. Steiner, H. X., McBean, G. J., Kohler, C., Roberts, P. J., and Schwarez, R. (1984) Ibotenate-induced neuronal degeneration in immature rat brain. *Brain Res.* **307,** 117–124.

165. Ferriero, D. M., Areavi, L. J., and Simon, R. P. (1990) Ontogeny of excitotoxic injury to nicotinamide adenine dinucleotide phosphate diaphorase reactive neurons in the neonatal rat striatum. *Neuroscience* **36,** 417–424.

14

Neurotoxiciy of NMDA Receptor Antagonists

Sunita Rajdev and Frank R. Sharp

1. INTRODUCTION

Glutamate, the major excitatory neurotransmitter in the mammalian nervous system, is also suggested to be the primary mediator of neurodegeneration in several neuropathological conditions ranging from acute stroke and epilepsy to chronic disorders, e.g., Alzheimer's disease and Huntington's chorea *(1–3)*. These neurotoxic effects of glutamate are thought to be mediated via the activation of *N*-methyl-D-aspartate (NMDA) subtypes of glutamate receptors. In the presence of excessive amounts of glutamate, the NMDA channel remains open and allows extracellular calcium to enter the cells *(4)*. Increased intracellular free calcium in turn triggers a broad spectrum of metabolic and ionic events, which, if uncontrolled, result in irreversible neuronal injury and death *(5)*. Therefore, it is important to develop strategies to block the excessive activation of NMDA receptors.

Pharmacological antagonists of the NMDA receptors, including dizocilpine (MK-801) and phencyclidine (PCP), have potent neuroprotective properties in both in vivo and in vitro models of neurotoxicity *(6–11)*. However, the potential usefulness of these agents is hampered by their side effects and paradoxical neurotoxic properties. Olney et al. *(12)* were the first to demonstrate that treatment of adult rats with noncompetitive NMDA antagonists causes morphological changes in selective neurons located in the posterior cingulate (PS) and retrosplenial (RS) cortices. Though various studies have investigated the potential mechanisms and ways of preventing the neurotoxicity of NMDA antagonists, many of the basic concerns evoked by Olney et al.'s study still remain unanswered. In addition, NMDA antagonists, especially PCP, also have psychotomimetic effects, which are indistinguishable from symptoms of schizophrenia. The behavioral effects of PCP are now being extensively studied in different species to develop an animal model of schizophrenia *(13–15)*. In

From: Highly Selective Neurotoxins: Basic and Clinical Applications *Edited by: R. M. Kostrzewa.*
Humana Press Inc., Totowa, NJ

this chapter we primarily focus on the neurotoxicity of noncompetitive NMDA receptor channel blocker PCP with appropriate references to the closely related agents ketamine and MK-801. These drugs are together referred to as PCP-like drugs.

2. BEHAVIORAL EFFECTS

PCP has been shown to act on multiple levels in the nervous system, resulting in widely different behaviors, depending on the dose and the species studied.

2.1. Animals

PCP and MK-801 have anticonvulsant and central depressant effects in animals that resemble barbiturates *(16,17)*. Systemic administration of PCP-like drugs, including MK-801 and ketamine, to rodents produces a motor syndrome characterized by increased locomotion, head weaving, body rolling, and sniffing progressing to ataxia at higher doses *(16–20)*. In pigeons, PCP and MK-801 produce a cataleptic state characterized by the loss of righting without eye closure and muscle relaxation. In monkeys, the noncompetitive NMDA receptor antagonists cause ataxia and nystagmus at low doses and anesthesia at higher doses. In many species other than rodents, including guinea pig, monkeys, and cats, PCP produces a state of tranquilization and reduced aggressiveness *(13)*. Since MK-801 produces similar effects, NMDA receptor blockade may be involved in this phenomenon. In addition to the central effects, PCP also affects the peripheral nervous system resulting in increased salivation, hypertension, and tachycardia at higher doses.

Competitive NMDA receptor antagonists, including AP5, CPP, CGS and 19755, also appear to exhibit similar motor incoordination in rodents, pigeons, and monkeys, though in some studies the symptoms were less intense than noncompetitive NMDA receptor antagonists *(19–22)*.

PCP induces a highly discriminative stimulus in rodents, to which other noncompetitive NMDA antagonists, like MK-801, ketamine, and TCP, fully cross-react *(23,24)*. In addition, benzomorphans like SKF 10,047 also induce PCP-like behavioral changes, crossgeneralize with PCP in drug discrimination paradigms, and antagonize NMDA responses *(18,25,26)*. Since their potency in behavioral animal models corresponds well to their potency in PCP binding assays, the NMDA receptor antagonism has been implicated in the discriminative stimulus properties.

PCP has been shown to interfere with the memory of previously learned responses and also with the acquisition of information learned under the influence of the drug *(27)*. It does not affect the ability of animals to learn, but rather affects the long-term storage of information. These effects are shared by SKF 10,047, ketamine, and MK-801, but not amphetamine, suggesting an involvement of NMDA receptors *(28–31)*. MK-801 has recently been shown to impair chronically the performance acquisition of mice on a spatial learning task *(31)*.

NMDA receptors play an important role in the process of long-term potentiation (LTP), implicated in learning and memory storage, and some studies suggest a role for the second messenger nitric oxide (NO) in this phenomenon *(32)*. PCP in low concentrations blocks LTP both in vivo and in vitro in pyramidal cells *(33)*. Since PCP has recently been reported to be a suicide inhibitor of NO synthase *(34)*, its effects on memory may be partly mediated by this mechanism, in addition to the blockade of NMDA receptors.

PCP is self-administered by animals *(35,36)*. Its reinforcing effects in monkeys are clear by the self-administration pattern that results in a state of continuous intoxication. In this respect, PCP resembles opiates and central nervous system (CNS) depressants rather than stimulants. Chronic administration in rats and monkeys results in dependence on some behavioral effects of PCP. Signs of PCP withdrawal in monkeys include frequent vocalization, fearfulness, oculomotor hyperactivity, somnolence, tremors, seizures, diarrhea, and piloerection, strongly suggesting that PCP is an addictive drug. Ketamine and low doses of MK-801 also facilitate the intracranial self-administration in rats *(37,38)*. PCP-like drugs have been shown to stimulate brain reward areas and thus facilitate self-stimulation in animal models, suggesting a profile of drugs abused by humans.

2.2. Humans

PCP has a wide range of behavioral symptoms, including anesthesia, analgesia, stimulant, depressant, and psychotomimetic effects. It was originally developed as an anesthetic agent (Sernyl). It proved analgesic in monkeys and humans and produced a state of dissociative anesthesia with little cardiovascular depression *(39)*. The term dissociative is derived from the fact that the users experience a sense of dissociation from the environment. Although during anesthesia the patient remains conscious with staring gaze, open mouth, flat expression, and rigid muscles, amnesia occurs for the entire surgical procedure. Unfortunately, in many patients, PCP produced an emergence reaction ranging from blurred vision, confusion, and hallucinations to more severe reactions, including paranoia, depersonalization, and bizarre behavior, with some symptoms persisting for 7–10 d after anesthesia *(13,40)*.

Though it was withdrawn from the clinical market in the mid-1960s, it rapidly gained popularity as a drug of abuse, commonly known as "angel dust." In small doses, PCP induces euphoria, stimulation, a sense of intoxication, staggered gait, slurred speech, nystagmus, and numbness of extremities, but at higher doses, depressive actions predominate. Depending on the dose, PCP may induce sweating, stereotyped movements, catatonia, blank stare, changes in body image, delusions, disorganized thoughts, drowsiness, apathy, agitation, delirium, and hostile and bizarre behavior usually accompanied by amnesia. Intoxication may result in stupor, respiratory depression, or coma with muscular rigidity, seizures, and occasionally death. Chronic PCP users reported persistent problems with recent memory and speech *(13,40)*.

Though the abuse of PCP is not widespread now, the research into its mechanism of action still continues, primarily prompted by the suggestion that PCP produces a state of psychosis closely resembling schizophrenia in humans. Both syndromes are characterized by altered sensory perception, thought disorder, and cognitive impairments. This notion is further supported by the fact that PCP exacerbates the symptoms of schizophrenia in stabilized patients and some PCP users develop schizophrenic behavior after chronic abuse *(13,40–42)*.

Ketamine, another dissociative anesthetic, less potent and shorter-acting than PCP, also produces symptoms of schizophrenia in healthy volunteers *(43)*. It is available for intravenous or intramuscular use as an anesthetic and is frequently used in pediatric cases. There are still worldwide reports of ketamine abuse especially among teenagers *(43,44)*.

The high-affinity antagonist MK-801 offered considerable promise for effective therapeutic intervention in conditions like stroke, with excellent preclinical results *(10,45)*. However, the product was not pursued beyond Phase I studies mostly because of the psychotomimetic side effects and likely abuse potential of this class of drugs *(37)*. Therefore, its effects in humans have not been investigated. The search for NMDA antagonists like memantine, which may decrease ischemic injury without producing severe side effects, still continues *(46)*.

3. INTERACTION WITH NEUROTRANSMITTERS

The sites and mechanisms of action by which PCP and its analogs can elicit schizophrenic-like symptoms in humans or function as a positive reinforcing agent with abuse liability remains uncertain. The discovery that specific PCP receptors are present in the mammalian brain, and that PCP selectively reduces the excitatory actions of glutamate and NMDA on spinal neurons greatly enhanced the understanding of the mechanism of action of PCP and its analogs *(47–49)*. Ketamine and MK-801 were also found to interact with the PCP receptor *(45,50)*. The variety of behavioral effects and dose-dependent differential effects suggests that PCP-like drugs may interact with more than a single class of receptors and that their behavioral effects are not mediated solely via its interaction with the NMDA receptor channel. The evidence from binding studies using more selective ligands indicates three primary recognition sites in the brain that recognize PCP with high affinity, i.e., the NMDA receptor channel, σ binding site, and dopamine transporter *(50–52)*. In addition, PCP has also been reported to act on K^+ channels *(53)*. Early pharmacological studies with PCP and analogs suggested a variety of possible neurochemical candidates for the induction of psychotomimetic effects, and have been reviewed recently by Johnson and Jones *(54)*. In the following paragraphs, we have briefly described some of the neurotransmitters affected by PCP-like drugs.

3.1. Glutamate

Glutamate receptors can be broadly classified into ionotropic and metabotropic receptors. The ionotropic receptors are further classified into three

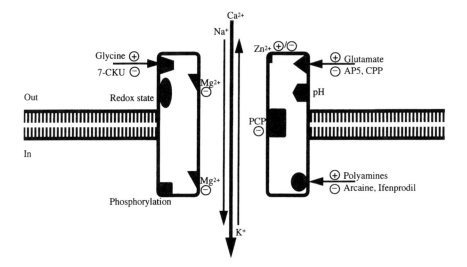

Fig. 1. Schematic model of NMDA receptor modulation. A plus sign indicates agonist and/or stimulatory and a minus sign indicates an antagonist and/or inhibitory modulation. Noncompetitive antagonists, such as PCP, MK-801, and ketamine, block the NMDA channel at the site labeled "PCP."

subtypes i.e., NMDA, α-amino-3-hydroxy-5-methyl-4-isoxaolepropionic acid (AMPA), and kainate receptors, originally based on the relative affinities of synthetic agonists. These findings have been recently confirmed by molecular biological studies *(55)*. Ionotropic receptors are linked to a cation channel that gates Na^+, K^+, and in some cases Ca^{2+} and are involved in most of the excitatory neurotransmission in the CNS. Whereas AMPA and kainate (together referred to as non-NMDA receptors) are primarily involved in fast depolarization at most glutamatergic synapses, NMDA receptors are more closely associated with the induction of synaptic plasticity, e.g., LTP and long-term depression. The metabotropic receptor, which is activated by trans-1-amino-1,3-cyclopentane dicarboxylate (ACPD), is linked by GTP binding protein to second messenger systems, including adenylate cyclase and protein kinase C.

Of these, the NMDA receptor is undoubtedly the best-characterized one *(56,57)*. The modulation of NMDA receptor channel by different agents is diagrammatically depicted in Fig. 1. Activation of NMDA receptors results in increased Ca^{2+}, Na^+ and K^+ conductances. Of particular importance is the calcium influx, which if uncontrolled, will eventually cause irreversible injury to the neurons *(5)*. Though many stimuli can increase intracellular free calcium, it is the NMDA receptor stimulation that appears to be particularly toxic to neurons *(58)*. In addition to NMDA (or glutamate, aspartate, quinolinate, and so forth), the NMDA receptor also requires glycine as a coagonist for optimal activation. NMDA receptors are competitively inhibited by several agents, including AP5, AP7, CPP, D-CPPene, and CGS 19755. Mg^{2+} also uniquely blocks the channel in a voltage-dependent manner both from an extracellular and intracellular site *(59)*. In addition, there is evidence that the receptor is regulated by zinc and other divalent cations, polyamines, phosphorylation, pH changes, and redox potential *(56,57)*.

The ion channel is also blocked by dissociative anesthetics at a site located within the channel pore. Agents like PCP, TCP, ketamine, MK-801, dextorphan, memantine, and remacemide all bind with varying affinities to this site widely known as the "PCP" site and block Ca^{2+} influx, with MK-801 being the most potent agent at this site. These drugs specifically antagonize the actions of NMDA with no effect on any other subtype of glutamate receptors. PCP-like drugs block the NMDA ion channel in a noncompetitive manner, i.e., inhibition cannot be overcome by increasing concentrations of the agonist, and the blockade is voltage- and use-dependent *(50,60,61)*.

The potency of PCP analogs and some σ-selective benzomorphans in displacing ^{3}H-PCP from these receptors correlated well with their relative potencies in behavioral tests, suggesting that PCP receptor plays an important role in the manifestation of the drugs' effects. Autoradiographic localization studies have demonstrated that PCP receptor may be superimposed onto NMDA receptors *(62)*.

3.2. Acetylcholine

PCP has been reported to have anticholinergic properties in a number of biochemical and behavioral studies. It exerts an inhibitory action at cholinergic receptors in the CNS and peripheral nervous system *(54,63,64)*. Physostigmine, a cholinomimetic, can reduce the psychotic symptoms induced by PCP *(65)*, suggesting that anticholinergic effects of PCP may be important in mediating PCP psychosis. Although it clearly inhibits binding to the muscarinic receptors its actions on nicotinic receptors are complex *(65)*. PCP blocks the activated state of nicotinic receptor in the electric organ of Torpedo and rat neuromuscular junction. However, it also appears to have cholinomimetic properties in other behavioral paradigms. In agreement with the latter, PCP has been shown to be a weak inhibitor of acetylcholinesterase and thus may potentiate the actions of acetylcholine after its release *(63,64)*. Furthermore, MK-801 sensitizes rats to pilocarpine-induced seizures. The interaction between PCP and cholinergic system differs depending on the brain region being studied. Although PCP analogs increase acetylcholine turnover in some brain areas, e.g., neocortex, there is no effect in other areas, such as the hippocampus and striatum *(64)*.

3.3. Dopamine

Because PCP produces amphetamine-like behavior in rodents and amphetamine-like psychosis in some humans, its effects on dopamine neurochemistry have been extensively studied *(54,66,67)*. PCP competitively inhibits the uptake of ^{3}H-dopamine into rat striatal slices and synaptosomes, thus prolonging the action of dopamine in the synaptic cleft. PCP is as potent as amphetamine and is 280-fold more potent than MK-801 as a dopamine uptake inhibitor. At higher concentrations, it also enhances the spontaneous efflux of ^{3}H-dopamine *(68–70)*. PCP-like drugs also increase the electrophysiological activity of dopaminergic neurons, whereas the competitive NMDA antagonists, like CPP and CGS 19755, are without effect *(71)*.

Numerous studies have shown that PCP and MK-801 can increase mesolimbic and mesocortical dopamine turnover and release, and that interference with dopaminergic system can reduce some behavioral effects of PCP-like drugs *(67,72,73)*. The decrease in D1- and D4-receptor is consistent with increased extracellular dopamine leading to downregulation of postsynaptic dopaminergic receptors *(74)*. Also, MK-801-induced facilitation of self-stimulation in rats was suppressed by pretreatment with haloperidol (D1/D2 antagonist) or SCH 23390 (D1 antagonist), suggesting an involvement of dopaminergic activity in this phenomenon *(75)*.

Compared to PCP (IC$_{50}$ = 0.46 μM), MK-801 (IC$_{50}$ = 129 μM) is only a weak blocker of dopamine uptake *(70)*. Dopamine uptake blockers that have no effect on NMDA receptor do not produce the full range of PCP-like behavior in animals *(76)* suggesting that elevated dopamine levels are unlikely to be the sole mediator of PCP psychosis.

3.4. Norepinephrine

Both PCP and MK-801 inhibit ^{3}H-norepinephrine uptake competitively, with MK-801 being slightly less potent than PCP *(70)*. Levels of norepinephrine have been reported to be unchanged in mouse and guinea pig forebrains, but to be decreased in rat whole brain following PCP *(54)*. It has been shown to reduce the firing rate of locus ceruleus neurons, and decrease apparent synthesis of norepinephrine at the level of tyrosine hydroxylase *(77,78)*. In cerebellar Purkinje cells and hippocampal pyramidal cells, local application of PCP produces depressant effects probably mediated via the release of presynaptic norepinephrine *(79)*. Although the reports of PCP and norepinephrine interactions are not consistent, norepinephrine does appear to play a role in PCP pharmacology. Many of the behavioral and biochemical effects of PCP are blocked by α-adrenergic antagonists. Furthermore, PCP withdrawal symptoms respond best to desipramine, a predominantly noradrenergic depressant *(79)*.

3.5. Opiate Receptors

Anesthetic effects of PCP and other dissociative anesthetics, especially ketamine, are the result of their opiate-like properties *(80)*. The psychotomimetic effects of PCP resemble those elicited by benzomorphans, such as SKF 10,047 *(81)*. PCP has modest affinity for σ-receptors. Until 1987, the ^{3}H-PCP binding site was commonly referred to as the PCP/σ-receptor. This idea was strengthened by the fact that a benzomorphan, SKF 10,047, an agonist for σ opiate receptor, has PCP-like discriminative stimulus properties *(25,26)*. However, recent data using compounds with selective action at either site (^{3}H-2DTG for σ site and ^{3}H-TCP for PCP site) has led to the recognition that the PCP and σ sites are not identical *(51)*. Both sites display affinities for PCP-like drugs and some benzomorphans, but the PCP site (associated with NMDA receptor channel) has a higher affinity for PCP analogs. The relative potencies are reversed at the σ site. Haloperidol, a dopamine antagonist, also has a significant affinity for σ-receptors, with no effect on the PCP receptor. The auto-radiographic studies using site-selective ligands have shown that the two sites have different distributions in rat brain *(82)*. Thus, the role of the σ-receptor in the effects of PCP is unresolved. It is now believed that the behavioral effects shared by PCP and opiates are mediated via a nonopiate receptor, probably the NMDA receptor-associated PCP site. MK-801, a potent NMDA receptor antagonist, has little affinity for σ-receptors, but shares many behavioral effects with PCP *(83)*.

PCP-like drugs also have some affinity for the μ-type receptors. The small contribution of these receptors to the behavioral effects of ketamine is consistent with the limited ability of naloxone to reverse the analgesic and behavioral effects of ketamine *(43)*. In addition, cross-tolerance between PCP and morphine has also been reported *(84)*.

3.6. γ-Aminobutyric Acid (GABA)

The effect of PCP-like drugs on the GABAergic system is controversial. It is believed that most initial effects of acute PCP intoxication can be blocked by benzodiazepines. The suggestion has been made that the effect of benzodiazepines, such as diazepam, may indirectly attenuate PCP actions by inducing a relative depressive state. GABA-transaminase inhibitors, which increase GABA levels, antagonize PCP-induced hyperactivity in mice and also decrease the latency for ketamine anesthesia *(85,86)*. High concentrations of ketamine block GABA uptake into cultured neurons, astrocytes, and synaptosomes *(87)*.

3.7. Serotonin (5HT)

Several studies have implicated (5HT) in actions of PCP *(88)*. PCP is a potent inhibitor of 5HT uptake into rat synaptosomes and also reduces the turnover of 5HT. Chronic treatment with PCP leads to increased striatal 5HT and metabolite 5HIAA levels, and significant reduction in the density of 5HT2 binding sites in rat cortex *(89,90)*. Turnover and microdialysis studies indicate that MK-801 also produces a large increase in the release of 5HT in the rat brain *(91)*. Presuming that this would increase 5HT2A-receptor stimulation and thus neuronal excitation, it has been suggested that increased 5HT would result in enhanced NMDA receptor activation for neurons possessing both NMDA and 5HT2A-receptors, owing to a reduction in Mg^{2+} blockade *(92)*. Since 5HT2A antagonists block many effects of PCP and MK-801, their potential therapeutic role as antipsychotic drugs is being investigated. The MK-801-stimulated release of dopamine from nucleus accumbens in rats was blocked by competitive NMDA receptor antagonists MDL 100,453 or the 5HT2A-receptor antagonist MDL 100,907, suggesting that interaction between glutamate and dopaminergic systems is regulated by the serotonergic system in the nucleus accumbens *(92)*.

3.8. Voltage-Gated Ion Channels

The effects of PCP on ion channels undoubtedly contribute to its effects on central neurotransmission. PCP is a potent inhibitor of K+ conductance of electrically excitable membranes and also blocks the ionic channel associated with the nicotinic acetylcholine receptor *(53,66)*. It prolongs the action potential and inhibits delayed rectification in frog skeletal muscle. Prolongation of action potential may be responsible for stimulatory effects of PCP on dopamine release from synaptosomes and excitatory amino acid release from the astrocytes *(53,93)*. It also inhibits a component of ^{86}Rb efflux in synaptosomes corresponding to the voltage-gated, noninactivating K+ channel. It has been suggested that the behavioral effects of PCP could be owing to modulation of neurotransmitter release at the central synapses, resulting from the blockade of K+ channels *(53)*.

PCP has also been reported to block Na+ channels in neuroblastoma cells *(94)*. In addition, PCP and its analogs inhibit the binding of dihydropyridine calcium channel antagonists *(95)*.

4. NEUROTOXICITY

As mentioned above, there is considerable interest in the potential clinical usefulness of noncompetitive NMDA receptor antagonists, since they were found to be very effective neuroprotectants in acute neurodegenerative conditions, such as hypoxia, ischemia, and seizures. Their neuroprotective action is primarily the result of their ability to block the excitotoxic actions of increased extracellular glutamate *(5–10)*. Unlike the available competitive NMDA receptor antagonists, like CPP and CGS 19755, which are highly hydrophilic, the noncompetitive NMDA antagonists, including PCP, MK-801, and ketamine are lipid-soluble drugs that rapidly cross the blood–brain barrier following peripheral administration and are concentrated in brain and adipose tissue *(43,80,96,97)*.

A major problem associated with the clinical use of PCP-like drugs is their serious side effects. In addition to their psychotomimetic properties, these compounds were also found to have neurotoxic effects. Olney et al. *(12)* first reported that PCP, MK-801, and ketamine induce a neurotoxic reaction in rats that consists of vacuolization of cytoplasm in selected pyramidal neurons of Posterior cingulate and retrosplenial cortices. Vacuole formation involved mitochondria and endoplasmic reticulum, and was similar to that seen after hypoxia/ischemia in hippocampal neurons. The order of potency of neurotoxic effect (MK-801 > PCP > ketamine) strongly suggested an involvement of the NMDA receptor-associated PCP site. Initial studies showed that following a single dose, the neurotoxic effects of PCP-like drugs were limited to pyramidal neurons in cortical layers III and IV of Posterior cingulate and retrosplenial cortices, and that the effect was transient. The vacuoles already detectable at 4 h were maximal at 12 h and gradually disappeared during the next 12–18 h. However, subsequent studies found that higher doses of MK-801 result in a small number of necrotic neurons, which can still be observed after 48 h, suggesting an irreversible effect *(98)*. Additional vacuolized cells have been observed in stratum pyramidal of hippocampus areas CA1, CA3, and subiculum following PCP administration to rats *(99)*. Similar neuronal swelling and vacuolization of Posterior cingulate and retrosplenial cortex neurons have been observed after the administration of competitive NMDA receptor antagonists, CPP and AP5, although at doses greater than those required to reduce ischemic damage *(100)*.

A recent study investigated the neurotoxic effects of PCP administered continuously over 5 d *(101)*. The neurotoxicity in this model was found to be more widespread in limbic regions and forebrain, and appeared to be principally owing to persistent blockade of NMDA receptors. In addition to Posterior cingulate and retrosplenial cortices, degenerating cells were also observed in olfactory regions, including olfactory tubercle and anterior olfactory nuclei and their projections, and parahippocampal and hippocampal regions, especially posterior and ventral entorhinal cortex, granule cells in ventral and posterior dentate, and scattered pyramidal cells in ventral CA3 and CA1 regions. The same pattern of degeneration has been observed following prolonged or repeated administration of MK-801, whereas nimodipine or DNQX did not produce this effect *(102,103)*. PCP-induced degeneration involved complete cells, with all their processes darkly stained, a pattern usually characteristic of excitotoxicity *(101)*.

Similar results were observed in vitro in a study by Mattson et al. *(104)*. Exposure of human fetal cortical cell cultures to 500 μ*M* PCP caused a progressive degeneration of neurons and astrocytes over a 2–8 d period characterized by early reversible vacuolization of soma and later irreversible neurite fragmentation and cell death. A sublethal concentration of PCP (100 μ*M*) suppressed axonal outgrowth. These neurotoxic effects of PCP appeared to be mediated via an inhibitory action on K+ channels and were reproduced by other K+ channel inhibitors.

4.1. Possible Mechanisms of Neurotoxicity

Though the pattern of NMDA antagonist neurotoxicity in vivo does not correlate with the NMDA receptor distribution, the fact that neurotoxic effects in rodents and psychotomimetic effects in human are reproduced by competitive NMDA receptor antagonists strongly implicates reduced functioning of the NMDA receptors as the underlying mechanism. Systematic pharmacological

evaluation showed that many drugs block the neurotoxicity of NMDA antagonists *(105–107)*. These include muscarinic receptor antagonists (atropine, scopolamine), GABA$_A$ receptor agonists (barbiturates and benzodiazepines), σ-receptor
antagonists (rimcazole), α-2 adrenergic-receptor agonists (clonidine), typical
antipsychotic drugs (haloperidol), and atypical antipsychotic drugs (clozapine).

These results led Olney et al. *(105)* to propose a model of NMDA antagonist
neurotoxicity composed of glutamatergic, GABAergic, and cholinergic neurons
in the cingulate cortex. In this model, stimulation of GABAergic interneurons by
pyramidal glutamatergic cingulate neurons via NMDA receptors is blocked by
PCP and analogs. This results in disinhibition of cholinergic terminals owing
to reduced GABA, which eventually results in increased release of acetylcholine
and thus damage of pyramidal glutamatergic neurons. The anticholinergics and
GABA-mimetic drugs would thus be effective against NMDA antagonist neurotoxicity by balancing the circuit. Drugs acting at σ-receptors like rimcazole
and haloperidol also block the neurotoxic actions of PCP-like drugs in rats. This
suggests an interaction between the σ site and muscarinic receptors. There is
evidence to suggest that stimulation of σ 1 site inhibits the M1-receptor activation-induced phosphoinositide metabolism *(108)*. Since noradrenergic agonists
prevent PCP neurotoxicity, glutamate may also control inhibitory noradrenergic neurons. Thus, by interfering with GABA and noradrenergic neurons,
NMDA antagonists would simultaneously abolish inhibitory control over multiple excitatory inputs (cholinergic, glutamatergic and neuropeptide Y) to neurons in posterior cingulate and retrosplenial cortices *(15,105)*.

A recent study demonstrated that high concentrations of MK-801 (>100 μ*M*),
which were toxic to cultured mouse cortical neurons, inhibit glutamate uptake in
both astrocytes and neuronal cultures *(93)*. Since glutamate uptake is electro-
genic and is inhibited by depolarization, MK-801 effects were explained by its
ability to depolarize astrocytes owing to K$^+$ channel blockade and thus reduced
membrane potential. MK-801-induced depolarization also caused an efflux of
excitatory amino acids from astrocytes. These results suggest that the neurotoxic
effects of NMDA receptor antagonists could partly result from the excitotoxicity
mediated by increased stimulation of glutamate receptors. The observations that
NMDA antagonist toxicity pattern resembles excitotoxicity and that PCP-
induced heat-shock protein 70 (hsp70) expression is inhibited by DNQX, a non-
NMDA receptor antagonist, are consistent with this hypothesis *(101,109)*.

However, the neurotoxicity may not extend to all the NMDA receptor antagonists. For example, SL 82,0715, an NMDA antagonist acting at the polyamine
site, does not injure cortical neurons *(110)*. The pathomorphological alterations
with NMDA antagonists seem to be related to the specific site of NMDA receptor with which they interact. Furthermore, the NMDA antagonist-induced neuronal injury has only been reported in rodents. Studies in other species,
especially primates, are required to address the potential relevance of the neurotoxicity to the possible therapeutic usefulness of these agents in various patho-

logical conditions. MK-801 at the dose of 1 mg/kg did not cause any evidence of neuronal damage in monkeys *(111)*.

4.2. Variables Affecting the Neurotoxicity of NMDA Antagonists

Adult female rats were found to be more susceptible to either PCP or MK-801-induced neuronal vacuolization than adult males of the matching age group *(98,112)*. In behavioral studies also, female rats were more responsive to MK-801, ketamine and PCP as evidenced by hyperactivity, stereotypy, motor incoordination, salivation, tremors, and ataxia *(113,114)*. Consistent with these observations, chronic testosterone treatment and ovariectomy reduced the effects of PCP in females, whereas chronic estradiol and castration increased the response of male rats to PCP *(114)*. The effects based on gender have not been well characterized in higher species. Increased neuronal NMDA sites and NMDA response with estradiol treatment provides a potential mechanism for enhanced sensitivity to NMDA effects in females *(115,116)*. However, the exact mechanism of these differences between males and females is not clear. Pharmacokinetic studies showed that the concentration of PCP in female rat brain regions were much higher than in males *(114)*. Moreover, the half-lives of PCP both in brain and plasma were longer in females than in males. Since the activity of hepatic cytochrome P450 system was lower in males than in females, a lower capability of female rats to metabolize PCP may explain these gender differences in PCP pharmacokinetics.

There are also considerable differences in the neurotoxic potencies of NMDA receptor antagonists based on age *(98,112)*. No vacuolization was observed following MK-801 treatment in the brains of 1-mo-old female rats. The effect was weak in 2-mo-old, but fully present in 3-mo-old rats. These effects may be partly related to incomplete development of neurotransmitter pathways or to changes in the cytochrome p450 system with age.

5. INDUCTION OF HSP70

The hsp70 and other heat-shock genes are induced in neurons by a variety of injurious stimuli, including ischemia, trauma, and seizures *(117)*. The study of these stress proteins is considered not only useful for quantifying the degree and extent of injury, but may also help in eventual understanding of the mechanism of injury in individual brain cells. Hsp70 is not detectable in neurons under normal conditions. It is believed that denatured proteins under the above-mentioned conditions activate heat-shock factor, which then acts on heat-shock elements in the promoter region of the heat-shock gene to initiate transcription. The hsp70 protein is thought to prevent partially denatured proteins from becoming irreversibly denatured, facilitate protein import into cellular organelles, and interact with newly synthesized proteins to ensure proper folding. There is evidence that prior stress inducing hsp70 can protect cells from subsequent, otherwise lethal injuries, suggesting a protective role for hsp70.

On this rationale, Sharp et al. studied the expression of hsp70 in rat brain following administration of PCP, MK-801, and ketamine *(118–120)*. Immuno-cytochemistry studies using a monoclonal antibody (MAb) against hsp70 revealed many hsp70 expressing pyramidal neurons in layer III of PC and layers III and V of RS cortices in rats injected with MK-801 (Fig. 2). Hsp70 protein was present in the cell body, dendrites, and axons, but not in the nucleolus of the neuron. Using electron microscopy, the vacuoles were detected in the cytoplasm of hsp70-positive neurons, which were similar to those described by Olney et al. *(12,105)*. The number of hsp70-positive neurons after 5 mg/kg MK-801 was less than that after 1 mg/kg dose. The fact that necrotic neurons are observed after 5, but not 1 mg/kg dose of MK-801 *(98)* suggests that the induction of hsp70 is primarily in neurons that are destined to survive. Ketamine (40–140 mg/kg) produced results similar to MK-801 *(118)*.

PCP administration, however, caused a more widespread induction of hsp70 *(119)*. At 10 mg/kg, PCP induced hsp70 primarily in pyramidal neurons in posterior cingulate and retrosplenial cortices. However, at 50 mg/kg dose, hsp70 was induced not only throughout the forebrain, including neocortex, piriform cortex, insular cortex, amygdala, and hippocampus, but also in cerebellar Purkinje cells *(121)*. *In situ* hybridization studies confirmed the presence of hsp70 mRNA in neurons in these same locations. The heat-shock protein induction by PCP-like drugs was maximal at 24 h and persisted up to 2 wk after the treatment.

The differential pattern of hsp70 induction with MK-801 and PCP may be owing to the abovementioned additional effects of PCP on σ site, K+ channels, or dopamine transporter, among others. Pretreatment with haloperidol, clozapine, SCH23390, and rimcazole blocked the hsp70 induction by PCP. Haloperidol (5 mg/kg) was most potent, followed by clozapine, SCH23390, and rimcazole *(119)*. Scopolamine also blocked the induction of hsp70 induction in PC by PCP. Muscimol attenuated the effects of PCP on hsp70 induction, whereas baclofen had no effect. These data suggest that D_1-D_2-D_4, and $GABA_A$ receptors may be involved in mediating the effects of PCP on hsp70 protein expression. If meso-cortical dopamine neurons also project onto GABA interneurons and inhibit them via D_2-receptors, then increased dopamine levels would also result in dis-inhibition of excitatory inputs to pyramidal PS or RS neurons. These results are interesting, since compounds that prevent PCP psychosis might do so by preventing or attenuating the neuronal injury by PCP. Since these same drugs prevent PCP neurotoxicity, heat-shock protein response may be triggered by the same toxic mechanism, and may thus serve as an easily identifiable and quantifiable marker for the injured neurons.

Although the hypothesis presented by Olney et al. *(105)* as explained in Section 4.1. explains how multiple excitatory inputs can worsen the injury and GABA agonists can attenuate injury, the actual mechanism of injury is still not clear. One possibility is that even if the NMDA receptors are blocked by PCP-like drugs, Ca^{2+} entry via other routes, e.g., voltage-gated calcium channels and

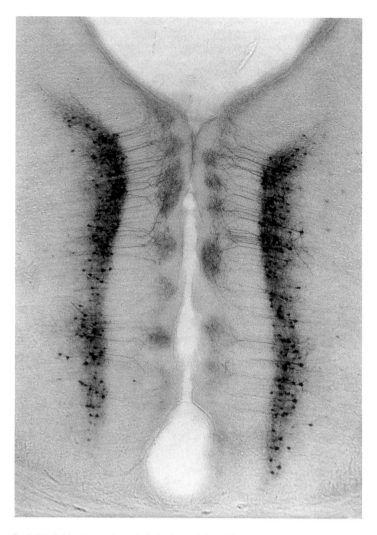

Fig. 2. MK-801-(1 mg/kg, ip) induced hsp70 protein in posterior cingulate cortex. The immunostaining was performed 24 h after the drug administration, using an MAb against hsp70. Most of the hsp70 immunoreactive neurons are pyramidal neurons.

non-NMDA ionotropic or metabotropic glutamate receptors, may be responsible for injuring the pyramidal neurons. Mattson et al. *(104)* reported increased intracellular free calcium concentrations in cultured fetal human cortical neurons chronically exposed to PCP. In accordance with this hypothesis, administration of nimodipine, an L-type Ca^{2+} channel antagonist, completely blocked the hsp70 induction in posterior cingulate and retrosplenial cortices as well as neocortex *(120)*. However, the hsp70 expression, though attenuated, was not completely blocked in piriform cortex and amygdala. This could be explained by the influx of Ca^{2+} through other types of voltage-gated Ca^{2+} channels or AMPA/kainate receptor subtypes of glutamate. Both L- and non-L-type calcium channel antagonists have been shown to modulate the behavioral effects of PCP in rats *(122,123)*. Furthermore, a recent study showed that DNQX, an antagonist of AMPA/kainate receptors, significantly reduced the hsp70 induction by PCP and ketamine in rat brain *(109)*.

6. ACTIVATION OF ASTROCYTES AND MICROGLIA

The brain usually responds to neuronal injury by reactive astrocytosis, i.e., by increasing the number and size of glial cells immunoreactive for the glial fibrillary acidic protein (GFAP) in the area of injury. Though the functional consequences of this phenomenon are not well understood, it is thought to participate in processes beneficial to the injured neurons *(124)*. Microglial activation has been well characterized as a marker of neuronal injury under a variety of pathological conditions and is considered a sensitive but relatively nonspecific indicator of cell damage *(125)*.

Fix et al. *(126)* characterized the spatial and temporal changes in astrocytes and microglia along with neurons after MK-801 treatment in rats. Beginning on d 3 after an injection of 10 mg/kg MK-801, GFAP immunoreactivity revealed hypertrophic astrocytes in a diffuse pattern throughout the region of neuronal necrosis (i.e., posterior cingulate and retrosplenial cortices). This change persisted through d 28 after treatment. Using GSA-lectin histochemistry, few reactive microglia were identified on d 1 in a multifocal pattern in PC/RS cortices and disappeared by d 28 after MK-801 treatment. Under the electron microscope, these microglia were found to be closely associated with necrotic neurons. Hsp70 induction, however, is maximal at 1 mg/kg dose *(118)*, suggesting that although hsp70 may indicate a sublethal stress, microglial activation may be associated with more severe lethal injury.

Using OX42 immunoreactivity as a marker, Nakki et al. also showed that PCP activates microglia, a response that was minimal at 10 mg/kg and increased with increasing doses *(121)*. Though most of the activated microglia were present in posterior cingulate and retrosplenial cortices, higher doses of PCP (50 mg/kg) also induced patches of activated microglia around the cerebellar Purkinje cell bodies (some of which were also hsp70 immunoreactive) and in the molecular layer, predominantly in cerebellar vermis. MK-801, on the other hand, did not induce any hsp70 or microglial activation in the cerebellum. Though haloperidol blocked ketamine and PCP-induced hsp70 expression and neuronal vacuolization induced by MK-801, it did not block the microglial activation by ketamine and PCP, suggesting that either haloperidol does not prevent NMDA antagonist-mediated cell death or that these two effects are mediated via different mechanisms *(106,127)*.

7. INDUCTION OF IMMEDIATE-EARLY GENES (IEGS)

IEGs are rapidly activated in brain following a variety of stimuli, including stress, trauma, depolarization, and toxins *(128)*. The most studied are the fos (c-*fos*, fosB, fra1 and fra2) and jun (c-*jun*, junB, and junD) gene families. The protein products of these genes form homo-(Jun-Jun) and hetero-(fos-jun) dimers, which in turn bind to the specific sequences of DNA on target genes and are thought to be involved in coupling neuronal excitation to changes in target gene expression. Other IEGs, e.g. NGF1-A (also known as krox24, egr-1 or zif/268), encode proteins that contain specific DNA binding zinc-finger motifs, which regulate the expression of late-response genes. Since depolarization and trans-synaptic activity can trigger IEG expression, these genes, especially c-*fos* expression, have been used to map the neural pathway following a specific stimulus *(129)*. Moreover, IEGs are also suggested to control the expression of genes involved in apoptosis as well as the expression of neuroprotective genes, such as NGF.

Since PCP-like drugs injure neurons in rat brain, the effects of PCP on IEG expression in rat brain were studied to gain further insight into the mechanism and neural pathways underlying PCP neurotoxicity *(130,131)*. An induction of all IEGs was observed in the cerebral cortex and hippocampus, within 30 min following PCP injection (10 and 50 mg/kg ip). A delayed induction (2–6 h following PCP injection) of c-*fos*, c-*jun*, NGFI-A, B, and C, and nurr1 mRNA was observed in layers II and III of PC cortex, the primary site of PCP toxicity. The results suggest a widespread neural activation by high doses of PCP, and the possibility that IEGs may be involved in deleterious (neuronal necrosis) or protective (hsp70 induction) cellular responses induced by PCP. High doses of PCP also induced fos in inferior olive, cerebellar granular layer, and deep cerebellar and vestibular nuclei, suggesting a role for the excitatory climbing fiber inputs from the inferior olive in cerebellar injury. Gao et al. *(132)* reported a selective induction of c-*fos* in the cingulate cortex following PCP administration in rats.

MK-801 also produced a time- and dose-dependent induction of IEGs, including c-*fos*, c-*jun*, jun-B, and krox-24, within 1–4 h of administration in posterior cingulate and retrosplenial neurons, that returned to baseline by 24 h *(133)*. In addition to IEGs, MK-801 also induced BDNF mRNA in the same neurons *(134)*. The induction of IEGs and BDNF could be blocked by atropine, suggesting the involvement of central muscarinic receptors in the induction of IEGs. The coordinate induction of fos and jun families suggests that MK-801 induces a transcriptional program in the same areas where its neurotoxic effects are most evident. Since IEG induction (1–3 h) preceded hsp70 expression (at 4 h) as well as pathomorphological changes in neurons, they appear to represent a rapid and sensitive indicator of neuronal injury.

8. METABOLIC ACTIONS

Autoradiography using ^{14}C-deoxyglucose allows a comprehensive anatomical assessment of function-related alterations in cerebral glucose utilization in vivo owing to a close coupling between neuronal activity and energy requirements. Many studies have reported the effects of NMDA antagonists on brain glucose metabolism *(135)*. MK-801 administration produced largest increases in both glucose utilization and blood flow in entorhinal and cingulate cortices, hippocampus, dentate gyrus *(136)*, and olfactory tubercle *(137)*, which reached a magnitude similar to that seen during seizures. The alterations in posterior regions of the above brain regions were significantly more than in the corresponding anterior structures, consistent with the degeneration pattern following MK-801. The glucose utilization was reduced in some regions, especially neocortex and inferior colliculus. These effects of MK-801 were only seen in conscious animals and were blocked by halothane. PCP and ketamine also produce similar changes in glucose utilization, particularly the metabolic activation within the limbic system *(138,139)*. This is interesting, since the limbic system has been implicated to be involved in psychotic conditions like schizophrenia *(140)*. However, there are distinct differences between the effects of PCP and MK-801, which could be explained by additional effects of PCP on σ and dopamine uptake sites. MK-801, on the other hand, has little affinity for σ-receptors and is only a weak blocker of dopamine uptake *(83)*. It is not clear if the increased metabolic rate occurs in the same cells that undergo degeneration or in other cortical elements, but the two phenomena may be related, since sustained stimulation of neuronal circuits is associated with hypermetabolism and alterations in cellular morphology, e.g., during seizures *(135)*.

Competitive antagonists of NMDA receptors have also been found to alter glucose utilization. At the highest dose tested, CPP increased glucose metabolism in entorhinal cortex and olfactory tubercle, but not in posterior cingulate cortex, the primary site of NMDA antagonist neurotoxicity *(141)*. Though competitive antagonists have been shown to induce comparative neurotoxicity to noncompetitive antagonists of NMDA receptors, the former are still believed to be more promising for clinical use. Some other drugs acting at different sites on NMDA receptor channel complex appear to be devoid of the abovementioned neurotoxic effects. SL 82,0715, acting at the polyamine site, and L 687,414, acting at the glycine site, are two such drugs *(110,142)*. At doses effective against ischemic neurotoxicity, they were found not to alter glucose metabolism, blood flow, or neuronal morphology.

Since psychotomimetic effects of PCP often have a delayed onset and can extend well beyond the peak plasma drug levels, Gao et al. *(99)* studied the delayed metabolic actions of PCP. A single high dose of PCP (8.6 mg/kg) produced an initial increase in glucose metabolism (3–6 h) and a later decrease in metabolic activity (24 h). The glucose metabolism did not return to baseline until 48 h. At lower doses of PCP (0.86 mg/kg), only a hypometabolic effect of

PCP could be observed and did not appear to be related to PCP-induced vac-uolization in RS neurons. The authors suggest that these delayed hypometabolic changes may be associated with the psychotomimetic effects of PCP.

9. PCP-LIKE DRUGS AS A MODEL OF SCHIZOPHRENIA

For the past two decades, the most widely investigated theory regarding the cause of schizophrenia was the "dopamine hypothesis" stating that schizophre-nia results from the abnormalities in the dopaminergic system *(143,144)*. The most compelling evidence for this hypothesis is the partial effectiveness of dopamine receptor antagonists in ameliorating symptoms of schizophrenia. However, a large body of evidence has accumulated that now complicates this hypothesis. Many schizophrenics, particularly those with negative symptoms, fail to respond to dopamine antagonist treatment. Recent studies have focused on an alteration in glutamatergic systems in schizophrenia *(13–15)*.

The principal symptoms of psychosis induced by amphetamine or cocaine are stimulant, and include stereotypy, agitation, paranoid delusions, and sensory hallucinations *(145)*. These symptoms are similar to the positive symptoms of schizophrenia. PCP-induced psychosis, on the other hand, more closely resem-bles the clinical picture of schizophrenia. The patients experience both the pos-itive and negative symptoms (alterations in body image, disorganization of thoughts, negativism, and apathy) of schizophrenia *(146)*. In schizophrenics, PCP produces even more striking effects, including exacerbation of thought dis-order and increased hostility. A single dose of PCP precipitated symptoms in stabilized schizophrenics that lasted for days to weeks *(147)*. On the other hand, stabilized schizophrenics are hyporesponsive to amphetamine *(148)* suggesting that PCP may be affecting the same neural pathways affected in patients with schizophrenia. PCP administration in animals is thought to be a useful model of schizophrenia *(149,150)*. Understanding the neurochemical basis of PCP-induced behaviors in animals may thus provide insight into the mechanism of schizophrenia.

The ability of PCP-like drugs to induce schizophrenic psychosis suggests that endogenous dysfunction of NMDA receptor neurotransmission may contribute to schizophrenia. The fact that competitive NMDA receptor antagonists CPP, CPPene, and CGS 19755 also produce PCP-like psychosis in humans further strengthens this hypothesis *(15)*. However, a potential interaction between dopaminergic and glutamatergic systems cannot be excluded. Though the block-ade of dopamine receptor (D2 subtype) partly explains the antipsychotic effect of haloperidol, it does not fully explain its clinical efficacy. A recent study showed that at clinically relevant concentrations, two antipsychotic agents, haloperidol and clozapine, had potent augmenting effects on the NMDA recep-tors in rat striatal slices *(151)*. The authors suggest that the action at NMDA receptors may contribute to their antipsychotic efficacy and further emphasize the importance of glutamatergic dysfunction in the etiology of schizophrenia.

A recent study reported the effects of subanesthetic doses of ketamine in normal healthy volunteers and concluded that ketamine produced symptoms resembling endogenous psychosis, particularly schizophrenia and dissociative states *(43)*. Lahti et al. *(152)* reported an activation of psychotic symptoms in schizophrenics with subanesthetic doses of ketamine. Ketamine increased the blood flow in anterior cingulate cortex, but decreased it in hippocampus. This is interesting, since PET studies with fluorodeoxyglucose probes revealed that glucose metabolism is decreased in schizophrenic subjects, primarily in hippocampus and anterior cingulate cortex *(153)*. PCP administration to rats caused an initial increase followed by a rebound decrease at 24–48 h in glucose metabolism in many structures, including the limbic system *(99)*. This effect of PCP was also present at low doses, which act specifically on the NMDA receptors, suggesting that similar neural pathways are affected during PCP administration and schizophrenia, and further support PCP-induced psychosis as a model of schizophrenia. One study has reported the loss of glutamatergic neurons in cingulate cortex of schizophrenic patients *(154)*. The location of these neurons was similar to those damaged by PCP-like drugs in rats. Therefore it is likely that the psychopathological effects of PCP might be interrelated to its morphopathological effects.

Symptoms of schizophrenia usually become evident around puberty. The induction of ketamine-induced emergence reaction is also age-dependent. Although reported rarely in children, the emergence reaction occurs in almost 50% of young and middle-aged users *(80)*. There are also reports of reduced susceptibility of children to PCP psychosis *(155)*. Furthermore, as described in Section 4.2., the neurotoxic effects of NMDA receptor antagonists are also age-dependent. It is important to understand the mechanisms undelying this phenomenon of age dependency.

The presence of specific PCP and σ-receptor sites implies the presence of endogenous ligands. In fact, a peptide, α-endopsychosin, isolated from porcine brain specifically inhibits the binding of ^{3}H-PCP in a concentration-dependent manner, mimics electrophysiological effects of PCP, and produces some PCP-like behavioral effects when injected into the brain *(156–158)*. Its distribution in vivo parallels that of NMDA receptors. Other endogenous ligands have been isolated from human brain with PCP-like biochemical properties and, therefore, may act as endogenous psychotogens. This is promising, since disturbance in such a system could be involved in schizophrenia.

Since NMDA receptors are modulated by a variety of agents, e.g., glycine and polyamines, a decrease in glutamate or any of these other factors could potentially result in reduced NMDA receptor function. Preliminary studies have shown beneficial effects of large doses of glycine in some schizophrenic subjects *(159)*. Efforts are under way to develop centrally active agents that can augment NMDA receptor-mediated neurotransmission and to evaluate their therapeutic potential for treating schizophrenia.

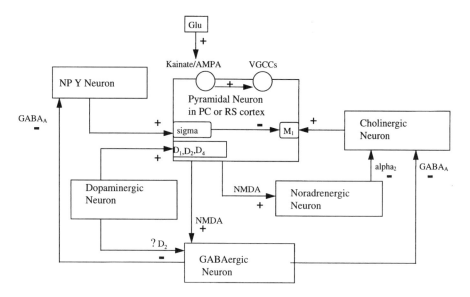

Fig. 3. Schematic hypothetical illustration of the possible circuits mediating the neurotoxic effects of NMDA antagonists. Blockade of NMDA receptors on inhibitory GABAergic and noradrenergic neurons would result in a disinhibition of multiple excitatory inputs on to the pyramidal cingulate neuron. Dopamine receptor antagonists (SCH 23390, haloperidol, and clozapine) also inhibit the neurotoxic effects of NMDA antagonists. Glutamate (Glu) released from the cingulate neuron or from other descending glutamatergic pathways could also injure the cingulate neuron by activating AMPA/kainate and voltage-gated calcium channels (VGCC). For more details, see Sections 4.2. and 5. A plus sign indicates stimulatory and a minus sign indicates an inhibitory input. M_1 = muscarinic receptors.

10. CONCLUSIONS

PCP-like drugs, though extremely useful neuroprotective agents against excitotoxicity, also have neurotoxic, psychomimetic, and reinforcing properties. Whether they would ever be used therapeutically is speculative. A single high dose of PCP and MK-801 causes neuronal vacuolization in posterior cingulate and retrosplenial cortices. Neurons in these same regions show an immediate increased metabolic activity, and induction of IEGs and hsp70. This response is associated with the induction of astrocytes and microglia. It is clear, however, that PCP-like drugs affect a variety of neurotransmitter pathways. Figure 3 highlights the proposed circuits that may be involved in the neurotoxicity of NMDA receptor antagonists.

The challenge now is to develop NMDA antagonists that are as effective as PCP or MK-801, but are devoid of the undesirable side effects or to find ways of blocking the toxicity of PCP and analogs. Recently, an anti-PCP MAb fragment was shown to reverse PCP-induced increases rapidly in locomotor activity and ataxia *(160)*. Though their neuroprotective effect appears to mediated via blockade of NMDA receptor channels, the molecular mechanisms underlying their psychomimetic or neurotoxic effects are unclear. More research is needed to clarify the mechanism of the behavioral effects of NMDA receptor antagonists. Unlike noncompetitive channel blockers or classic competitive NMDA antagonists, drugs acting at the glycine or polyamine site to reduce NMDA receptor activation appear to be devoid of these undesirable neurotoxic and psychotomimetic side effects. They are currently being investigated as a potential treatment for stroke *(110,142)*.

Regardless of their therapeutic potential, noncompetitive NMDA receptor antagonists provide us with a comprehensive model of schizophrenia closely mimicking the pathology at both behavioral and anatomic levels. Further studies to clarify the mechanism of PCP neurotoxicity and psychosis may result in better understanding of the pathology underlying schizophrenia.

REFERENCES

1. Rothman, S. M. and Olney, J. W. (1986) Glutamate and the pathophysiology of hypoxic-ischemic brain damage. *Ann. Neurol.* **19,** 105–111.
2. Olney, J. W. (1989) Excitatory amino acids and neuropsychiatric disorders. *Biol. Psychiatr.* **26,** 505–525.
3. Lipton, S. A. and Rosenberg, P. A. (1994) Excitatory amino acids as a final common pathway for neurologic disorders. *N. Engl. J. Med.* **330,** 613–622.
4. MacDermott, A. B., Mayer, M. L., Westbrook, G. L., Smith, S. J., and Barker, J. L. (1986) NMDA-receptor activation increases cytoplasmic calcium concentration in cultured spinal cord neurons. *Nature* **321,** 519–522.
5. Verity, M. A. (1992) Ca^{2+} dependent processes as mediators of neurotoxicity. *Neurotoxicol.* **13,** 139–148.
6. Wieloch, T. (1985) Hypoglycemia-induced neuronal damage prevented by an *N*-methyl-D-aspartate antagonist. *Science* **230,** 681–683.
7. Weiss, J., Goldberg, M. P., and Choi, D. W. (1986) Ketamine protects cultured neocortical neurons from hypoxic injury. *Brain Res.* **380,** 186–190.
8. Hayes, R. L., Chapouris, R., Lyeth, B. G., Jenkins, L., Robinson, S. E., Young, H. F., and Marmarou, A. (1987) Pretreatment with phencyclidine (PCP) attenuates long-term behavioral deficits following concussive brain injury in the rat. *Soc. Neurosci. Abstracts* **13,** 1254.
9. Choi, D. W., Koh, J. Y., and Peters, S. (1988) Pharmacology of glutamate neurotoxicity in cortical cell culture: attenuation by NMDA antagonists. *J. Neurosci.* **8,** 185–196.
10. Park, C. K., Nehls, D. G., Graham, D. I., Teasdale, G. M., and McCulloch, J. (1988) The glutamate antagonist MK-801 reduces focal ischemic brain damage in the rat. *Ann. Neurol.* **24,** 543–551.
11. Clifford, D. B., Olney, J. W., Benz, A. M., Fuller, T. A., and Zorumski, C. F. (1990) Ketamine, phencyclidine, and MK-801 protect against kainic acid-induced seizure-related brain damage. *Epilepsia* **31,** 382–390.

12. Olney, J. W., Labruyere, J., and Price, M. T. (1989) Pathological changes induced in cerebrocortical neurons by phencyclidine and related drugs. *Science* **244,** 1360–1362.

13. Javitt, D. C. and Zukin, S. R. (1991) Recent advances in the phencyclidine model of schizophrenia. *Am. J. Psychiatr.* **148,** 1301–1308.

14. Ellison, G. (1995) The N-methyl-D-aspartate antagonists phencyclidine, ketamine and dizocilpine as both behavioral and anatomical models of the dementias. *Brain Res. Rev.* **20,** 250–267.

15. Olney, J. W. and Farber, N. B. (1995) Glutamate receptor dysfunction and schizophrenia. *Arch. Gen. Psychiatr.* **52,** 998–1007.

16. Koek, W., Woods, J. H., and Winger, G. D. (1988) MK-801, a proposed noncompetitive antagonist of excitatory amino acid neurotransmission, produces phencyclidine-like behavioral effects in pigeons, rats and rhesus monkeys. *J. Pharmacol. Exp. Ther.* **245,** 969–974.

17. Balster, R. L. and Chait, L. D. (1978) The behavioral effects of phencyclidine in animals. *NIDA Res. Monogr.* **21,** 53–65.

18. Contreras, P. C., Contreras, M. L., O'Donohue, and Lair, C. C. (1988) Bicohemical and behavioral effects of sigma and PCP ligands. *Synapse* **2,** 240–243.

19. Tricklebank, M. D., Singh, L., Oles, R. J., Preston, C., and Iversen, S. D. (1989) The behavioral effects of MK-801: a comparison with antagonists acting non-competitively and competitively at the NMDA receptor. *Eur. J. Pharmacol.* **167,** 127–135.

20. Willets, J., Balster, R. L., and Leander, J. D. (1990) The behavioral pharmacology of NMDA receptor antagonists. *Trends Pharmacol. Sci.* **11,** 423–428.

21. Koek, W., Kleer, E., Mudar, P. J., and Woods, J. H. (1986) Phencyclidine-like catalepsy induced by the excitatory amino acid antagonist DL-2-amino-5-phosphonovalerate. *Behav. Brain Res.* **19,** 257–259.

22. Bennett, D. A., Bernard, P. S., Amrick, C. L., Wilson, D. E., Liebman, J. M., and Hutchison, A. J. (1989) Behavioral pharmacological profile of CGS 19755, a competitive antagonist at N-methyl-D-aspartate receptors. *J. Pharmacol. Exp. Ther.* **250,** 454–460.

23. Tricklebank, M. D., Singh, L., Oles, R. J., Wong, E. H., and Iversen, S. D. (1987) A role for receptors of N-methyl-D-aspartic acid in the discriminative stimulus properties of phencyclidine. *Eur. J. Pharmacol.* **141,** 497–501.

24. Willets, J. and Balster, R. L. (1988) Phencyclidine-like discriminative stimulus properties of MK-801 in rats. *Eur. J. Pharmacol.* **146,** 167–169.

25. Brady, K. T., Balster, R. L., and May, E. L. (1982) Stereoisomers of N-allylnormetazocine: phencyclidine-like behavioral effects in squirrel monkeys and rats. *Science* **215,** 178–180.

26. Shannon, H. E. (1983) Pharmacological evaluation of N-allylnormetazocine (SKF 10047) on the basis of its discriminative stimulus properties in the rat. *J. Pharmacol. Exp. Ther.* **225,** 144–152.

27. Handelmann, G. E., Contreras, P. C., and O'Donahue, T. L. (1987) Selective memory impairment by phencyclidine in rats. *Eur. J. Pharmacol.* **140,** 69–73.

28. Moershbaecher, J. M. and Thompson, D. M. (1983) Differential effects of prototype opioid agonists on the acquisition of conditional discriminations in monkeys. *J. Pharmacol. Exp. Ther.* **226,** 738–748.

29. Thompson, D. M. and Moershbaecher, J. M. (1984) Phencyclidine in combination with *D*-amphetamine: differential effects on acquisition and performance of response chains in monkeys. *Pharm. Biochem. Behav.* **20,** 619–627.

30. Thompson, D. M., Winsauer, P. J., and Mastropaolo, J. (1987) Effects of phencyclidine, ketamine and MDMA on complex operant behavior in monkeys. *Pharm. Biochem. Behav.* **26,** 401–405.

31. Wozniak, D. F., Brosnan-Watters, G., Nardi, A., McEwen, M., Corso, T. D., Olney, J. W., and Fix, A. S. (1996) MK-801 neurotoxicity in male mice—histologic effects and chronic impairment in spatial learning. *Brain Res.* **707,** 165–179.

32. Schuman, E. M. and Madison, D. V. (1991) A requirement for the intracellular messenger nitric oxide in long term potentiation. *Science* **254,** 1503–1506.

33. Stringer, J. L., Greenfield, L. J., Hackett, J. T., and Guyenet, P. G. (1983) Blockade of long-term potentiation by phencyclidine and sigma opiates in the hippocampus in vivo and in vitro. *Brain Res.* **280,** 127–138.

34. Osawa, Y. and Davila, J. C. (1993) Phencyclidine, a psychotomimetic agent and drug of abuse, is a suicide inhibitor of brain nitric oxide synthase. *Biochem. Biophys. Res. Commun.* **194,** 1435–1439.

35. Balster, R. L. and Woolverton, W. L. (1980) Continuous access phencyclidine self-administration by rhesus monkeys leading to physical dependence. *Psychopharmacology* **70,** 5–10.

36. Marquis, K. L. and Moreton, J. E. (1987) Animal models of intravenous phencyclidine self-administration. *Pharm. Biochem. Behav.* **27,** 385–389.

37. Corbett, D. (1989) Possible abuse potential of the NMDA antagonist MK-801. *Behavioral Brain Res.* **34,** 239–246.

38. Herberg, L. J. and Roce, I. C. (1989) The effect of MK-801 and other antagonists of NMDA-type glutamate receptors on brain stimulation reward. *Psychopharmacology* **99,** 87–90.

39. Geifenstein, F. E., Devault, M., Yoshitake, J., and Gajewski, J. E. (1958) A study of 1-arylcyclohexyl-amine for anesthesia. *Anesth. Analg.* **37,** 283.

40. Burns, R. S. (1981) The effects of phencyclidine in man: a review, in *PCP (Phencyclidine): Historical and Current Perspectives* (Domino, E. F., ed.), NPP Books, Ann Arbor, MI, pp. 449–464.

41. Allen, R. M. and Young, S. J. (1978) Phencyclidine-induced psychosis. *Am. J. Psychiatr.* **135,** 1081–1084.

42. Erard, R., Luisada, P. V. and Peele, R. (1980) The PCP psychosis: prolonged intoxication or drug-precipitated functional illness? *J. Psychiatr. Drugs* **12,** 235–251.

43. Krystal, J. H., Karper, L. P., Seibyl, J. P., Freeman, G. K., Delaney, R., Bremner, J. D., Heninger, G. R., Bowers, M. B. Jr., and Charney, D. S. (1994) Subanesthetic effects of the noncompetitive NMDA antagonist, ketamine, in humans. Psychotomimetic, perceptual, cognitive, and neuroendocrine responses. *Arch. Gen. Psychiatr.* **51,** 199–214.

44. Dalgarno, P. J. and Shewan, D. (1996) Illicit use of ketamine in Scotland. *J. Psychol.* **28,** 191–199.

45. Wong, E. H. F., Kemp, J. A., Priestley, T., Knight, A. R., Woodruff, G. N., and Iverson, L. L. The anticonvulsant MK-801 is a potent *N*-methyl-D-aspartate antagonist. *Proc. Natl. Acad. Sci. USA* **83,** 7104–7108.

46. Kornhuber, J., Weller, M., Schoppmeyer, K., and Reiderer, P. (1994) Amantidine and memantine are NMDA receptor antagonists with neuroprotective properties. *J. Neural Transm.* **43 (Suppl.)**, 91–104.

47. Lodge, D. and Anis, N. A. (1982) Effects of phencyclidine on excitatory amino acid activation of spinal interneurons in the cat. *Eur. J. Pharmacol.* **77**, 203–204.

48. Vincent, J. P., Kartalovski, B., Geneste, P., Kamenka, J. M., and Lazdunski, M. (1979) Interaction of phencyclidine ("angel dust") with a specific receptor in rat brain membranes. *Proc. Natl. Acad. Sci. USA* **6**, 4678–4682.

49. Zukin, S. R. and Zukin, R. S. (1979) Specific [^{3}H]phencyclidine binding in rat central nervous system. *Proc. Natl. Acad. Sci. USA* **76**, 5372–5376.

50. Lodge, D. and Johnson, K. M. (1990) Noncompetitive excitatory amino acid receptor antagonists. *Trends Pharamacol. Sci.* **11**, 81–86.

51. Sonders, M. S., Keana, J. F., and Weber, E. (1988) Phencyclidine and psychotomimetic sigma opiates: recent insights into their biochemical and physiological sites of action. *Trends Neurosci.* **1**, 37–40.

52. Akunne, H. C., Reid, A. A., Thurkauf, A., Jacobson, A. E., deCosta, B. R., Rice, K. C., Heyes, M. P., and Rothman, R. B. (1991) ^{3}[H]-1-[1-(2-thienyl)cyclohexyl] piperidine labels two high affinity binding sites associated with biogenic amine reuptake complex. *Synapse* **8**, 289–300.

53. Albuquerque, E. X., Aguayo, L. G., Warnick, J. E., Weinstein, H., Glick, S. D., Maayani, S., Ickowicz, R. K., and Blaustein, M. P. (1981) The behavioral effects of phencyclidines may be due to their blockade of potassium channels. *Proc. Natl. Acad. Sci. USA* **78**, 7792–7796.

54. Johnson, K. M. and Jones, S. M. (1990) Neuropharmacology of phencyclidine: basic mechanisms and therapeutic potential. *Ann. Rev. Pharmacol. Toxicol.* **30**, 707–750.

55. Schoepfer, R., Monyer, H., Sommer, B., Wisden, W., Sprengel, R., Kuner, T., Lomeli, H., Herb, A., Kohler, M., and Burnashev, N. (1994) Molecular biology of glutamate receptors. *Prog. Neurobiol.* **42**, 353–357.

56. Reynolds, I. J. (1990) Modulation of NMDA receptor responsiveness by neurotransmitters, drugs and chemical modification. *Life Sci.* **47**, 1785–1792.

57. Bigge, C. F. (1993) Structural requirements for the development of potent *N*-methyl-D-aspartic acid (NMDA) receptor antagonists. *Biochem. Pharmacol.* **45**, 1547–1561.

58. Rajdev, S. and Reynolds, I. J. (1994) Glutamate-induced intracellular calcium changes and neurotoxicity in cortical neurons in vitro: effect of chemical ischemia. *Neuroscience* **62**, 667–679.

59. Reynolds, I. J. and Rajdev, S. (1994) Calcium, magnesium and glutamate receptors, in *Direct and Allosteric Control of Glutamate Receptors.* (Palfreyman, M. G., Reynolds, I. J., and Skolnick, P., eds.), CRC, Boca Raton, FL, pp. 39–56.

60. Honey, C. R., Miljkovic, Z., and MacDonald, J. F. (1985) Ketamine and phencyclidine cause a voltage-dependent block of responses to L-aspartic acid. *Neurosci. Lett.* **61**, 135–139.

61. MacDonald, J. F., Miljkovic, Z., and Pennefather, P. (1987) Use-dependent block of excitatory amino acid currents in cultured neurons by ketamine. *J. Neurophysiol.* **58**, 251–266.

62. Maragos, W. F., Penney, J. B. and Young, A. B. (1988) Anatomic correlation of NMDA and ^{3}H-TCP-labeled receptors in rat brain. *J. Neurosci.* **8,** 493–501.

63. Maayani, S., Weinstein, H., Ben-Zvi, N., Cohen, S., and Sokolovsky, M. (1974) Psychotomimetics as anticholinergic agents. I. 1-Cyclohexylpiperidine derivatives: anticholinesterase activity and antagonistic activity to acetylcholine. *Biochem. Pharmacol.* **23,** 1263–1281.

64. Murray, T. F. (1983) A comparison of phencyclidine with other psychoactive drugs on cholinergic dynamics in the rat brain, in *Phencyclidine and Related Arylcyclohexylamines.* (Kamenka, J. M., Domino, E. F., and Geneste, P., eds.), NPP Books, Ann Arbor, MI, pp. 547–567.

65. Castellani, S., Adams, P. M, and Giannini, A. J. (1982) Physostigmine treatment of acute phencyclidine intoxication. *J. Clin. Psychiatr.* **43,** 10–11.

66. Albuquerque, E. X., Tsai, M. C., Aronstam, R. S., Eldefrawi, A. T., and Eldefrawi, M. E. (1980) Sites of action of phencyclidine. II. Interaction with the ionic channel of the nicotinic receptor. *Mol. Pharmacol.* **18,** 167–178.

67. Greenberg, B. D. and Segal, D. S. (1985) Acute and chronic behavioral interactions between phencyclidine (PCP) and amphetamine: evidence for a dopaminergic role in some PCP-induced behaviors. *Pharm. Biochem. Behav.* **23,** 99–105.

68. Ary, T. E. and Komiskey, H. L. (1982) Phencyclidine-induced release of ^{3}H-dopamine from chopped striatal tissue. *Neuropharmacology* **21,** 639–645.

69. Bowyer, J. E., Spuhler, K. P., and Weiner, N. (1984) Effects of phencyclidine, amphetamine and related compounds on dopamine release from and uptake into striatal synaptosomes. *J. Pharmacol. Exp. Ther.* **229,** 671–680.

70. Snell, L. D., Yi, S. J., and Johnson, K. M. (1988) Comparison of the effects of MK-801 and phencyclidine on catecholamine uptake and NMDA-induced norepinephrine release. *Eur. J. Pharmacol.* **145,** 223–226.

71. French, E. D., Ferkany, J., Abreu, M., and Levenson, S. (1991) Effects of competitive NMDA antagonists on midbrain dopamine neurons: an electrophysiological and behavioral comparison to phencyclidine. *Neuropharmacology* **30,** 1039–1046.

72. Deutch, A. Y., Tam, S. Y., Freeman, A. S., Bowers, M. B. Jr., and Roth, R. H. (1987) Mesolimbic and mesocortical dopamine activation induced by phencyclidine: contrasting patterns to striatal response. *Eur. J. Pharmacol.* **134,** 257–264.

73. Liljequist, S., Ossowska, K., Grabowskaanden, M., and Anden, N. E. (1991) Effect of the NMDA receptor antagonist, MK-801, on locomotor activity and on the metabolism of dopamine in various brain areas of mice. *Eur. J. Pharmacol.* **195,** 55–61.

74. Healy, D. J. and Meadorwoodruff, J. H. (1996) Dopamine receptor gene expression in hippocampus is differentially regulated by the NMDA receptor antagonist MK-801. *Eur. J. Pharmacol.* **306,** 257–264.

75. Olds, M. E. (1996) Dopaminergic basis for the facilitation of brain stimulation reward by the NMDA receptor antagonist, MK-801. *Eur. J. Pharmacol.* **306,** 23–32.

76. Koek, W., Colpaert, F. C., Woods, J. H., and Kamenka, J. M. (1989) The phencyclidine (PCP) analog *N*-[1-(2-benzo(B)thiophenyl) cyclohexyl]piperidine shares cocaine-like but not other characteristic behavioral effects with PCP, ketamine and MK-801. *J. Pharmacol. Exp. Ther.* **250,** 1019–1027.

77. Lacey, M. G. and Henderson, G. (1986) Actions of phencyclidine on rat locus coeruleus neurons in vitro. *Neuroscience* **17,** 485–494.

78. Hitzemann, R. J., Loh, H. H., and Domino, E. F. (1973) Effect of phencyclidine on the accumulation of ^{14}C-catecholamines formed from ^{14}C-tyrosine. *Arch. Int. Pharmacodyn.* **202,** 252–258.

79. Palmer, M. P., Bickford, P. C., Hoffer, B. J., and Freedman, R. (1983) Electrophysiological evidence for presynaptic actions of phencyclidine on noradrenergic transmission in rat cerebellum and hippocampus, in *Phencyclidine and Related Arylcyclohexylamines: Present and Future Applications.* (Kamenka, J. M., Domino, E. F. Geneste, P., eds.), NPP Books, Ann Arbor, MI, pp. 443–470.

80. Hass, D. A. and Harper, D. G. (1992) Ketamine: a review of its pharmacological properties and use in ambulatory anesthesia. *Anesth. Prog.* **39,** 61–68.

81. Vaupel, D. B. (1983) Naltrexone fails to antagonize the sigma effects of PCP and SKF 10,047 in the dog. *Eur. J. Pharmacol.* **92,** 269–274.

82. Largent, B. L., Gundlach, A. L., and Snyder, S. H. (1986) Pharmacological and autoradiographic discrimination of sigma and phencyclidine receptor binding sites in brain with (+)-[^{3}H]SKF 10,047, (+)-[^{3}H]-3-[3-hydroxyphenyl]-*N*-(1-propyl)piperidine and [^{3}H]-1-[1-(2-thienyl)cyclohexyl]piperidine. *J. Pharmacol. Exp. Ther.* **238,** 739–748.

83. Wong, E. H. and Nielsen, M. (1989) The *N*-methyl-D-aspartate receptor channel complex and the sigma site have different target sizes. *Eur. J. Pharmacol.* **172,** 493–496.

84. Fidecka, S. (1987) Opioid mechanisms of some behavioral effects of ketamine. *Pol. J. Pharmacol. Pharm.* **39,** 353–360.

85. Seiler, N. and Grauffel, C. (1992) Antagonism of phencyclidine-induced hyperactivity in mice by elevated brain GABA concentrations. *Pharm. Biochem. Behav.* **41,** 603–606.

86. Myslobodsky, M. S., Ackerman, V., Golovchinsky, V., and Engel, J., Jr. (1979) Ketamine-induced rotation: interaction with GABA-transaminase inhibitors and picrotoxin. *Pharm. Biochem. Behav.* **11,** 483–486.

87. Wood, J. D. and Hert, L. (1990) Ketamine-induced changes in GABA system of mouse brain. *Neuropharmacology* **19,** 805–808.

88. Nabeshima, T., Yamaguchi, K., Hiramatsu, M., Amano, M., Furukawa, H., and Kameyama, T. (1984) Serotonergic involvement in phencyclidine-induced behaviors. *Pharm. Biochem. Behav.* **21,** 401–408.

89. Nabeshima, T., Hiramatsu, M., Furukawa, H., and Kameyama, T. (1985) Effects of acute and chronic administrations of phencyclidine on the levels of serotonin and 5-hydroxy-indoleacetic acid in discrete brain areas of mouse. *Life Sci.* **36,** 939–946.

90. Nabeshima, T., Noda, Y., Yamaguchi, K., Ishikawa, K., Furukawa, H., and Kameyama, T. (1985) Acute and chronic phencyclidine administration changes serotonin receptors in rat brain. *Eur. J. Pharmacol.* **109,** 129,130.

91. Whitton, P. S., Biggs, C. S., Pearce, B. R., and Fowler, L. J. (1992) MK-801 increases extracellular 5-hydroxytryptamine in rat hippocampus and striatum in vivo. *J. Neurochem.* **58,** 1573–1575.

92. Schmidt, C. J. and Fadayel, G. M. (1996) Regional effects of MK-801 on dopamine release: effects of competitive NMDA or 5-HT$_{2A}$ receptor blockade. *J. Pharmacol. Exp. Ther.* **277,** 1541–1549.

93. Longuemare, M. C., Keung, E. C., Chun, S., Sharp, F. R., Chan, P. H., and Swanson, R. A. (1996) MK-801 reduces uptake and stimulates efflux of excitatory amino acids via membrane depolarization. *Am. J. Physiol.* **39,** C1398–C1404.

94. Tourneur, Y., Romey, G., and Lazdunski, M. (1982) Phencyclidine blockade of sodium and potassium channels in neuroblastoma cells. *Brain Res.* **245,** 154–158.

95. Bolger, G. T., Rafferty, M. F., and Skolnick, P. (1986) Enhancement of brain calcium antagonist binding by phencyclidine and related compounds. *Pharm. Biochem. Behav.* **24,** 417–423.

96. Misra, A. L., Pontani, R. B., and Bartolomeo, J. (1979) Persistence of phencyclidine (PCP) and metabolites in brain and adipose tissue and implications for long-lasting behavioral effects. *Res. Commun. Chem. Pathol. Pharmacol.* **24,** 431–445.

97. Wallace, M. C., McCormack, A., and McCulloch, J. (1989) ^{3}H-MK-01: an in vivo ligand for glutamate mechanisms in the normal and ischemic brain. *J. Cereb. Blood Flow Metab.* **9,** S745.

98. Fix, A. S., Wozniak, D. F., Truex, L. L., McEwen, M., Miller, J. P., and Olney, J. W. (1995) Quantitative analysis of factors influencing neuronal necrosis induced by MK-801 in the rat posterior cingulate/retrosplenial cortex. *Brain Res.* **696,** 194–204.

99. Gao, X. M., Shirakawa, O., Du, F., and Tamminga, C. A. (1993) Delayed regional metabolic actions of phencyclidine. *Eur. J. Pharmacol.* **241,** 7–15.

100. Ellison, G. (1994) Competitive and non-competitive NMDA antagonists induce similar limbic degeneration. *Neuroreport* **5,** 2688–2692.

101. Ellison, G. and Switzer, R. C. (1993) Dissimilar patterns of degeneration in brain following four different addictive stimulants. *Neuroreport* **5,** 17–20.

102. Corso, T., Neafsy, E. L., and Collins, M. (1992) Ethanol-induced degeneration of dentate gyrus, entorhinal cortex and other olfactory related areas in rats, effects of co-administration of MK-801, DNQX or nimodipine. *Soc. Neurosci. Abstracts* **18,** 540.

103. Horvath, Z. and Buzsaki, G. (1993) MK-801-induced neuronal damage in normal rats. *Soc. Neurosci. Abstracts* **19,** 354.

104. Mattson, M. P., Rychlik, B., and Cheng, B. (1992) Degenerative and axon outgrowth-altering effects of phencyclidine in human fetal cerebral cortical cells. *Neuropharmacol.* **31,** 279–291.

105. Olney, J. W., Labruyere, J., Wang, G., Wozniak, D. F., Price, M. T., and Sesma, M. A. (1991) NMDA antagonist neurotoxicity: mechanism and prevention. *Science* **254,** 1515–1518.

106. Farber, N. B., Price, M. T., Labruyere, J., Nemnich, J., St. Peter, H., Wozniak, D. F., and Olney, J. W. (1993) Antipsychotic drugs block phencyclidine receptor-mediated neurotoxicity. *Biol. Psychiatr.* **34,** 119–121.

107. Farber, N. B., Foster, J., Duhan, N. L., and Olney, J. W. (1995) Alpha 2-adrenergic agonists prevent MK-801 neurotoxicity. *Neuropsychopharmacology* **12,** 347–349.

108. Bowen, W. D., Kirschner, B. N., Newman, A. H., and Rice, K. C. (1988) Sigma receptors negatively modulate agonist-stimulated phosphoinositide metabolism in rat brain. *Eur. J. Pharmacol.* **149,** 399,400.

109. Sharp, J. W., Petersen, D. L., and Langford, M. T. (1995) DNQX inhibits phencyclidine (PCP) and ketamine induction of the hsp70 heat shock gene in the rat cingulate and retrosplenial cortex. *Brain Res.* **687,** 114–124.

110. Duval, D., Roome, N., Gauffeny, C., Nowicki, J. P., and Scatton, B. (1992) SL 82.0715, an NMDA antagonist acting at the polyamine site, does not induce neurotoxic effects on rat cortical neurons. *Neurosci. Lett.* **137,** 193–197.

111. Auer, R. N. (1994) Assessing structural changes in the brain to evaluate neurotoxicological effects of NMDA receptor antagonists. *Psychopharmacol. Bull.* **30,** 585–591.

112. Auer, R. N. (1996) Effect of age and sex on N-methyl-D-aspartate antagonist-induced neuronal necrosis in rats. *Stroke* **27,** 743–746.

113. Honack, D. and Loscher, W. (1993) Sex differences in NMDA receptor mediated responses in rats. *Brain Res.* **620,** 167–170.

114. Nabeshima, T., Yamaguchi, K., Yamada, K., Hiramatsu, M., Kuwabara, Y., Furukawa, H., and Kameyama, T. (1984) Sex-dependent differences in the pharmacological actions and pharmacokinetics of phencyclidine in rats. *Eur. J. Pharmacol.* **97,** 217–227.

115. Smith, S. S. (1989) Estrogen administration increases neuronal response to excitatory amino acids as a long term effect. *Brain Res.* **503,** 354–357.

116. Weiland, N. G. (1992) Estradiol selectively regulates agonist binding sites on the N-methyl-D-aspartate receptor complex in the CA1 region of the hippocampus. *Endocrinology* **131,** 622–628.

117. Massa, S. M., Swanson, R. A., and Sharp, F. R. (1996) The stress gene response in brain. *Cerebrovas. Brain Metabol. Rev.* **8,** 95–158.

118. Sharp, F. R., Jasper, P., Hall, J., Noble, L., and Sagar, S. M. (1991) MK-801 and ketamine induce heat shock protein HSP72 in injured neurons in posterior cingulate and retrosplenial cortex. *Ann. Neurol.* **30,** 801–809.

119. Sharp, F. R., Butman, M., Wang, S., Koistinaho, J., Graham, S. H., Sagar, S. M., Noble, L., Berger, P., and Longo, F. M. (1992) Haloperidol prevents induction of the hsp70 heat shock gene in neurons injured by phencyclidine (PCP), MK801, and ketamine. *J. Neurosci. Res.* **33,** 605–616.

120. Sharp, F. R., Butman, M., Koistinaho, J., Aardalen, K., Nakki, R., Massa, S. M., Swanson, R. A., and Sagar, S. M. (1994) Phencyclidine induction of the hsp 70 stress gene in injured pyramidal neurons is mediated via multiple receptors and voltage gated calcium channels. *Neuroscience* **62,** 1079–1092.

121. Nakki, R., Koistinaho, J., Sharp, F. R., and Sagar, S. M. (1995) Cerebellar toxicity of phencyclidine. *J. Neurosci.* **15,** 2097–2108.

122. Popoli, P., Pezola, A., Benedetti, M., and Scotti-de-Caroles, A. (1992) Verapamil and flunarizine inhibit phencyclidine-induce defects: an EEG and behavioral study in rats. *Neuropharmacology* **31,** 1185–1191.

123. Popoli, P., Pezola, A., and Scotti-de-Caroles, A. (1993) Influence of non L-type calcium channel antagonists on phencyclidine-induced effects in rats. *Life Sci.* **52,** 2055–2061.

124. Eddleston, M. and Mucke, L. (1993) Molecular profile of reactive astrocytes—implications for their role in neurologic disease. *Neuroscience* **54,** 15–36.

125. Thomas, W. E. (1992) Brain macrophages: evaluation of microglia and their function. *Brain Res. Rev.* **17,** 61–74.

126. Fix, A. S., Ross, J. F., Stitzel, S. R., and Switzer, R. C. (1996) Integrated evaluation of central nervous system lesions: stains for neurons, astrocytes, and microglia reveal the spatial and temporal features of MK-801-induced neuronal necrosis in the rat cerebral cortex. *Toxicol. Pathol.* **24,** 291–304.

127. Nakki, R., Nickolenko, J., Chang, J., Sagar, S. M., and Sharp, F. R. (1996) Haloperidol prevents ketamine- and phencyclidine-induced HSP70 protein expression but not microglial activation. *Exp. Neurol.* **137,** 234–241.

128. Hughes, P. and Dragunow, M. (1995) Induction of immediate-early genes and the control of neurotransmitter-regulated gene expression within the nervous system. *Pharmacol. Rev.* **47,** 133–178.

129. Sagar, S. M., Sharp, F. R., and Curran, T. (1988) Expression of c-fos protein in brain: metabolic mapping at the cellular level. *Science* **240,** 1328–1331.

130. Nakki, R., Sharp, F. R., Sagar, S. M., and Honkaniemi, J. (1996) Effects of phencyclidine on immediate early gene expression in the brain. *J. Neurosci. Res.* **45,** 13–27.

131. Nakki, R., Sharp, F. R., and Sagar, S. M. (1996) Fos expression in the brainstem and cerebellum following phencyclidine and MK801. *J. Neurosci. Res.* **43,** 203–212.

132. Gao, X. M., Hoshomoto, T., and Tamminga, C. A. (1994) Time dependent changes in expression of immediate early genes in rat brain after phencyclidine. *Soc. Neurosci. Abstracts* **20,** 732.

133. Gass, P., Herdegen, T., Bravo, R., and Kiessling, M. (1993) Induction and suppression of immediate early genes in specific rat brain regions by the non-competitive *N*-methyl-D-aspartate receptor antagonist MK-801. *Neuroscience* **53,** 749–758.

134. Hughes, P., Dragunow, M., Beilharz, E., Lawlor, P., and Gluckman, P. (1993) MK801 induces immediate-early gene proteins and BDNF mRNA in rat cerebrocortical neurons. *Neuroreport* **4,** 183–186.

135. McCulloch, J. and Iversen, L. L. (1991) Autoradiographic assessment of the effects of *N*-methyl-D-aspartate (NMDA) receptor antagonists in vivo. *Neurochem. Res.* **16,** 951–963.

136. Nehls, D. G., Kurumaji, A., Park, C. K., and McCulloch, J. (1988) Differential effects of competitive and non-competitive *N*-methyl-D-aspartate antagonists on glucose use in the limbic system. *Neurosci. Lett.* **91,** 204–210.

137. Kurumaji, A. and McCulloch, J. (1989) Effects of MK-801 upon local cerebral glucose utilization in conscious rats and in rats anaesthetized with halothane. *J. Cereb. Blood Flow Metabol.* **9,** 786–794.

138. Meibach, R. C., Glick, S. D., Cox, R., and Maayani, S. (1979) Localization of phencyclidine-induced changes in brain energy metabolism. *Nature* **282,** 625–626.

139. Crosby, G., Crane, A. M., and Skoloff, L. (1982) Local changes in cerebral glucose utilization during ketamine anesthesia. *Anesthesiology* **56,** 437–443.

140. Tamminga, C. A., Thaker, G. K., Alphs, L. D., Buchanan, R. W., Kirkpatrick, B., Carpenter, W. T., and Chase, T. N. (1992) Limbic system abnormalities identified in schizophrenia using positron emission tomography with fluorodeoxyglucose and neocortical alterations with deficit syndrome. *Arch. Gen. Psychiatr.* **49,** 522–530

141. Kurumaji, A., Nehls, D. G., Park, C. K., and McCulloch, J. (1989) Effects of NMDA antagonists MK-801 and CPP, upon local cerebral glucose use. *Brain Res.* **496,** 268–284.

142. Hargreaves, R. J., Rigby, M., Smith, D., and Hill, R. G. (1993) Lack of effect of L-687,414 ((+)-cis-4-methyl-HA-966), an NMDA receptor antagonist acting at the glycine site, on cerebral glucose metabolism and cortical neuronal morphology. *Br. J. Pharmacol.* **110,** 36–42.

143. Sayed, Y. and Garrison, J. M. (1983) The dopamine hypothesis of schizophrenia and the antagonistic action of neuroleptic drugs—a review. *Psychopharmacol. Bull.* **19,** 283–288.

144. Carlsson, A. (1988) The current status of the dopamine hypothesis of schizophrenia. *Neuropsychopharmacology* **1**, 179–186.

145. Angrist, B., Sathananthan, G., Wilk, S., and Gershon, S. (1974) Amphetamine psychosis: behavioral and biochemical aspects. *J. Psychiatr. Res.* **11**, 13–23.

146. Kay, S. R., Fiszbein, A., and Opler, L. A. (1987) The positive and negative syndrome scale (PANSS) for schizophrenia. *Schizophr. Bull.* **13**, 261–276.

147. Itil, T., Keskiner, A., Kiremitci, N., and Holden, J. M. (1967) Effect of phencyclidine in chronic schizophrenics. *Can. Psychiatr. Assoc. J.* **12**, 209–212.

148. Kornetsky, C. (1976) Hyporesponsivity of chronic schizophrenic patients to dextroamphetamine. *Arch. Gen. Psychiatr.* **33**, 1425–1428.

149. Ogawa, S., Okuyama, S., Araki, H., Nakazato, A., and Otomo, S. (1994) A rat model of phencyclidine psychosis. *Life Sci.* **55**, 1605–1610.

150. Noda, Y., Yamada, K., Furukawa, H., and Nabeshima, T. (1995) Enhancement of immobility in a forced swimming test by subacute or repeated treatment with phencyclidine: a new model of schizophrenia. *Br. J. Pharmacol.* **116**, 2531–2537.

151. Banerjee, S. P., Zuck, L. G., Yablonsky-Alter, E., and Lidsky, T. I. (1995) Glutamate agonist activity: implications for antipsychotic drug action and schizophrenia. *Neuroreport* **6**, 2500–2504.

152. Lahti, A. C., Holcomb, H. H., Medoff, D. R., and Tamminga, C. A. (1995) Ketamine activates psychosis and alters limbic blood flow in schizophrenia. *Neuroreport* **6**, 869–872.

153. Tamminga, C. A., Thaker, G. K., Buchanan, R., Kirkpatrick, B., Alphs, L. D., Chase, T. N., and Carpenter, W. T. (1992) Limbic system abnormalities identified in schizophrenia using positron emission tomography with fluorodeoxyglucose and neocortical alterations with deficit syndrome. *Arch. Gen. Psychiatr.* **49**, 522–530.

154. Squires, R. F., Lajtha, A., Saederup, E., and Palkovits, M. (1993) Reduced [^{3}H]flunitrazepam binding in cingulate cortex and hippocampus of postmortem schizophrenic brains: is selective loss of glutamatergic neurons associated with major psychoses? *Neurochem. Res.* **18**, 219–223.

155. Welch, M. J. and Correa, G. A. (1980) PCP intoxication in young children and infants. *Clin. Pediatr.* **19**, 511–514.

156. Quirion, R., DiMaggio, D. A., French, E. D., Contreras, P. C., Shiloach, J., Pert, C. B., Everist, H., Pert, A., and O'Donohue T. L. (1984) Evidence for an endogenous peptide ligand for the phencyclidine receptor. *Peptides* **5**, 967–973.

157. Contreras, P. C., DiMaggio, D. A., and O'Donohue, T. L. (1987) An endogenous ligand for the sigma opioid binding site. *Synapse* **1**, 57–61.

158. Zukin, S. R., Zukin, R. S., Vale, W., Rivier, J., Nichtenhauser, R., Snell, L. D., and Johnson, K. M. (1987) An endogenous ligand of the brain sigma/PCP receptor antagonizes NMDA-induced neurotransmitter release. *Brain Res.* **416**, 84–89.

159. Javitt, D. C., Zylberman, I., Zukin, S. R., Heresco-Levy, U., and Lindenmayer, J. P. (1994) Amelioration of negative symptoms in schizophrenia by glycine. *Am. J. Psychiatr.* **151**, 1234–1236.

160. Valentine, J. L., Mayersohn, M., Wessinger, W. D., Arnold, L. W., and Owens, S. M. (1996) Antiphencyclidine monoclonal Fab fragments reverse phencyclidine-induced behavioral effects and ataxia in rats. *J. Pharmacol. Exp. Ther.* **278**, 709–716.

15
Toxic Vanilloids

Arpad Szallasi

1. FOREWORD: ARE VANILLOIDS REALLY TOXIC?

The best-known vanilloid (Table 1), capsaicin (*see* Fig. 1 for structure), is most commonly recognized as the irritant principle in hot peppers and is consumed on a daily basis by one-quarter of the world's population *(1)*, most likely including several readers of this chapter. Other popular vanilloids include piperine, the active ingredient in black pepper, as well as zingerone, responsible for the picancy of ginger (Table 1). Therefore, it is important to emphasize: do not panic by the title of this chapter; culinary use of vanilloids is probably safe; at least, there are no data to suggest otherwise. The absorption of vanilloids in the gastrointestinal tract appears to be limited (as the Hungarian folk wisdom goes: "Hot pepper bites twice!"), and although there are reports that chili peppers may be carcinogenic in rodents, it is likely that such actions are owing to contaminants (e.g., aflatoxin) in pepper preparations rather than to vanilloids *(2)*.

2. INTRODUCTION AND HISTORICAL PERSPECTIVES

It is still a puzzle why the very same burning sensation produced in the tongue by capsaicin, which repels animals (despite several experimental efforts to model human chili preference, rats fail to develop a liking for hot peppers) is found pleasant by so many humans. Capsaicin, as first noted by Högyes *(3)* over a century ago, lowers body temperature in animals. Connoisseurs of hot, spicy food experience a similar cooling effect owing to "gustatory sweating" *(4)*. Whereas this effect of capsaicin on thermoregulation, along with an antidiarrheic activity, may explain the popularity of hot peppers in hot climates, where the normal dietary intake of *Capsicum* may be as high as 2.5 g per diem (corresponding to approx 1 mg capsaicin per os/kg body wt, daily), in temperate climates, hot pepper preference is supposed to be a hereditary trait reflecting the "masochism of everyday life" *(5)*.

From: Highly Selective Neurotoxins: Basic and Clinical Applications *Edited by: R. M. Kostrzewa.*
Humana Press Inc., Totowa, NJ

Table 1
Notable Vanilloids

Capsaicinoids
   Natural: capsaicin (in red and green hot peppers of the genus *Capsicum*);
      piperine (in black pepper, *Piper nigrum*); zingerone (in ginger root, *Zingiber officinale*)
   Synthetic: nonivamide ("synthetic capsaicin"); civamide (cis-isomer of
      capsaicin); olvanil and nuvanil (orally active analogs, Procter & Gamble)

Resiniferanoids
   Natural: resiniferatoxin and tinyatoxin (in several member of the genus
      Euphorbia)
   Synthetic: 12-deoxyphorbol 13-phenylacetate 20-homovanillate (fails to induce
      hypothermia); phorbol 12-phenylacetate 13-acetate 20-homovanillate (binds
      noncooperatively).

Pungent diterpenes with a 1,4-unsaturated dialdehyde moiety
   For example, velleral and isovelleral (in the mushroom *Lactarius vellereus*),
      warburganal (in the East African *Warburgia* plant)

Preceding the pioneering work of the late Nicholas Jancsó, who demonstrated that sensory nerves subserving pain and neurogenic inflammation can be stimulated and desensitized by capsaicin application *(6,7)*, the sole interest in capsaicin congeners had been generated by a search for improved pepper-flavored food additives. This search led to the synthesis of nonivamide, erroneously called "synthetic capsaicin," and the worldwide practice of adulterating capsaicin with this cheap substitute.

In the 1970s, capsaicin was introduced to neuropharmacology as a remarkably selective tool to identify an otherwise both functionally and morphologically diverse subpopulation of primary sensory neurons *(8,9; also see* Table 2). In the 1980s, capsaicin was found to be beneficial in neuropathic pain syndromes *(10,11)* and capsaicin-containing creams (e.g., Axsain, Zostrix) became commercially available. In the late 1980s, capsaicin research was given impetus by the discovery of ultrapotent capsaicin analogs *(12,13;* Table 1), the paradigm of which is resiniferatoxin (RTX; structure shown in Fig. 1), a phorbol-related, extremely irritant diterpene isolated from several members of the genus *Euphorbia*. Since capsaicinoids and resiniferanoids differ dramatically in the rest of the molecule, but share a (homo)vanillyl substituent (compare structures in Fig. 1) critical for biological activity *(13)*, these two classes of compounds were collectively referred to as "vanilloids" *(12)*. In keeping with this terminology, the long sought after capsaicin receptor, the actual existence of which was first demonstrated by the specific binding of [3H]RTX (Fig. 2), was termed the "vanilloid receptor" *(12–14)*.

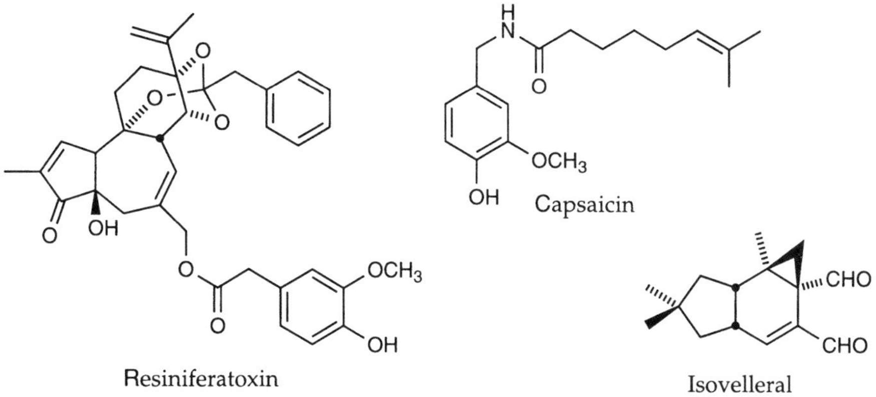

Fig. 1. Vanilloid structures: capsaicin, resiniferatoxin, and isovelleral, representing the three known major classes of naturally occurring vanilloid agonists (capsaicinoids, resiniferanoids, and pungent diterpenes with a 1,4-unsaturated dialdehyde moiety, respectively).

Nowadays vanilloids are in the focus of interest (1) as a promising approach to mitigate itching, neuropathic pain, and a variety of pathobiological conditions in which mediators released from vanilloid-sensitive neurons are thought to play a major role, and (2) as ingredients in "cop-in-a-can" sprays *(15)*. It can easily be predicted that in the near future, an increasing number of people will be exposed to high topical doses of vanilloids, either during therapy or police intervention, which will raise public concern regarding the toxicity of vanilloids. Mysterious deaths following the use of capsaicin sprays for "aerosol restraint" have already been reported. Thus it would not be surprising at all if "Sergeant Pepper" stood trials shortly in US courts.

3. VANILLOID-SENSITIVE NEURONS

3.1. Morphological/Neurochemical Markers

In the adult rat, approx 60% of primary sensory neurons with somata in dorsal root ganglia (DRG) or trigeminal ganglia are activated by capsaicin *(16)*. In general, these are small (diameters in the range of 20–50 µm) neurons, giving rise to either unmyelinated (C)-, or thin myelinated (Aδ)-fibers *(8,9; also see* Table 2). Peripheral endings comprise polymodal nociceptors, which respond to a range of stimuli, from noxious heat to chemical irritants *(17; also see* Table 2). Central fibers enter the spinal cord via the dorsal roots, and generally synapse with second-order neurons in layers I, II, and V of the dorsal horn. Most of these neurons are killed by neonatal capsaicin administration *(18)*, and some of them (about 10%) presumably also degenerate in the adult *(19)*, if sufficiently high doses are used. At present, it is unclear whether this is a qualita-

Table 2
Some Characteristics of Vanilloid-Sensitive Neurons[a]

Anatomical location: sensory (dorsal root as well as trigeminal) ganglia

Morphology: small (20–50 μm) pseudounipolar cells with unmyelinated (C)-fibers, possessing chemonociceptors or C-mechanoheat-sensitive nociceptors; also, some large neurons with thin-myelin (Aδ)-fibers, which carry A-mechanoheat-sensitive nociceptors

Mediators: glutamate (rapid, central transmitter); neurokinins (notably substance P); CGRP; galanin; VIP; somatostatin; secretoneurin; cholecystokinin

Enzymes: NOS; fluorid-resistant acid phosphatase

Receptors: vanilloid receptors; cholecystokinin B receptors; GABA receptors; serotonin receptors

[a]There are vanilloid-sensitive sensory neurons in nodose ganglia, too, and in several brain nuclei (e.g., locus ceruleus, preoptic area).

tive difference between newborn and mature animals. Whereas probably all the vanilloid-sensitive neurons use glutamate as the rapid central neurotransmitter *(20)*, they show considerable heterogeneity in expression of neuropeptides (Table 2). The tachykinin substance P (SP), present in approx 20% of small DRG neurons, is thought to represent the main excitatory sensory peptide, the effect of which is modulated by concomitantly released calcitonin gene-related peptide (CGRP) *(20,21)*. CGRP itself, however, is expressed in about 50% of neurons of all sizes. Somatostatin, which does not colocalize with SP, is another peptide present in vanilloid-sensitive neurons *(9,20,21)*. Other peptide markers of vanilloid-sensitive neurons include cholecystokinin, bombesin, vasoactive intestinal polypeptide (VIP), and galanin *(8,9;* Table 1). Although these peptides are normally present in a very small number of neurons, they become abundant and gain special importance following mechanical *(21,22)* or chemical *(22)* nerve damages *(see* Section 3.4.). Apparently, there is no peptide that either alone or in combination with another marker would label every vanilloid-sensitive neuron. Worse still, the expression of these peptides is not restricted to vanilloid-sensitive neurons.

In addition to dorsal root as well as trigeminal ganglia, the nodose ganglia also contain neurons sensitive to capsaicin *(9)*. Furthermore, there are capsaicin-sensitive structures in the central nervous system (CNS) *(23)*, including the preoptic area, which is believed to mediate the above-mentioned thermoregulatory action of vanilloids *(24)*.

Finally, it has to be kept in mind that if sufficiently high doses of capsaicin are used, the selectivity for primary sensory neurons is no longer observed *(9; also see* Table 3). At high concentrations, capsaicin is capable of inhibiting a

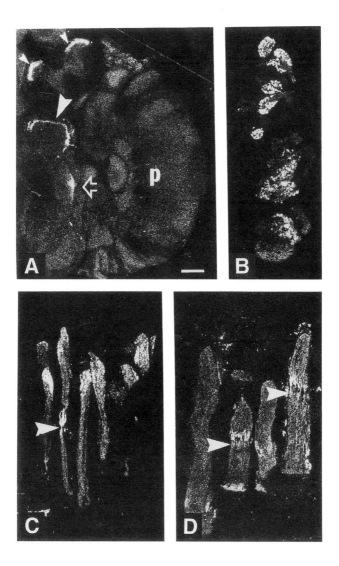

Fig. 2. Autoradiographic visualization of vanilloid receptors (specific [3H]resiniferatoxin binding sites) in the rat: DRG (**B**) and nodose (**C**) ganglia, containing cell bodies of vanilloid-sensitive neurons; spinal cord (**A**, small arrowheads) and brainstem (A, big and open arrowheads), the main termination site for central fibers; as well as vagal (**C**) and sciatic (**D**) nerves, in which peripheral axons traverse. Note that specific [3H]resiniferatoxin binding sites accumulate proximal to ligation of nerves (big arrowheads in C and D), indicating intra-axonal receptor transport to the periphery. Reprinted with permission from ref. *(43)*.

Table 3
Vanilloid Actions

1. Specific, neuronal effects (excitation, desensitization, neurotoxicity) can be detected in vanilloid receptor-bearing neurons of DRG, trigeminal, and nodose ganglia, as well as in certain brain nuclei, such as the locus ceruleus.
2. Nonspecific neurotoxicity can be observed through the neuroaxis of the rat if sufficiently high capsaicin doses are used.
3. Other nonspecific (i.e., not vanilloid receptor-mediated) actions:
   a. Direct interaction at other receptors/channels (e.g., voltage-gated K+-channels in the heart or melanotrophs of the pituitary; nicotinic acetylcholine receptors in pheochromocytoma cells).
   b. Inhibition of enzymes (e.g., NADH-coenzyme Q oxidoreductase of the mitochondrial respiratory chain).
   c. Changes in membrane fluidity.
   d. Binding to the plasma protein orosomucoid (also known as α_1-acid glycoprotein.

variety of enzymes and of changing the fluidity of membrane lipids *(9)*. Even more disturbing, it shows activity in several assays of mutageneicity *(2)*. Although it is consoling that McMahon and coworkers *(25)* did not observe signs of toxicity in rats receiving a daily treatment with capsaicin for months, the therapeutic window of capsaicin seems to be worrisomely narrow.

3.2. Physiological Functions of Vanilloid-Sensitive Neurons and Their Contribution to Disease States

Little is known about the function of "resting" vanilloid-sensitive neurons. A sustained release of neuropeptides is thought to have important trophic, immuno-, and vasoregulatory functions *(26)*. For example, rats lacking vanilloid-sensitive neurons owing to neonatal capsaicin treatment are known to develop skin lesions *(27)*. When activated, vanilloid-sensitive neurons transmit nociceptive information into the CNS (afferent function), whereas peptides released from their peripheral endings initiate the cascade of inflammatory events, collectively referred to as neurogenic inflammation *(6–9)*. Abnormal impulse generation in these nerves may be perceived as neuropathic pain or result in hyperreflexia (e.g., neuropathic bladder), whereas enhanced peptide release in the periphery is implicated in symptoms like vasomotor rhinitis, arthritis, or cluster headache *(11,28)*.

3.3. Vanilloid Actions: Excitation, Desensitization, and Neurotoxicity

Among the multitude of xenobiotics known to activate primary sensory neurons *(29)* vanilloids are unique in that, after an initial period of excitation, they render the neurons insensitive to further stimuli *(6–9)*. This phenomenon is traditionally referred to as desensitization *(6,7)*. It has been suggested that vanilloid receptors, believed to be ligand-gated nonselective cation channels *(30)*, cycle between closed and open states via intermediate states *(31)*. According to this model, pharmacological desensitization can be thought of as the time vanilloid receptors spend in intermediate state(s), unable to bind vanilloids. This receptor-level desensitization does not require the presence Ca^{2+} *(9)*, and is likely to stem from a ligand-induced conformational change. However, since the extracellular fluid is not Ca^{2+}-free by any means, the opening of vanilloid receptors invariably leads to an increase in intracellular Ca^{2+} and a resulting activation of Ca^{2+}-dependent enzymes of which Ca^{2+}- and calmodulin-dependent protein phosphatase 2B (calcineurin) is believed to play a central role in desensitization by dephosphorylating the vanilloid receptor/channel complex *(30)*. In this state, referred to as tachyphylaxis, vanilloid-sensitive neurons no longer respond to a subsequent vanilloid administration, but are sensitive to other stimuli *(8,9)*. Further increasing Ca^{2+} concentrations impairs neuronal functions by blocking voltage-sensitive Ca^{2+}-channels (which are responsible for nerve-evoked neurotransmitter release) as well as other, as yet unidentified, Ca^{2+}-dependent processes *(30)*. This state, in which neurons no longer respond to any stimuli, is usually termed functional desensitization *(9)*. Ca^{2+} is probably also a key player in vanilloid-induced neurotoxicity (there are clear parallelisms with toxicity by excitatory amino acids), as is the osmotic damage that follows Na^+ influx *(9,12,30)*.

3.4. Vanilloid-Induced Messenger Plasticity in Sensory Neurons

Following vanilloid administration, the expression of peptides in the affected neurons is dramatically altered *(32,33)*. For example, SP is downregulated, whereas VIP or galanin is enhanced. The expression of the enzyme nitric oxide synthase (NOS) is also increased *(32)*. These changes in peptide expression, collectively referred to as messenger plasticity, mimic those seen after mechanical nerve damage, such as axotomy *(21)*. Messenger plasticity in axotomized animals is believed to represent adaptive responses promoting neuronal survival *(21)*. Its underlying mechanism(s) is (are) poorly understood, but a deprivation of neurotrophic factors, such as nerve growth factor (NGF), transported intraxonally from the periphery, where they are synthesized, to the somata of neurons is an obvious candidate mechanism *(21)*. At present it is unclear whether messenger plasticity seen after vanilloid treatment reflects a "chemical injury" to the neurons *(22)*, or stems only from the well-documented ability of vanilloids to block intra-axonal transport *(8,9)*. Probably, it is a dose-dependent combination of these two mechanisms. In either case, it has to be clearly distinguished from neurodegeneration when peptides are indiscriminately depleted *(8,9)*. Of the changes in peptide expression, enhanced galanin is of special importance: galanin is believed to inhibit excitatory transmission in the spinal cord *(34)*. Thus, it may represent a neurochemical basis for vanilloid-induced analgesia *(33)*.

4. VANILLOID RECEPTORS

4.1. The Vanilloid Receptor/Channel Complex

The vanilloid receptor is either a ligand-gated, nonselective cation channel *per se* or at least closely coupled to one *(9,30)*. Its molecular target size, as determined by radiation inactivation, is approx 300 kDa *(35)*, suggesting a multisubunit structure. Moreover, it comprises at least two interacting ligand recognition sites displaying positive cooperativity *(14)*. It was speculated that positive cooperativity may serve as an amplification mechanism to enhance the efficacy of endogenous ligands *(36)*. Therefore, the positive cooperativity of vanilloid receptors implies the existence of endogenous vanilloids. However, cooperative binding does not appear to be an inherent property of vanilloid receptors, but rather a ligand-induced feature *(37)*. Thus, it is premature to jump to any conclusions regarding its function.

Pharmacological receptors are preferably named after their endogenous ligands. The vanilloid receptor is, however, still an orphan receptor, and it is not known whether it belongs to a receptor family. Among the known endogenous activators of vanilloid-sensitive neurons, protons have repeatedly been suggested to act as endogenous vanilloids *(9,39,38)*. However, most recently Oh and colleagues *(39)* failed to observe an activation by low pH of vanilloid-operated conductances in patch-clamped trigeminal ganglion neurons. Thus, the question whether vanilloids and protons gate identical, or only similar, conductances is still open.

There is experimental evidence to suggest that the maintenance of vanilloid sensitivity requires a continuous supply of NGF in a subpopulation of primary sensory neurons *(11,30)*. The role of NGF in the regulation of vanilloid receptor expression is, however, hampered by the finding that axotomy fails to downregulate specific [^{3}H]RTX binding sites in the rat *(33)*. It is also known that a subset of vanilloid-sensitive neurons (for example, those that express somatostatin) do not possess *trk*-receptors for NGF.

4.2. Vanilloid Receptors: Heterogeneity and Species-Related Differences

Vanilloids show striking differences in relative potencies for different biological end points *(9,12,13)*, which is difficult to reconcile with the original model of a single class of shared receptors *(30)*, but is entirely consistent with the existence of multiple receptor classes *(33)*. Structure/activity relations for receptor binding and Ca^{2+}-uptake are different, suggesting that these assays detect distinct vanilloid receptor classes *(40)*. Moreover, vanilloid-induced cation fluxes show an amazing variability in their gating properties *(31)*, indicating further heterogeneity of the receptor detected by Ca^{2+}-uptake. Also, there is evidence that whereas certain conductances are activated by both capsaicin and RTX, others recognize either ligand only *(31)*. Compared to capsaicin, RTX tends to evoke slowly activating, but lasting currents *(31)*, which, in turn, favor Ca^{2+} accumulation (and hence desensitization) to the activation of voltage-sensitive Na$^+$-channels (impulse generation). It may represent the neurophysiological basis to explain why RTX is more potent to desensitize than to evoke pain responses. Conversely, piperine and zingerone, representing pungent, but nondesensitizing capsaicinoids, predominantly cause rapidly activating, but transient currents *(40a)*.

Given the striking species-related differences in capsaicin sensitivity *(8,9)*, it is hardly unexpected that the expression of vanilloid receptors is likewise species-dependent *(14,41)*. In general, nonmammals do not respond to capsaicin (that is why capsaicin can be used to produce "squirrel-free" bird feed), and they do not express vanilloid receptors *(14,41)*. However, among mammals, the connection between vanilloid receptors and vanilloid sensitivity is complicated by the well-known species-related differences in the expression of sensory neuropeptides and their receptors *(14)*.

Whereas the presence of vanilloid receptors in human tissues obtained postmortem is unambiguous *(42,43)*, at present it is unclear to what extent the observed differences between the properties of human and animal vanilloid receptors may reflect postmortem tissue autolysis.

4.3. The Diversity of Ligands for Vanilloid Receptors

4.3.1. Agonists

Capsaicinoids (Table 1) comprise a family of structurally closely related compounds *(44)*. Natural capsaicinoids include capsaicin, piperine and zingerone. Of synthetic capsaicinoids, olvanil and nuvanil are notable, partly because they are relatively well absorbed following *per os* administration, and partly because they tend to activate preferentially central endings of vanilloid-sensitive neurons *(11)*. Resiniferanoids (Table 1) include the naturally occurring RTX and tinyatoxin (which differs from RTX only in the absence of the 3-methoxy group), as well as a number of synthetic ligands obtained by the esterification of an RTX-like diterpene skeleton (e.g., phorbol, 12-deoxyphorbol, and so forth) at C20 position *(12–14,37)*.

Resiniferanoids are of particular interest because of the following reasons:

1. Their ultrapotency: for example, RTX is three to four orders of magnitude more potent than capsaicin for most vanilloid-like responses *(12,13)*;
2. Their selectivity: for example, RTX, unlike capsaicin, desensitizes pulmonary J1 receptors without prior excitation *(12,13)*; 12-deoxyphorbol 13-phenylacetate 20-homovanillate fails to induce any hypothermia response at doses at which it effectively desensitizes against neurogenic inflammatory agents *(12,13)*; and
3. Their effect on cooperativity: for example, phorbol 12-phenylacetate 13-acetate 20-homovanillate binds noncooperatively. Furthermore it abolishes positive cooperativity of binding to other ligands as well *(37)*.

In the case of capsaicinoids, the presence of a (homo)vanillyl group is essential for biological activity *(6,9,44)*, a rule later confirmed by resiniferanoids

(12,13). However, most recently it has been found that pungent diterpenes with a 1,4-unsaturated dialdehyde moiety (Table 1), as exemplified by isovelleral *(see* structure in Fig. 1) isolated from a pungent mushroom, are able to activate capsaicin-sensitive neurons in a vanilloid receptor-mediated fashion *(33)*. This finding suggests that structure/activity relationships for vanilloid-like activity are much more complex than thought previously *(33)*.

4.3.2. Antagonists

The competitive vanilloid receptor antagonist capsazepine came from an extensive synthetic program at the Sandoz Institute for Medical Research in London, aimed at understanding the structure/activity relationship of capsaicinoids *(44)*. The potency of capsazepine is species-dependent. Unfortunately, the micromolar concentrations of capsazepine, which are frequently required to block vanilloid receptors, may also inhibit voltage-sensitive Ca^{2+}-channels *(45)*. Alternatively, ruthenium red, an inorganic dye thought to block the vanilloid channel pore without interfering with ligand binding, may be used as a "functional vanilloid antagonist" *(46)*. However, ruthenium red is not selective for vanilloid receptors.

4.3.3. Modulators

Little is known about modulators of vanilloid receptors, but their very existence is exciting because it implies that at least in theory, compounds can be synthesized that do not open the channel (hence, they do not cause irritation), but induce the conformational changes leading to desensitization. At present, compounds that are able to modulate vanilloid receptors include some heavy metals, such as zinc, sulfhydryl-reactive agents, and phenothiazine drugs *(43)*. Zinc is of interest as a potential endogenous modulator of vanilloid receptor functions, and it is tempting to speculate that the well-documented ability of phenothiazines to ameliorate neuropathic pain is somehow related to their interaction at vanilloid receptors *(43,47)*.

5. VANILLOIDS: CURRENT CLINICAL USES, FUTURE THERAPEUTIC POTENTIAL

Generally speaking, vanilloids comprise a family of potential nonsteroidal, anti-inflammatory, nonnarcotic analgesic drugs *(10,11,33,48)*. Since vanilloid receptors are expressed almost exclusively by primary sensory neurons, a vanilloid receptor-selective ligand should have virtually no nonneuronal actions. Moreover, it seems to be a lucrative approach to block vanilloid-sensitive neurons (and thereby prevent mediator release), rather than to target the receptors for the multitude of mediators released from such neurons. Furthermore, vanilloids because of their ability to stop intra-axonal transport may be used to prevent recurrent herpes diseases *(49)*. Finally, perineural injection of high, neurotoxic doses of capsaicin may be used to mitigate otherwise intractable neuropathic pain *(50)*. Capsaicin preparations have been tried in a variety of disease

Table 4
Current Therapeutic Uses of Capsaicin*a*

Chronic pain of neuropathic origin
  Postherpetic neuralgia; diabetic neuropathy; postmastectomy pain syndrome; stump pain; reflex sympathetic dystrophy; trigeminal neuralgia; osteoarthritis; rheumatoid arthritis; cluster headache; fibromyalgia; Guillain-Barré syndrome

Pruritus
  Psoriasis; hemodialysis; notalgia paraesthetica; brachioradial pruritus; aquagenic pruritus; vulvar vestibulitis

Symptoms associated with nerve hyperactivity
  Vasomotor rhinitis; bladder hyperreflexia

Symptoms of unknown etiology
  Apocrin chromhydrosis

aSee refs. *(10,11,28,33,48,50)* for further details.

states (as summarized in Table 4), but with some exceptions, such as vasomotor rhinitis, they did not live up to expectations *(11,33)*. Capsaicin creams suffer from two major drawbacks: first, they are very irritant (hence the high dropout rate plaguing clinical trials), and second, their efficacy is limited *(48)*. Efforts to synthesize a capsaicinoid better than capsaicin have been disappointing. At present, RTX is undergoing clinical trials in Arizona to mitigate postherpetic neuralgia, as well as pain related to diabetic polyneuropathy. Although RTX outperforms capsaicin in animal experimentation *(48)*, it remains to be seen whether it is really a better drug in humans.

6. CONCLUDING REMARKS AND FUTURE RESEARCH DIRECTIONS

Vanilloids show remarkable selectivity for a subset of primary sensory neurons involved in nociception and neurogenic inflammation. Vanilloids represent an invaluable research tool to identify such neurons, and have a clear, yet unfulfilled, therapeutic potential to mitigate neuropathic pain and pathological conditions in which neuropeptides released from vanilloid-sensitive neurons play a pivotal role. There is now good evidence that the membrane recognition site for vanilloids, the vanilloid receptor, is not a single receptor, but a family of receptors. The heterogeneity of vanilloid receptors implies that receptor subtype-selective ligands may be synthesized, devoid of the undesirable side effects of existing vanilloids. Furthermore, vanilloid receptors are apparently much broader in their ligand selectivity (for example, they bind pungent dialdehydes as well as neuroleptic drugs) than thought previously. At present, the vanilloid receptor is still an orphan receptor, although several lines of indirect evidence suggest the existence of endogenous vanilloids. In addition, it is not known whether there is a lone receptor or a receptor family for vanilloids. Molecular cloning of the vanilloid receptor will, no doubt, clarify this question. Stay tuned: rapid advances are anticipated in this field!

REFERENCES

1. Govindarajan, V. S. (1986) Capsicum—production, technology, chemistry, and quality. II. Processed products, standards, world production, and trade. *CRC Crit. Rev. Food Sci. Nutr.* **23,** 207–288.
2. Surh, Y.-J. and Lee, S. S. (1996) Capsaicin in hot chili pepper: carcinogen, co-carcinogen or anticarcinogen? *Food Chem. Toxicol.* **34,** 313–316.
3. Högyes, E. (1878) Beiträge zur physiologischen Wirkung der Bestandteile des Capsicum annuum. *Arch. Exp. Pathol. Pharmakol.* **9,** 117–130.
4. Lee, T. S. (1954) Physiological gustatory sweating in warm climate. *J. Physiol. (Lond.)* **124,** 528–542.
5. Rozin, P. and Schiller, D. (1980) The nature and acquisition of a preference for chili pepper by humans. *Motivation Emotion* **1,** 77–101.
6. Jancsó, N. (1960) Role of nerve terminals in the mechanism of inflammatory reactions. *Bull. Millard. Fillmore Hosp. Buffalo, NY* **7,** 53–77.
7. Jancsó, N. (1968) Desensitization with capsaicin and related acylamides as a tool for studying the function of pain receptors, in *Pharmacology of Pain* (Lin, K., Armstrong, D., and Pardo, E. G., eds.), Pergamon, Oxford, pp. 33–55.
8. Buck, S. H. and Burks, T. F. (1986) The neuropharmacology of capsaicin: a review of some recent observations. *Pharmacol. Rev.* **38,** 179–226.
9. Holzer, P. (1991) Capsaicin: cellular targets, mechanisms of action, and selectivity for thin sensory neurons. *Pharmacol. Rev.* **43,** 144–201.
10. Rumsfield, J. A. and West, D. P. (1991) Topical capsaicin in dermatologic and peripheral pain disorders. *DICP Ann. Pharmacother.* **25,** 381–387.
11. Winter, J., Bevan, S., and Campbell, E. A. (1995) Capsaicin and pain mechanisms. *Br. J. Anaesthesia* **75,** 157–168.
12. Szallasi, A. and Blumberg, P. M. (1990) Resiniferatoxin and its analogs provide novel insights into the pharmacology of vanilloid (capsaicin) receptors. *Life Sci.* **47,** 1399–1408.
13. Blumberg, P. M., Szallasi, A., and Acs, G. (1993) Resiniferatoxin—an ultrapotent capsaicin analogue, in *Capsaicin in the Study of Pain* (Wood, J. N., ed.), Academic, New York, pp. 45–62.
14. Szallasi, A. (1994) The vanilloid (capsaicin) receptor: receptor types and species differences. *Gen. Pharmacol.* **25,** 223–243.
15. Hyder, K. (1996) Is CS the wrong solution? *New Scientist* **149,** 12,13.
16. Liu, L., Wang, Y., and Simon, S. A. (1996) Capsaicin activated currents in rat dorsal root ganglion cells. *Pain* **64,** 191–195.
17. Szolcsányi, J. (1993) Actions of capsaicin on sensory receptors, in *Capsaicin in the Study of Pain* (Wood, J. N., ed.), Academic, New York, pp. 1–26.
18. Jancsó, G., Király, E., and Jancsó-Gábor, A. (1978) Pharmacologically induced selective degeneration of chemosensitive sensory neurones. *Nature* **270,** 741–743.

19. Jancsó, G., Király, E., Joó, F., Such, G., and Nagy, A. (1985) Selective degeneration by capsaicin of a subpopulation of primary sensory neurons in the adult rat. *Neurosci. Lett.* **59**, 209–214.

20. Urbán, L., Thompson, S. W. N., and Dray, A. (1994) Modulation of spinal excitability: co-operation between neurokinin and excitatory amino acid transmitters. *Trends Neurosci.* **17**, 432–438.

21. Hökfelt, T., Zhang, X., and Wiesenfeld-Hallin, Z. (1994) Messenger plasticity in primary sensory neurons following axotomy and its functional implications. *Trends Neurosci.* **17**, 22–30.

22. Jancsó, G. (1992) Pathobiological reactions of C-fibre primary sensory neurones to peripheral nerve injury. *Exp. Physiol.* **77**, 405–431.

23. Ritter, S. and Dinh, T. T. (1993) Capsaicin-induced degeneration in rat brain and retina, in *Capsaicin in the Study of Pain* (Wood, J. N., ed.), Academic, New York, pp. 105–138.

24. Szolcsányi, J., Joó, F., and Jancsó-Gábor, A. (1971) Mitochondrial changes in preoptic neurones after capsaicin desensitization of the hypothalamic thermodetectors in rats. *Nature* **299**, 116,117.

25. McMahon, S. B., Lewin, G., and Bloom, S. R. (1991) The consequences of long-term capsaicin application in the rat. *Pain* **44**, 301–310.

26. Holzer, P. (1988) Local effector functions of capsaicin-sensitive sensory nerve endings: involvement of tachykinins, calcitonin gene-related peptide and other neuropeptides. *Neuroscience* **24**, 739–768.

27. Thomas, D. A., Dubner, R., and Ruda, M. A. (1994) Neonatal capsaicin treatment in rats results in scratching behavior with skin damage: potential model of non-painful dysesthesia. *Neurosci. Lett.* **171**, 101–104.

28. Maggi, C. A. and Meli, A. (1988) The sensory-efferent function of capsaicin-sensitive sensory neurons. *Gen. Pharmacol.* **19**, 1–43.

29. Rang, H. P., Bevan, S., and Dray, A. (1994) Nociceptive peripheral neurons: cellular properties, in *Textbook of Pain* (Wall, P. D. and Melzack, R., eds.), Churchill Livingstone, Edinburgh, pp. 57–78.

30. Bevan, S. and Docherty, R. J. (1993) Cellular mechanisms of action of capsaicin, in *Capsaicin in the Study of Pain* (Wood, J. N., ed.), Academic, New York, pp. 27–44.

31. Liu, L. and Simon, S. A. (1996) Capsaicin-induced currents with distinct desensitization and Ca^{2+} dependence in rat trigeminal ganglion cells. *J. Neurophysiol.* **75**, 1503–1514.

32. Farkas-Szallasi, T., Lundberg, J. M., Wiesenfeld-Hallin, Z., and Szallasi, A. (1995) Increased levels of GMAP, VIP and nitric oxide synthase, and their mRNAs, in lumbar dorsal root ganglia of the rat following systemic resiniferatoxin treatment. *NeuroReport* **6**, 2230–2234.

33. Szallasi, A. and Blumberg, P. M. (1996) Vanilloid receptors: new insights enhance potential as a therapeutic target. *Pain* **68**, 195–208.

34. Yanagisawa, M., Yagi, N., Otsuka, M., Yanaihara, A., and Yanaihara, N. (1986) Inhibitory effects of galanin on the isolated spinal cord of the newborn rat. *Neurosci. Lett.* **70**, 278–282.

35. Szallasi, A. and Blumberg, P. M. (1991) Molecular target size of the vanilloid (capsaicin) receptor in pig dorsal root ganglia. *Life Sci.* **48**, 1863–1869.

36. Maderspach, K. and Fajszi, C. (1982) β-Adrenergic receptors of brain cells. Membrane integrity implies apparent positive cooperativity and higher affinity. *Biochim. Biophys. Acta.* **692,** 469–478.

37. Szallasi, A., Acs, G., Cravotto, G., Blumberg, P. M., Lundberg, J. M., and Appendino, G. (1996) A novel agonist, phorbol 12-phenylacetate 13-acetate 20-homovanillate, abolishes positive cooperativity of binding by the vanilloid receptor. *Eur. J. Pharmacol.* **299,** 221–228.

38. Bevan, S. and Geppetti, P. (1994) Protons: small stimulants of capsaicin-sensitive sensory nerves. *Trends Neurosci.* **17,** 509–512.

39. Oh, U., Hwang, S. W., and Kim, D. (1996) Capsaicin activates a nonselective cation channel in cultured neonatal rat dorsal root ganglion neurons. *J. Neurosci.* **16,** 1659–1667.

40. Acs, G., Lee, J., Marquez, V. E., and Blumberg, P. M. (1996) Distinct structure-activity relations for stimulation of ^{45}Ca uptake and for high affinity binding in cultured rat dorsal root ganglion neurons and dorsal root ganglion membranes. *Mol. Brain Res.* **35,** 173–182.

40a. Liu, L. and Simon, S. A. (1996) Similarities and differences in the currents activated by capscicin, piperine, and zingerone in rat trigeminal ganglion cells. *J. Neurophysiol.* **71,** 1858–1869.

41. Szallasi, A. (1996) The vanilloid receptor, in *Neurogenic Inflammation* (Geppetti, P. and Holzer, P., eds.), CRC, Boca Raton, FL, pp. 43–52.

42. Acs, G., Palkovits, M., and Blumberg, P. M. (1994) [^{3}H]Resiniferatoxin binding by the human vanilloid (capsaicin) receptor. *Mol. Brain Res.* **23,** 185–190.

43. Szallasi, A. (1995) Autoradiographic visualization and pharmacological characterization of vanilloid (capsaicin) receptors in several species, including man. *Acta. Physiol. Scand.* **155 (Suppl 629),** 1–68.

44. Walpole, C. S. J. and Wrigglesworth, R. (1993) Structural requirements for capsaicin agonists and antagonists, in *Capsaicin in the Study of Pain* (Wood, J. N., ed.), Academic, New York, pp. 63–82.

45. Docherty, R. J., Yeats, J. C., and Bevan, S. (1993) Non-specific block of voltage-activated calcium channels by capsazepine, an antagonist of capsaicin. *XXXII Congress Int. Union Physiol. Sci.* **256,** 7.

46. Amann, R. and Maggi, C. A. (1991) Ruthenium red as a capsaicin antagonist. *Life Sci.* **49,** 849–856.

47. Acs, G., Palkovits, M., and Blumberg, P. M. (1995) Trifluoperazine modulates [^{3}H]resiniferatoxin binding by human and rat vanilloid (capsaicin) receptors and affects ^{45}Ca uptake by adult rat dorsal root ganglion cultures. *J. Pharmacol. Exp. Ther.* **274,** 1090–1098.

48. Szallasi, A. and Blumberg, P. M. (1993) Mechanisms and therapeutic potential of vanilloids (capsaicin-like molecules). *Adv. Pharmacol.* **24,** 123–155.

49. Stanberry, L. R. (1990) Capsaicin interferes with the centrifugal spread of virus in primary and recurrent genital herpes simplex virus infections. *J. Infect. Dis.* **162,** 29–34.

50. Szolcsányi, J. (1991) Perspectives of capsaicin-type agents in pain therapy and research, in *Contemporary Issues in Chronic Pain Management* (Parris, W. C. V., ed.), Kluwer Academic, Boston, pp. 97–124.

Index